Yukio Ishida and Toshio Yamamoto

Linear and Nonlinear Rotordynamics

Related Titles

Yukio Ishida and Toshio Yamamoto

Linear and Nonlinear Rotordynamics

A Modern Treatment with Applications

2nd Enlarged and Improved Edition

WILEY-VCH Verlag GmbH & Co. KGaA

The Authors

Prof. Yukio Ishida
308 Nenokami, Nagakuto
Aichi-ken 480-1141
Japan

Prof. Toshio Yamamoto
414 Takama-cho, Meito-ku
Nagoya 465-0081
Japan

Library of Congress Card No.: applied for

British Library Cataloguing-in-Publication Data
A catalogue record for this book is available from the British Library.

Bibliographic information published by the Deutsche Nationalbibliothek
The Deutsche Nationalbibliothek lists this publication in the Deutsche Nationalbibliografie; detailed bibliographic data are available on the Internet at <http://dnb.d-nb.de>.

Cover Design Adam-Design, Weinheim
Typesetting Laserwords Private Limited, Chennai, India
Printing and Binding Markono Print Media Pte Ltd, Singapore

Printed in Singapore
Printed on acid-free paper

Print ISBN: 978-3-527-40942-6

ePDF ISBN: 978-3-527-65192-4

ePub ISBN: 978-3-527-65191-7

mobi ISBN: 978-3-527-65190-0

oBook ISBN: 978-3-527-65189-4

Dedicated to the memory of Professor Yamamoto

Contents

Foreword to the First Edition

The dynamics of rotating machinery has been extensively investigated during the past century. Many people in England, Germany, and the United States, in the late nineteenth and early twentieth centuries, studied the fundamental concepts associated with rotordynamic systems and investigated the effects of many types of mechanisms on these systems. The published material in this area diminished significantly between the start of World War I and the end of World War II. With the development of the gas turbine as a commercially viable engine after World War II, the need to better understand the dynamics of high-speed rotating systems became critical. Subsequent development of the digital computer assisted the development of many highly sophisticated procedures for analyzing, simulating, designing, and testing rotor systems. Many talented and dedicated people, throughout the world, have contributed to a better understanding of the dynamics of high-speed machinery during the last half of the twentieth century.

During the approximately 20-year interval defined by the end of World War II and the beginning of the digital computer revolution, three people stand out as significant contributors to the rotordynamic literature. F. M. Dimentberg in Russia documented his work in 1961 with the publication of a book entitled *Flexural Vibrations of Rotating Shafts*. In Czechoslovakia, Alex Tondl documented his studies in 1965 with the publication of a book entitled *Some Problems of Rotor Dynamics*. During the same period, Toshio Yamamoto worked independently and conscientiously at his Nagoya University laboratory on numerous topics related to rotordynamic systems. His work focused on fundamental concepts related to the dynamics of high-speed machinery and included meticulous laboratory test rigs to back up his analytical predictions. All of this work was done without the benefit of a high-speed digital computer or sophisticated electronic test equipment. Some of his work was documented in 1954 and 1957 with the publication of two *Nagoya University Memoirs* entitled "On the Critical Speed of a Shaft" and "On the Vibrations of a Rotating Shaft." These works and other publications of Dr Yamamoto, were not, however, circulated widely outside Japan. Thus it has taken the engineering world a little longer to recognize the genius of this dedicated and talented researcher. This book includes presentations on several of the original topics investigated by Dr Yamamoto and Dr Ishida, and also many important contributions by several other rotating machinery analysts and researchers around the world.

In 1975, Yukio Ishida graduated as one of Toshio Yamamoto's doctoral students at Nagoya University. Since that time, Dr Ishida has both independently and in collaboration with Dr Yamamoto made many additional and significant contributions to the area of rotordynamics. In particular, he and Dr Yamamoto have paid special attention to the effects of various nonlinear mechanisms on the dynamic behavior of rotor systems. Dr Ishida has also extensively investigated the use of modern digital signal processing as a valuable tool for the analytical and experimental investigation of vibrations in rotordynamic systems. Several topics associated with these more recent studies are included in this work. It is highly fitting that Dr Yamamoto and Dr Ishida are finally documenting a significant portion of their half a century of rotordynamic expertise with the publication of this book.

Professor and Chair
Department of Engineering
Texas Christian University
1991–1998

Harold D. Nelson, Ph.D.

Preface to the First Edition

Rotating machinery, such as steam turbines, gas turbines, internal combustion engines, and electric motors, are the most widely used elements in mechanical systems. As the rotating parts of such machinery often become the main source of vibrations, correct understanding of the vibration phenomena and sufficient knowledge of rotordynamics are essential for considering adequate means to eliminate vibrations. However, compared to rectilinear vibrations, the whirling motions of rotors seem difficult for students and engineers to understand in the beginning.

Studies of rotordynamics started more than 100 years ago. At that early stage, the primary concern was prediction of the resonance rotational speeds called *critical speeds*. As the normal operation speed increased above the critical speeds, engineers encountered various kinds of new problems, and rotordynamics developed through their efforts to overcome these technical difficulties. It is a very difficult task to master the entire range of rotordynamics by reading various technical papers. Although many standard books on vibrations contain a chapter explaining rotordynamics, the content is insufficient for practicing engineers. Recently, many excellent books on rotordynamics have been published by distinguished researchers. However, some of them are too practically oriented and others contain only recent technical topics.

The scope of this book includes most branches of rotordynamics. But it is intended especially to provide a detailed explanation of the basic concepts of rotordynamics because correct understanding becomes a strong tool with which to tackle vibration problems.

In Chapter 1, a classification of rotating shaft systems and a historical perspective are given.

In Chapter 2, the fundamentals of rotordynamics are explained using rotor models consisting of a massless elastic shaft and a rigid disk. The key ideas common to all branches of rotordynamics, such as critical speed, gyroscopic moment, whirling motion, and frequency diagrams, are explained. Balancing a rigid rotor is also discussed.

In Chapter 3, the dynamic analysis of a rotor with distributed mass is presented. Gyroscopic moment inertia and rotary are considered in the equation of the motion. Balancing a flexible rotor is also explained.

In Chapter 4, vibrations of an asymmetrical shaft with unequal stiffness and an asymmetrical rotor with unequal moments of inertia are discussed. Such systems with rotating asymmetry in stiffness or inertia belong to the category of parametrically excited systems, and the coefficients of the equations of motion are functions of time. The most distinguished feature of these systems is the appearance of an unstable zone at the major critical speed. Analysis of free vibrations, forced vibrations, and unstable vibrations is explained in this chapter.

In Chapter 5, various types of nonlinear resonances are considered. Rotating shaft systems have many elements that can cause nonlinearity in the shaft restoring forces. Several methods to obtain solutions for subharmonic resonances, combination resonances, and chaotic vibrations are presented. Nonlinear phenomena, such as jump phenomena, hysteresis phenomena, and period doublings, are shown. The dynamic behavior of a cracked shaft, which has attracted the interest of researchers in the field of vibration diagnosis, is also explained.

In Chapter 6, the effects of internal damping on shaft stability are explained. Owing to internal friction in the shaft material and dry friction between rotating components, self-excited vibrations appear in a wide speed range above the major critical speed. Expressions for such internal damping forces and the characteristics of self-excited vibrations are investigated.

In Chapter 7, nonstationary vibrations during transition through critical speeds are explained. Such nonstationary phenomena become a matter of great interest when the rated speed of a rotating machine is above the major critical speeds. Nonstationary phenomena are highly dependent on the magnitude of the driving torque of a motor. Interaction of the oscillating system with the energy source occurs, especially when the driving torque is small.

In Chapter 8, vibrations due to various machine elements are outlined. Ball bearings, bearing pedestals, universal joints, and couplings are the elements widely used in rotating machinery and may cause vibrations. In practical operations, an understanding of the characteristics of these machine elements is indispensable.

In Chapter 9, various kinds of flow-induced vibrations are explained. Oil whip in journal bearings, steam whirls in turbines, and vibrations of a hollow rotor partially filled with liquid are typical phenomena caused by liquid flow.

In Chapter 10, the finite element method is explained. In the analysis of a complex-shaped rotor in a practical machine, theoretical derivation of the equations of motion is impossible. Therefore, numerical procedures suitable for computer calculation are adopted instead. One of the most widely used means in the analysis of complex structures is the finite element method. This chapter describes how to apply this method to analyze rotating shaft systems.

In Chapter 11, another representative computational method called the *transfer matrix method* is explained. This method is especially suitable for the analysis of rotor systems.

In Chapter 12, a digital signal processing technique called complex fast Fourier transform (complex-FFT) is explained. In contrast to the ordinary FFT, this technique can distinguish between forward and backward whirling motions. The complex-FFT method is useful especially for the experimental study of rotor whirling motions.

Nagoya, Japan
January 2001

Toshio Yamamoto
Yukio Ishida

Preface to the Second Edition

Ten years have passed since the first edition of this book was published in 2001. I was very pleased to learn that the first edition has come to be used as a text in some graduate schools, and I am grateful to have received valuable comments and questions from many readers about the contents. The highly constructive advice from readers has helped us to polish the interpretations and explanations contained in this text and motivated us to publish this second edition. The major objective of our revisions is to first discuss additional subject matter of interest to engineers in the field of rotordynamics. Furthermore, we also decided to include additional case histories in order to illustrate the importance of the contents; exercises have also been included to aid readers' understanding of the text. The additional chapters cover topics such as balancing (Chapter 4), vibration suppression (Chapter 11), steam turbines and windmills (Chapter 12), cracked rotors (Chapter 13), and magnetic bearings (Chapter 17). In addition, Chapter 9 on vibrations of mechanical elements is extensively revised. It is hoped that these revisions will assist readers in their endeavor to develop a holistic understanding of rotordynamics.

In the second half of the twentieth century, the power of turbines increased very rapidly, and accordingly, the theory of rotordynamics developed remarkably during this time. Now, many of the researchers and engineers who first engaged in the design of these turbines are gradually retiring. I am afraid that their technology and "know-how," which have been accumulated through many failures and successes, may be lost over time. Owing to the commercial demand in recent times for the shortening of the development period, many engineers have come to depend on commercial software, and thus opportunities to learn from practice seem to me to be decreasing. It is our greatest wish that this book should help to transfer the contributions of our predecessors to the next generation of rotordynamists.

Finally, Professor Toshio Yamamoto, the coauthor of the first edition of this book, regrettably passed away in 2007 at the age of 86. I believe that he would be pleased at the publication of this new edition, although he was unable to make these changes by his own hand. With this in mind, I dedicate this book to Professor Yamamoto with respect to and in memory of his great achievements in the field of rotordynamics.

Nagoya, Japan
January, 2012

Yukio Ishida

Acknowledgments

The first edition and this second edition were developed out of lecture notes used in our undergraduate and postgraduate courses. The information contained in this book is based not only on our own work but also on the work of many excellent pioneers and leaders in the field of rotordynamics. We hereby record our sincere gratitude to all these distinguished scholars. We also wish to express our special thanks to Professor Harold D. Nelson, formerly of Texas Christian University, for writing the Foreword and to Professor Ali H. Nayfeh, of Virginia Polytechnic Institute and State University, for recommending and acting as series editor for this publication. Further, we would like to thank Professor Takashi Ikeda of Hiroshima University, Professor Imao Nagasaka of Chubu University, Professor Takao Torii of Shizuoka University, Professor Tsuyosi Inoue of Nagoya University and Dr. Kentaro Takagi of Nagoya University for their review and kind suggestions. Finally, we would like to thank our former and present students for their enthusiasm and contributions to our research work at Nagoya University.

1
Introduction

1.1
Classification of Rotor Systems

In general, rotating machinery consists of disks of various shapes, shafts whose diameters change depending on their longitudinal position, and bearings situated at various positions.[1] In vibration analyses, such a complex rotor system is simplified and a suitable mathematical model is adopted. In this modeling process, we must know which parameters are important for the system.

Rotating machines are classified according to their characteristics as follows: If the deformation of the rotating shaft is negligible in the operating speed range, it is called a *rigid rotor*. If the shaft deforms appreciably at some rotational speed in the operating speed range, it is called a *flexible rotor*. We cannot determine to which of these categories the rotor system belongs by considering only its dimensions. In rotordynamics, the rotating speeds that produce resonance responses due to mass eccentricity are called *critical speeds*. The deformation of a rotor becomes highest in the vicinity of the critical speed. Therefore, the range of the rated speed relative to these critical speeds determines whether the rotor is rigid or flexible.

Figure 1.1 called a *critical speed diagram* shows variations of critical speeds and vibration modes versus the stiffness of the supports for a symmetrical rotor. The left part of this figure represents values for rotors that are supported softly. In the first and second modes, the rotors do not deform appreciably but the supporting parts deflect. In this case, the rotor is considered to be a rigid rotor. As the stiffness of the supports decreases, the natural frequency of these modes approaches zero. In the third mode, the rotor deforms and it is considered to be a flexible rotor. Depending on the type of mode to be discussed, the same system may be considered as a rigid or a flexible rotor. On the right part of this figure, deformation occurs in all three modes and therefore the rotor is considered to be flexible in every mode.

In some models, disks are considered to be rigid and the distributed mass of an elastic shaft is concentrated at the disk positions. Such a model is called a

1) Rotor is often used as the general term for the rotating part of a rotating machine. The opposite term is *stator*, which means the static part of the machine.

Linear and Nonlinear Rotordynamics: A Modern Treatment with Applications, Second Edition.
Yukio Ishida and Toshio Yamamoto.

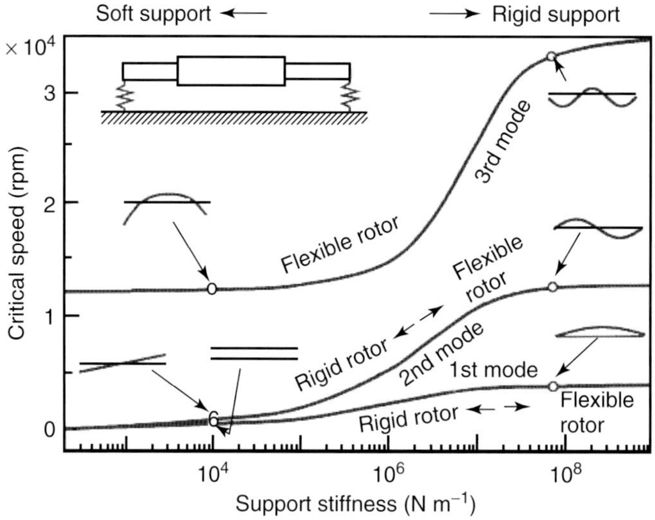

Figure 1.1 Critical speeds and mode shape versus the stiffness of the bearing support.

lumped-parameter system. If a flexible rotor with distributed mass and stiffness is considered, this model is called a *distributed-parameter system* or *continuous rotor system*. The mathematical treatment of the latter is more difficult than that of the former because it is governed by the partial differential equations.

Rotors are sometimes classified into vertical shaft systems and horizontal shaft systems. We mainly discuss the former model, however, the latter model tends to be considered in cases in which we must clarify the effect of gravity.

Rotors are sometimes classified as *high-speed rotors* or *low-speed rotors*. In this case, speed means the angular velocity or the peripheral velocity. Since high angular velocity often causes vibration, we use the term in association with the first definition in this book. The boundary between high and low is not clear and it differs depending on the situation. In some cases, the major critical speed is considered as the boundary. In ball bearing engineering, the term refers to the latter definition because it determines heating due to friction. The dimensionless parameter called *DN value* is used as an index related to the peripheral velocity. This value is defined as the product of the shaft diameter (mm) and the rotational speed (rpm). However, the unit symbol (mm · rpm) is omitted from the result of the calculation. Concerning high peripheral velocity, ball bearings and roller bearings of a main shaft of an aircraft engine have attained velocities of as much as $DN = 3 \times 10^6$ in a laboratory setting, though in practice these bearings are generally used at approximately 2.2×10^6 (Zaretsky, 1998). With regard to high angular velocity bearings, the spindle of a drill for dental use operates at approximately 50×10^4rpm. However, since its shaft diameter is small, its *DN* value is consequently relatively small. For example, a bearing with an inner diameter of 3.175 mm used in dental drilling operates at approximately $DN = 1.6 \times 10^6$.

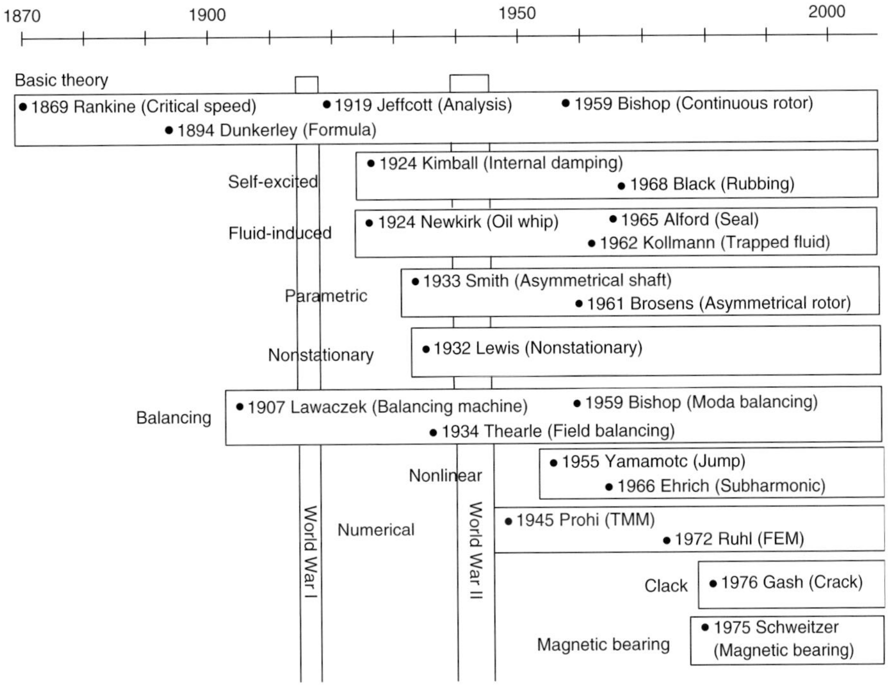

Figure 1.2 History of rotordynamics.

1.2 Historical Perspective

The evolution of research in the field of rotordynamics is shown in Figure 1.2. Research on rotordynamics spans at least a 140-year history, starting with Rankine's paper on whirling motions of a rotor in 1869. Rankine discussed the relationship between centrifugal and restoring forces and concluded that operation above a certain rotational speed is impossible. Although this conclusion was wrong, his paper (refer to the "Topic: The First Paper on Rotordynamics") is important as the first publication on rotordynamics. The research progressed significantly at the end of the nineteenth century with contributions by de Laval and others. De Laval, an engineer in Sweden, invented a one-stage steam turbine and succeeded in its operation. He first used a rigid rotor, but later used a flexible rotor and showed that it was possible to operate above the critical speed by operating at a rotational speed about seven times the critical speed (Stodola, 1924).

In the early days, the major concern for researchers and designers was to predict the critical speed, because the first thing that had to be done in designing rotating machinery was to avoid resonance. Dunkerley (1894) derived an empirical formula that gave the lowest critical speed for a multirotor system. He was the first to use the term "*critical speed*" for the resonance rotational speed. The word "*critical*," which refers to a state or a value at which an abrupt change in a quality or state occurs, was coined possibly based on Rankine's conclusion mentioned above. Holzer (1921) proposed an approximate method to calculate the natural frequencies and mode shapes of torsional vibrations.

The first recorded fundamental theory of rotordynamics can be found in a paper written by Jeffcott (1919) . We can appreciate Jeffcott's great contributions if we recall that a shaft with a disk at the midspan is called the *Jeffcott rotor*, especially among researchers in the United States. This simplified fundamental rotor system is also called the *Laval rotor*, named after de Laval.

The developments made in rotordynamics up to the beginning of the twentieth century are detailed in the masterpiece written by Stodola (1924). This superb book explains nearly the entire field related to steam turbines. Among other things, this book includes the dynamics of elastic shafts with disks, the dynamics of continuous rotors without considering the gyroscopic moment, the balancing of rigid rotors, and methods for determining approximate values of critical speeds of rotors with variable cross sections.

Thereafter, the center of research shifted from Europe to the United States, and the scope of rotordynamics expanded to consider various other phenomena. Campbell (1924) at General Electric investigated vibrations of steam turbines in detail. His diagram, representing critical speed in relation to the cross points of natural frequency curves and the straight lines proportional to the rotational speed, is now widely used and referred to as the *Campbell diagram*. As the rotational speed increased above the first critical speed, the occurrence of self-excited vibrations became a serious problem. In the 1920s, Newkirk (1924) and Kimball (1924) first recognized that internal friction of shaft materials could cause an unstable whirling motion. These phenomena, in which friction that ordinarily dampens vibration causes self-excited vibration, attracted the attention of many researchers. Newkirk and Taylor (1925) investigated an unstable vibration called *oil whip*, which was due to an oil film in the journal bearings. Rotor is generally surrounded by a stator such as seals with a small clearance. Newkirk (1926) showed a forward whirl induced by a hot spot on the rotor surface, which was generated by the contact of the rotor and the surroundings. This hot spot instability is called the *Newkirk effect*.

About a decade later, the study of asymmetrical shaft systems and asymmetrical rotor systems began. The former are systems with a directional difference in shaft stiffness, and the latter are those with a directional difference in rotor inertia. Two pole generator rotors and propeller rotors are examples of such systems. As these directional differences rotate with the shaft, terms with time-varying coefficients appear in the governing equations. These systems therefore fall into the category of parametrically excited systems. The most characteristic property of asymmetrical systems is the appearance of unstable vibrations in some rotational speed ranges. Smith (1933)'s report is a pioneering work on this topic. Various phenomena related to the asymmetries of rotors were investigated actively in the middle of the twentieth century by Taylor (1940) and Foote, Poritsky, and Slade (1943), Brosens and Crandall (1961), and Yamamoto and Ota (1963a, 1963b, 1964).

Nonstationary phenomena during passage through critical speeds have been studied since Lewis reported his investigation on the Jeffcott rotor in 1932.

Numerous reports on this topic are classified into two groups. One group classifies nonstationary phenomena that occur in a process with a constant acceleration and the other classifies phenomena that occur with a limited driving torque. In the latter case, mutual interaction between the driving torque and the shaft vibration must be considered. As the theoretical analysis of such transition problems is far more difficult than that of stationary oscillations, many researchers adopted numerical integrations. The asymptotic method developed by the Russian school of Krylov and Bogoliubov (1947) and Bogoliubov and Mitropol'skii (1958) considerably boosted the research on this subject.

The vibrations of rotors with continuously distributed mass were also studied. The simplest continuous rotor model corresponding to the Euler beam was first studied in the book by Stodola (1924). In the 1950s and 1960s, Bishop (1959), Bishop and Gladwell (1959), and Bishop and Parkinson (1965) reported a series of papers on the unbalance response and the balancing of a continuous rotor. Eshleman and Eubanks (1969) derived more general equations of motion considering the effects of rotary inertia, shear deformation, and gyroscopic moment, and investigated these effects.

The most important and fundamental procedure to reduce unfavorable vibrations is to eliminate geometric imbalance in the rotor. The balancing technique for a rigid rotor was established relatively early. A practical balancing machine based on this technique was invented by Lawaczeck in 1907 (Miwa and Shimomura, 1976). In 1925, Suehiro invented a balancing machine that conducts balancing at a speed in the postcritical speed range (Miwa and Shimomura, 1976). And in 1934, Thearle developed the two-plane balancing (Thearle, 1934). The arrival of high-speed rotating machines made it necessary to develop a balancing technique for flexible rotors. Two representative theories were proposed. One was *the modal balancing method* proposed in the 1950s by Federn (1957) and Bishop and Gladwell (1959). The other was *the influence coefficient method* proposed in the early 1960s and developed mainly in the United States alone with the progress of computers. Goodman (1964) improved this method by taking into the least square methods.

In the latter half of the twentieth century, various vibrations due to fluid were studied. The above-mentioned oil whip is a representative flow-induced vibration of rotors. In the middle of the twentieth century, Hori (1959) succeeded in explaining various fundamental characteristics of oil whip by investigating the stability of shaft motion and considering pressure forces due to oil films. At almost the same time, other types of flow-induced vibrations attracted the attention of many researchers. Seals that are used to reduce the leakage of working fluids through the interface between rotors and stators sometimes induce unstable vibrations. In 1964, Alford reported accidents due to labyrinth seals. Another one was a self-excited vibration called *the steam whirl*. The mechanism of this vibration in turbines was explained by Thomas (1958) and that in compressors was explained by Alford (1965). These phenomena are still attracting the interest of many researchers for practical importance. The vibration of a hollow rotor containing fluid is a relatively new problem of flow-induced vibrations. In 1967, Ehrich reported that fluid trapped in engine shafts induced asynchronous vibrations. A noteworthy paper on this phenomenon is that of Wolf (1968). He succeeded in explaining the appearance of an unstable speed range in a postcritical region of a rotor system containing inviscid fluid.

As rotors became lighter and their operational speeds higher, the occurrence of nonlinear resonances such as subharmonic resonances became a serious problem. Yamamoto (1955, 1957a) studied various kinds of nonlinear resonances after he reported on subharmonic resonances due to ball bearings, in 1955. He discussed systems with weak nonlinearity that can be expressed by a power series of low order. Aside from subharmonic resonances, he also investigated combination resonances (he named them summed-and-differential harmonic oscillations) and combination tones. In the 1960s, Tondl (1965) studied nonlinear resonances due to oil films in journal bearings. Ehrich (1966) reported subharmonic resonances observed in an aircraft gas turbine with squeeze-film damper bearings. The cause of strong nonlinearity in aircraft gas turbines is the radial clearance of squeeze-film damper bearings. Later, Ehrich (1988, 1991) reported the occurrence of various types of subharmonic resonances up to a very high order and also chaotic vibrations in practical engines.

In the practical design of rotating machinery, it is necessary to know accurately the natural frequencies, modes, and forced responses to unbalances in complex-shaped rotor systems. The representative techniques used for this purpose are the transfer matrix method and the finite-element method. Prohl (1945) used the transfer matrix method in the analysis of a rotor system by expanding the method originally developed by Myklestad (1944). This analytical method is particularly useful for multirotor-bearing systems and has developed rapidly since the 1960s by the contribution of many researchers such as Lund and Orcutt (1967) and Lund (1974). The finite-element method was first developed in structural dynamics and then used in various technological fields. The first application of the finite-element method to a rotor system was made by Ruhl and Booker (1972). Then, Nelson and McVaugh (1976) generalized it by considering rotating inertia, gyroscopic moment, and axial force.

From the 1950s, cracks were found in rotors of some steam turbines (Ishida, 2008). To prevent serious accidents and to develop a vibration diagnosis system for detecting cracks, research on vibrations of cracked shafts began. In the 1970s, Gasch (1976) and Henry and Okah-Avae (1976) investigated vibrations, giving consideration to nonlinearity in stiffness due to open–close mechanisms. They showed that an unstable region appeared or disappeared at the major critical speed, depending on the direction of the unbalance. The research is still being developed and various monitoring systems have been proposed.

The latest topics in rotordynamics are magnetic bearings that support a rotor without contacting it and active control. This study has received considerable attention since Schweitzer (1975) reported his work in 1975. Nonami (1985) suppressed an unbalance response of a rotor controlling the bearing support actively using the optimal regulator theory.

In this chapter, the history of rotordynamics has been summarized briefly. The authors recommend the readers to read excellent introductions on the history of rotordynamics by Miwa (1991), Dimarogonas (1992), and Nelson (1994, 1998). The following are representative books on rotordynamics, some of which have detailed bibliographies: Stodola (1924), Kimball (1932), Biezeno and Grammel (1939), Dimentberg (1961), Tondl (1965), Gunter (1966), Roewy and Piarulli (1969), Eshleman, Shapiro, and Rumabarger (1972), Gasch and Pfutzner (1975), Miwa and Shimomura (1976), Dimarogonas and Paipetis (1983), Vance (1988), Darlow (1989), Lalanne and Ferraris (1990), Zhang (1990), Rao (1991), Ehrich (1992), Lee (1993), Childs (1993), Kramer (1993), Gasch, Nordmann, and Pfützner (2002), Genta (2005), Muszynska (2005), and Bachschmid, Pennacchi, and Tanzi (2010).

Topic: The First Paper on Rotordynamics

The first paper on rotordynamics is attributed to William John Macquorn Rankine (1820–1872), a professor at the University of Glasgow. He is well known as a scientist and engineer, particularly for his development of the Rankine cycle in the theory of heat engines. However, his interests were extremely varied. In 1869, he wrote an article entitled "On the Centrifugal Force of Rotating Shafts" in "The Engineer." Figure 1.3 shows a part of this article. He published his analysis of spinning shafts and wrote as follows: "For a shaft of a given length, a diameter, and material, there is a limit of

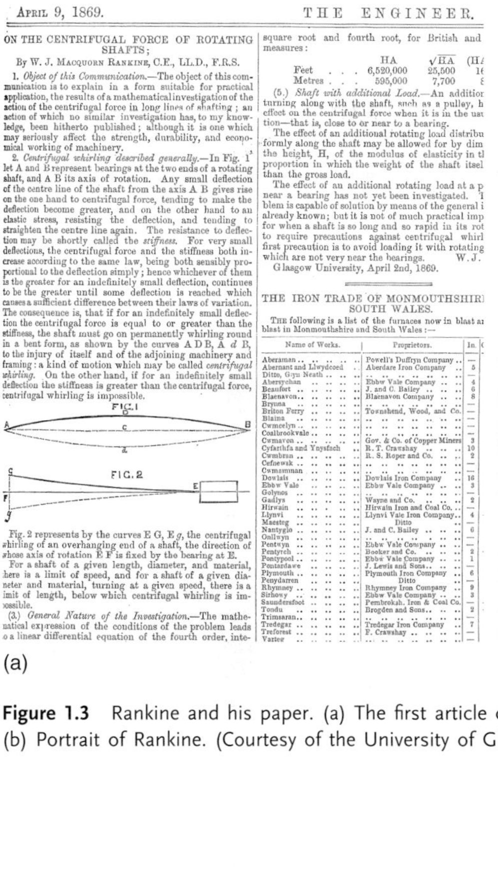

April 9, 1869. THE ENGINEER.

ON THE CENTRIFUGAL FORCE OF ROTATING SHAFTS;

By W. J. Macquorn Rankine, C.E., LL.D., F.R.S.

1. *Object of this Communication.*—The object of this communication is to explain in a form suitable for practical application, the results of a mathematical investigation of the action of the centrifugal force in long lines of shafting; an action of which no similar investigation has, to my knowledge, been hitherto published; although it is one which may seriously affect the strength, durability, and economical working of machinery.

2. *Centrifugal whirling described generally.*—In Fig. 1 let A and B represent bearings at the two ends of a rotating shaft, and A B its axis of rotation. Any small deflection of the centre line of the shaft from the axis A B gives rise on the one hand to centrifugal force, tending to make the deflection become greater, and on the other hand to an elastic stress, resisting the deflection, and tending to straighten the centre line again. The resistance to deflection may be shortly called the *stiffness*. For very small deflections, the centrifugal force and the stiffness both increase according to the same law, being both sensibly proportional to the deflection simply; hence whichever of them is the greater for an indefinitely small deflection, continues to be the greater until some deflection is reached which causes a sufficient difference between their laws of variation. The consequence is, that if for an indefinitely small deflection the centrifugal force is equal to or greater than the stiffness, the shaft must go on permanently whirling round in a bent form, as shown by the curves A D B, A *d* B, to the injury of itself and of the adjoining machinery and framing: a kind of motion which may be called *centrifugal whirling*. On the other hand, if for an indefinitely small deflection the stiffness is greater than the centrifugal force, centrifugal whirling is impossible.

Fig. 2 represents by the curves E G, E *g*, the centrifugal whirling of an overhanging end of a shaft, the direction of whose axis of rotation E F is fixed by the bearing at E.

For a shaft of a given length, diameter, and material, there is a limit of speed, and for a shaft of a given diameter and material, turning at a given speed, there is a limit of length, below which centrifugal whirling is impossible.

(3.) *General Nature of the Investigation.*—The mathematical expression of the conditions of the problem leads to a linear differential equation of the fourth order, inte-

square root and fourth root, for British and measures:

	HA	$\sqrt{HA}$	(HA
Feet . . .	6,520,000	25,500	16
Metres . . .	595,000	7,700	8

(5.) *Shaft with additional Load.*—An additior turning along with the shaft, such as a pulley, h effect on the centrifugal force when it is in the us tion—that is, close to or near to a bearing.

The effect of an additional rotating load distribu formly along the shaft may be allowed for by dim the height, H, of the modulus of elasticity in t proportion in which the weight of the shaft itsel than the gross load.

The effect of an additional rotating load at a p near a bearing has not yet been investigated. T blem is capable of solution by means of the general i already known; but it is not of much practical imp for when a shaft is so long and so rapid in its rot to require precautions against centrifugal whirl first precaution is to avoid loading it with rotating which are not very near the bearings. W. J.

Glasgow University, April 2nd, 1869.

THE IRON TRADE OF MONMOUTHSHIRE SOUTH WALES.

The following is a list of the furnaces now in blast a blast in Monmouthshire and South Wales:—

Name of Works.	Proprietors.	In.
Aberaman	Powell's Duffryn Company ..	—
Abernant and Llwydcoed ..	Aberdare Iron Company ..	5
Ditto, Glyn Neath	.. ..	—
Abersychan	Ebbw Vale Company	4
Beaufort	J. and C. Bailey	6
Blaenavon..	Blaenavon Company	8
Brynna	.. ..	—
Briton Ferry	Townshend, Wood, and Co.	—
Blaina	.. ..	—
Cwmcelyn	.. ..	—
Coalbrookvale	.. ..	—
Cwmavon	Gov. & Co. of Copper Miners	3
Cyfarthfa and Ynysfach ..	R. T. Crawshay	10
Cwmbran	R. S. Roper and Co.	2
Cefnewsk	.. ..	—
Cwmamman	.. ..	—
Dowlais	Dowlais Iron Company ..	16
Ebbw Vale	Ebbw Vale Company	3
Golynos	.. ..	—
Gadlys	Wayne and Co.	2
Hirwain	Hirwain Iron and Coal Co. ..	—
Llynvi	Llynvi Vale Iron Company..	4
Maesteg	Ditto ..	—
Nantyglo	J. and C. Bailey	6
Onllwyn	.. ..	—
Pentwyn	Ebbw Vale Company	—
Pentyrch	Booker and Co.	2
Pontypool	Ebbw Vale Company	1
Pontardawe	J. Lewis and Sons..	—
Plymouth	Plymouth Iron Company ..	6
Penydarren	Ditto ..	—
Rhymney	Rhymney Iron Company ..	9
Sirhowy	Ebbw Vale Company	3
Saundersfoot	Pembroksh. Iron & Coal Co.	—
Tondu	Brogden and Sons..	2
Trimsaran..	.. ..	—
Tredegar	Tredegar Iron Company ..	7
Treforest	F. Crawshay	—
Varteg	.. ..	—

(a)

(b)

Figure 1.3 Rankine and his paper. (a) The first article on rotordynamics (Rankine, (1869)). (b) Portrait of Rankine. (Courtesy of the University of Glasgow.)

speed, and for a shaft of a given diameter and material, turning a given speed, there is ***a limit of length***, below which centrifugal whirling is impossible." This limit of length corresponds to the critical speed and he gave the correct formula calculating critical speed. Although his prediction that the supercritical operation is impossible is not correct, his analysis is still worth mentioning.

References

Alford, J.S. (1964) Protection of labyrinth seals from flexural vibration. *Trans. ASME, J. Eng. Power*, **86** (2), 141–148.

Alford, J.S. (1965) Protecting turbomachinery from self-excited rotor whirl. *Trans. ASME, J. Eng. Power*, **87** (4), 333–344.

Bachschmid, N., Pennacchi, P., and Tanzi, E. (2010) *Cracked Rotors*, Springer.

Biezeno, C.B. and Grammel, R. (1939) *Technische Dynamik*, Julius Springer, Berlin. (English translation (1954) *Engineering Dynamics*, Steam Turbines, Vol. 111, Blackie & Son, Glasgow).

Bishop, R.E.D. (1959) Vibration of rotating shafts. *J. Mech. Eng. Sci.*, **1** (1), 50–65.

Bishop, R.E.D. and Gladwell, G.M.L. (1959) The vibration and balancing of an unbalanced flexible rotor. *J. Mech. Eng. Sci.*, **1** (l), 66–77.

Bishop, R.E.D. and Parkinson, A.G. (1965) Second order vibration of flexible shafts. *Philos. Trans. R. Soc. Lond., Ser. A*, **259** (1095), 1–31.

Bogoliubov, N.N. and Mitropol'skii, Y.A. (1958) *Asymptotic Methods in the Theory of Nonlinear Oscillations*, Gordon and Breath, New York. (English translation, 1961, Gordon and Breath, New York), (in Russian).

Brosens, S.H. and Crandall, S.H. (1961) Whirling of unsymmetrical rotors. *Trans. ASME, J. Appl. Mech.*, **28** (3), 355–362.

Campbell, W. (1924) The Protection of steam-turbine disk wheels from axial vibration. *Trans. ASME*, **46**, 31–160.

Childs, D. (1993) *Turbomachinery Rotordynamics*, John Wiley & Sons, Inc., New York.

Darlow, M.S. (1989) Balancing of High-Speed Machinery, *Springer-Verlag*, New-York.

Dimarogonas, A.D. (1992) *A Brief History of Rotor Dynamics*, Rotordynamics, Vol. **92**, Springer-Verlag, New York, pp. 1–10.

Dimarogonas, A.D. and Paipetis, S.A. (1983) *Analytical Methods in Rotor Dynamics*, Applied Science Publishers, Barking, Essex.

Dimentberg, F.M. (1961) *Flexural Vibrations of Rotating Shafts*, Butterworths, London.

Dunkerley, S. (1894) On the whirling and vibration of shaft. *Philos. Trans. R. Soc. Lond., Ser. A*, **185**, 279–359.

Ehrich, F.F. (1966) *Subharmonic Vibration of Rotors in Bearing Clearance*, ASME, New York, ASME paper 66-MD-I.

Ehrich, F.F. (1967) The influence of trapped fluids on high speed rotor vibration. *Trans. ASME, J. Eng. Ind.*, **89B** (4), 806–812.

Ehrich, F.F. (1988) High order subharmonic response of high speed rotors in bearing clearance. *Trans. ASME, J. Vib. Acoust. Stress Reliab. Des.*, **113** (1), 50–56.

Ehrich, F.F. (1991) Some observations of chaotic vibration phenomena in high-speed rotordynamics. *Trans. ASME, J. Vib. Acoust. Stress Rehab. Des.*, **113** (1), 50–56.

Ehrich, F.F. (1992) *Handbook of Rotordynamics*, McGraw-Hill, New York.

Eshleman, R.L. and Eubanks, R.A. (1969) On the critical speeds of a continuous rotor. *Trans. ASME, J. Eng. Ind.*, **91** (4), 1180–1188.

Eshleman, R., Shapiro, W., and Rumabarger, J.H. (1972) *Flexible Rotor-Bearing System Dynamics*, I: Critical Speeds and Response of Flexible Rotor Dynamics Analysis, II. Bearing Influence and Representation in Rotor Dynamics Analysis, ASME.

Federn, K. (1957) Grundlagen einer systematischen Schwingungsentstörung wellenelastischer Rotoren. *VDI Ber.*, **24**, 9–25.

Foote, W.R., Poritsky, H., and Slade, J.J. (1943) Critical speeds of a rotor with unequal shaft flexibilities, mounted in bearings of unequal flexibility I. *Trans. ASME, J. Appl. Mech.*, **10** (2), 77–84.

Gasch, R. (1976) Dynamic behaviour of a simple rotor with a cross-sectional crack, *Proceedings of the International Conference on Vibrations in Rotating Machinery*, Institute of Mechanical Engineers, New York, pp. 123–128.

Gasch, R., Nordmann, R., and Pfützner, H. (2002) *Rotordynamik*, 2 Auflage, Springer.

Gasch, R. and Pfutzner, H. (1975) *Rotordynamik: Eine Einfuhrung*, Springer-Verlag, Berlin.

Genta, G. (2005) *Dynamics of Rotating Systems*, Springer.

Goodman, T.P. (1964) A least-squares method for computing balance corrections. *Trans. ASME, J. Eng. Ind.*, **86** (3), 273–279.

Gunter, E.J. (1966) *Dynamic Stability of Rotor-Bearing Systems*, National Aeronautics and Space Administration, Washington, DC.

Henry, T.A. and Okah-Avae, B.E. (1976) *Proceedings of the International Conference on Vibrations in Rotation Machinery*, Institute of Mechanical Engineers, New York, pp. 15–17.

Holzer, H. (1921) *Die Berechnung der Drehschwingungen*, Springer-Verlag, Berlin.

Hori, Y. (1959) A theory of oil whip. *Trans. ASME, J. Appl. Mech.*, **26** (2), 189–198.

Ishida, Y. (2008) Cracked rotors: industrial machine case histories and nonlinear effects shown by simple Jeffcott rotor. *Mech. Syst. Signal Process.*, **22** (4), 805–817.

Jeffcott, H.H. (1919) The lateral vibration of loaded shafts in the neighborhood of a whirling speed: the effect of want of balance. *Philos. Mag. A*, **37**, 304–315.

Kimball, A.L. (1924) Internal friction theory of shaft whirling. *Gen. Electric Rev.*, **27** (4), 244–251.

Kimball, A.L. (1932) *Vibration Prevention in Engineering*, John Wiley & Sons, Ltd, London.

Kramer, E. (1993) *Dynamics of Rotors and Foundations*, Springer-Verlag, New York.

Krylov, N. and Bogoliubov, N. (1947) *Introduction to Non-Linear Mechanics*, Translation from Russian Monographs, Oxford University Press.

Lalanne, M. and Ferraris, G. (1990) *Rotordynamics: Prediction in Engineering*, John Wiley & Sons, Inc., New York.

Lee, C.W. (1993) *Vibration Analysis of Rotors*, Kluwer Academic, Norwell, MA.

Lewis, F.W. (1932) Vibrations during acceleration through a critical speed. *Trans. ASME*, **54** (3), 253–261.

Lund, J.W. (1974) Stability and damped critical speed of a flexible rotor in fluid-film bearings. *Trans. ASME, J. Eng. Ind.*, **96** (2), 509–517.

Lund, J.W. and Orcutt, F.K. (1967) Calculation and experiments on the unbalance response of a flexible rotor. *Trans. ASME, J. Eng. Ind.*, **89** (4), 785–795.

Miwa, S. (1991) The dawn of machine dynamics: the balancing of wheels and the critical speed of rotating shafts. *Trans. JSME*, **57** (541), 3063–3070. (in Japanese).

Miwa, S. and Shimomura, G. (1976) *Balancing of Rotating Machinery*, Corona Publishing Co., Tokyo. (in Japanese).

Muszynska, A. (2005) *Rotordynamics*, Taylor & Francis.

Myklestad, N.O. (1944) A new method for calculating natural modes of uncoupled bending vibrations of airplane wings and other types of beams. *J. Aeronaut. Sci.*, **II** (2), 153–162.

Nelson, H.D. (1994) Modeling, analysis, and computation in rotordynamics: a historical perspective. *Proceedings of the 4th IFToMM International Conference on Rotor Dynamics*, pp. 171–177.

Nelson, H.D. (1998) Rotordynamic modeling and analysis procedures: a review. *JSME Int.J. Ser. C*, **41** (1), 1–12.

Nelson, H.D. and McVaugh, J.M. (1976) The dynamics of rotor bearing systems, using finite elements. *Trans. ASME, J. Eng. Ind.*, **98** (2), 593–600.

Newkirk, B.L. (1924) Shaft whipping. *Gen. Electr. Rev.*, **27** (3), 169–178.

Newkirk, B.L. (1926) Shaft rubbing. *Mech. Eng.*, **48** (8), 830–832.

Newkirk, B.L. and Taylor, H.D. (1925) Shaft whirling due to oil action in journal bearings. *Gen. Electr. Rev.*, **28** (7), 559–568.

Nonami, K. (1985) Vibration control of a rotor by active bearings. *Trans. JSME*, **31** (470), 2463–2472. (in Japanese).

Prohl, M.A. (1945) A general method for calculating critical speeds of flexible rotors. *J. Appl. Mech.*, **12** (3), 142–148.

Rankine, W.J.M. (1869) On the centrifugal force of rotating shafts. *Engineering*, **27**, 249–249.

Rao, J.S. (1991) *Rotor Dynamics*, 2nd edn, John Wiley & Sons, Inc., New York.

Roewy, R.G. and Piarulli, V.J. (1969) *Dynamics of Rotating Shafts*, The Shock and Vibration Information Center, U.S. Department of Defense, Washington, DC.

Ruhl, R.L. and Booker, J.F. (1972) A finite element model for distributed parameter turbo rotor system. *Trans. ASME, J. Eng. Ind.*, **94** (1), 126–132.

Schweitzer, G. (1975) Stabilization of self-excited rotor vibrations by active dampers, *Dynamics of Rotors*, Springer-Verlag, New York, pp. 472–493.

Smith, D.M. (1933) The motion of a rotor carried by a flexible shaft in flexible bearings. *Proc. R. Soc. Lond., Ser. A*, **142**, 92–118.

Stodola, A. (1924) *Dampf und Gas-Turbinen*, Verlag von Julius Springer, Berlin. (English translation (1927) *Steam and Gas Turbines*, McGraw-Hill, New York).

Taylor, H.D. (1940) Critical speed behavior of unsymmetrical shafts. *J. Appl. Mech.*, **7** (2), 71–79.

Thearle, E.L. (1934) Dynamic balancing in the field. *Trans. ASME, J. Appl. Mech.*, **56** (10), 745–753.

Thomas, J.J. (1958) *Instabile Eigenschwingungeri von Turbinenlaufern, Angefacht durch die Spaltstromungen, in Stoptbuchsen und Beschauflungen*, AEG-Sonderdruck, pp. 1039–l063.

Tondl, A. (1965) *Some Problems of Rotor Dynamics*, Czechoslovak Academy of Sciences, Prague.

Vance, J.H. (1988) *Rotordynamics of Turbomachinery*, John Wiley & Sons, Inc., New York.

Wolf, J.A. (1968) Whirl dynamics of a rotor partially filled with liquid. *Trans. ASME, J. Appl. Mech.*, **35** (4), 676–682.

Yamamoto, T. (1955) On the critical speed of a shaft of sub-harmonic oscillation. *Trans. JSME*, **21** (111), 853–858. (in Japanese).

Yamamoto, T. (1957a) On the vibrations of a rotating shaft. Chapter II: non-linear and non-symmetrical spring characteristics of the shaft supported by single-row radial ball bearings; Chapter III: On the critical speed of a shaft of sub-harmonic oscillation and on sub-harmonic oscillation on rectilinear vibrations. *Mem. Fac. Eng., Nagoya Univ.*, **9** (1), 25–40.

Yamamoto, T. and Ota, H. (1963a) On the vibrations of a shaft carrying an unsymmetrical rotating body. *Bull. JSME*, **6** (21), 29–36.

Yamamoto, T. and Ota, H. (1963b) Unstable vibrations of the shaft carrying an unsymmetrical rotating body. *Bull. JSME*, **6** (23), 404–411.

Yamamoto, T. and Ota, H. (1964) On the dynamically unstable vibrations of a shaft carrying an unsymmetrical rotating body. *Bull. JSME*, **31** (3), 515–5522.

Zaretsky, E.V. (1998) Quest for the 3 million DN-bearing – A history, NTN Technical Review, No.67, 7–19 (in Japanese)

Zhang, W. (1990) *Fundamental Theory of Rotordynamics*, Chinese Science Academy, Beijing. (in Chinese).

2
Vibrations of Massless Shafts with Rigid Disks

2.1
General Considerations

In practical rotating machinery, various disks and blades are mounted on a shaft whose diameter changes in its longitudinal direction. In vibration analysis, such a rotating machine is replaced by a simple mathematical model. Figure 2.1 illustrates several mathematical models that are often adopted in the analyses of rotor vibrations. In Figure 2.1, a–c are lumped-parameter systems in which a rigid disk is mounted on a massless elastic shaft with a circular cross section. When a rotor is mounted at the center of an elastic shaft supported at both ends, the lateral deflection r and the inclination θ of the rotor are independent of each other and therefore this system can be decomposed into two rotor systems with two degrees of freedom (2 DOF). Figure 2.1a shows a 2 DOF model executing a deflection motion. This simplified mathematical model called *the Jeffcott rotor* is the most widely used model in the theoretical analysis of rotors. Figure 2.1b is a 2 DOF model executing an inclination motion. Different from the system shown in Figure 2.1a, the gyroscopic moment acts in this system and, as a result, natural frequencies change as a function of the rotational speed. This is the simplest model that has characteristics unique to rotor systems. For some types of vibration problems, this model is more suitable than the Jeffcott rotor. Figure 2.1c shows 4 DOF models where the deflection motion and the inclination motion couple with each other. Figure 2.1d shows a rigid rotor that is supported elastically. The movement of this rotor is governed by the same equations of motion as those for the model in Figure 2.1c. In this chapter, the fundamentals of rotordynamics are explained using these mathematical models.

2.2
Rotor Unbalance

The term *unbalance* refers to the nonuniform distribution of the mass of a rotor about its axis of rotation. In ISO1925 (ISO, International Organization for Standardization), *unbalance* is defined as the "condition which exists in a rotor when vibratory force or motion is imparted to its bearings as a result of centrifugal

Linear and Nonlinear Rotordynamics: A Modern Treatment with Applications, Second Edition.
Yukio Ishida and Toshio Yamamoto.

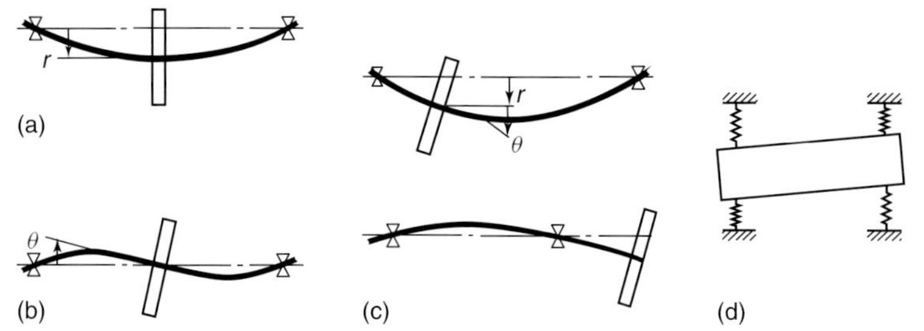

Figure 2.1 Lumped-parameter rotor models: (a) 2 DOF model (deflection), (b) 2 DOF model (inclination), (c) 4 DOF model, (d) rigid-rotor model.

force." The main cause of vibrations is the excitation due to inevitable mass unbalance of rotors. We cannot avoid residual unbalance because it occurs owing to various causes such as manufacturing error, thermal deformation, material inhomogeneity, wear, and corrosion. Unbalance also occurs owing to stuck up of tolerance in assembly. When a well-balanced rotor is mounted on a well-balanced shaft, the necessary assembly tolerances permit radial displacement of the rotor, and, as a result, unbalance occurs.

In a rigid rotor, unbalance is produced when the principal axis of inertia (refer to Appendix A) does not coincide with the axis of rotation. As shown in Figure 2.2, two types of error produce unbalance. One is the case that, as shown in Figure 2.2a, the principal axis of inertia is displaced parallel to the axis of rotation. In this case, the center of gravity G of a rotor deviates from the axis of rotation by e. We call this e *"the eccentricity."* This eccentricity produces a centrifugal force proportional to the square of the rotational speed. The eccentricity can be detected without operating the rotor because it always directs downward if the shaft is supported horizontally by bearings with little friction. The other is the case that the principal axis of inertia is displaced not parallel to but intersecting with the axis of rotation at the center of gravity G. We call this angle τ between these two axis *"the skew angle."* This

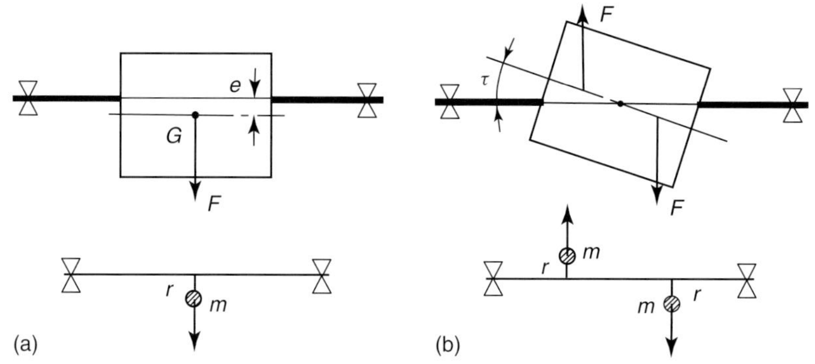

Figure 2.2 Unbalances in a lumped-parameter rotor system: (a) eccentricity and (b) skew angle.

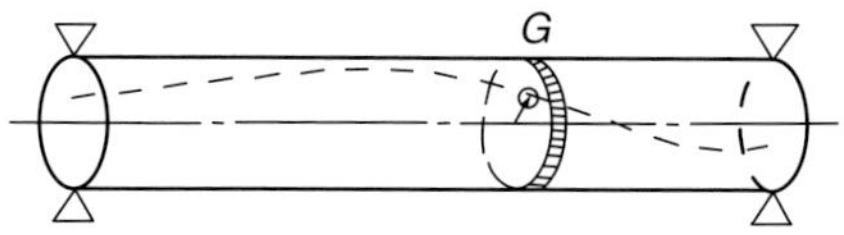

Figure 2.3 Unbalance in a continuous rotor.

skew angle cannot be detected without rotating the shaft. When the rotor rotates, the centrifugal force at the center of gravity of the left half and that of the right half works in the opposite direction, and, as a result, a moment proportional to the square of the rotational speed works. Generally, the direction of the eccentricity and the skew angle is not the same. As shown in Figure 2.2, the unbalance due to the eccentricity is represented by a model with one concentrated mass, while the unbalance due to a skew angle is represented by two concentrated masses.

The eccentricity and skew angle are also referred to as the *static* and *dynamic unbalance*, respectively (Loewy and Piarulli, 1969; Ehrich, 1992). However, these definitions are not used widely. In the field of balancing technology, unbalance has been classified into three categories: *static unbalance, couple unbalance,* and *dynamic unbalance.* These widely accepted definitions are explained in Chapter 4.

A model of a flexible rotor with distributed mass is shown in Figure 2.3. Consider a sliced element that is perpendicular to the axis of rotation. This element has the eccentricity. The magnitude and direction of this eccentricity change depending on the position of the element.

2.3 Lateral Vibrations of an Elastic Shaft with a Disk at Its Center

2.3.1 Derivation of Equations of Motion

Figure 2.4 shows a vertical rotor executing a lateral whirling motion and a coordinate system. For simplicity, it is assumed that the rotor is vertical and the gravity force is not considered. The z-axis of the coordinate system O-xyz coincides with the bearing centerline, which connects the centers of the upper and lower bearings.

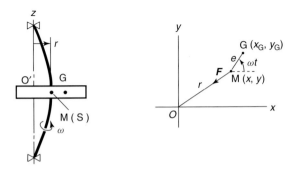

Figure 2.4 2 DOF rotor system for lateral vibration.

The disk has mass m, and its center of gravity $G(x_G, y_G)$ deviates slightly from the geometrical center M. Owing to errors in assembly, this point M does not always coincide with the point $S(x, y)$ through which the shaft centerline passes. The eccentricity $\overline{SG} = e$ produces the unbalance. In the following, it is assumed, for simplicity, that the geometrical center M coincides with the shaft center S, and the eccentricity is given by $\overline{MG} = e$.

Assume that the rotor spinning with an angular speed ω [rad/s] is executing a planar motion. Let the radial deflection of the rotor be r. Such an orbital motion is called a *whirl* in rotordynamics. An elastic restoring force $\boldsymbol{F}(F_x, F_y)$ works on a rotor at point S. The components of this elastic restoring force are expressed by $F_x = -kx$ and $F_y = -ky$, where k is a spring constant. Suppose that a damping force does not work. The equations of motion are given by

$$\left.\begin{aligned} m\ddot{x}_G &= -kx \\ m\ddot{y}_G &= -ky \end{aligned}\right\} \tag{2.1}$$

If we take $t = 0$ at the instance that $\overline{SG}$ coincides with the positive direction of the x-axis, the direction angle is represented by ωt. Substituting the relationships

$$\left.\begin{aligned} x_G &= x + e\cos\omega t \\ y_G &= y + e\sin\omega t \end{aligned}\right\} \tag{2.2}$$

into Eq. (2.1), we obtain

$$\left.\begin{aligned} m\ddot{x} + kx &= me\omega^2\cos\omega t \\ m\ddot{y} + ky &= me\omega^2\sin\omega t \end{aligned}\right\} \tag{2.3}$$

Equation (2.3) can be rearranged in the form

$$\left.\begin{aligned} \ddot{x} + p_0^2 x &= e\omega^2\cos\omega t \\ \ddot{y} + p_0^2 y &= e\omega^2\sin\omega t \end{aligned}\right\} \tag{2.4}$$

where $p_0 = \sqrt{k/m}$.

Angular speed is a magnitude of a vector "*angular velocity*." The terms *angular frequency* and *circular frequency* are also used instead of angular speed. In the field of rotordynamics, the terminology *rotational speed* is often used. Rotational speed means the number of revolution per unit time. The unit "rpm" expressing revolutions per minute is more common than the unit "rps" expressing revolutions per second. We can convert angular speed ω (rpm) to rotational speed by $\omega/(2\pi)$ [rps] or $60\omega/(2\pi)$ [rpm].

2.3.2
Free Vibrations of an Undamped System and Whirling Modes

Here, we consider a free whirling motion in the absence of unbalance and damping forces. It is governed by the following equations of motion obtained by substituting $e = 0$ in Eq. (2.4).

$$\left.\begin{aligned} \ddot{x} + p_0^2 x &= 0 \\ \ddot{y} + p_0^2 y &= 0 \end{aligned}\right\} \tag{2.5}$$

There is no coupling term between these two equations, and therefore each expression can be solved as a 1 DOF system. The response to the initial excitation is given by

$$\left.\begin{aligned} x &= A\sin(p_0 t + \alpha) \\ y &= B\sin(p_0 t + \beta) \end{aligned}\right\} \tag{2.6}$$

Each expression represents a free undamped vibration. The vibrations in the x- and y-directions are independent of each other but have the same angular speed p_0. The shape of this orbit in the xy-plane is generally an ellipse. When $\alpha = \beta$ holds, Eq. (2.6) represents a linear motion through the origin. In the analysis of a whirling motion, it is essential to interpret it as a superposition of circular motions. Indeed, the elliptical motion given by Eq. (2.6) can be expressed as a summation of two circular motions that whirl in directions opposite to each other. We can understand this by the following example:

Exercise 2.1: Decompose the planer motion given by Eq. (2.6) for $\alpha = \pi/2$ and $\beta = 0$ into circular motions.

Answer: In this case, Eq. (2.6) becomes

$$\left.\begin{aligned} x &= A\cos p_0 t \\ y &= B\sin p_0 t \end{aligned}\right\} \tag{1}$$

This represents the elliptical motion illustrated on the left side in Figure 1. By rearranging Eq. (1), we obtain

$$\left.\begin{aligned} x &= \frac{1}{2}(A+B)\cos p_0 t + \frac{1}{2}(A-B)\cos(-p_0 t) \\ y &= \frac{1}{2}(A+B)\sin p_0 t + \frac{1}{2}(A-B)\sin(-p_0 t) \end{aligned}\right\} \tag{2}$$

From this expression, we see that this elliptical motion is composed of two circular motions with the same angular speed p_0, but moving in opposite directions. The whirl in a counterclockwise direction has the radius $(A+B)/2$, and that in a clockwise direction $|A-B|/2$.

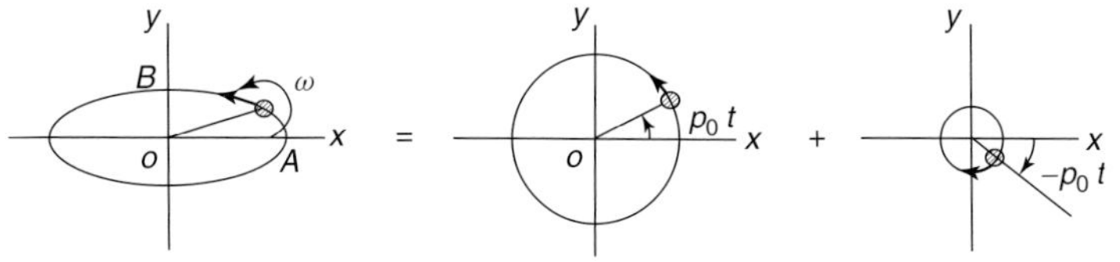

Figure 1 Decomposition of a whirling motion.

A planer motion of a rotor is called a *whirling motion* or a *whirl*. And a circular whirl in the same direction as the shaft rotation is called a *forward whirl*, and that in the opposite direction is termed a *backward whirl*. The angular speed p of a free whirling motion is called a *natural angular frequency* (or shortly *natural frequency*). Instead of natural angular frequencies, *natural frequencies* given by $f = p/(2\pi)$ [rps, Hz] or $f = 60p/(2\pi)$ [rpm] are also used.

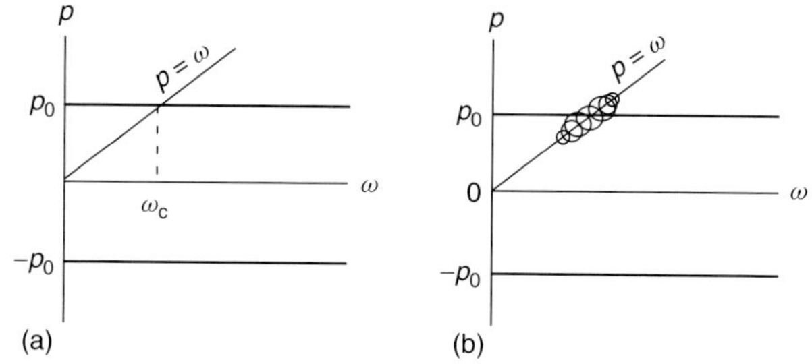

Figure 2.5 Natural frequency diagram (Campbell diagram).

Plot of these natural angular frequencies versus the rotational speed is called a *natural angular frequency diagram* (or shortly *natural frequency diagram*) or a $p - \omega$ diagram. Positive and negative values of p correspond to forward and backward whirls, respectively. Diagram of natural frequencies f versus the rotational speed n is also used.

In the Jeffcott rotor discussed in this section, its natural frequencies have values $p = \pm p_0 = \pm\sqrt{k/m}$, which do not depend on the angular speed ω. Its natural frequency diagram is illustrated in Figure 2.5a where natural angular frequencies become horizontal straight lines in the diagram. Resonance mentioned later occurs at the angular frequency ω_c given by the cross point of this straight line and the line $p = \omega$. A forced oscillation becomes large in the neighborhood of this resonance frequency. Sometimes, radii of these whirling motions are represented by relative sizes of circles in the diagram as shown in Figure 2.5b. These diagrams are called *Campbell diagram*.

Note: Campbell Diagram

Wilfred Campbell was born in Manchester, England, in 1884. When he served an apprenticeship with the Lancashire and Yorkshire Railway, he was selected as one of two to attend the Manchester Municipal School of technology. In 1907, he went to America. In 1909, he was employed by the General Electric Company as a draftsman. From 1919 on, he worked as an engineer in the Turbine Engineering Department of GE and made outstanding contributions to turbine design. He used the frequency versus rotational speed plots extensively in his work and Figure 2.6 is an example. This type of diagram has been named the *Campbell diagram* after him. He passed away on July 7, 1924, after an attack of acute appendicitis. (This biographical note is reprinted partly through the courtesy of the General Electric Co., GER, 1924.)

2.3.3 Synchronous Whirl of an Undamped System

A whirling motion governed by Eq. (2.3) or (2.4) is discussed. Since each equation has the same form of a rectilinear system of 1 DOF, it can be solved in the similar way.

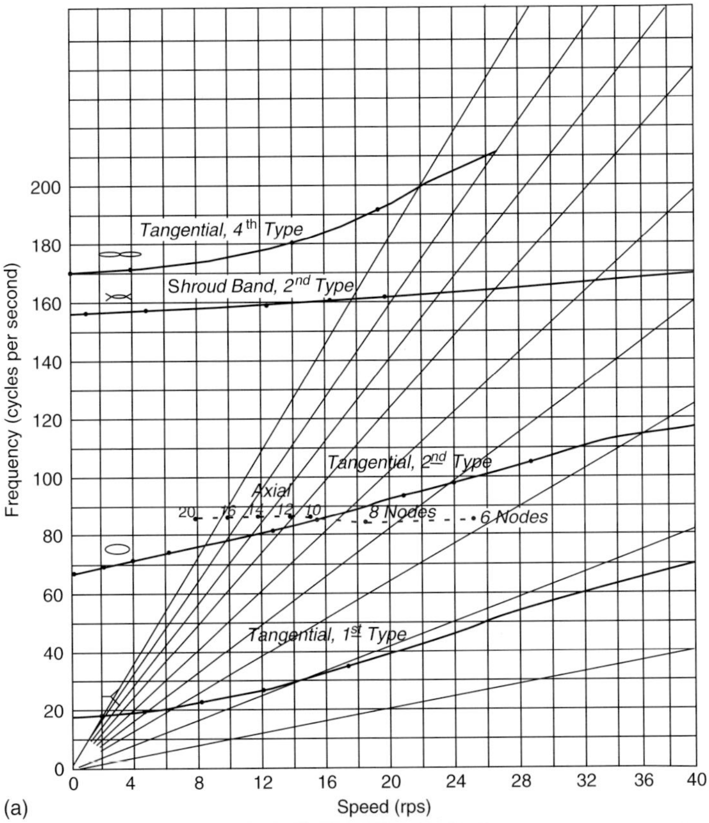

(a)

(b)

Figure 2.6 Tangential and axial vibration of turbine wheel and bucket: (a) Campbell diagram (Campbell and Heckman, 1925 courtesy of ASME) and (b) Wilfred Campbell (courtesy of GE).

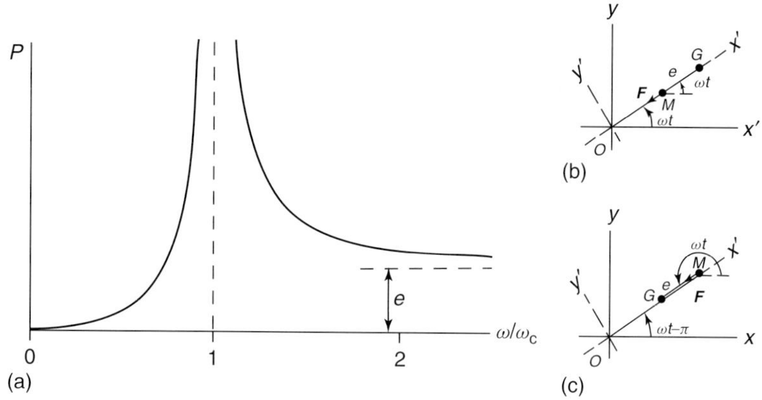

Figure 2.7 Response curve and configuration in a whirling motion: (a) response curve, (b) $\omega < \omega_c$ and (c) $\omega > \omega_c$.

The general solution of Eq. (2.4) is obtained by adding a particular solution to the solution of its homogeneous equation as follows:

$$\left.\begin{aligned} x &= A_1 \cos p_0 t + B_1 \sin p_0 t + \frac{e\omega^2}{p_0{}^2 - \omega^2} \cos \omega t \\ y &= A_2 \cos p_0 t + B_2 \sin p_0 t + \frac{e\omega^2}{p_0{}^2 - \omega^2} \sin \omega t \end{aligned}\right\} \tag{2.7}$$

Since the damping always exists in a practical system, the first two components diminish and the steady-state response is given by the third term as follows:

$$\left.\begin{aligned} x &= P\cos(\omega t + \beta) \\ y &= P\sin(\omega t + \beta) \end{aligned}\right\} \tag{2.8}$$

where

$$\left.\begin{aligned} &P = \frac{e\omega^2}{\left|p_0{}^2 - \omega^2\right|} \\ &\beta = 0 \;(\text{for } \omega < p_0), \quad \beta = -\pi \;(\text{for } \omega > p_0) \end{aligned}\right\} \tag{2.9}$$

This result is illustrated in Figure 2.7a. Equation (2.8) represents a *synchronous whirling motion* with a constant radial deflection. Figure 2.7(a) shows that the deflection P becomes ∞ at $\omega = p_0$. This resonance rotational speed ω is called the *critical speed* or, more specifically, *the major critical speed*, and it is denoted by ω_c.

The bearing centerline O, the geometrical center M, and the center of gravity G have the relative positions shown in Figure 2.7b,c. In a very high rotational speed region, the deflection becomes $P \approx e$ and the center of gravity G almost coincides with the bearing center O. This characteristic is called *the self-centering*.

Note: Centrifugal Force and the Coriolis Force

Equation (2.3) is expressed in the static coordinates (or the inertia coordinates) O−xy. Now, we transform this into the expression in the rotating coordinate O−$x'y'$ which rotates with a constant angular speed ω. The transformation is given by

$$\left.\begin{aligned} x &= x' \cos \omega t - y' \sin \omega t \\ y &= x' \sin \omega t + y' \cos \omega t \end{aligned}\right\} \tag{2.10}$$

Substituting this into Eq. (2.3), we obtain

$$\left.\begin{aligned} m\ddot{x}' &= -kx' + 2m\dot{y}'\omega + m(x' + e)\omega^2 \\ m\ddot{y}' &= -ky' - 2m\dot{x}'\omega + my'\omega^2 \end{aligned}\right\} \tag{2.11}$$

The motion of the center of gravity is given by $x'_G = x' + e$ and $y'_G = y'$. This equation shows that, when we investigate the motion in the rotating coordinate system, we must consider apparent forces $(2m\dot{y}'\omega, -2m\dot{x}'\omega)$ and $(mx'_G\omega^2, my'_G\omega^2)$ in addition to the real restoring force $\boldsymbol{F}(-kx', -ky')$ of the spring. The former is called *Coriolis force* and the latter is called *the centrifugal force*.

Note: Stability in the Speed Range Higher than the Major Critical Speed

If we observe the steady-state whirling motion in the rotating coordinate system $O\text{-}x'y'$ rotating with the angular speed ω, the centrifugal force $\boldsymbol{C}$ appears and works on G as mentioned in the previous "Note: Centrifugal Force and the Coriolis Force." However, since the Coriolis force is proportional to the velocity, it does not appear in this steady-state motion. In the speed range lower than the major critical speed, substituting the steady-state solutions $x = P\cos\omega t$ and $y = P\sin\omega t$ into Eq. (2.10), we obtain $x' = P$ and $y' = 0$. Then from Eq. (2.11), we obtain $kP = m(P + e)\omega^2$. This expression shows that the spring force $F = kP$ and the centrifugal force $C = m(P + e)\omega^2$ are in balance. Similarly, in a high speed range, the solution is $x = -P\cos\omega t$ and $y = -P\sin\omega t$ and $x' = -P$ and $y' = 0$. Then, we see that $F = kP$ and $C = m(P - e)\omega^2$ are in balance. These forces are illustrated in Figure 2.8.

In Figure 2.8b for the range above the major critical speed, the centrifugal force C acting at G and the restoring force F acting at M work in the directions opposite to and facing one another. Therefore, this relative position seems to be impossible to reach. As introduced in Section 1.2, it was believed that the operation above the major critical speed is impossible in the early days of the history of rotordynamics. This incorrect understanding was probably based on such an idea. The possibility of stable operation above the major critical speed must be investigated by a theoretical stability analysis. It was proved that this

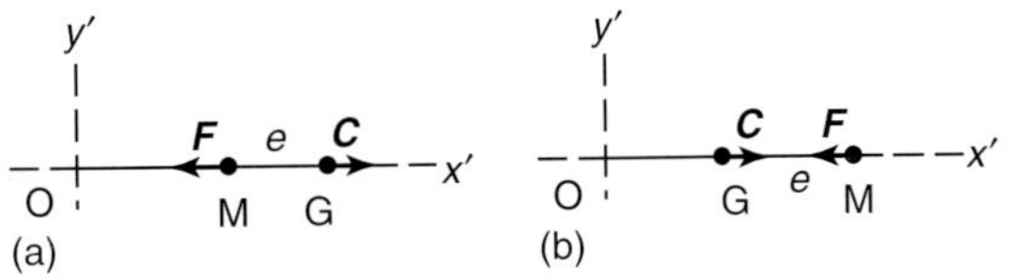

Figure 2.8 Forces in the rotating coordinate system: (a) $\omega < \omega_c$ and (b) $\omega > \omega_c$.

equilibrium state is stable under the action of the Coriolis force (Stodola, 1924). Since the rotor has velocity when it deviates slightly from the equilibrium state, the Coriolis force appears and this force restores the condition. A similar proof is given in Appendix B.

Equation (2.9) is not valid at the major critical speed where $\omega = p_0$ holds. Therefore, we derive the solution at the major critical speed as follows. When $\omega = p_0$, Eq. (2.4) becomes

$$\left.\begin{aligned} \ddot{x} + p_0^2 x &= e p_0^2 \cos p_0 t \\ \ddot{y} + p_0^2 y &= e p_0^2 \sin p_0 t \end{aligned}\right\} \tag{2.12}$$

The particular solution for the initial condition $x(0) = y(0) = 0$ and $\dot{x}(0) = \dot{y}(0) = 0$ can be assumed to have the form $x(t) = r(t)\sin p_0 t$ and $y(t) = r(t)\cos p_0 t$. Substituting this solution into Eq. (2.12) and equating the coefficients of $\cos p_0 t$ and $\sin p_0 t$, respectively, on both sides, we get $\ddot{r} = 0$ and $\dot{r} = ep_0/2$. Then, we can write the solution in the following form:

$$\left.\begin{aligned} x &= \frac{ep_0}{2} t \sin \omega t = \frac{ep_0}{2} t \cos\left(\omega t - \frac{\pi}{2}\right) \\ y &= -\frac{ep_0}{2} t \cos \omega t = \frac{ep_0}{2} t \sin\left(\omega t - \frac{\pi}{2}\right) \end{aligned}\right\} \tag{2.13}$$

The deflection $r(t) = (ep_0/2)t$ diverges, tracing a spiral orbit as shown in Figure 2.9. The phase of the eccentricity $\overline{\text{MG}}$ is always 90° ahead of the phase of the deflection $\overline{\text{OM}}$, and the kinetic and potential energies of the system increases continuously.

2.3.4
Synchronous Whirl of a Damped System

Figure 2.10 shows a damped rotor system. If the rotational speed is constant, the angular position of the rotor is expressed by ωt. Assume that the viscous damping force $\boldsymbol{D}(D_x, D_y)$ acts in the direction opposite to the velocity of the disk. Adding the damping terms $D_x = -c\dot{x}$ and $D_y = -c\dot{y}$ to Eq. (2.3), we obtain the equations of motion:

$$\left.\begin{aligned} m\ddot{x} + c\dot{x} + kx &= me\omega^2 \cos \omega t \\ m\ddot{y} + c\dot{y} + ky &= me\omega^2 \sin \omega t \end{aligned}\right\} \tag{2.14}$$

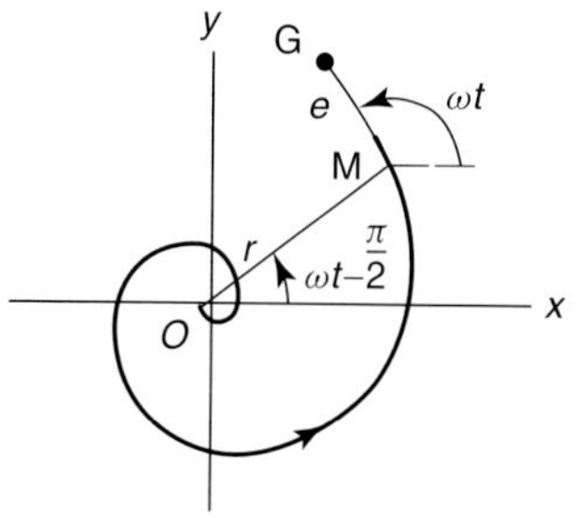

Figure 2.9 Spiral orbit at the major critical speed.

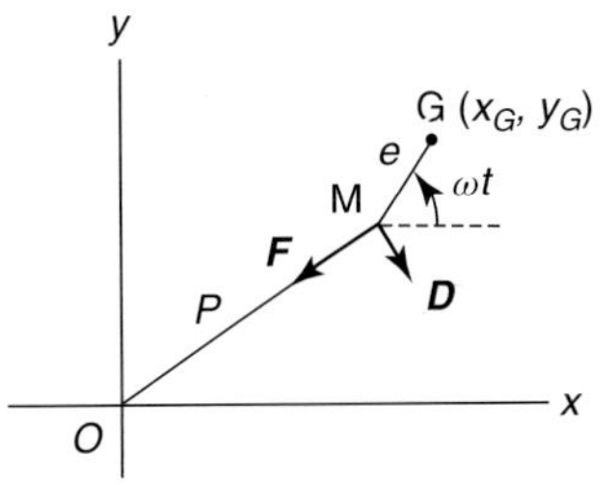

Figure 2.10 Damped 2 DOF system.

We assume a steady-state solution in the same form as Eq. (2.8) as follows:

$$\left.\begin{aligned} x &= P\cos(\omega t + \beta) \\ y &= P\sin(\omega t + \beta) \end{aligned}\right\} \tag{2.15}$$

Substituting these expressions into Eq. (2.14) and equating the coefficients of $\cos(\omega t + \beta)$ and $\sin(\omega t + \beta)$, respectively, on both sides, we obtain the expressions for the amplitude P and the phase angle β:

$$\left.\begin{aligned} (k - m\omega^2)P &= me\omega^2 \cos\beta \\ -cP\omega &= me\omega^2 \sin\beta \end{aligned}\right\} \tag{2.16}$$

From this, we obtain the expressions for the amplitude P and the phase angle β as follows:

$$\left.\begin{aligned} P &= \frac{e\omega^2}{\sqrt{(p_0^2 - \omega^2)^2 + (e\omega/m)^2}} \\ \tan\beta &= -\frac{c\omega/m}{p_0^2 - \omega^2} \end{aligned}\right\} \tag{2.17}$$

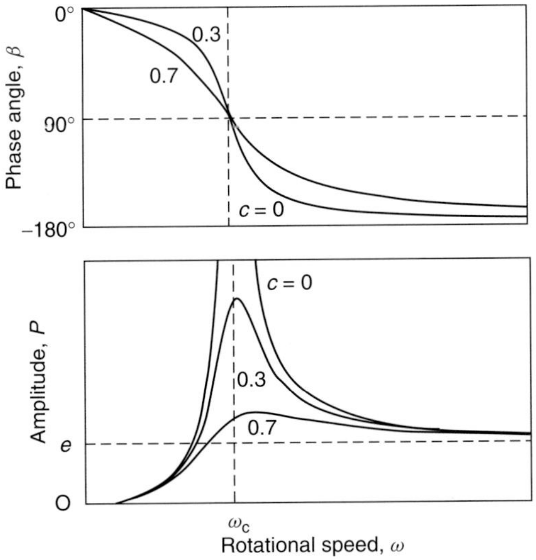

Figure 2.11 Effect of damping to the response curve.

Figure 2.11 shows responses for various values of the damping coefficient c. The damping diminishes the maximum amplitude in the vicinity of the major critical speed ω_c. In cases where damping is present, the phase angle changes continuously and always has the value $-\pi/2$ at the major critical speed ω_c.

2.3.5
Energy Balance

In this section, the energy balance of a disk during a synchronous whirling motion is considered. Let us assume that the disk moves along a circular orbit shown by the broken line in Figure 2.11. In addition to the spring force $\boldsymbol{F} = (-kx, -ky)$ and the damping force $\boldsymbol{D} = (-c\dot{x}, -c\dot{y})$, a torque T_s works on the disk from the shaft. If a resisting torque due to the bearings is neglected, this torque T_s is equal to the driving torque T_d applied by the motor. The restoring force $\boldsymbol{F}$ does not work at the disk because it is perpendicular to the orbit. The damping force $\boldsymbol{D}$ works in the direction opposite to the movement and therefore it dissipates the energy of the rotor. The work W_{out} dissipates during the period T corresponding to one cycle, which is given as follows:

$$\begin{aligned} W_{out} &= -\int_0^T \mathbf{D} \cdot \mathrm{d}\mathbf{r} = -\int_0^T (D_x \mathrm{d}x + D_y \mathrm{d}y) \\ &= \int_0^T (c\dot{x} \cdot \dot{x} + c\dot{y} \cdot \dot{y})\mathrm{d}t = 2\pi c\omega P^2 \end{aligned} \tag{2.18}$$

The equations of motion for the rotation of the disk around the center of gravity is

$$I_p\ddot{\psi} = (F_x + D_x)e \sin\psi - (F_y + D_y)e \cos\psi + T_s \tag{2.19}$$

where I_p is the polar moment of inertia of the disk. If the angular speed $\dot{\psi} = \omega$ is constant, we obtain

$$0 = (-kx - c\dot{x})e \sin\omega t - (-ky - c\dot{y})e \cos\omega t + T_d \tag{2.20}$$

Substituting the relationship of Eq. (2.15), we obtain

$$M = -keP \sin\beta - ceP\omega \cos\beta = c\omega P^2 \tag{2.21}$$

The work W_{in} done by the torque T_d during the period T is shown by

$$W_{in} = \int_0^T T_d \mathrm{d}\psi = \omega \int_0^T T_d \mathrm{d}t = 2\pi c\omega P^2 \tag{2.22}$$

From the comparison of Eqs. (2.18) and (2.22), we know that the dissipated energy is supplied by the driving torque of the motor.

2.4
Inclination Vibrations of an Elastic Shaft with a Disk at Its Center

2.4.1
Rotational Equations of Motion for Single Axis Rotation

Before we discuss the derivation of the equations of motion for inclination, let us review the basic rotational equations of motion. For the sake of simplicity, we consider the case that a disk is rotating around an axis in a fixed direction as shown in Figure 2.12a. In the translational motion, Newton's second law of motion holds that the time derivative of momentum is equal to the force applied to the mass. Similarly, the time derivative of the angular momentum $L = I\omega$ is equal to the moment M applied to the rotor. This relationship is expressed as follows:

$$\frac{d}{dt}(I\omega) = M \tag{2.23}$$

where I is the moment of inertia and $\omega = \dot{\theta}$ is the angular speed.

Since the momentum and the angular momentum can be expressed by vectors as shown in Figure 2.12b, we represent Eq. (2.23) by vectors and transform it as follows:

$$d\boldsymbol{L} = \boldsymbol{M}dt \tag{2.24}$$

This expression can be interpreted as follows: the moment $\boldsymbol{M}$ working on the body during the time interval dt produces the increment d$\boldsymbol{L}$ given by Eq. (2.24) in the same direction as $\boldsymbol{M}$. This vector relationship holds also in a three-dimensional space treated in the following.

2.4.2
Equations of Motion

Figure 2.13 shows an inclination system where a disk is mounted at the center of a flexible shaft. This 2 DOF system is the simplest rotor system that has a gyroscopic effect. In this section we derive the equations of motion for the inclination vibration on the assumption that the inclination angle of the disk is small.

We use the rectangular coordinate system O-xyz whose z-axis coincides with the bearing center line. Suppose that the rotor undergoes inclination vibrations only

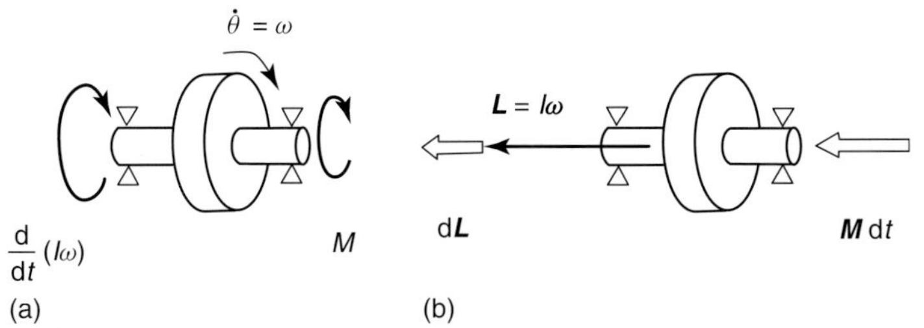

Figure 2.12 Relationship between the moment and the increment of angular momentum.

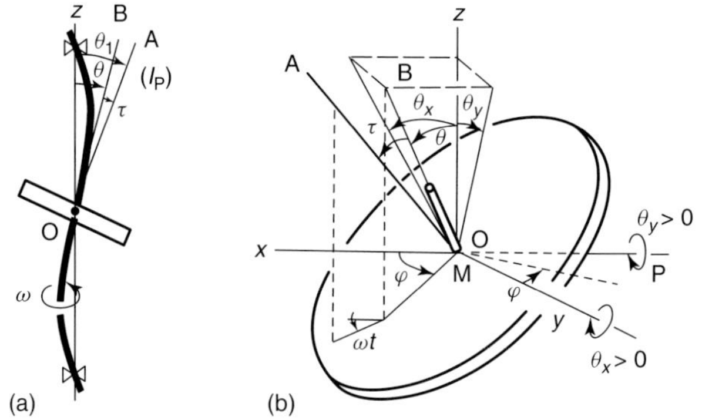

Figure 2.13 2 DOF rotor for an inclination oscillation: (a) rotor model and (b) coordinate system.

without lateral deflection and that the geometrical center M of the disk always stays on the bearing centerline. The inclination angle of the shaft at the position of the disk is denoted by θ and the disk centerline OA inclines from this tangential direction OB by τ. These two angles θ and τ are generally not in the same direction. The inclination angle of the centerline OA, which coincides with the principal axis of the polar moment of inertia of the disk I_p, is denoted by θ_1. The projections of the shaft inclination angle θ onto the xz- and yz-planes are denoted by θ_x and θ_y, and those of the rotor inclination angle θ_1 are denoted by θ_{1x} and θ_{1y}, respectively. If the angle θ is small, the relationships

$$\theta_x \approx \theta \cos\varphi, \quad \theta_y \approx \theta \sin\varphi \tag{2.25}$$

hold approximately, where φ is the direction in which the axis OB inclines.

In the derivation of the equations of motion, the use of Eulerian angles (explained in detail in Appendix C) is recommended as the most reliable means. However, here we adopt a procedure that is more suitable for grasping the physical meaning of each quantity.

The angular momentum vectors of the rotor shown in Figure 2.13 are shown in Figure 2.14a. Let the diametral moment of inertia and the polar moment of inertia

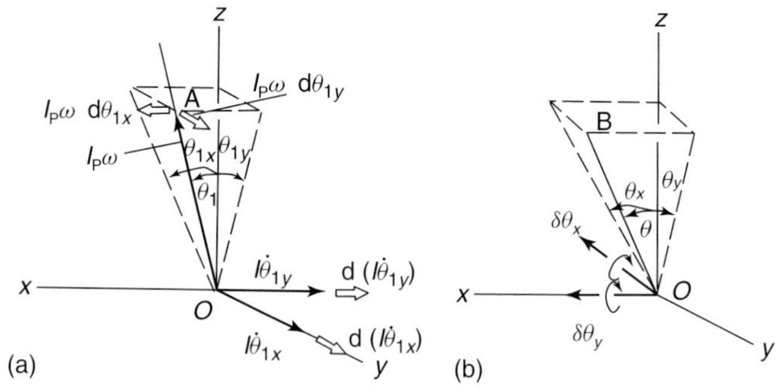

Figure 2.14 Angular momentum vectors and restoring moment vectors: (a) angular momentum vectors and (b) restoring moments.

be I and I_p, respectively. The unit vectors in the x- and y-directions are denoted by $\mathbf{e}_x$ and $\mathbf{e}_x$, respectively. The diametrical direction of the rotor is perpendicular to the direction OA; however, we can take the principal axes to the direction in the x- and y-directions approximately under the assumption that the rotor inclination angle θ_1 is small. When a nonrotating rotor inclines, the angular momentum vector $\boldsymbol{L}$ has the component $I\dot{\theta}_{1x}\mathbf{e}_y$ in the positive y-direction and $-I\dot{\theta}_y\mathbf{e}_x$ in the negative x-direction. If this disk is rotating around the axis OA with the angular speed ω, the angular momentum vector of magnitude $I_p\boldsymbol{\omega}$ is added in the direction OA. In order to derive the equations of motion, we use Eq. (2.24) for $\boldsymbol{M}$ and $\boldsymbol{L}$. In order to apply this relationship, we must first obtain the decrements of these vectors. Suppose that the angles θ_{1x} and θ_{1y} changes by $\mathrm{d}\theta_{1x}$ and $\mathrm{d}\theta_{1y}$, respectively in a small time $\mathrm{d}t$, then the variations shown in Figure 2.14a appear in the angular momentums. Specifically, an increment $-\mathrm{d}(I\dot{\theta}_{1y})\mathbf{e}_x$ appears in the negative x-direction because of the change in the angular speed $\dot{\theta}_{1y}$, and $\mathrm{d}(I\dot{\theta}_{1x})\mathbf{e}_y$ appears in the positive y-direction because of the change in $\dot{\theta}_{1x}$. In addition, as the direction of the angular momentum vector $I_p\boldsymbol{\omega}$ moves when the magnitude of the angles θ_{1x} and θ_{1y} changes, the increments $+I_p\omega\mathrm{d}\theta_{1x}\mathbf{e}_x$ and $+I_p\omega\mathrm{d}\theta_{1y}\mathbf{e}_y$ appear in the positive x- and y-directions, respectively. Therefore, the total increment during a small time $\mathrm{d}t$ in the negative x-direction is $\mathrm{d}(I\dot{\theta}_{1y}) - I_p\omega\mathrm{d}\theta_{1x}$, and that in the positive y-direction is $\mathrm{d}(I\dot{\theta}_{1x}) + I_p\omega\mathrm{d}\theta_{1y}$. And the increment of the angular momentum during the small time duration $\mathrm{d}t$ is given by

$$\mathrm{d}\boldsymbol{L} = \left\{-\mathrm{d}(I\dot{\theta}_{1y}) + I_p\omega\mathrm{d}\theta_{1x}\right\}\mathbf{e}_x + \left\{\mathrm{d}(I\dot{\theta}_{1x}) + I_p\omega\mathrm{d}\theta_{1y}\right\}\mathbf{e}_y \tag{2.26}$$

These changes are caused by the restoring moment of the elastic shaft. From Figure 2.14b, the components of this restoring moment are $-\delta\theta_x\mathbf{e}_y$ in the negative y-direction and $\delta\theta_y\mathbf{e}_x$ in the positive x-direction, where δ is the spring constant of the elastic shaft for the inclination. Therefore, the restoring moment vector is expressed by

$$\boldsymbol{M} = \delta\theta_y\mathbf{e}_x - \delta\theta_x\mathbf{e}_y \tag{2.27}$$

Substituting Eqs. (2.26) and (2.27) into Eq. (2.24) and comparing the coefficients of the same unit vectors, we have

$$\left.\begin{aligned} I\ddot{\theta}_{1x} + I_p\omega\dot{\theta}_{1y} &= -\delta\theta_x \\ I\ddot{\theta}_{1y} - I_p\omega\dot{\theta}_{1x} &= -\delta\theta_y \end{aligned}\right\} \tag{2.28}$$

Selecting the time $t = 0$ at the instant that the direction of τ coincides with the x-direction in Figure 2.13, we have

$$\left.\begin{aligned} \theta_{1x} &= \theta_x + \tau\cos\omega t \\ \theta_{1y} &= \theta_y + \tau\sin\omega t \end{aligned}\right\} \tag{2.29}$$

Substituting this into Eq. (2.28), we get

$$\left.\begin{aligned} I\ddot{\theta}_x + I_p\omega\dot{\theta}_y + \delta\theta_x &= (I - I_p)\tau\omega^2\cos\omega t \\ I\ddot{\theta}_y - I_p\omega\dot{\theta}_x + \delta\theta_y &= (I - I_p)\tau\omega^2\sin\omega t \end{aligned}\right\} \tag{2.30}$$

Exercise 2.2: Figure 1a illustrates a top called a *gyroscope*. When a gyroscope spinning at a high speed is placed on the top of the column, it begins to whirl slowly. Such a motion is referred to as *precession*. Suppose that the axis of spin is horizontal. Let the spin speed of the gyroscope be $\Omega = 600$ rpm, the mass m and radius R of the disk be 100 g and 3.5 cm, respectively, and the distance a between the center of gravity and the end of the shaft be 4 cm. Let us consider the number of times this gyroscope whirls per minute around the point O. The polar moment of inertia of the disk is given by $I_p = mR^2/2$ and the mass of the frame (gimbal) can be neglected.

Answer: Let us assume that a vertical plane P containing the shaft rotates with the angular speed p. The angle between plane P and the x-axis is represented by $\varphi = pt$. The angular momentum vector $\boldsymbol{L}$ with magnitude $L = I_p\Omega$ is inside this plane. The gravitational force mg and the normal force $N(= mg)$ work on the gyroscope as shown in the figure above. The friction acting in the horizontal direction is neglected in the figure. The magnitude of the moment due to the gravitational force is expressed by mga. The moment vector (not illustrated) perpendicular to the plane P produces the increment $\mathrm{d}\boldsymbol{L}$ of the angular momentum vector in the same direction, perpendicular to the plane P. The magnitude of this increment is $I_p\Omega\,\mathrm{d}\varphi$. Therefore, from Eq. (2.24), we have

$$I_p\Omega\,\mathrm{d}\varphi = mga\,\mathrm{d}t \tag{1}$$

Since $\mathrm{d}\varphi/\mathrm{d}t = p$ and $I_p = mR^2/2$, this expression becomes

$$(mR^2/2)\Omega p = mga \tag{2}$$

Substituting the concrete values of each parameter, we have

$$p = \frac{2ga}{\Omega R^2} = \frac{2 \times (9.8\ \mathrm{m\ s^{-2}}) \times (0.04\ \mathrm{m})}{(2 \times 3.14 \times 600/60\ \mathrm{s^{-1}}) \times (0.035\ \mathrm{m})^2} = 10.2\ \mathrm{rad\ s^{-1}}$$
$$= 97\ \mathrm{rpm} \tag{3}$$

Namely, the gyroscope rotates 97 times per minute.

[note] The meaning of the notation M_g will be explained in the NOTE in Section 2.4.4.

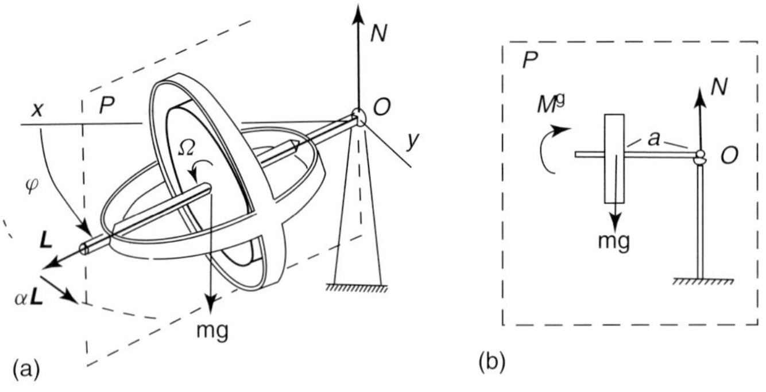

Figure 1 Precessional motion.

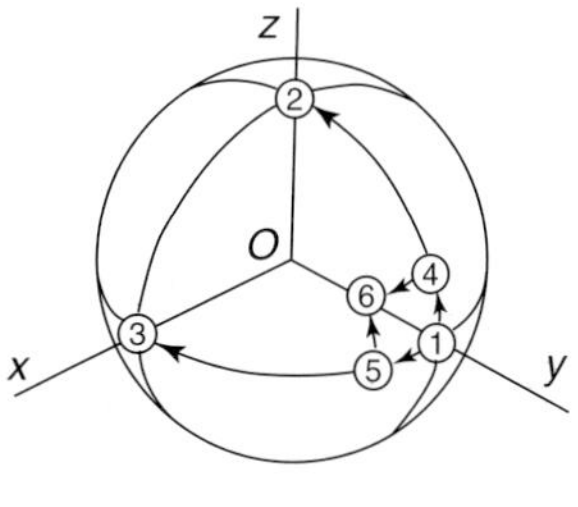

Figure 2.15 Rotation of a rigid body.

Note: Is Rotation a Vector?

In Figure 2.13b, the angle θ is the rotation around the axis OP. Therefore, if we apply *the right-hand grip rule*, it seems that the angle can be expressed by a rotation "vector" that has the magnitude θ and the direction $O \rightarrow P$. However, not all quantities that have well-defined magnitudes and directions are necessarily vectors. Fitzpatrick (2006) explained this problem as follows: One of the properties of vectors is the commutative law. For example, if $\boldsymbol{a}$ and $\boldsymbol{b}$ are vectors, the relationship $\boldsymbol{a} + \boldsymbol{b} = \boldsymbol{b} + \boldsymbol{a}$ holds. This means that it is independent of the order of addition. Let us check whether the rotation angles have this property. Figure 2.15 shows a sphere that can rotate in any direction. *Case 1*: When the sphere rotates by 90° about the x-axis and then rotates by -90° about the z-axis, the point ① on the sphere moves as ① → ② → ②. *Case 2*: If the steps of Case 1 are executed vice versa, the point ① moves as ① → ③ → ③. Comparing these two cases, we see that the commutative law does not hold in this case. This means that the rotation angle cannot be represented by a vector. Next, let us consider the case that all the rotation angles are small enough. *Case 3*: When the sphere is rotated by 1° about the x-axis and then is rotated by -1° about the z-axis, the point ① on the sphere moves as ① → ④ → ⑥. *Case 4*: If the steps of Case 3 are done vice versa, the point ① moves as ① → ⑤ → ⑥. These two cases give the same result. Namely, the rotations with small angles are commutative. In other words, although a rotation angle θ is not a vector but an infinitesimal rotation angle $\delta\theta$ around an axis can be represented by a vector. Since the angular speed is defined by a ratio of a small rotation angle $\delta\theta$ to an infinitesimal time interval dt, the angular speed $\omega = \delta\theta/\mathrm{d}t$ is represented by a vector $\boldsymbol{\omega}$.

Now, let the unit vectors in the x, y, z-directions in the rectangular coordinate system O-xyz be e_x, e_y, and e_z, respectively. When a rigid body is rotating about the x-axis at an angular speed ω_x, about the y-axis at ω_y and about the z-axis at ω_z, the angular speed of the body is expressed by $\boldsymbol{\omega} = \omega_x \boldsymbol{e}_x + \omega_y \boldsymbol{e}_y + \omega_z \boldsymbol{e}_z$. However, the rotation angle cannot be expressed in this way.

2.4.3
Free Vibrations and Natural Angular Frequency

In this section, we discuss a free whirling motion. In the absence of the skew angle τ, Eq. (2.30) becomes

$$\left.\begin{aligned} I\ddot{\theta}_x + I_p\omega\dot{\theta}_y + \delta\theta_x = 0 \\ I\ddot{\theta}_y - I_p\omega\dot{\theta}_x + \delta\theta_y = 0 \end{aligned}\right\} \tag{2.31}$$

Differing from the case described in Eq. (2.5), these two equations are coupled because of the existence of the first derivative terms called *the gyroscopic terms*.

Suppose that the solution in the θ_x-direction has the form $\theta_x = A\cos(pt + \beta')$. Then, inspection of Eq. (2.31) reveals that the solution in the θ_y-direction must have the form $\theta_y = R_y \sin(pt + \beta')$ and the relationship $R_x = R_y$ holds. Therefore, we can assume that

$$\left.\begin{aligned} \theta_x = A\cos(pt + \beta') \\ \theta_y = A\sin(pt + \beta') \end{aligned}\right\} \tag{2.32}$$

Substituting this into Eq. (2.31), we get the following frequency equation:

$$Ip^2 - I_p\omega p - \delta = 0 \tag{2.33}$$

The roots of this equation, p_f and p_b, are given by

$$\left.\begin{aligned} p_f = \frac{I_p\omega + \sqrt{(I_p\omega)^2 + 4\delta I}}{2I}(> 0) \\ p_b = \frac{I_p\omega - \sqrt{(I_p\omega)^2 + 4\delta I}}{2I}(< 0) \end{aligned}\right\} \tag{2.34}$$

This result shows that the natural frequencies change as a function of the rotational speed ω. If ω is assumed to be positive, p_f is always positive and p_b is negative. This means that p_f is the natural frequency of a forward whirling mode and p_b is that of a backward whirling mode. Adopting the terminology used in physics to explain the motion of spinning tops, the former whirling mode is sometimes called *a forward precession* and the latter *a backward precession*.

Typical plots of the natural frequencies p_f and p_b versus the rotational speed ω are shown in Figure 2.16. Figure 2.16a is an example of a disk-shaped rotor

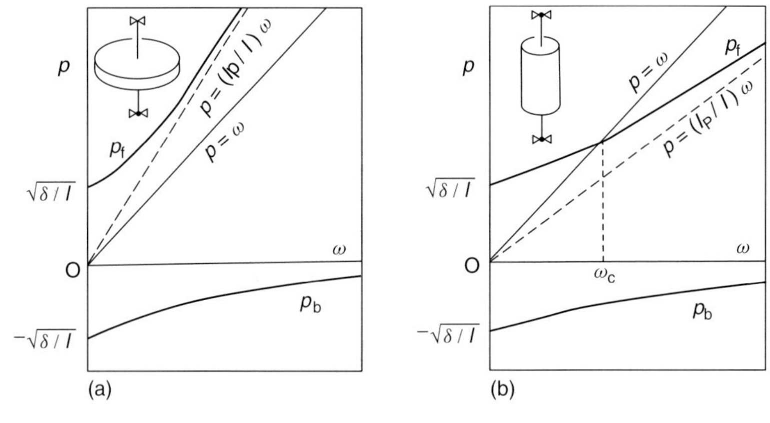

Figure 2.16 Natural frequency diagram: (a) disk-shaped rotor ($I_p > I$) and (b) cylindrical rotor ($I_p < I$).

and Figure 2.16b is that of a cylindrical rotor. As the rotational speed ω increases, the natural frequency of a forward whirling mode p_f increases and approaches the straight line $p = (I_p/I)\omega$ asymptotically. Meanwhile, the absolute value of the natural frequency of a backward whirling mode p_b decreases and approaches the value zero. Resonance occurs at w_c given by the cross point of the straight line $p = \omega$ and the curve p_f.

2.4.4
Gyroscopic Moment

In Figure 1a in Exercise 2.2, if we observe the top from the vertical plane P, moving together with the axis of the top, the top appears to stand in the same position with a constant inclination angle of $\theta = \pi/2$ as shown in Figure 1b in Exercise 2.2. Since the gravitational force mga is at work, we must consider that a countermoment with the same magnitude is working in the opposite direction. In this case, Eq. (2.28) becomes

$$-\frac{dL}{dt} + mga = 0 \tag{2.35}$$

This apparent (or inertia/fictitious) moment $M_g = -dL/dt$ working in the opposite direction is called a *gyroscopic moment*. In Exercise 2.2, it was given by the expression $M_g = -I_p\Omega p$.

In Figure 2.8, in which a whirling motion of a Jeffcott rotor is depicted with rotating coordinates, the spring force and the centrifugal force (which is an inertia force) are in balance. The gyroscopic moment is the inertia couple corresponding to the inertia force in a deflection motion. In Eq. (2.35), the moment due to gravity and the gyroscopic moment are in balance. When we hold the gyroscope shown in Figure 1 in Exercise 2.2 and turn it toward the direction of φ in the horizontal plane, the gyroscopic moment that tends to turn the gyroscope upward can be felt. Conversely, when we turn the top in the opposite direction, this gyroscopic moment works downward.

The above mentioned explanation is a special case for $\theta = \pi/2$ treated in Exercise 2.2. In the general case of arbitrary inclination angles, the expression differs. In the following, we derive the general expression for the gyroscopic moment. Here, the inclination θ has no restrictions in terms of its smallness.

In Figure 2.17a, a rotor is spinning around the axis OA with an angular speed Ω, and this axis OA executes a precession with an angular speed p keeping the inclination angle θ constant. The shaft is connected with the axis of the motor rotating with the angular speed ω by a spring. It is assumed that this spring can transmit the driving torque but does not exert a restoring moment. The motion of a rotor is governed by the equations of motion for the translation and those for the rotation around the center of gravity. In the following, we discuss the latter case. Now, we use the subscripts 1, 2, and 3 for the quantities along each principal axis of the rotor. Notation ① represents the direction of the symmetrical axis and notations ② and ③ represent the diametral directions. The direction ③ is not depicted in the Figure. For the sake of convenience, we use a fictitious model shown in Figure 2.17b and c. In this model, a rotor whose center of gravity is on the z-axis can rotate smoothly in the pipe. This pipe has a rigid vertical shaft that is supported by two bearings. In Figure 2.17b, a rotor rotates with the angular speed Ω in the pipe keeping the direction of OA constant. In this condition, the angular velocity vector $\boldsymbol{\Omega}$ is pointing toward the direction of ①. In Figure 2.17c, the pipe rotates around the rigid shaft with the angular speed p but the rotor is not rotating inside the pipe ($\Omega = 0$). The angular velocity vector $\boldsymbol{p}$, which has the same direction as the rigid axis, can be decomposed into the components in the directions of the principal axes of inertia. The component in the direction ① is $p\cos\theta$ and that in the direction ② is $p\sin\theta$. The rotor shown in Figure 2.17a executes these two motions simultaneously. Therefore, its angular velocity vector is obtained by summing up the angular velocity vectors of these two motions.

The terminology "rotational speed" used hereon corresponds to the angular velocity ω of the motor shown in Figure 2.17a. The angular speed Ω is relative to the rotating plane P, in other words it is the angular speed which is observed by a person from the plane P. The following expression illustrates the relationship between the two angular velocities:

$$\omega = \Omega + p \tag{2.36}$$

For example, in Figure 2.17a, if the rotor turns once keeping the plane P in the same direction, the axis of the motor also turns once. Next, the plane P turns around once keeping the rotor fixed to the plane, the axis of the motor then turns once further.

The components ω'_1, ω'_2, and ω'_3 of the angular velocity vector in each direction of the principal axes are given by the following expressions:

$$\left.\begin{aligned} \omega'_1 &= p\cos\theta + \Omega = p\cos\theta + \omega - p, \\ \omega'_2 &= p\sin\theta \\ \omega'_3 &= 0 \end{aligned}\right\} \tag{2.37}$$

The angular velocity vector $\boldsymbol{\omega}'$ with these components has the direction shown in Figure 2.17a.

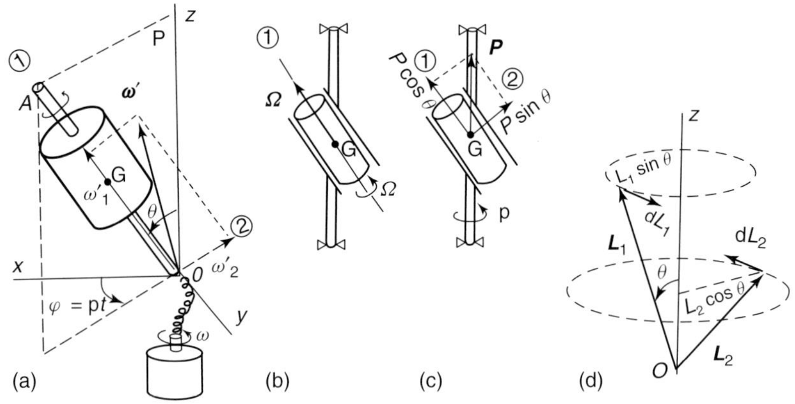

Figure 2.17 Derivation of the gyroscopic moment.

The angular momentum vectors in the directions ①, ②, and ③ are obtained by multiplying the angular speed and the principal moment of inertia in each direction as follows:

$$L_1 = I_p(p\cos\theta + \omega - p), \quad L_2 = Ip\sin\theta, \quad L_3 = 0 \tag{2.38}$$

The results are shown in Figure 2.17d. The angular momentum vector $\boldsymbol{L}$ is obtained by

$$\boldsymbol{L} = L_1\boldsymbol{e}'_1 + L_2\boldsymbol{e}'_2 + L_3\boldsymbol{e}'_3 \tag{2.39}$$

where $\boldsymbol{e}'_1$, $\boldsymbol{e}'_2$, and $\boldsymbol{e}'_3$ are the unit vectors in each principal axis. The direction of $\boldsymbol{L}$ is different from that of $\boldsymbol{\omega}'$.

In Figure 2.17a, the gravitational force creates a moment. This moment produces an increment $\mathrm{d}\boldsymbol{L}$ of the angular momentum. Correspondingly, the center line of the rotor changes its direction by $\mathrm{d}\varphi$. The vector $\mathrm{d}\boldsymbol{L}$ is perpendicular to the plane P and, from Figure 2.1d, its magnitude is

$$\begin{aligned} dL &= L_1\sin\theta\mathrm{d}\varphi - L_2\cos\theta\mathrm{d}\varphi \\ &= \{I_p p\cos\theta + I_p(\omega - p)\}\sin\theta\mathrm{d}\varphi - Ip\sin\theta\cos\theta\mathrm{d}\varphi \end{aligned} \tag{2.40}$$

Therefore, the apparent moment is

$$M_g = -\frac{dL}{dt} = -\{I_p p\cos\theta + I_p(\omega - p) - Ip\cos\theta\}p\sin\theta \tag{2.41}$$

This is the general expression for any inclination angle θ.

Now, let us consider several special cases. When $\theta = \pi/2$, since $\sin\theta = 1$ and $\cos\theta = 0$, Eq. (2.38) becomes

$$L_1 = I_p(\omega - p) = I_p\Omega, \quad L_2 = Ip, \quad L_3 = 0 \tag{2.42}$$

and from Eq. (2.41), the gyroscopic moment is

$$M_g = -I_p(\omega - p)p = -I_p\Omega p \tag{2.43}$$

This result agrees with the result of Exercise 2.2.

When θ is a small angle, Eq. (2.38) becomes

$$L_1 \approx I_p\omega, \quad L_2 \approx Ip\theta, \quad L_3 = 0 \tag{2.44}$$

and, from Eq. (2.41), the gyroscopic moment is

$$M_g = -(I_p\omega - Ip)p\theta \tag{2.45}$$

The precise expression for the momentum around the symmetrical axis of the rotor is given by Eq. (2.38). However, in Figure 2.14, we started by assuming that L_1 is $I_p\omega$ because the known value was ω. Here, we confirmed from Eq. (2.44) that, under the assumption that the inclination angle θ is small, that expression is approximately correct.

Note: Two Definitions of the Eyroscopic Moment

Once again, we discuss the case that the inclination angle θ is small as shown in Figure 2.13. Let us consider the case that the disk is whirling with a constant angular speed p around the z-axis with the constant angle θ. Now, we observe this motion on the plane BOz which is rotating around the z-axis. Substituting $\theta_x \approx \theta \cos pt$, and $\theta_y \approx \theta \sin pt$ into Eq. (2.31), we can obtain the following expression after rearrangement:

$$-(I_p\omega\theta p - I\theta p^2) + (-\delta\theta) = 0 \tag{2.46}$$

The first two terms correspond to the gyroscopic moment given by Eq. (2.45). This expression shows that, as illustrated in Figure 2.18a, the restoring moment $-\delta\theta$ works to reduce θ and the gyroscopic moment $M_g = -(I_p\omega - Ip)\theta p$ works to increase θ. Since the relationship $p_f > (I_p/I)\omega > 0 > p_b$ holds in Figure 2.15, $M_g = I(p - I_p\omega/I)\theta p > 0$ is positive and acts in the same direction for both $p = p_f$ and $p = p_b$.

From Eq. (2.46), we obtain the following expression:

$$-Ip^2\theta + (\delta + I_p\omega p)\theta = 0 \tag{2.47}$$

If we compare this with the corresponding expression of nonrotating shafts $-Ip^2\theta + \delta\theta = 0$, we see that the spring constant changes from δ to $\delta + I_p\omega p$. In other words, the term $I_p\omega p$ is added to a spring constant. In the case of a forward whirl, since ω and p have the same sign and $\omega p > 0$, the equivalent spring constant increases. While, in the case of a backward whirl, since the signs of ω and p are different and $\omega p < 0$, it decreases. As a result, in Figure 2.16, the forward natural frequency increases as the rotational speed increases, and the absolute value of the backward natural frequency decreases. Some researchers call $-I_p\omega p\theta$ the gyroscopic moment (Gash, 1975). Here, we represent this expression utilizing the notation M_g'. The direction of M_g' changes depending on the direction of the whirling motion as shown in Figure 2.18b and c, because $p = p_f > 0$ and $p = p_b < 0$.

In order to clarify the difference of these two quantities, $M_g = -(I_p\omega - Ip)\theta p$ is sometimes called the *gyroscopic moment in a broad definition* and $M_g' (= -I_p\omega p\theta)$ is called the *gyroscopic moment in a narrow definition*. In other words, the top can

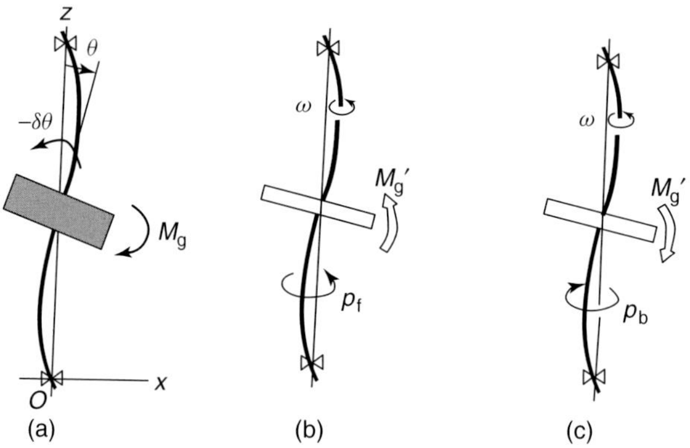

Figure 2.18 Directions of the gyroscopic moment.

whirl without falling because of the gyroscopic moment in a broad definition and the natural frequencies change as a function of the rotational speed because of the gyroscopic moment in a narrow definition.

2.4.5
Synchronous Whirl

In the following, synchronous whirling caused by skew angles is discussed. The equations of motion of a damped system are given as follows:

$$\left.\begin{aligned} I\ddot{\theta}_x + I_p\omega\dot{\theta}_y + c\dot{\theta}_x + \delta\theta &= (I - I_p)\tau\omega^2\cos\omega t \\ I\ddot{\theta}_y - I_p\omega\dot{\theta}_x + c\dot{\theta}_y + \delta\theta &= (I - I_p)\tau\omega^2\sin\omega t \end{aligned}\right\} \tag{2.48}$$

From the term on the right-hand side, we see that the unbalance force does not work if the relation $I = I_p$ holds. For example, $I = I_p$ holds in the case of a cylindrical rotor whose length is $\sqrt{3}/2$ times its diameter (see Appendix A). Such a cylindrical rotor is dynamically equal to a sphere. The direction of moments are different between a cylindrical rotor ($I > I_p$) and a disk-shaped rotor. The solution of Eq. (2.48) can be assumed in the following form:

$$\left.\begin{aligned} \theta_x &= P\cos(\omega t + \beta) \\ \theta_y &= P\sin(\omega t + \beta) \end{aligned}\right\} \tag{2.49}$$

Substituting this solution into Eq. (2.25) and equating the coefficients of the samc trigonomctric functions on both sides, we can obtain the equations for the amplitude P and the phase β. For example, in the case of no damping ($c = 0$), they are given by

$$\left.\begin{aligned} &P = \frac{\left|(I - I_p)\tau\omega^2\right|}{\left|\delta - (I - I_p)\omega^2\right|} \\ &\beta = 0 \ (\text{for } \omega < \omega_c), \quad \beta = -\pi \ (\text{for } \omega > \omega_c) \end{aligned}\right\} \tag{2.50}$$

The major critical speed ω_c is the rotational speed at which the amplitude becomes infinite, and it is given by $\omega_c = \sqrt{\delta/(I - I_p)}$ which exists only for a cylindrical rotor ($I > I_p$). This corresponds to the fact that, in Figure 2.16, the cross point between the straight line $p = \omega$ and the curve p_f exists only in the case of $I > I_p$ because the curve p_f asymptotically approaches the straight line $p = (I_p/I)\omega$.

Figure 2.19 illustrates the configurations of the rotor at various rotational speeds in the rotating coordinate system. Note how the relative positions of the skew angle τ and the inclination angle θ change according to the rotational speed. First, the disc ($I < I_p$) is divided into two equal parts (left and right) by the diameter as depicted in Figure 2.19a. The cylinder ($I > I_p$) is divided into two parts (upward and downward) by the plane perpendicular to its center line as depicted in Figure 2.19b. Then, each part is replaced by a concentrated mass. When the rotor rotates, two centrifugal forces F work on these masses and they constitute a couple. In each state, this couple and the restoring moment are in balance. In the case of a disk,

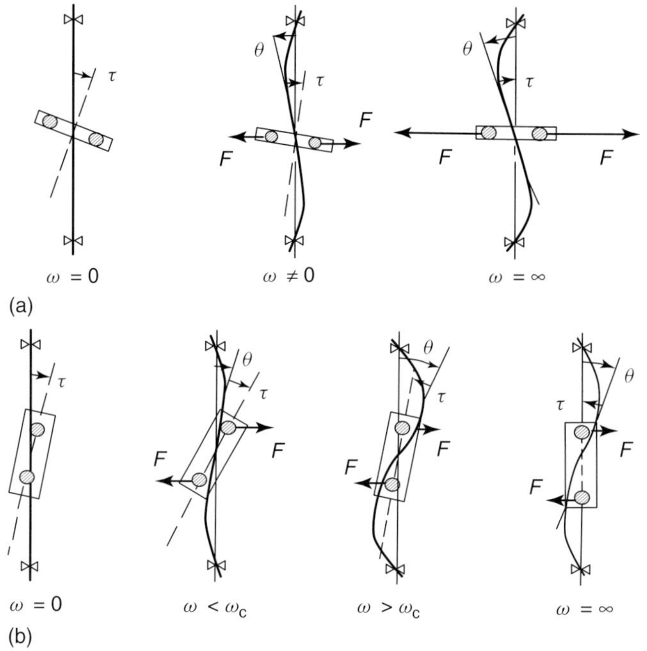

Figure 2.19 Relative positions of angles τ and θ: (a) sase of $I < I_p$ and (b) case of $I > I_p$.

as the rotor speed increases, the inclination angle of the disk decreases, keeping the relative position of θ and τ the same. While, in the case of a cylinder, first the inclination angle θ increases, the relative position of θ and τ changes at the major critical speed, and then the angle θ decreases. In both cases, the principal axis of inertia of the rotor coincides with the bearing center line at $\omega = \infty$, and the rotor becomes still when the rotational speed becomes infinite. This shows that the self-centering effect appears in an inclination model.

2.5 Vibrations of a 4 DOF System

2.5.1 Equations of Motion

Figure 2.20 shows a vertical rotor system in which a disk is mounted on a massless elastic shaft. As the disk position is not at the center of the shaft, the lateral motion and the inclination motion couple with each other. This rotor has eccentricity e and skew angle τ. The rectangular coordinate system O-xyz, whose z-axis coincides with the bearing centerline, is fixed in space. Generally, a rigid body in space has 6 DOF. However, in a rotating machinery in which the disk rotates with a constant speed ω and performs a planer motion (i.e., a motion in which case the deflection

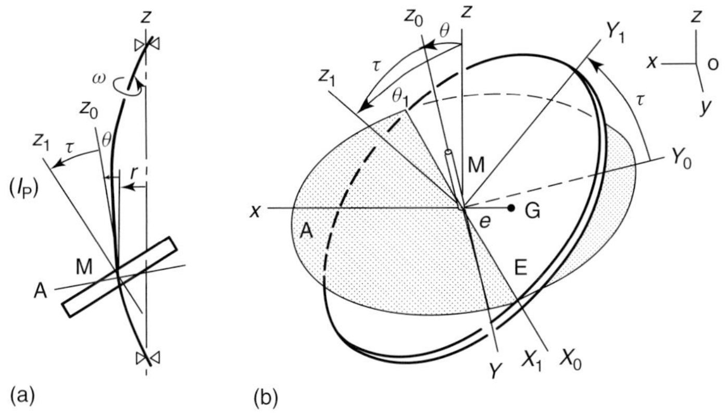

Figure 2.20 4 DOF rotor system: (a) rotor system and (b) coordinate system.

in the direction of the bearing centerline is fairly small compared to that in the lateral direction), the disk has 4 DOF. In this section, the whirling motions of such a rotor system are discussed under the assumption that the inclination angle of the rotor is small.

It is assumed that the centerline of the elastic shaft passes through the geometrical center M of the disk. When the shaft is not rotating ($\omega = 0$), this point M is located at the origin O. The coordinate systems used in the following explanations are shown in Figure 2.20. The coordinate system M-XYZ is a system where the X-, Y-, and Z-axes are in parallel with the x-, y-, and z-axes, respectively. The shaded plane A is a virtual disk perpendicular to and rotating with the shaft. The disk is inclined from A by a small angle τ. The rotating coordinate system M-$X_0Y_0Z_0$ has its X_0-axis at the line of node ME (i.e., the intersection of plane A and the disk) and its Z_0-axis in the tangential direction of the elastic shaft. And, the rotating coordinate system M-$X_1Y_1Z_1$ has its Z_1-axis on the principal axis of the polar moment of inertia, and its X_1-axis at the line of node ME. In the following, we derive the equations of motion by two ways.

2.5.1.1 **Derivation by Using the Results of 2 DOF System**

The equations of motion for such a 4 DOF system are obtained from Eq. (2.1) for a deflection motion and Eq. (2.28) for an inclination motion by replacing their restoring forces with the expression for this system. Suppose that a deflection x, y and an inclination θ_x, θ_y appear when a force F'_x, F'_y and a moment M'_{xz}, M'_{yz} work. Here, F'_x and F'_y are the components working in the positive directions of the x- and y-axes, and M'_{xz} and M'_{yz} are the components working in the positive directions of θ_x and θ_y in the xz- and yz-planes, respectively. Then the following relationships hold:

$$\left.\begin{array}{l} F'_x = \alpha x + \gamma\theta_x \\ F'_y = \alpha y + \gamma\theta_y \end{array}\right\} \quad \left.\begin{array}{l} M'_{xz} = \gamma x + \delta\theta_x \\ M'_{yz} = \gamma y + \delta\theta_y \end{array}\right\} \qquad (2.51)$$

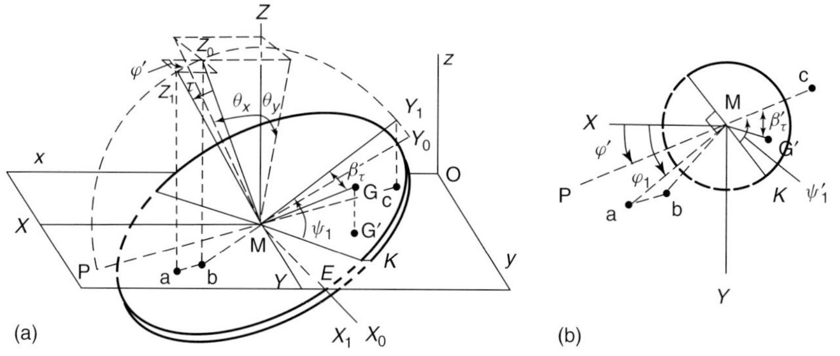

Figure 2.21 Relationship between θ_{1x}, θ_{1y} and θ_x, θ_y.

where α, γ, and δ are spring constants. The concrete forms of these constants can be derived using the theory of strength of materials. The restoring force F_x, F_y and the restoring moment M_{xz}, M_{yz} are given by changing the sign of the expression in Eq. (2.51). From Eqs. (2.1) and (2.28), we obtain

$$\left.\begin{aligned} m\ddot{x}_G &= -(\alpha x + \gamma\theta_x) \\ m\ddot{y}_G &= -(\alpha y + \gamma\theta_y) \end{aligned}\right\} \tag{2.52}$$

$$\left.\begin{aligned} I\ddot{\theta}_{1x} + I_p\omega\dot{\theta}_{1y} &= -(\gamma x + \delta\theta_x) \\ I\ddot{\theta}_{1y} - I_p\omega\dot{\theta}_{1x} &= -(\gamma y + \delta\theta_y) \end{aligned}\right\} \tag{2.53}$$

Next, we derive the relationship between θ_{1x}, θ_{1y} and θ_x, θ_y. In Figure 2.21 that corresponds to Figure 2.20, the intersection line of the disk and the horizontal plane is MK. Let the angle between this cross line and the Y_1-axis be ψ_1. The points a, b, and c are the projections of points on the axes MZ_1, MZ_0, MY_1 onto the horizontal plane XMY, respectively. Let the direction of the center line (the Z_1-axis) of the disk be φ_1 and we put $\Theta_1 = \varphi_1 + \psi_1$. From Eq. (2.36), we have $\dot{\Theta}_1 = \omega$. (Note that angle φ_1 and angle ψ_1 are on the different planes). Let the projections of the center of gravity G, the angle β_τ between the direction of G and the Y_1-axis and the angle between the direction of the skew angle (the direction Mc) and MK onto the horizontal plane XMY be G$'$, β'_τ, and ψ'_1, respectively (Figure 2.21). Then, the relationships $\beta'_\tau \approx \beta_{\tau 1}$ and $\psi'_1 \approx \psi$ hold. Therefore, we obtain

$$\angle \mathrm{XMG}' = \varphi_1 + 90^\circ + \psi'_1 - \beta'_\tau \approx \varphi_1 + 90^\circ + \psi_1 - \beta_\tau = \Theta_1 + 90^\circ - \beta_\tau \tag{2.54}$$

Since $\dot{\Theta}_1 = \omega$, if we represent $\angle XMG' = \omega t$ by selecting the time origin properly, the coordinates of the center of gravity G$'$ are given by

$$x_G = x + e\cos\omega t, \quad y_G = y + e\sin\omega t \tag{2.55}$$

Substituting this relationship into Eq. (2.52) yields the following equations of motion for the lateral oscillations:

$$\left.\begin{aligned} m\ddot{x} + \alpha x + \gamma\theta_x &= me\,\omega^2\cos\omega t \\ m\ddot{y} + \alpha y + \gamma\theta_y &= me\,\omega^2\sin\omega t \end{aligned}\right\} \tag{2.56}$$

Let the direction of the skew angle τ measured from the X-axis be φ'. Since the relationship $\varphi' = \angle XMG' + \beta'_\tau - 180^\circ \approx \omega t + \beta_\tau - 180^\circ$ holds in Figure 2.21b, we obtain from Eq. (2.29) as

$$\left.\begin{aligned}\theta_{1x} &= \theta_x + \tau\cos\varphi' = \theta_x - \tau\cos(\omega t + \beta_\tau)\\ \theta_{1y} &= \theta_y + \tau\sin\varphi' = \theta_y - \tau\sin(\omega t + \beta_\tau)\end{aligned}\right\} \tag{2.57}$$

Substituting this expression into Eq. (2.53), we have the following equations of motion for the inclination oscillation:

$$\left.\begin{aligned}I\ddot{\theta}_x + I_p\omega\dot{\theta}_y + \gamma x + \delta\theta_x &= (I_p - I)\tau\omega^2\cos(\omega t + \beta_\tau)\\ I\ddot{\theta}_y - I_p\omega\dot{\theta}_x + \gamma y + \delta\theta_y &= (I_p - I)\tau\omega^2\sin(\omega t + \beta_\tau)\end{aligned}\right\} \tag{2.58}$$

Combining Eqs. (2.56) and (2.58) and considering damping, we can finally obtain the following equations of motion for the 4 DOF system.

$$\left.\begin{aligned}&m\ddot{x} + c_{11}\dot{x} + c_{12}\dot{\theta}_x + \alpha x + \gamma\theta_x = me\omega^2\cos\omega t\\ &m\ddot{y} + c_{11}\dot{y} + c_{12}\dot{\theta}_y + \alpha y + \gamma\theta_y = me\omega^2\sin\omega t\\ &I\ddot{\theta}_x + I_p\omega\dot{\theta}_y + c_{21}\dot{x} + c_{22}\dot{\theta}_x + \gamma x + \delta\theta_x = (I_p - I)\tau\omega^2\cos(\omega t + \beta_\tau)\\ &I\ddot{\theta}_y - I_p\omega\dot{\theta}_x + c_{21}\dot{y} + c_{22}\dot{\theta}_y + \gamma y + \delta\theta_y = (I_p - I)\tau\omega^2\sin(\omega t + \beta_\tau)\end{aligned}\right\} \tag{2.59}$$

Note: Euler Angle

Although they already appeared in Figure 2.20, we sometimes use *Euler angles* that describe the orientation of a rigid body. For example, in Figure 2.22, the position of the disk is given by the Euler angles θ_1, φ_1, and ψ_1, which specify the orientation of the coordinate system M-$X_1Y_1Z_1$, where θ_1 is the angle representing an inclination of the Z_1-axis (i.e., a zenith angle), φ_1 is the angle designating the direction into which the Z_1-axis inclines (i.e., a precessional angle), and ψ_1 is the angle representing the rotation about the Z_1-axis (i.e., a spinning angle).

2.5.1.2 Derivation by Lagrange's Equations

For more complicated systems, such as an asymmetrical shaft and asymmetrical rotor that are mentioned in Chapter 4, the Lagrange's equation is more convenient.

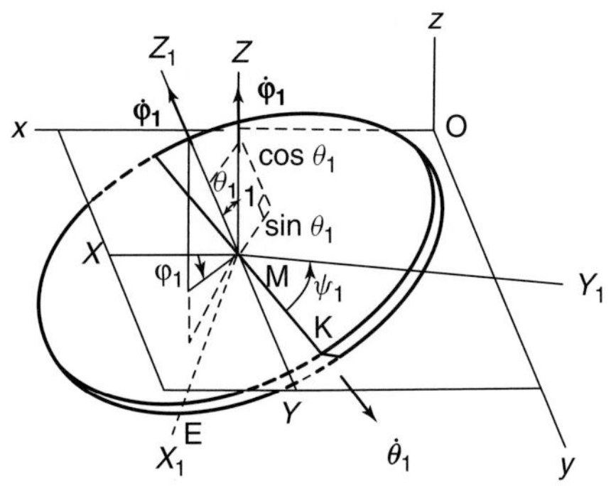

Figure 2.22 Euler angles.

Table 2.1 Direction cosines for MK, MZ, and MZ_1.

	X_1	Y_1	Z_1
$MK(\dot{\theta}$-direction)	$\sin\psi_1$	$\cos\psi_1$	0
$MZ(\psi_1$-direction)	$\sin\theta_1\cos\psi_1$	$\sin\theta_1\cos\psi_1$	$\cos\theta_1$
$MZ_1(\dot{\psi}_1$-direction)	0	0	1

Here, the derivation of equations of motion of a 4 DOF symmetrical rotor system by Lagrange's equations is explained.

First, we derive the expression for the kinetic energy. Let the center of gravity of the rotor be $(x_G, y_G, 0)$ in Figure 2.20. The deflection is neglected. The kinetic energy T_1 for the deflection oscillation is expressed by

$$T_1 = \frac{1}{2}m(\dot{x}_G^2 + \dot{y}_G^2) \tag{2.60}$$

Substituting Eq. (2.55) into this expression, we obtain the following expression by neglecting higher order terms

$$T_1 = \frac{1}{2}m(\dot{x}^2 + \dot{y}^2 + e^2\omega^2 - 2e\omega\dot{x}\sin\omega t + 2e\omega\dot{y}\cos\omega t) \tag{2.61}$$

Let the principal moment of inertia around the principal axis X_1, Y_1, Z_1 of the disk be I_p, I, I and the components of the angular speed be $\omega_{X1}, \omega_{Y1}, \omega_{Z1}$, respectively. Then the kinetic energy T_2 for the inclination motion is given by

$$T_2 = \frac{1}{2}(I_p\omega_{Z1}{}^2 + I\omega_{Y1}{}^2 + I\omega_{X1}{}^2) \tag{2.62}$$

In Figure 2.22, let the Euler angles representing the coordinate system $M\text{-}X_1Y_1Z_1$ fixed to the rotor be $\theta_1, \varphi_1, \psi_1$. Then, the angular speeds are given by $\dot{\theta}_1, \dot{\varphi}_1, \dot{\psi}_1$. Let the line of node of the disk and the horizontal plane (xy-plane) be MK. The vectors $\dot{\boldsymbol{\theta}}_1, \dot{\boldsymbol{\varphi}}_1, \dot{\boldsymbol{\psi}}_1$ are in the same directions as MK, MZ, and MZ_1, respectively.

The directions of MK, MZ, and MZ_1 relative to $M\text{-}X_1Y_1Z_1$, whose axes coincide with the principal axes, are given by the direction cosines shown in Table 2.1. For example, the direction cosines of MZ are obtained by deriving the components in the X_1, Y_1, Z_1-directions of the unit vector in the MZ-direction in Figure 2.22. By referring to this table, the components of the angular speed in the directions of the principal axes MX_1, MY_1, and MZ_1 become

$$\left.\begin{aligned}
\omega_{X1} &= \dot{\theta}_1\sin\psi_1 - \dot{\varphi}_1\sin\theta_1\cos\psi_1 \\
\omega_{Y1} &= \dot{\theta}_1\cos\psi_1 + \dot{\varphi}_1\sin\theta_1\sin\psi_1 \\
\omega_{Z1} &= \dot{\varphi}_1\cos\theta_1 + \dot{\psi}_1
\end{aligned}\right\} \tag{2.63}$$

Substituting this into Eq. (2.62), we have

$$T_2 = \frac{1}{2}\left\{I_p(\dot{\varphi}_1\cos\theta_1 + \dot{\psi}_1)^2 + I(\dot{\theta}_1^2 + \dot{\varphi}_1^2\sin^2\theta_1)\right\} \tag{2.64}$$

If the angle θ is small, the relationships $\sin\theta \approx \theta_1$ and $\cos\theta \approx 1 - \theta_1^2/2$ hold approximately. Substituting these relationships into Eq. (2.64) and neglecting small quantities of the third order and more, we obtain

$$\begin{aligned} T_2 &= \frac{1}{2}\left\{I_p\{(\dot{\varphi}_1 + \dot{\psi}_1)^2 - (\dot{\varphi}_1 + \dot{\psi}_1)\dot{\varphi}_1\theta_1^2\} + I(\dot{\theta}_1^2 + \dot{\varphi}_1^2\theta_1^2)\right\} \\ &= \frac{1}{2}\left\{I_p(\omega^2 - \omega\dot{\varphi}_1\theta_1^2) + I(\dot{\theta}_1^2 + \dot{\varphi}_1^2\theta_1^2)\right\} \end{aligned} \tag{2.65}$$

where the above-mentioned relationships $\Theta_1 = \varphi_1 + \psi_1$ and $\dot{\Theta}_1 = \omega$ are used. From $\theta_{1x} = \theta_1 \cos\varphi_1, \theta_{1y} = \theta_1 \sin\varphi_1$, we have

$$\left.\begin{aligned} &\dot{\theta}_{1x} = \dot{\theta}_1\cos\varphi_1 - \theta_1\dot{\varphi}_1\sin\varphi_1, \quad \dot{\theta}_{1y} = \dot{\theta}_1\sin\varphi_1 + \theta_1\dot{\varphi}_1\cos\varphi_1, \\ &\theta_{1x}\dot{\theta}_{1y} - \dot{\theta}_{1x}\theta_{1y} = \theta_1^2\dot{\varphi}_1, \quad \dot{\theta}_{1x}^2 + \dot{\theta}_{1y}^2 = \dot{\theta}_1^2 + \theta_1^2\dot{\varphi}_1^2 \end{aligned}\right\} \tag{2.66}$$

Substituting these relationships into Eq. (2.65), we have becomes

$$\begin{aligned} T_2 &= \frac{1}{2}\left[I_p\omega\{\omega + (\dot{\theta}_{1x}\theta_{1y} - \theta_{1x}\dot{\theta}_{1y})\} + I(\dot{\theta}_{1x}{}^2 + \dot{\theta}_{1y}{}^2)\right] \\ &= \frac{1}{2}\omega^2\left\{I_p(1-\tau^2) + I\tau^2\right\} + \frac{1}{2}I_p\omega(\dot{\theta}_x\theta_y - \theta_x\dot{\theta}_y) + \frac{1}{2}I(\dot{\theta}_x{}^2 + \dot{\theta}_y{}^2) \\ &+ \frac{1}{2}(2I - I_p)\tau\omega\{\dot{\theta}_x\sin(\omega t + \beta_\tau) - \dot{\theta}_y\cos(\omega t + \beta_\tau)\} \\ &+ \frac{1}{2}I_p\tau\omega^2\{\theta_y\sin(\omega t + \beta_\tau) + \theta_x\cos(\omega t + \beta_\tau)\} \end{aligned} \tag{2.67}$$

The kinetic energy is obtained by substituting Eqs. (2.61) and (2.67) into

$$T = T_1 + T_2 \tag{2.68}$$

The potential energy is given by

$$\begin{aligned} V &= \frac{1}{2}(\alpha r^2 + 2\gamma r\theta + \delta\theta^2) \\ &= \frac{1}{2}\left\{\alpha(x^2 + y^2) + 2\gamma(x\theta_x + y\theta_y) + \delta(\theta_x^2 + \theta_y^2)\right\} \end{aligned} \tag{2.69}$$

Substituting these expressions into Lagrange's equation

$$\frac{d}{dt}\left(\frac{\partial T}{\partial \dot{q}}\right) - \frac{\partial T}{\partial q} + \frac{\partial V}{\partial q} = 0 \tag{2.70}$$

where q is the general coordinates, we can obtain the equations of motion:

$$\left.\begin{aligned} &m\ddot{x} + \alpha x + \gamma\theta_x = me\omega^2\cos\omega t \\ &m\ddot{y} + \alpha y + \gamma\theta_y = me\omega^2\sin\omega t \\ &I\ddot{\theta}_x + I_p\omega\dot{\theta}_y + \gamma x + \delta\theta_x = (I_p - I)\tau\omega^2\cos(\omega t + \beta) \\ &I\ddot{\theta}_y - I_p\omega\dot{\theta}_x + \gamma y + \delta\theta_y = (I_p - I)\tau\omega^2\sin(\omega t + \beta) \end{aligned}\right\} \tag{2.71}$$

This agrees with the expression that is obtained by neglecting damping in Eq. (2.60).

We derived equations of motion in two different ways. The third method utilizing the relationship "the time derivatives of angular momentum expressed by Euler angles are equal to moments" is explained in Appendix C.

2.5.2
Free Vibrations and a Natural Frequency Diagram

Free vibrations in a system with no external force due to unbalance and no damping are governed by the following equations of motion, which are given by putting $e = \tau = 0, c_{ij} = 0\ (i, j = 1, 2)$ into Eq. (2.59) or Eq. (2.71):

$$\left.\begin{array}{l} m\ddot{x} + \alpha x + \gamma\theta_x = 0 \\ m\ddot{y} + \alpha y + \gamma\theta_y = 0 \\ I\ddot{\theta}_x + I_p\omega\dot{\theta}_y + \gamma x + \delta\theta_x = 0 \\ I\ddot{\theta}_y - I_p\omega\dot{\theta}_x + \gamma y + \delta\theta_y = 0 \end{array}\right\} \tag{2.72}$$

In this expression, four variables, x, y, θ_x, and θ_y, are coupled with each other. From the observation of this expression, we know that x and θ_x are in the same phase, as y and θ_y are, and x and y are out of phase by 90′, as θ_x and θ_y are. Similarly, x and y and θ_x and θ_y have the same amplitude, respectively. Therefore, we can assume the solutions for free oscillations to be

$$\left.\begin{array}{ll} x = A\cos(pt + \beta'), & \theta_x = B\cos(pt + \beta') \\ y = A\sin(pt + \beta'), & \theta_y = B\sin(pt + \beta') \end{array}\right\} \tag{2.73}$$

Substituting this into Eq. (2.72) and comparing the coefficients of the terms on the right- and left-hand sides, we get

$$\left.\begin{array}{l} (\alpha - mp^2)A + \gamma B = 0 \\ \gamma A + (\delta + I_p\omega p - Ip^2)B = 0 \end{array}\right\} \tag{2.74}$$

By eliminating A and B, we obtain the following frequency equation:

$$f(p) \equiv (\alpha - mp^2)(\delta + I_p\omega p - Ip^2) - \gamma^2 = 0 \tag{2.75}$$

where the expression on the left-hand side is represented by $f(p)$. We can solve this quartic equation numerically and get a natural frequency diagram such as Figure 2.23. The four natural frequencies $p_1, \ldots, p_4$ of this system have the following characteristics regardless of the dimensions of the rotor:

1) The relationship $p_1 > \sqrt{\alpha/m} > p_2 > 0 > p_3 > -\sqrt{\alpha/m} > p_4$ holds. The positive p_1 and p_2 have forward whirling modes, and p_3 and p_4 have backward whirling modes.
2) The natural frequencies $p_1, \ldots, p_4$ change as a function of the rotational speed. As $\omega \to \infty$, they approach as $p_1 \to (I_p/I)\omega, p_2 \to \sqrt{\alpha/m}, p_3 \to 0, p_4 \to -\sqrt{\alpha/m}$, asymptotically.
3) When $\omega = 0$, the relationships $p_1 = -p_4$ and $p_2 = -p_3$ hold.

These characteristics are easily confirmed by transforming Eq. (2.75) into

$$\omega = \frac{mIp^4 - (\alpha I + \delta m)p^2 + (\alpha\delta - \gamma^3)}{mI_p(p^2 - \alpha/m)p} \tag{2.76}$$

and considering ω as a function of p. For example, asymptotes can be obtained as follows: Setting the denominator equal to zero, we know that $\omega = \infty$ at $p = 0$ and $\pm\sqrt{\alpha/m}$. In the range of high frequency p, the terms of the highest order of p

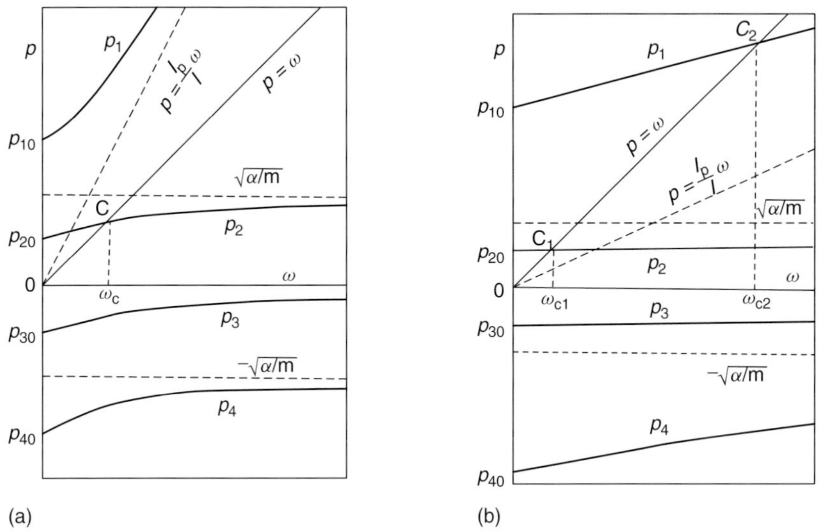

Figure 2.23 Natural frequency diagram of 4 DOF rotor system: (a) case of $I_p > I$ and (b) case of $I_p < I$.

become predominant in both the numerator and the denominator. Therefore, we get the approximate expression $\omega \approx mIp^4/mI_pp^3 = (I/I_p)p$ there. This means that p approaches the straight line $p = (I_p/I)\omega$ as ω increases. Furthermore, by setting the numerator equal to zero, we obtain the points of intersection of the ordinate and the natural frequency curves in Figure 2.23.

Note: Spring Constants of Elastic Shafts

The concrete forms of spring constants of an elastic shaft α, γ, δ can be derived using the theory of strength of materials. For example, in the case of a rotor system where both ends of the shaft are supported simply, as shown in Figure 2.24a, these constants are given by

$$\alpha = 3EI_0\frac{(a^2 - ab + b^2)l}{a^3b^3}, \quad \gamma = 3EI_0\frac{(a-b)l}{a^2b^2}, \quad \delta = 3EI_0\frac{l}{ab} \tag{2.77}$$

where a and b are the distances of the upper and lower bearings from the disk center, $l(= a + b)$ is the shaft length, E is Young's modulus, and I_0 is the

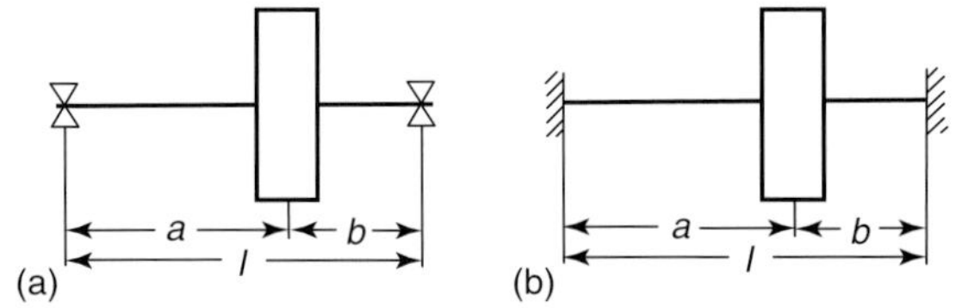

Figure 2.24 Rotor with various supports: (a) free-free supports and (b) fixed-fixed support.

cross-sectional area moment of inertia of the shaft. In the case of a shaft with a circular cross section with the diameter d it is given by

$$I_0 = \frac{\pi d^4}{64} \tag{2.78}$$

In the case of a rotor whose both ends are fixed, as shown in Figure 2.24b, these constants are given by

$$\alpha = 12EI_0\frac{(a^2 - ab + b^2)l}{a^3b^3}, \quad \gamma = 6EI_0\frac{(b-a)l}{a^2b^2}, \quad \delta = 4EI_0\frac{l}{ab} \tag{2.79}$$

Exercise 2.3: A disk with radius $R = 25$ cm and thickness $h = 0.55$ cm is mounted on an elastic shaft with length $l = 70$ cm and diameter $d = 12$ nm at the position 14 cm from one end. The specific gravity of the disk material is 7.8 and Young's modulus of the shaft material is $E = 206$ GPa. The shaft is supported simply at both ends. Illustrate a natural frequency diagram in the speed range 0–4000 rpm.

Answer: Using the given parameters, we can obtain the mass as $m = 8.423$ kg. From Appendix A, the diametral and polar moments of inertia of the rotor are $I = m\left(R^2/4 + h^2/12\right) = 0.263$ kg m^2 and $I_p = (m/2)R^2 = 0.132$ kg m^2, respectively, From Eq. (2.77) we obtain $\alpha = 2.328 \times 10^5$N $\cdot$ m^{-1}, $\gamma = 3.009 \times 10^4$N, and $\delta = 5.616 \times 10^3$N $\cdot$ m. Substituting these values into Eq. (2.76), we can obtain Figure 1.

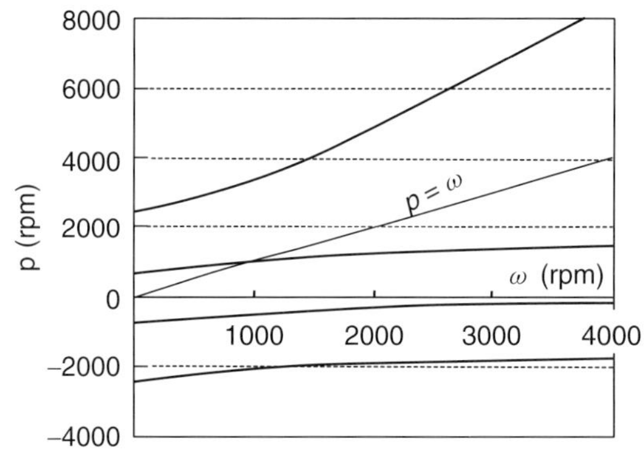

Figure 1 Natural frequency diagram.

2.5.3 Synchronous Whirling Response

When unbalances exist, a resonance occurs at the major critical speed ω_c, and a forward whirling motion with large amplitude occurs. For simplicity, let us consider the case governed by Eq. (2.71) with no damping. Because this equation is linear, its solution is given by superimposing the response for a force due to the eccentricity e and that for the skew angle τ. As the response is not delayed in the system with no damping, the first terms in the following equations are assumed for the former and the second terms for the latter. Substituting the assumed solution into the equations of motion (Eq. (2.71)) and comparing the coefficients of the terms on both sides, we get

$$\left.\begin{array}{l} x = P\cos\omega t + Q\cos(\omega t + \beta_\tau) \\ y = P\sin\omega t + Q\sin(\omega t + \beta_\tau) \\ \theta_x = P'\cos\omega t + Q'\cos(\omega t + \beta_\tau) \\ \theta_y = p'\sin\omega t + Q'\sin(\omega t + \beta_\tau) \end{array}\right\} \tag{2.80}$$

where

$$\left.\begin{array}{l} P = \dfrac{m\omega^2\left\{\delta - (I - I_p)\,\omega^2\right\}e}{f(\omega)}, \quad Q = \dfrac{-(I - I_p)\omega^2\gamma\tau}{f(\omega)} \\ P' = \dfrac{-m\omega^2\gamma e}{f(\omega)}, \quad Q' = \dfrac{(I - I_p)\omega^2(\alpha - m\omega^2)\tau}{f(\omega)} \end{array}\right\} \tag{2.81}$$

The term $f(\omega)$ is given by putting $p = \omega$ in Eq. (2.75). The major critical speed $\omega = \omega_c$ is obtained by setting the denominator zero in this expression. From $f(\omega) = 0$, we obtain

$$\omega_c^2 = \frac{\{\alpha(I - I_p) + \delta m\} \pm \sqrt{\{\alpha(I - I_p) + \delta m\}^2 - 4m(I - I_p)(\alpha\delta - \gamma^2)}}{2m(I - I_p)} \tag{2.82}$$

As the relationship $\alpha\delta - \gamma^2 > 0$ holds, the number for the major critical speed is one for $I_p > I$ and two for $I_p < I$. These major critical speeds are given by the cross points of the straight line $p = \omega$ and the p_1 and p_2 curves in Figure 2.23.

In the case with damping, we can obtain the solution by assuming them to be in the same form as Eq. (2.15), which take into account the phase delay.

2.6 Vibrations of a Rigid Rotor

2.6.1 Equations of Motion

Until now, we discussed rotor systems with a disk on an elastic shaft. However, as explained in Figure 1.1, the rotor must be considered rigid when its elastic deformation is negligible in the operation range. If the rigidity of the support is comparatively small, the system is expressed by a model shown in Figure 2.25. Many practical machines can be modeled by such a rigid-rotor model. Suppose that a rotor considered rigid is supported by springs with stiffness constants k_1 and k_2, as shown in Figure 2.25. Damping in the bearings is represented by damping coefficients c_1 and c_2. The rectangular coordinate system O-xyz has its axis at the

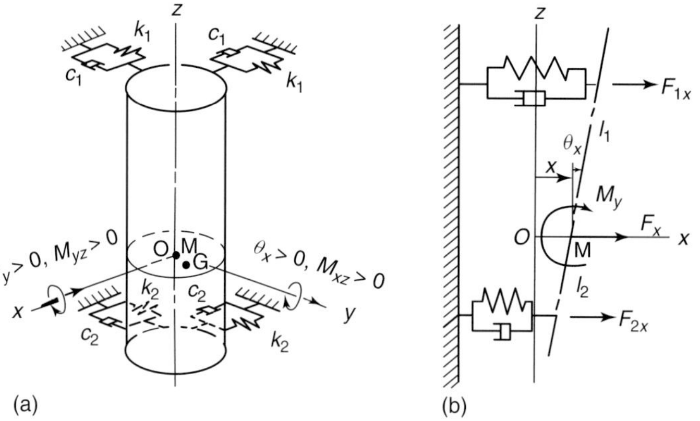

Figure 2.25 Model of a rigid rotor: (a) theoretical model and (b) *xz*-plane.

centerline of the rotor at rest and has its origin at the position of the geometrical center M of the cross section with the center of gravity G. This rotor has unbalances because of an eccentricity e and the skew angle τ. Let us assume that center of gravity G is located at distances l_1 and l_2 from the upper and lower bearings.

Let the deflection of the rotor during a whirling motion be (x_G, y_G) and the inclination of the principal axis of the rotor be $(\theta_{1x}, \theta_{1y})$. The changes in the momentum and in the angular momentum of the rotor per unit time are represented by the left-hand sides of Eqs. (2.1) and (2.28). Owing to the spring forces and to the damping forces at the supports, a resultant force (F_x, F_y) and a resultant moment (M_x, M_y) work. If we represent the moment by $M_{xz}(= M_y)$ and $M_{yz}(= -M_x)$, the equations of motion are given by

$$\left.\begin{array}{l} m\ddot{x}_G = F_x \\ m\ddot{y}_G = F_y \\ I\ddot{\theta}_{1x} + I_p\omega\dot{\theta}_{1y} = M_{xz} \\ I\ddot{\theta}_{1y} - I_p\omega\dot{\theta}_{1x} = M_{yz} \end{array}\right\} \tag{2.83}$$

Let the deflections of the upper and lower ends of the rotor be (x_1, y_1) and (x_2, y_2), respectively. A spring force $(-k_1x_1, -k_1y_1)$ and a damping force $(-c_1\dot{x}_1, c_1\dot{y}_1)$ work at the upper end, and $(-k_2x_2, -k_2y_2)$ and $(-c_2\dot{x}_2, -c_2\dot{y}_2)$ work at the lower end. Let the summation of the forces working at the upper end be (F_{1x}, F_{1y}) and that at the lower end be (F_{2x}, F_{2y}). From Figure 2.25b, we obtain

$$\left.\begin{array}{l} F_x = F_{1x} + F_{2x} = -k_1x_1 - c_1\dot{x}_1 - k_2x_2 - c_2\dot{x}_2 \\ F_y = F_{1y} + F_{2y} = -k_1y_1 - c_1\dot{y}_1 - k_2y_2 - c_2\dot{y}_2 \\ M_{xz} = l_1F_{1x} - l_2F_{2x} = -l_1(k_1x_1 + c_1\dot{x}_1) + l_2(k_2x_2 + c_2\dot{x}_2) \\ M_{yz} = l_1F_{1y} - l_2F_{2y} = -l_1(k_1y_1 + c_1\dot{y}_1) + l(k_2y_2 + c_2\dot{y}_2) \end{array}\right\} \tag{2.84}$$

Using the relationships

$$\left.\begin{array}{ll} x_1 = x + l_1\theta_x & y_1 = y + l_1\theta_y \\ x_2 = x - l_2\theta_x & y_2 = y - l_2\theta_y \end{array}\right\} \tag{2.85}$$

we represent Eq. (2.84) by (x, y), which is the deflection of M, and (θ_x, θ_y) which is the inclination of the centerline. Let the angle between the direction of the center of gravity G and the x-axis be ωt, and the angle between the direction in which the principal axis inclines and the direction of G be β_τ (see Figure 2.21). Then the relationships

$$\left. \begin{aligned} x_G &= x + e\cos\omega t, \quad \theta_{1x} = \theta_x - \tau\cos(\omega t + \beta_\tau) \\ y_G &= y + e\sin\omega t, \quad \theta_{1y} = \theta_y - \tau\sin(\omega t + \beta_\tau) \end{aligned} \right\} \tag{2.86}$$

hold. Substituting these into Eq. (2.83), we get the following equations of motion.

$$\left. \begin{aligned} &m\ddot{x} + c_{11}\dot{x} + c_{12}\dot{\theta}_x + \alpha x + \gamma\theta_x = me\omega^2\cos\omega t \\ &m\ddot{y} + c_{11}\dot{y} + c_{12}\dot{\theta}_y + \alpha y + \gamma\theta_y = me\omega^2\sin\omega t \\ &I\ddot{\theta}_x + I_p\omega\dot{\theta}_y + c_{21}\dot{x} + c_{22}\dot{\theta}_x + \gamma x + \delta\theta_x = (I_p - I)\tau\omega^2\cos(\omega t + \beta_\gamma) \\ &I\ddot{\theta}_y - I_p\omega\dot{\theta}_x + c_{21}\dot{y} + c_{22}\dot{\theta}_y + \gamma y + \delta\theta_y = (I_p - I)\tau\omega^2\sin(\omega t + \beta_\tau) \end{aligned} \right\} \tag{2.87}$$

where

$$\left. \begin{aligned} &c_{11} = c_1 + c_2, \quad c_{12} = c_{21} = c_1 l_1 - c_2 l_2, \quad c_{22} = c_1 l_1^2 + c_2 l_2^2 \\ &\alpha = k_1 + k_2, \quad \gamma = k_1 l_1 - k_2 l_2, \quad \delta = k_1 l_1^2 + k_2 l_2^2 \end{aligned} \right\} \tag{2.88}$$

The expression in Eq. (2.87) has the same form as Eq. (2.59) for an elastic shaft and a disk.

2.6.2 Free Whirling Motion and Whirling Modes

When an unbalance force and a damping force do not work, Eq. (2.87) becomes

$$\left. \begin{aligned} &m\ddot{x} + \alpha x + \gamma\theta_x = 0 \\ &m\ddot{y} + \alpha y + \gamma\theta_y = 0 \\ &I\ddot{\theta}_x + I_p\omega\dot{\theta}_y + \gamma x + \delta\theta_x = 0 \\ &I\ddot{\theta}_y - I_p\omega\dot{\theta}_x + \gamma y + \delta\theta_y = 0 \end{aligned} \right\} \tag{2.89}$$

The solution for this equation was obtained in Section 2.5.2. As $I > I_p$ holds in this system, the natural frequency diagram becomes as shown in Figure 2.26 and has two major critical speeds, ω_{c1} and ω_{c2}. When $k_1 l_1 - k_2 l_2 = 0$, the coupling terms disappear ($\gamma = 0$) and the deflection and inclination motions become independent of each other. From Eq. (2.75), the frequency equations for this case are given by

$$\left. \begin{aligned} &\alpha - mp^2 = 0 \\ &\delta + I_p\omega p - Ip^2 = 0 \end{aligned} \right\} \tag{2.90}$$

The vibrations for p_2 and p_3 obtained from the former have the parallel mode shown in Figure 2.27a and those for p_1 and p_4 obtained from the latter have the conical mode shown in Figure 2.27b. In the coupled system ($\gamma \neq 0$) shown in Figure 2.26, every vibration for $p_1, \ldots, p_4$ becomes a conical mode whose node is different from G.

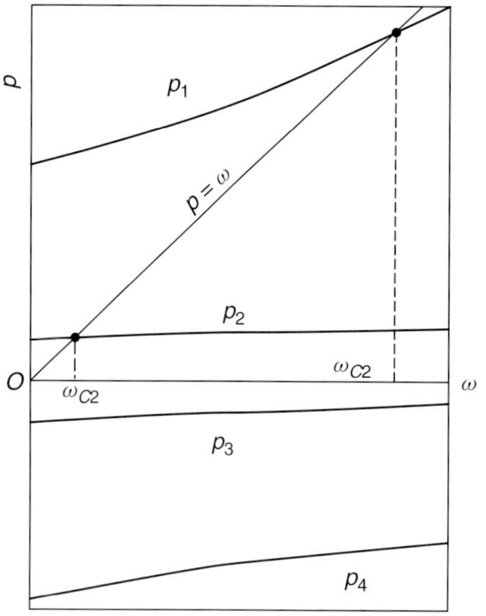

Figure 2.26 Natural frequency diagram of a rigid rotor.

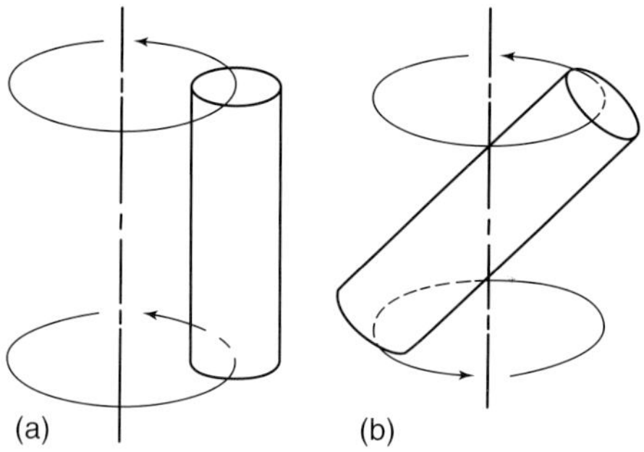

Figure 2.27 Vibration modes of an uncoupled system ($\gamma = 0$): (a) parallel mode and (b) conical mode.

2.7
Approximate Formulas for Critical Speeds of a Shaft with Several Disks

Practical rotating machines often have several disks. To calculate the critical speeds of these systems, we must solve theoretically in a manner similar to the one mentioned above or apply some numerical procedure such as the finite-element method (Chapter 14) or the transfer matrix method (Chapter 15). However, much labor is necessary when using these procedures. Instead, some simple approximate methods can be used practically. Although the accuracy of these approximate methods is not very high, and some of them give only the lowest critical speed, they are often used because they are simple and convenient. These methods were proposed originally for the calculation of a nonrotating system such as a beam. However, we can apply these approximate methods to rotor systems with a small gyroscopic moment to estimate their fundamental natural frequencies because their natural frequencies do not change as a function of the rotational speed and are almost equal to those of nonrotating systems. In this section, two classical approximate methods, *Rayleigh's method* and *Dunkerley's formula*, are presented.

2.7.1
Rayleigh's Method

This method is based on the following principle: "*In a conservative system vibrating freely at its fundamental natural frequency, the maximum potential energy stored in the spring is equal to the maximum kinetic energy that appears in the system.*" On the basis of this principle, the fundamental frequency is obtained using the following procedure.

First, we assume the shape of a shaft deflection and imagine that a nonrotating shaft oscillates sinusoidally between the two extremes of this assumed shape. Next, we calculate the maximum kinetic energy T_{max} and the maximum potential energy V_{max}. Setting them equal, we can obtain the approximate value of the fundamental natural frequency. If the assumed shape is not correct, a value a little larger than the exact frequency is obtained. It is recommended that a static deflection curve be used as the assumption because it is known empirically that we can generally obtain the value with minor error of about several percentage.

Here, we explain this procedure using a rotor model with n disks, shown in Figure 2.28. Let the mass and the static deflection of the ith disk be m_i and δ_i, respectively. If we assume that the shaft oscillates sinusoidally with the frequency p keeping the same shape as this static deflection curve, the deflection of the ith disk is $y_i = \delta_i \sin pt$. The maximum potential energy is

$$V_{max} = \frac{1}{2}\left(\sum_{i=1}^{n} m_i g \delta_i\right) \tag{2.91}$$

Since the maximum velocity of the ith disk is $p\delta_i$, the maximum kinetic energy is

$$T_{max} = \frac{p^2}{2}\left(\sum_{i-1}^{n} m_i \delta_i^2\right) \tag{2.92}$$

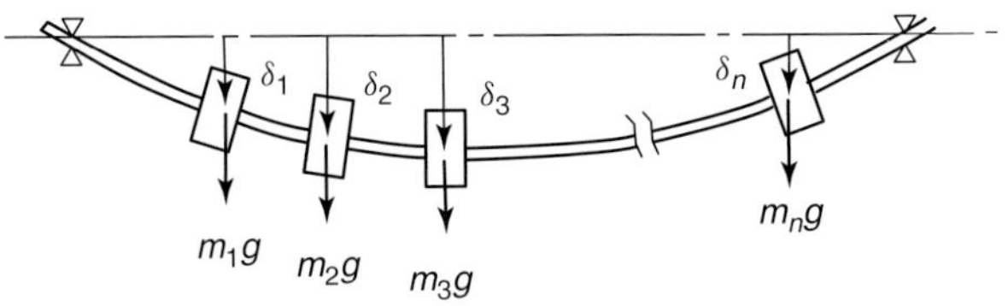

Figure 2.28 Static deflection of the rotor.

Setting $V_{max} = T_{max}$ based on the above mentioned principle, we get the fundamental natural frequency, that is, the critical speed ω_c as follows:

$$\omega_c = \sqrt{\frac{g\sum_{i=1}^{n} m_i\delta_i}{\sum_{i=1}^{n} m_i\delta_i}} \tag{2.93}$$

2.7.2 Dunkerley's Formula

Using a formula found experimentally by Dunkerley (1894), we can get an approximate value of the fundamental natural frequency from the related simple rotor systems. In the case of a multidisk system shown in Figure 2.28, we consider n sets of rotors that have only one disk. The ith set has a disk at the ith position of the original system. Let the spring constant and the disk mass of the ith set be k_i and m_i, respectively. The critical speed ω_i is given by

$$\omega_i = \sqrt{\frac{k_i}{m_i}} \tag{2.94}$$

From these critical speeds of rotors with one disk, the critical speed ω_{c1} of the original multidisk system is given as follows:

$$\frac{1}{\omega_{c1}^2} = \sum_{i=1}^{n} \frac{1}{\omega_i^2} \tag{2.95}$$

If the mass of the shaft is not negligible, the term $1/\omega_s^2$, where ω_s is the critical speed of the shaft alone, is added on to the right-hand side.

It is known that Rayleigh's method gives a value larger than the exact value and that Dunkerley's formula gives a smaller one. Therefore, it is recommended that both methods are used to estimate the critical speed.

References

Campbell, W. and Heckman, W.C. (1925) Tangential vibration of steam turbine buckets. *Trans. ASME*, **46**, 643–671, paper no. 1975.

Dunkerley, S. (1894) On the whirling and vibration of shaft. *Philos. Trans. R. Soc. Lond., Ser. A*, **185**, 279–359.

Ehrich, F.F. (1992) *Handbook of Rotordynamics*, McGraw-Hill, New York.

Fitzpatrick, R. (2012) Is Rotation a Vector? *http://farside.ph.utexas.edu/teaching/301/lectures/node100.html.*

GER (1924) A Special Publication of the General Electric Co., Published in Honor of Wilfred Campbell Shortly After his Untimely Death in 1924, From GER-800.

Loewy, R.G. and Piarulli, V.J. (1969) *Dynamics of Rotating Shafts*, The Shock and Vibration Information Center, U.S. Department of Defense, Washington, DC.

Stodola, A. (1924) *Dampf und Gas-Turbinen*, Verlag von Julius Springer, Berlin.

3
Vibrations of a Continuous Rotor

3.1
General Considerations

A distributed parameter model whose mass is distributed along the shaft is sometimes used as an analytical model for a practical rotor with a complex shape. For example, an elastic rotor whose diameter does not vary much or an elastic shaft that has many disks with almost the same size at positions with almost the same spacing is modeled appropriately by such a distributed parameter model. In the following, we call this model, which has distributed mass, distributed stiffness, and distributed damping, a *continuous rotor*. Analysis of continuous rotors is based on the theory of beam lateral vibrations. The most fundamental theory is the Bernoulli–Euler beam theory which is based on the assumption called *Bernoulli–Euler's hypothesis*. It states that the cross section remains as a flat plane and is perpendicular to the centerline during vibration. The latter assumption means that the deformation due to the shear force can be neglected and the former assumption makes it possible to calculate the strain at an arbitrary position in the beam. In this theory, the bending moment at an arbitrary position of the shaft is proportional to the inverse of the radius of curvature of the centerline of the shaft. This model is called *the Bernoulli–Euler beam* or *classical beam*. In this model, only the lateral motion is considered. In the equations of motion derived under this assumption, the equations in the x- and y-directions perpendicular to the bearing center line do not couple. If the rotor is slender, this model well represents the motion of the rotor. For example, Bishop (1959) and Bishop and Parkinson (1965) used this model for their study of rotor vibrations.

It is known that when the wave length of the vibration mode becomes small relative to the beam thickness, the effect of the inertia for rotation, called *rotary inertia*, and that of the deformation due to shearing forces, called *shear deformation*, appear. A beam model taking these two effects into account is called the *Timoshenko beam*. In a rotor system, when the diameter-to-length ratio increases, these two effects appear in addition to the effect of the gyroscopic moment.

Dimentberg (1961) and Tondl (1965) analyzed continuous rotor systems considering the effects of the gyroscopic moment and the rotary inertia. Eshleman and Eubanks (1969) derived the equations of motion considering gyroscopic moment,

Linear and Nonlinear Rotordynamics: A Modern Treatment with Applications, Second Edition.
Yukio Ishida and Toshio Yamamoto.

rotary inertia, shear deformation, and axial torque, and discussed in detail the effects of these secondary factors.

In this chapter the vibrations of a continuous rotor with uniform cross section are explained. The gyroscopic moment and the rotary inertia are considered in this model.

3.2 Equations of Motion

Figure 3.1 shows the theoretical model of an elastic continuous rotor with a circular cross section. The length of this uniform rotor is l. As the rotor is supported vertically, gravitational force does not work. For simplicity, shear deformation is not considered.

The rectangular coordinate system O-xys is fixed in space. The axis along the bearing centerline is represented by s instead of z because we use z as a complex variable representing the deflection of the rotor in the following. The deflections in the x- and y-directions at the position s are denoted by $u(s, t)$ and $v(s, t)$, respectively. The inclination angle $\theta(s, t)$ of the tangent to the rotor deflection curve is represented by two components $\theta_x(s, t)$ and $\theta_y(s, t)$, which are the projections of θ to the xs- and ys-planes, respectively, and are given by

$$\theta_x = \frac{\partial u}{\partial s}, \quad \theta_y = \frac{\partial v}{\partial s} \tag{3.1}$$

Let the moments working in the xs- and ys-planes be M_{xs} and M_{ys}, respectively. From the Bernoulli–Euler beam theory, we get the following relationships between M_{xs}, M_{ys}, and θ_x, θ_y:

$$M_{xs} = EI\frac{\partial \theta_x}{\partial s}, \quad M_{ys} = EI\frac{\partial \theta_y}{\partial s} \tag{3.2}$$

where E is Young's modulus and I is the cross-sectional area moment of inertia. The signs of the bending moments M_{xs} and M_{ys} are defined as positive when they work to increase θ_x and θ_y as s increases, as shown in Figure 3.1.

To derive the equations of motion, we consider a sliced element with thickness ds at s as shown in Figure 3.2. The polar moment of inertia dI_{p} and the diametral moment of inertia dI_{d} of this element are given by (refer to Appendix A)

$$dI_{\mathrm{p}} = (\rho A ds)\frac{R^2}{2}, \quad dI_{\mathrm{d}} \approx \frac{dI_{\mathrm{p}}}{2} = (\rho A ds)\frac{R^2}{4} \tag{3.3}$$

where R is the radius of the rotor, A the cross-sectional area, and ρ the density of the shaft material.

We pay attention to the correspondence between this sliced element and the disk in Section 2.4. Suppose that the rotor is executing a whirling motion and the inclination of the sliced element changes from θ_x to $\theta_x + d\theta_x$ and from θ_y to $\theta_y + d\theta_y$, respectively, during time duration dt. Referring to Figure 2.14a, we know that changes in the angular momentums about the principal axes of the moment of

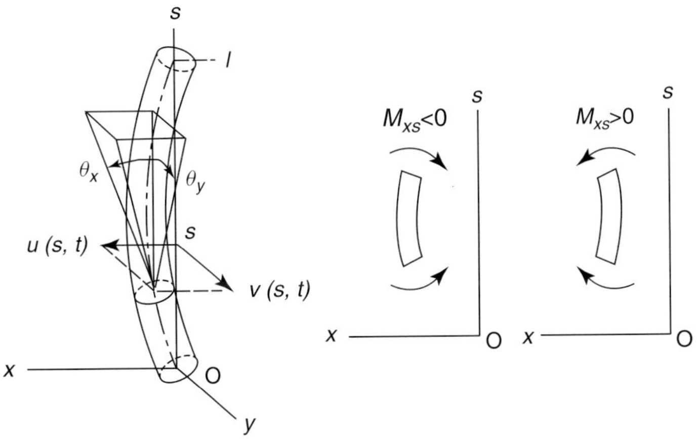

Figure 3.1 Continuous rotor and coordinates.

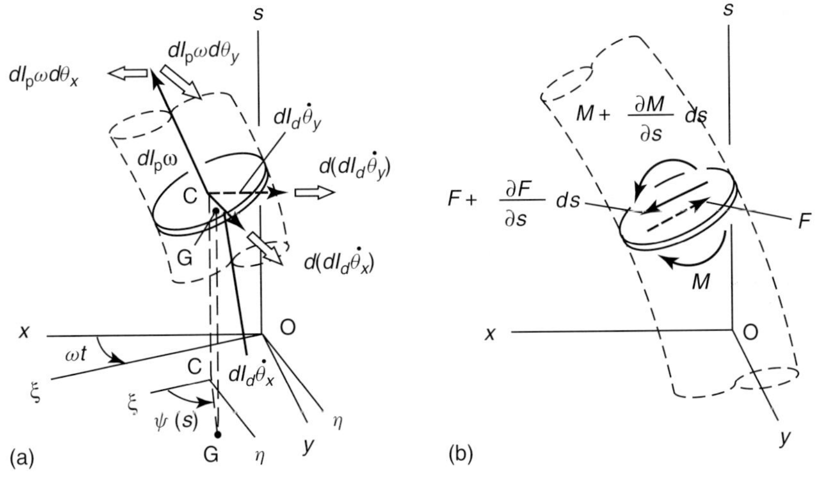

Figure 3.2 Sliced element: (a) increments of angular momentum and (b) shearing forces and moments.

inertia are those shown in Figure 3.2a. The shearing force $F(s,t)$ and the moment $M(s,t)$ are shown in Figure 3.2b. The equations of motion are obtained from the relationships among momentum and angular momentum changes, and shearing forces and moments, as follows.

In the first place, we consider the lateral motion of a sliced element. Let the viscous damping coefficient per unit length of the rotor be c, the deflection of the center of gravity at the position s be (u_G, v_G), and the components of a force in the x- and y-directions be F_x and F_y, respectively. Newton's second law

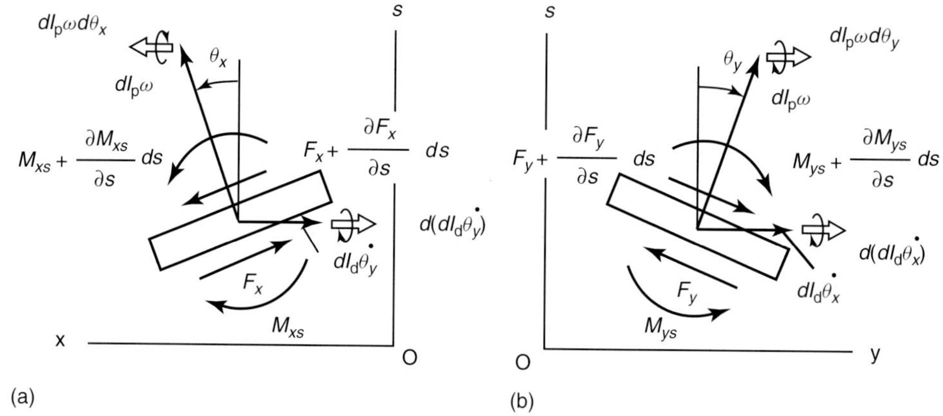

Figure 3.3 Components in the *xs*- and *ys*-planes: (a) *xs*-plane and (b) *ys*-plane.

gives

$$\left.\begin{aligned}(\rho A ds)\frac{\partial^2 u_G}{\partial t^2} &= -F_x + \left(F_x + \frac{\partial F_x}{\partial s} ds\right) - c\frac{\partial u}{\partial t} ds \\ (\rho A ds)\frac{\partial^2 v_G}{\partial t^2} &= -F_y + \left(F_y + \frac{\partial F_y}{\partial s} ds\right) - c\frac{\partial v}{\partial t} ds\end{aligned}\right\} \tag{3.4}$$

After rearrangement, we get

$$\rho A\frac{\partial^2 u_G}{\partial t^2} + c\frac{\partial u}{\partial t} = \frac{\partial F_x}{\partial s}, \quad \rho A\frac{\partial^2 v_G}{\partial t^2} + c\frac{\partial v}{\partial t} = \frac{\partial F_y}{\partial s} \tag{3.5}$$

Next, we consider the inclination motion. The components in the *xs*- and *ys*-planes are shown in Figure 3.3. These components are represented by the vectors following the rule shown in Figure 2.14. The moment vectors in the *x*- and *y*-directions are composed of the moments working on the element and the moments due to the shearing forces. From the relationships between the variations in angular momentum during dt and the applied moments in the *x*- and *y*-directions, we get the following expressions:

$$\left.\begin{aligned}\frac{d(dI_d\dot{\theta}_x) + dI_p\omega d\theta_y}{dt} &= -M_{xs} + \left(M_{xs} + \frac{\partial M_{xs}}{\partial s} ds\right) + F_x ds \\ \frac{-d(dI_d\dot{\theta}_y) + dI_p\omega d\theta_x}{dt} &= M_{ys} - \left(M_{ys} + \frac{\partial M_{ys}}{\partial s} ds\right) - F_y ds\end{aligned}\right\} \tag{3.6}$$

where $F_x ds$ and $F_y ds$ represent moments due to the shearing forces. Substituting these into Eq. (3.1), we obtain

$$\left.\begin{aligned}\left(\frac{dI_d}{ds}\right)\frac{\partial^3 u}{\partial s\partial t^2} + \left(\frac{dI_p}{ds}\right)\omega\frac{\partial^2 v}{\partial s\partial t} &= \frac{\partial M_{xs}}{\partial s} + F_x \\ \left(\frac{dI_d}{ds}\right)\frac{\partial^3 v}{\partial s\partial t^2} - \left(\frac{dI_p}{ds}\right)\omega\frac{\partial^2 u}{\partial s\partial t} &= \frac{\partial M_{ys}}{\partial s} + F_y\end{aligned}\right\} \tag{3.7}$$

The first and second terms appear because of the rotary inertia and the gyroscopic moment, respectively. By differentiating these equations by s and substituting Eq. (3.5) into them to eliminate F_x and F_y, and by using Eqs. (3.1) and (3.2) and expressing the components M_{xs} and M_{ys} as u and v, we get the following equations of motion:

$$\left.\begin{aligned} EI\frac{\partial^4 u}{\partial s^4}+\rho A\frac{\partial^2 u_G}{\partial t^2}-\frac{\rho A R^2}{4}\left[\frac{\partial^4 u}{\partial s^2\partial t^2}+2\omega\frac{\partial^3 v}{\partial s^2\partial t}\right]+c\frac{\partial u}{\partial t}=0\\ EI\frac{\partial^4 v}{\partial s^4}+\rho A\frac{\partial^2 v_G}{\partial t^2}-\frac{\rho A R^2}{4}\left[\frac{\partial^4 v}{\partial s^2\partial t^2}-2\omega\frac{\partial^3 u}{\partial s^2\partial t}\right]+c\frac{\partial v}{\partial t}=0 \end{aligned}\right\} \tag{3.8}$$

where dI_{d} and dI_{p} are replaced by the expression of Eq. (3.3).

Since the unbalance, that is, the magnitude and the direction of the eccentricity of the center of gravity G, changes as a function of s as shown in Figure 2.3, we represent the deflection of the center of gravity G as follows:

$$\left.\begin{aligned} u_G(s,t)&=u(s,t)+e(s)\cos\{\omega t+\varphi(s)\}\\ &=u(s,t)+e_\xi(s)\cos\omega t-e_\eta\sin\omega t\\ v_G(s,t)&=v(s,t)+e(s)\sin\{\omega t+\varphi(s)\}\\ &=v(s,t)+e_\xi(s)\sin\omega t+e_\eta(s)\cos\omega t \end{aligned}\right\} \tag{3.9}$$

where $e_\xi(s)=e(s)\cos\phi$, $e_\eta(s)=e(s)\sin\varphi$, and φ is the angular position of the center of gravity G relative to the ξ-axis in the coordinate system O-$\xi\eta s$ (refer to Figure 3.2), which rotates with the rotor. The position of the coordinate system O-$\xi\eta s$ on the rotor is arbitrary. Substituting Eq. (3.9) into Eq. (3.8) gives

$$\left.\begin{aligned} &EI\frac{\partial^4 u}{\partial s^4}+\rho A\frac{\partial^2 u}{\partial t^2}-\frac{\rho A R^2}{4}\left[\frac{\partial^1 u}{\partial s^2\partial t^2}+2\omega\frac{\partial^3 v}{\partial s^2\partial t}\right]+c\frac{\partial u}{\partial t}\\ &=\rho A\omega^2\{e_\xi(s)\cos\omega t-e_\eta(s)\sin\omega t\}\\ &EI\frac{\partial^4 v}{\partial s^4}+\rho A\frac{\partial^2 v}{\partial t^2}-\frac{\rho A R^2}{4}\left[\frac{\partial^4 v}{\partial s^2\partial t^2}-2\omega\frac{\partial^3 u}{\partial s^2\partial t}\right]+c\frac{\partial v}{\partial t}\\ &=\rho A\omega^2\{e_\xi(s)\sin\omega t+e_\eta(s)\cos\omega t\} \end{aligned}\right\} \tag{3.10}$$

These are the equations of motion governing the motion of a continuous rotor. These equations are also represented by complex numbers as follows:

$$\begin{aligned} &EI\frac{\partial^4 z}{\partial s^4}+\rho A\frac{\partial^2 z}{\partial t^2}-\frac{\rho A R^2}{4}\left[\frac{\partial^4 z}{\partial s^2\partial t^2}-2i\omega\frac{\partial^3 z}{\partial s^2\partial t}\right]+c\frac{\partial z}{\partial t}\\ &=\rho A\omega^2\bar{e}(s)\exp(i\omega t) \end{aligned} \tag{3.11}$$

where $\bar{e}(s)=e_\xi(s)+ie_\eta(s)$.

In the following, we use a prime for differentiation by s and a dot for differentiation by t. This notation is used for both ordinary and partial differentiations.

Note: Effects of Various Factors

Eshleman and Eubanks (1969) obtained more precise equations of motion, which included gyroscopic moments, rotary inertia, transverse shear, and externally applied torque T, and investigated their effects on critical speeds. Shear deformation means the deformation of a shaft element due to transverse shear forces, as shown in Figure 3.4. Since the element that has the shape shown by the dashed lines deforms to the one shown by the solid lines because of shearing forces, the normal line of the cross section (with gradient ψ) shifts from the centerline of the rotor (with the gradient $\partial u/\partial s$) by β. This angle β ($= \partial u/\partial s - \varphi$) is called the *shear angle*. Shear force F_x and shear angle β have the relationship $\beta = F_x/(kAG)$, where A is the area of the cross section of the rotor and G is the shear modulus. The constant k is determined by the shape of the cross section, for example, $k = 2/3$ for a rectangle and $k = 3/4$ for a circle.

Eshleman and Eubanks derived the following equations of motion:

$$\begin{aligned} & EIz'''' + \rho A\ddot{z} + \frac{EI\rho}{kG}\ddot{z}'' - iTz''' + i\frac{T\rho}{kG}\ddot{z}' \\ & \quad - \frac{\rho AR^2}{4}\left[\ddot{z}'' - \frac{\rho}{kG}\ddddot{z} + 2i\omega\left(-\dot{z}'' + \frac{\rho}{kG}\dddot{z}\right)\right] = 0 \end{aligned} \tag{3.12}$$

Comparing this expression with Eq. (3.11), we find that some new terms are included in the equation. We can guess the causes of these terms from their parameters. The third term is caused by the shear deformation, the fourth by the axial torque, the fifth by the interaction of shear deformation and axial torque, the seventh by the interaction of shear deformation and rotary inertia, and the ninth by the interaction of shear deformation and gyroscopic moment.

The following explanation on the effects of secondary factors is based on the investigation by Eshleman and Eubanks (1969). Figure 3.5 shows a comparison of the first critical speed of a rotor calculated by various theoretical models. Both ends of these models are supported by short bearings that give a condition of simple support. These critical speeds change as a function of the slenderness ratio $\bar{r} = R/(2l)$. Since the scale of the ordinate is nondimensionalized by a proper quantity, we should note the relative change of their values.

The dashed line (– – –) represents the critical speed of a rotor corresponding to the Bernoulli–Euler beam. Since it does not depend on the slenderness ratio, it has a constant value. The double-dotted curve (-•-•-) is for a rotor in which

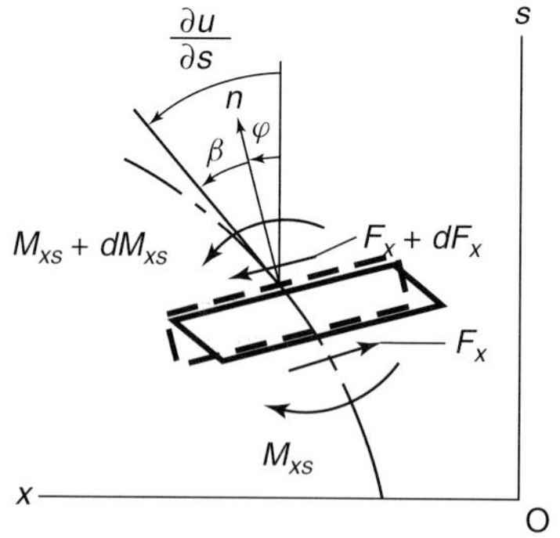

Figure 3.4 Shear deformation.

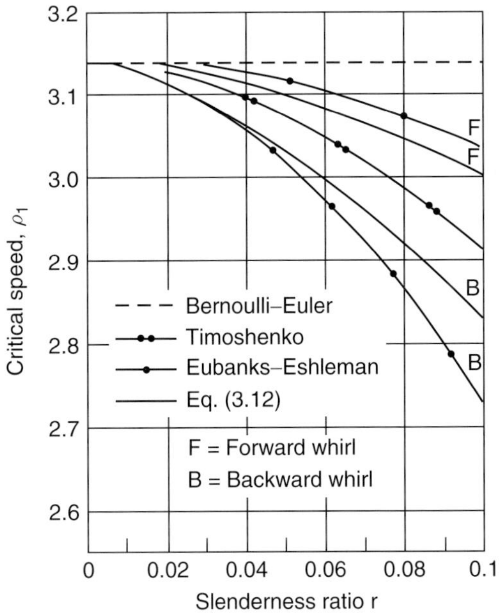

Figure 3.5 Relationship between slenderness ratio and first critical speed for various methods. (From Eshleman and Eubanks (1961), Courtesy of ASME.)

the effects of shear deformation and rotary inertia are considered. This rotor corresponds to the Timoshenko beam. The critical speed of this rotor decreases as the slenderness ratio increases. The single-dotted curve (-•-) is the case where the gyroscopic moment is considered in addition to shear deformation and rotary inertia. However, the interaction of shear deformation and rotary inertia and that of shear deformation and gyroscopic moment are not considered. Since the gyroscopic moment raises a forward critical speed and lowers a backward critical speed, the critical speed separates into two values. The curve without dots (—) is for a rotor governed by Eq. (3.12). This model, which takes axial torque and the interactions of various factors into consideration, gives the most precise result. The effects of these factors become more remarkable as the order of critical speed increases.

It is also known that concerning an axial force, the tension raises the critical speeds and the compression lowers them (Ziegler, 1977).

3.3
Free Whirling Motions and Critical Speeds

The following analysis can be executed either by solving Eq. (3.10) expressed in real numbers or Eq. (3.11) expressed in complex numbers. We adopt the latter method because of its simplicity of expression.

The equations of motion governing undamped free whirling motions are obtained by setting $c = 0$ and $\bar{e} = 0$ in Eq. (3.11) as follows:

$$EIz'''' + \rho A\ddot{z} - \frac{\rho AR^2}{4}(\ddot{z}'' - 2i\omega\dot{z}'') = 0 \tag{3.13}$$

Here, we discuss the case of a continuous rotor that is supported simply at both ends. This boundary condition is given by the following expressions, which denote that the deflections and moments are zero at both ends

$$\left.\begin{aligned} z(0,t) = 0, \quad EIz''(0,t) = 0 \\ z(l,t) = 0, \quad EIz''(l,t) = 0 \end{aligned}\right\} \tag{3.14}$$

Let the mass per unit length of the rotor be m $(= \rho A)$ and the quantity $\rho AR^2/4$ relating to the moment of inertia be mr^2 where r is the radius of gyration. Then, Eq. (3.13) becomes

$$EIz'''' + m\ddot{z} - mr^2(\ddot{z}'' - 2i\omega\dot{z}'') = 0 \tag{3.15}$$

3.3.1
Analysis Considering Only Transverse Motion

The equation of motion corresponding to the Bernoulli–Euler beam is given by putting $mr^2 = 0$ in Eq. (3.15):

$$\ddot{z} + {c_0}^2 z'''' = 0 \tag{3.16}$$

where ${c_0}^2 = EI/m = EI/(\rho A)$. We look for a solution of the form

$$z(s,t) = \varphi(s)\psi(t) \tag{3.17}$$

This solution is a product of functions, each of which depends on only one of the independent variables. When Eq. (3.17) is substituted into Eq. (3.16), we get

$$\frac{\ddot{\psi}}{\psi} = -{c_0}^2 \frac{\varphi''''}{\varphi} (= -p^2) \tag{3.18}$$

Since the left-hand side is a function of only t and the right-hand side is a function of only s, this equation can hold only if both sides are constant. We denote this constant by $-p^2$ $(p > 0)$. (If we denote it by $+p^2$, the result would be divergent solutions that are apparently inappropriate for an isolated system.) Then, the following two ordinary differential equations are obtained:

$$\ddot{\psi}(t) + p^2\psi(t) = 0 \tag{3.19}$$

$$\varphi''''(s) - \frac{p^2}{{c_0}^2}\varphi(s) = 0 \tag{3.20}$$

From Eq. (3.19) we get

$$\psi(t) = A_1 \cos pt + A_2 \sin pt \tag{3.21}$$

and from Eq. (3.20), we get

$$\begin{aligned}\varphi(s) &= B_1 e^{i\nu s} + B_2 e^{-i\nu s} + B_3 e^{\nu s} + B_4 e^{-\nu s} \\ &= C_1 \sin \nu s + C_2 \cos \nu s + C_3 \sinh \nu s + C_4 \cosh \nu s\end{aligned} \tag{3.22}$$

where $\nu = \sqrt{p/c_0}$. Boundary conditions of Eq. (3.14) give

$$\left.\begin{aligned}\varphi(0) = 0, \quad \varphi''(0) = 0 \\ \varphi(l) = 0, \quad \varphi''(l) = 0\end{aligned}\right\} \tag{3.23}$$

Applying these conditions to Eq. (3.22), we get

$$C_2 = C_3 = C_4 = 0, \quad \sin \nu l = 0 \tag{3.24}$$

From the latter, we know that $\nu l = n\pi$ must hold. This means that ν takes discrete values $\nu_n(= n\pi/l)$ corresponding to integer n. Then, the natural frequency p_n results in

$$p_n = c_0 {\nu_n}^2 = \left(\frac{n\pi}{l}\right)^2 \sqrt{\frac{EI}{\rho A}} \tag{3.25}$$

The mode shape of order n corresponding to the natural frequency p_n is expressed by

$$\varphi_n(s) = C_n \sin \nu_n s = C_n \sin \frac{n\pi s}{l} \tag{3.26}$$

Here, C_n represents the amplitude of the nth mode. The function φ_n is called the *eigenfunction* corresponding to the *eigenvalue* ν_n. Normalizing this function by putting $C_n = 1$, we get normalized mode function $\overline{\varphi}_n(s)$ given by

$$\overline{\varphi}_n(s) = \sin \nu_n s = \sin \frac{n\pi s}{l} \tag{3.27}$$

This mode shape is the sinusoidal deflection curve shown in Figure 3.6. Finally, we get the following general solution satisfying the boundary condition of Eq. (3.14):

$$z(s,t) = \sum_{n=1}^{\infty} \overline{\varphi}_n(s)(A_{1n} \cos p_n t + A_{2n} \sin p_n t) \tag{3.28}$$

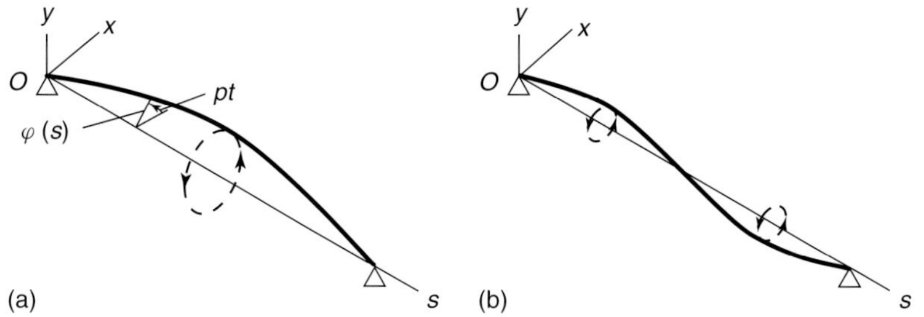

Figure 3.6 Vibration modes of a rotor corresponding to the Bernoulli–Euler beam: (a) first mode $p_1 = \left(\frac{\pi}{l}^2 \sqrt{\frac{EI}{\rho A}}\right)$ and $\overline{\varphi_1}(s) = \sin \frac{\pi}{l} s$; (b) second mode $p_2 = 2^2 p_1$ and $\overline{\varphi_2}(s) = \sin \frac{2\pi}{l} s$.

Thus, the free whirling motion of a continuous rotor is composed of motions that whirl like a skipping rope with the frequency p_n given by Eq. (3.25), and with the mode shape φ_n given by Eq. (3.26). As shown in Eq. (3.28), forward and backward whirling motions exist for each mode.

3.3.2
Analysis Considering the Gyroscopic Moment and Rotary Inertia

The equations of motion, including the effects of the gyroscopic moment and rotary inertia, are given by Eq. (3.15). Assuming the solution in the form of Eq. (3.17) and substituting it into Eq. (3.15) gives

$$EI\varphi''''\psi + m\varphi\ddot{\psi} - mr^2(\varphi''\ddot{\psi} - 2i\omega\varphi''\dot{\psi}) = 0 \tag{3.29}$$

This expression cannot be separated into the function with only s and the function with only t, like Eq. (3.18). Therefore, we solve this equation in a different way on the basis of physical considerations. It was shown that the deflection curve of a rotor treated in Section 3.3.1 was sinusoidal. However, since the gyroscopic moment, which differs depending on the position of the element, works in the rotor treated in this section, the deflection curve $\varphi(s)$ may differ from a sinusoidal curve, and the concrete form is unknown at this stage. However, it is reasonable to assume that the shaft whirls with frequency p, keeping mode shape $\varphi(s)$ constant in each mode. From such a consideration, we assume the solution in the following form, which is more concrete than Eq. (3.17):

$$z(s,t) = \varphi(s)e^{ipt} \tag{3.30}$$

Substituting this into Eq. (3.29) and dividing it by e^{ipt} gives

$$EI\varphi'''' - mr^2p(2\omega - p)\varphi'' - m\varphi p^2 = 0 \tag{3.31}$$

Assuming the mode shape in the form of $\varphi = Ae^{\lambda s}$ and substituting it into this equation, we get a quadratic equation on λ^2. Then, we get four roots $\lambda_1, \ldots, \lambda_4$, which include the variable p. The roots λ_1 and λ_2 are pure imaginary numbers, and λ_3 and λ_4 are real numbers that have the same value but with opposite signs. Therefore, we get the solution in the same form as Eq. (3.22). From the boundary condition of Eq. (3.23), it becomes

$$\varphi(s) = C\sin\nu s \tag{3.32}$$

Furthermore, from the boundary condition $\varphi(s) = 0$ at $s = l$, we get the relationship $\nu l = n\pi\,(n = 1, 2, 3, \ldots)$. Let ν satisfying this relationship be $\nu_n(= n\pi/l)$. Substituting the corresponding mode shape $\varphi_n(s) = C_n\sin\nu_n s$ into Eq. (3.31), we get the expression

$$EI{\nu_n}^4 + mr^2p_n(2\omega - p_n){\nu_n}^2 - m{p_n}^2 = 0 \tag{3.33}$$

where $p = p_n$ is the natural frequency corresponding to the nth mode. This equation gives

$$p_n = \frac{r^2{\nu_n}^2\omega \pm \sqrt{(r^2{\nu_n}^2\omega)^2 + {c_0}^2(1 + r^2{\nu_n}^2){\nu_n}^4}}{1 + r^2{\nu_n}^2} \tag{3.34}$$

where $c_0 = \sqrt{EI/m}$. The positive root $p = p_{fn}$ represents a frequency of a forward whirling mode, while the negative root $p = p_{bn}$ represents that of a backward whirling mode.

A natural frequency diagram is shown in Figure 3.7. In such a case of a slender uniform shaft with no disk, the effect of the gyroscopic moment does not appear appreciably in the diagram and the natural frequencies do not change much. In order to show the change in natural frequencies in the wide range of the rotational speed, a frequency diagram of the type shown in Figure 3.8 is sometimes used (Tondl, 1965).

In this figure, forward and backward natural frequencies of up to the sixth mode are illustrated.

3.3.3 Major Critical Speeds

The intersections of the straight line $p = \omega$ and the curves p_{fn} in Figures 3.7 and 3.8 give the major critical speeds ω_n. Setting $p_n = \omega = \omega_n$ in Eq. (3.33), we get

$$\omega_n = \frac{c_0 \nu_n{}^2}{\sqrt{1 - r^2 \nu_n{}^2}} = \frac{(n\pi/l)^2 \sqrt{EI/\rho A}}{\sqrt{1 - r^2 (n\pi/l)^2}} \tag{3.35}$$

Since ω_n is real only when the quantity under the radical sign in the denominator is positive, the major critical speed does not exist for higher modes with large n.

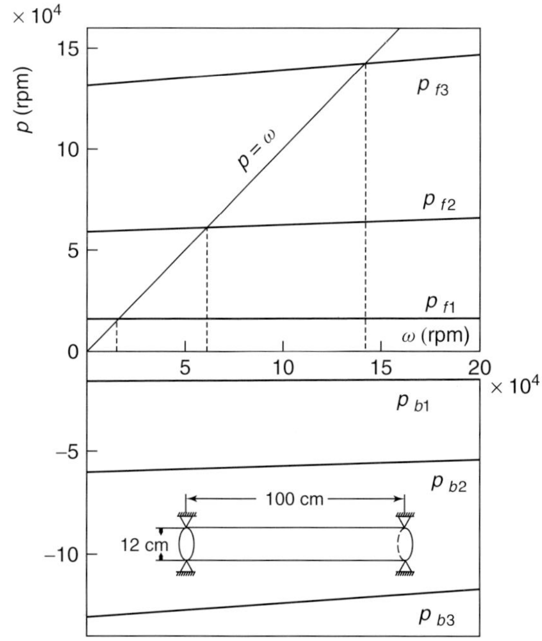

Figure 3.7 Natural frequency diagram of a continuous rotor (linear scale).

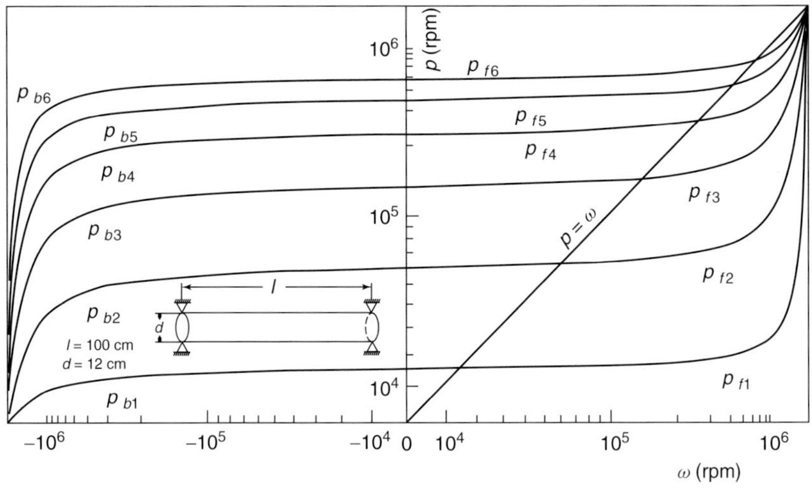

Figure 3.8 Natural frequency diagram of a continuous rotor (logarizmic scale).

This means that the number of the major critical speeds is finite, although the number of the degrees of freedom is infinite.

3.4 Synchronous Whirl

The response to the unbalance force in the vicinity of the major critical speed is obtained from the equations of motion, Eq. (3.10) or (3.11). Although we can analyze using either real numbers or complex numbers, we execute the analysis with real numbers.

As shown in Figure 3.9, the eccentricity of the center of gravity forms a three-dimensional curve whose direction and distance from the origin change

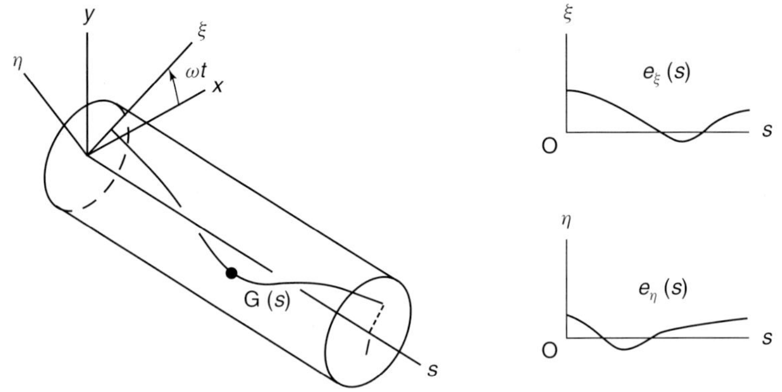

Figure 3.9 Distribution of eccentricity and coordinates.

depending on position s in the coordinate system O-$\xi\eta s$ rotating with the angular velocity ω.

This curve is represented by the function $e_\xi(s)$ in the ξs-plane and $e_\eta(s)$ in the ηs-plane, and these functions are expanded by the eigenfunctions. The eigenfunctions of a continuous rotor have the following characteristics, called the *orthogonality*:

$$\int_0^l \rho(s)\overline{\varphi}_n(s)\overline{\varphi}_m(s)ds = \begin{cases} 0 & (n \neq m) \\ \text{constant} & (n = m) \end{cases} \tag{3.36}$$

where $\rho(s)$ is mass per unit length. The eigenfunctions for a uniform shaft are obtained as Eq. (3.26) and are given by

$$\overline{\varphi}_n(s) = \sin \nu_n s = \sin\left(\frac{n\pi}{l}\right)s \quad (n = 1, 2, \ldots) \tag{3.37}$$

These eigenfunctions have the following characteristics corresponding to Eq. (3.36):

$$\int_0^l \overline{\varphi}_n(s)\overline{\varphi}_m(s)ds = \begin{cases} 0 & (n \neq m) \\ l/2 & (n = m) \end{cases} \tag{3.38}$$

The functions $e_\xi(s)$ and $e_\eta(s)$ can be expanded in eigenfunction series as

$$e_\xi(s) = \sum_{n=1}^{\infty} e_{\xi\eta}\overline{\varphi}_n(s), \quad e_\eta(s) = \sum_{n=1}^{\infty} e_{\eta n}\overline{\varphi}_n(s) \tag{3.39}$$

where each coefficient can be determined from

$$e_{\xi n} = \frac{2}{l}\int_0^l e_\xi(s)\overline{\varphi}_n(s)ds, \quad e_{\eta n} = \frac{2}{l}\int_0^l e_\eta(s)\overline{\varphi}_n(s)ds \tag{3.40}$$

These quantities $e_{\xi n}$ and $e_{\eta n}$ are called *nth modal eccentricity*. Since the rotor system treated here is linear, the deflection of the rotor is represented by the summations of the response to each unbalance force due to modal eccentricity. Therefore, we assume the solutions of Eq. (3.10) in the forms

$$\left.\begin{aligned} u &= \sum_{n=1}^{\infty} \overline{\varphi}_n(s)(P_{xcn}\cos\omega t - P_{xsn}\sin\omega t) \\ v &= \sum_{n=1}^{\infty} \overline{\varphi}_n(s)(P_{ysn}\sin\omega t + P_{ycn}\cos\omega t) \end{aligned}\right\} \tag{3.41}$$

Substituting these into Eq. (3.10) and equating the coefficients of the right- and left-hand sides concerning $\cos\omega t$ and $\sin\omega t$ in each mode, we get

$$\left.\begin{aligned} A_n(\omega)P_{xcn} - \overline{c}\omega P_{xcn} + B_n(\omega)P_{ysn} &= \omega^2 e_{\xi n} \\ \overline{c}\omega P_{xcn} + A_n(\omega)P_{xsn} + B_n(\omega)P_{ycn} &= \omega^2 e_{\eta n} \\ B_n(\omega)P_{xcn} + A_n(\omega)P_{ysn} - \overline{c}\omega P_{ycn} &= \omega^2 e_{\xi n} \\ B_n(\omega)P_{xsn} + \overline{c}\omega P_{ysn} + A_n(\omega)P_{ycn} &= \omega^2 e_{\eta n} \end{aligned}\right\} \tag{3.42}$$

where

$$\left.\begin{aligned} A_n(\omega) &= (EI/\rho A)\nu_n^4 - \omega^2 - r^2\omega^2\nu_n^2 \\ B_n(\omega) &= 2r^2\omega^2\nu_n^2, \quad \bar{c} = c/\rho A \end{aligned}\right\} \tag{3.43}$$

Using Cramer's rule, we can solve Eq. (3.42) in the form

$$\left.\begin{aligned} P_{xcn} = P_{ysn} &= \frac{\omega^2}{D}\begin{vmatrix} e_{\xi n} & -\bar{c}\omega & B_n & 0 \\ e_{\eta n} & A_n & 0 & B_n \\ e_{\xi n} & 0 & A_n & -\bar{c}\omega \\ e_{\eta n} & B_n & \bar{c}\omega & A_n \end{vmatrix} \\ P_{xsn} = P_{ycn} &= \frac{\omega^2}{D}\begin{vmatrix} A_n & e_{\xi n} & B_n & 0 \\ \bar{c}\omega & e_{\eta n} & 0 & B_n \\ B_n & e_{\xi n} & A_n & -\bar{c}\omega \\ 0 & e_{\eta n} & \bar{c}\omega & A_n \end{vmatrix} \end{aligned}\right\} \tag{3.44}$$

where

$$D = \begin{vmatrix} A_n & -\bar{c}\omega & B_n & 0 \\ \bar{c}\omega & A_n & 0 & B_n \\ B_n & 0 & A_n & -\bar{c}\omega \\ 0 & B_n & \bar{c}\omega & A_n \end{vmatrix} \tag{3.45}$$

Then we can obtain the amplitude P_n and the phase β_n of the nth mode as follows:

$$\left.\begin{aligned} &{P_n}^2 = {P_{xn}}^2 = {P_{xcn}}^2 + {P_{xsn}}^2 = {P_{yn}}^2 = {P_{ysn}}^2 + {P_{ycn}}^2 \\ &\beta_n = \tan^{-1}(P_{xsn}/P_{xcn}) = \tan^{-1}(P_{ycn}/P_{ysn}) \end{aligned}\right\} \tag{3.46}$$

Summing these solutions for the nth mode, we arrive at the response to the unbalance of Eq. (3.39):

$$\left.\begin{aligned} u(s,t) &= \sum_{n=1}^{\infty} \overline{\varphi}_n(s) P_n \cos(\omega t + \beta_n) \\ v(s,t) &= \sum_{n=1}^{\infty} \overline{\varphi}_n(s) P_n \sin(\omega t + \beta_n) \end{aligned}\right\} \tag{3.47}$$

These are the steady-state solutions of forced oscillations due to unbalance. These expressions mean that the deflection curve of a rotor whirling in the forward direction with the frequency ω is the superposition of curves $\overline{\varphi}_1(s)P_1, \overline{\varphi}_2(s)P_2$, and so on. Resonance phenomena appear when the denominator of Eq. (3.44) becomes small.

Here, we investigate the resonance speeds where the amplitude P_n of the nth mode becomes large for the case of no damping. From Eq. (3.44) with $\bar{c} = 0$, we have the following solutions:

$$P_{xcn} = P_{ysn} = \frac{\omega^2 e_{\xi n}}{A_n + B_n}, \qquad P_{xsn} = P_{ycn} = \frac{\omega^2 e_{\eta n}}{A_n + B_n} \tag{3.48}$$

Therefore, when

$$A_n + B_n = \left(\frac{EI}{\rho A}\right) {\nu_n}^4 - (1 - r^2{\nu_n}^2)\omega^2 = 0 \tag{3.49}$$

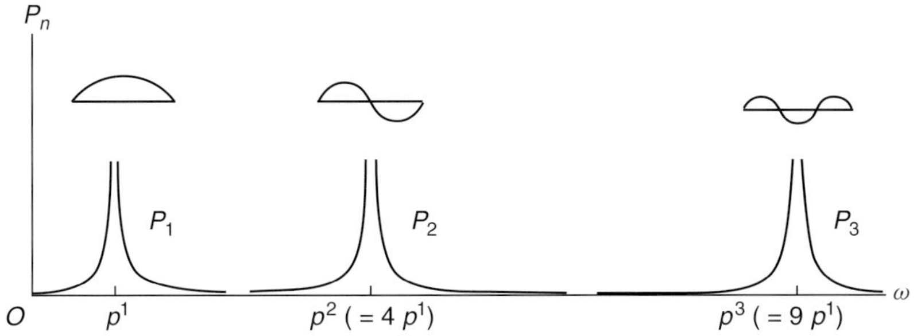

Figure 3.10 Resonance curves and shapes of rotor deflection (case with small gyroscopic moment).

holds, the amplitude becomes infinite. The rotational speeds satisfying this expression are clearly the same as the critical speed of a forward whirling mode given by Eq. (3.35).

For example, in the case of a system with no gyroscopic moment treated in Section 3.3.1, we have the following expression by setting $r = 0$ in Eqs. (3.48) and (3.49):

$$P_n = \frac{\omega^2 e_n}{p_n^2 - \omega^2} \tag{3.50}$$

where $e_n = \sqrt{e_{\xi n}{}^2 + e_{\eta n}{}^2}$ and p_n is given by Eq. (3.25). The amplitude variation and the shapes of deflection at each resonance are shown in Figure 3.10.

Topic: Supercritical-Speed Helicopter, Power-Transmission Shaft

A practical example of the continuous rotor is shown in Figure 3.11. Prause, Meacham, and Voorhees (1967) developed viscous dampers to suppress vibrations of the supercritical-speed power transmission shaft of Chinook helicopter shown in Figure 3.11a. The length, the outer diameter, and the inner diameter of the tube shaft are 338.50, 4.50, and 4.26 in., respectively. This shaft is connected to the couplings at both ends. The damper unit is installed at the location 30 in. from one end. The damper is composed of a bearing mounted on the shaft, a thin disc supported by this bearing, and two plates fixed on the body. The thin disk is inserted between plates and the damper oil fills the narrow gap. The response curves for three different damping coefficients are shown in Figure 3.11a. Since a vibration pickup at one location is used, maximum amplitudes at all critical speeds were not measured. The first, second (not measured), third, and the fourth critical speeds are observed. It was impossible to observe from the fourth resonance in the case of the maximum damping. (The reason is explained in the referenced paper.) The numerically obtained mode shapes up to the sixth critical speeds are shown in Figure 3.11b.

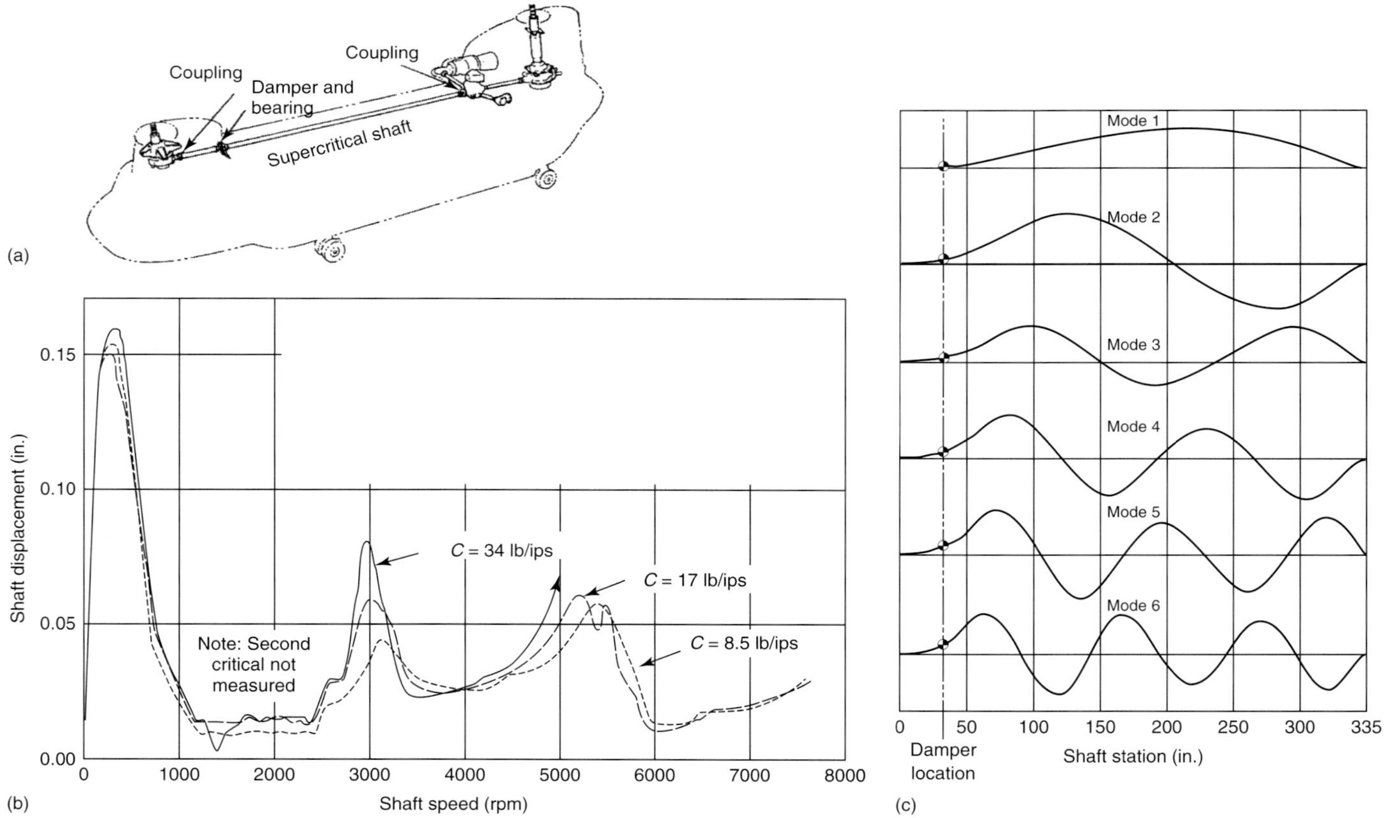

Figure 3.11 Boeing/Vertol Chinook helicopter: (a) drive shaft with a damper, (b) amplitudes for three damping coefficients, and (c) damped vibration mode shape. (Prause, Meacham and Voorhees, 1967, Courtesy of ASME.)

References

Bishop, R.E.D. (1959) The vibration of rotating shafts. *J. mech. Eng. Sci.*, **1** (1), 50–65.

Bishop, R.E.D. and Parkinson, A.G. (1965) Second order vibration of flexible shafts. *Philos. Trans. R. Soc. Lond., Ser. A*, **259** (1095), 1–31.

Dimentberg, F.M. (1961) *Flexural Vibrations of Rotating Shafts*, Butterworths, London.

Eshleman, R.L. and Eubanks, R.A. (1969) On the critical speeds of a continuous rotor. *Trans. ASME, J. Eng. Ind.*, **91** (4), 1180–1188.

Prause, R.H., Meacham, H.C., and Voorhees, J.E. (1967) The design and evaluation of a supercritical-speed helicopter power-transmission shaft. *Trans. ASME, J. Eng. Ind.*, 719–728.

Tondl, A. (1965) *Some Problems of Rotor Dynamics*, Czechoslovak Academy of Sciences, Prague.

Ziegler, H. (1977) *Principles of Structural Stability*, 2nd edn, Birkhauser Verlag, Basel, p. 87.

4
Balancing

4.1
Introduction

As mentioned in Section 2.2, the principal axis of inertia of a rotor deviates from the axis of rotation owing to errors that occur during the manufacturing process and also during assemblage. This deviation creates a centrifugal force that causes vibration and noise. The first consideration in reducing vibration is the elimination of such deviations. *Balancing* refers to the identification of the size and phase angles of these errors (an eccentricity of mass and angular deviation), then placing weights on the rotor to counter the unbalance. We must remember that we cannot eliminate unbalance totally but can bring it down to a tolerable level. In this chapter, methods to balance rotors are explained.

4.2
Classification of Rotors

In Figure 1.1, we showed that rotors are classified into two categories, namely rigid rotors and flexible rotors. A rotor is generally considered to be rigid when it is operated at the speed below the first critical speed (for example, lower than about 70%) and its shape does not change, while a rotor is called a *flexible rotor* when it operates close to or above the first critical speed and generates a slight elastic deformation. However, it is possible to break down the classification of rotors into more precise classes. The ISO has further classified rotors into the following five categories from the view point of balancing.

- **Class 1: Rigid rotor**
  - The unbalance of this type of rotor can be corrected in any two arbitrarily selected planes (two-plane balancing) at low speed. After balancing, the balancing condition does not change up to the operating speed.
  - Example: Gear wheel (Figure 4.1a).

Linear and Nonlinear Rotordynamics: A Modern Treatment with Applications, Second Edition.
Yukio Ishida and Toshio Yamamoto.

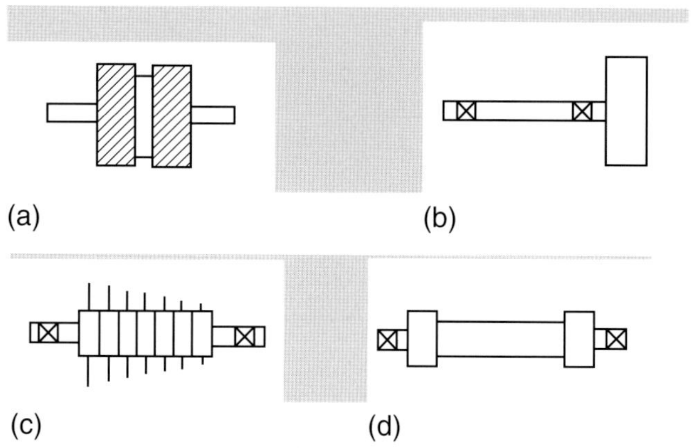

Figure 4.1 Example of rotors and their classes: (a) gear wheel (Class 1), (b) grinder wheel (Class 2), (c) jet engine compressor (Class 2), and (d) generator rotor (Class 3).

- **Class 2: Quasi-flexible rotor**
  - This type of rotor is flexible, but can be balanced by using a low-speed balancing machine. After balancing, the balancing condition does not change up to the operating speed.
  - Example: Grinding wheel (Figure 4.1b, single-plane balancing), jet engine compressor (Figure 4.1c, first each disc is balanced individually and then assembled).
- **Class 3: Flexible rotor**
  - This type of rotor requires high-speed balancing, which takes into account the deformation of the shaft due to centrifugal force.
  - Generally, the methods of modal balancing, influence coefficient method, and so on, are used in these cases.
  - Example: Generator rotor (Figure 4.1d).
- **Class 4: Flexible-attachment rotor**
  - This type of rotor belongs to Class 1, 2, or 3 but flexible components are attached.
  - Example: Rotor with rubber blades whose balancing changes depending on the operation speed.
- **Class 5: Single-speed flexible rotor**
  - This type of rotor belongs to Class 3. But, since it is used at one specific service speed and owing to economic reasons, it is balanced only at that single service speed.
  - Example: High-speed motor.

Among these categories, Classes 1 and 3 are of greatest importance.

4.3 Balancing of a Rigid Rotor

4.3.1 Principle of Balancing

In Section 2.2, we studied that the eccentricity of the center of gravity and the skew angle of the principal axis of inertia are the causes of unbalance in a rigid rotor. In order to eliminate these errors, generally two correction planes are necessary. Therefore, the following *two-plane balancing* is adopted.

4.3.1.1 Two-Plane Balancing

Firstly, the mass eccentricity is eliminated as follows. Here, we use vectors to represent the eccentricities and centrifugal forces, for which both magnitude and direction are necessary. Figure 4.2a shows a rotor with static unbalance and its equivalent model with concentrated masses. Balancing is attained if the centrifugal force $\boldsymbol{F} = m\boldsymbol{e}\omega^2$ due to the eccentricity is canceled by the centrifugal forces due to *correction weights.* In practical machines, the positions of the correction planes (I and II) are determined from the shape of the rotor. Balancing is done by removing some parts of the rotor or by attaching correction masses in planes I and II. In practice, removing some parts is done by drilling, milling, or grinding. Addition of weight would require the use of wire solders, bolted or riveted washers, and welded weights. Suppose that masses m_1 and m_2 are attached to the surface with radii a_1 and a_2, respectively. In order to cancel the unbalance force $\boldsymbol{F}$ by the centrifugal forces $\boldsymbol{F}_{\mathrm{I}} = m_1\boldsymbol{a}_1 w^2$ and $\boldsymbol{F}_{\mathrm{II}} = m_2\boldsymbol{a}_2 w^2$, the following relationships must hold among the forces shown in Figure 4.2a:

$$F_I + F_{II} = F, \quad F_I a = F_{II} b \tag{4.1}$$

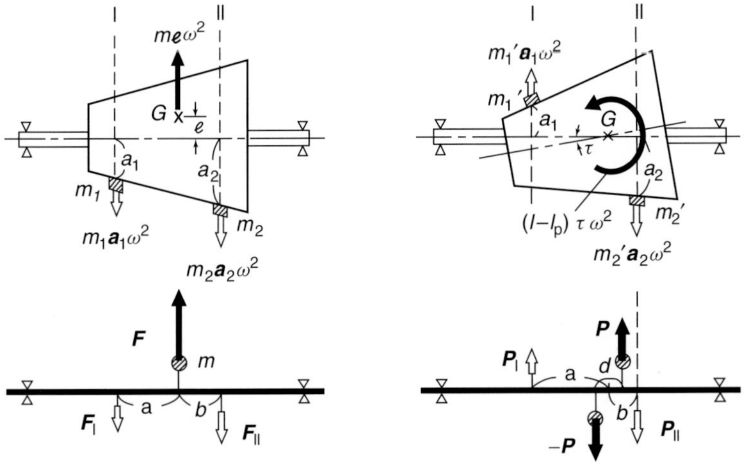

Figure 4.2 Elimination of unbalances in a rigid rotor: (a) eccentricity and (b) skew angle.

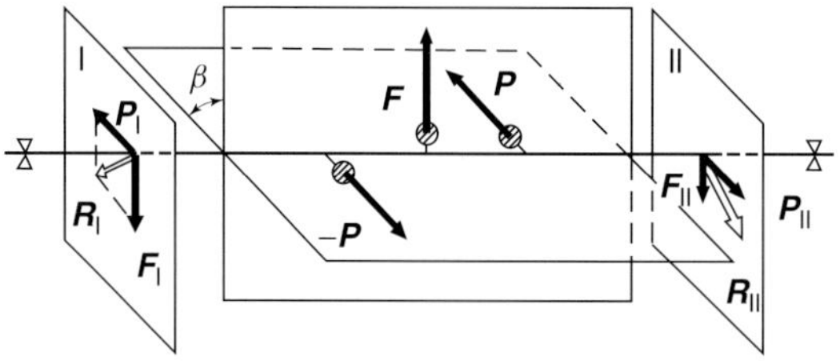

Figure 4.3 Composition of balancing weights in correction planes.

Then, we have

$$F_I = \frac{bF}{a+b}, \quad F_{II} = \frac{aF}{a+b} \tag{4.2}$$

Next, we consider the elimination of the skew angle shown in Figure 4.2b. The moment $M = (I - I_p)\tau\omega^2$ due to this skew angle can be equivalently replaced by a couple of forces $P(= M/d)$, which are in the same plane as τ and separated by the distance d. We add correction masses m_1' and m_2' to cancel the moment M by the centrifugal forces $\boldsymbol{P}_I = m_1'\mathbf{a}_1\omega^2$ and $\boldsymbol{P}_{II} = m_2'\mathbf{a}_2\omega^2$. In this case, the following relationships must hold:

$$P_I a + P_{II} b = M, \quad P_I = P_{II} \tag{4.3}$$

The latter is the condition to prevent a new eccentricity due to the addition of m_1' and m_2'. In general expression, including the directions, the condition is expressed by $\boldsymbol{P}_I = -\boldsymbol{P}_{II}$. From Eq. (4.3), we obtain

$$P_I = P_{II} = \frac{M}{a+b} \tag{4.4}$$

Figure 4.3 shows a general case where the eccentricity and the skew angle coexist. The balancing is attained by adding balancing weights in correction planes I and II so as to produce the resultant forces $\boldsymbol{R}_I$ and $\boldsymbol{R}_{II}$ determined by the vector relationships:

$$\boldsymbol{R}_I = \boldsymbol{F}_I + \boldsymbol{P}_I, \quad \boldsymbol{R}_{II} = \boldsymbol{F}_{II} + \boldsymbol{P}_{II} \tag{4.5}$$

The balancing method mentioned above is called the *two-plane balancing*.

4.3.1.2 Single-Plane Balancing

When a rotor is thin and the disk is perpendicularly attached to the shaft, the balancing is attained practically by adding a correction weight in one plane. In such a case, the eccentricity is detected by some means and a proper correction weight with mass m_1 is attached to the opposite direction as shown in Figure 4.4. This method is called the *single-plane balancing*.

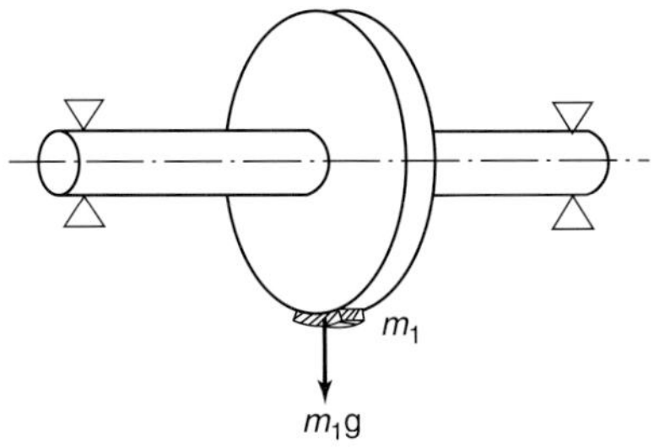

Figure 4.4 Single-plane balancing.

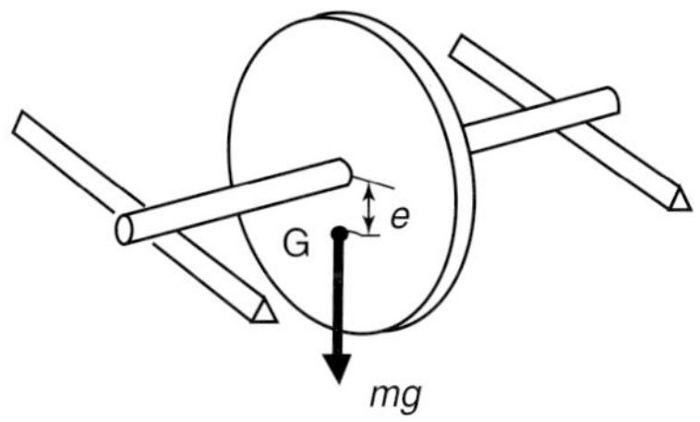

Figure 4.5 Static balancing machine.

4.3.2 Balancing Machine

The above explanation describes the principle of balancing. Since we cannot know both magnitudes of unbalances (that is, F and M) and the location of the center of gravity (that is, a and b) in practical terms, we cannot use Eqs. (4.2) and (4.4). Instead, we must balance the rotor utilizing other experimentally measured data.

In factories, rotors are balanced at the final stage of production. This is called *factory balancing* or *production balancing*. Factory balancing is performed using a *balancing machine*. Balancing machines are classified into *static balancing machines* that utilize gravitational force and *dynamic balancing machines* that utilize centrifugal force. Generally, balancing machines are used in order to balance rigid rotors.

4.3.2.1 Static Balancing Machine

When a rotor is supported by ball bearings with little friction or placed on knife edges as shown in Figure 4.5, the rotor stops when the eccentricity is located downward. Static balancing machine utilizes this phenomenon and this method is referred to as *static balancing*. As mentioned previously, the skew angle cannot be determined utilizing this method.

4.3.2.2 Dynamic Balancing Machine

The balancing procedure which identifies unbalance by rotating the rotor is called *dynamic balancing*. In this dynamic balancing machine, the magnitude and angular position of the eccentricity can be simultaneously detected. Dynamic balancing machines are classified into two categories as shown in Figure 4.6. One is the *hard-bearing balancing machine* in which the bearing support is hard. In this type of machine, its natural frequencies are higher than the rotor's operation range and forces (or accelerations) are detected at the bearings. The machine is rigid and,

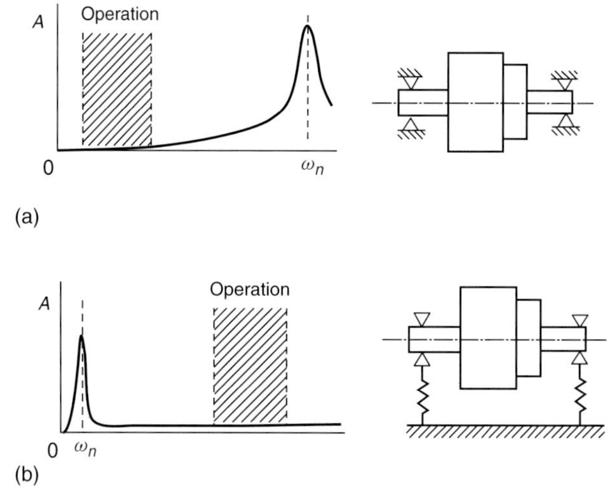

Figure 4.6 Dynamic balancing machine: (a) hard-bearing type and (b) soft-bearing type.

comparatively, can detect large unbalance. This type of balancing machine can be used for a wide range of mass distributions. The other is the *soft-bearing balancing machine* in which the bearings are supported softly. Most soft-bearing balancing machines detect vibrations (deflection or velocity). These machines are suitable for precise balancing of small machines, such as a small motor. Although they are called *hard-bearing* and *soft-bearing*, the difference exists not in the bearings themselves, but in their suspension.

Furthermore, considering their operation speed relative to the major critical speed of the rotor, balancing is also classified into the two categories of low-speed balancing and high-speed balancing. The hard-bearing type belongs to the former and the soft-bearing type belongs to the latter. Elastic continuous shaft systems are balanced utilizing high-speed balancing.

Principles of Hard-Bearing Balancing Machines Figure 4.7 shows a model of a hard-bearing balancing machine. It is assumed that correction planes I and II are taken at both sides of the cylinder. In this machine, the magnitude and angular position of an eccentricity and a skew angle can be detected. Let the distance between

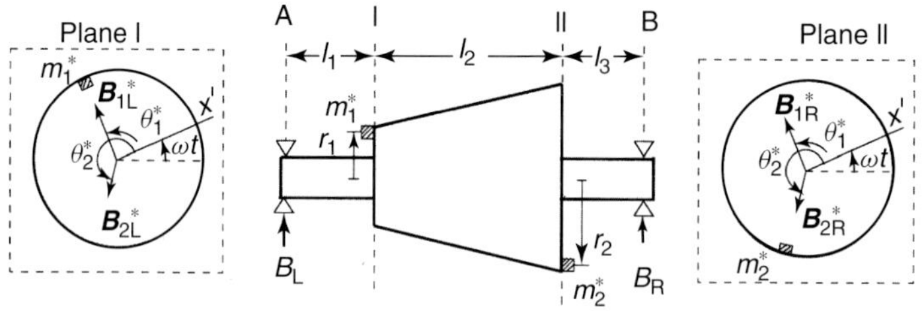

Figure 4.7 Hard-bearing balancing machine and forces at the bearings.

the left bearing and correction plane I be l_1, the distance between correction planes I and II be l_2, and the distance between correction plane II and the right bearing be l_3. From Figure 4.3, we see that the eccentricity and the skew angle produce forces $-\boldsymbol{R}_1$ and $-\boldsymbol{R}_2$. We eliminate these forces utilizing the following process:

- **Step 1:** First, we determine the reference direction on the side surfaces of the cylinder and take the x'-axis to be in this direction. Let r_1 and r_2 be radii whose equivalent masses are mounted on both sides. Suppose that mass m_1^* at the angle θ_1^* (at radius r_1) in correction plane I produces the force $-\boldsymbol{R}_1$. However, the values of m_1^* and θ_1^* are the unknown parameters at this stage. Hereon, the superscript $*$ is used to denote unknown values. When this rotor is operated at an angular speed ω, forces with magnitudes $B_{1\mathrm{L}}^*$ and $B_{1\mathrm{R}}^*$ work in the direction θ_1^* on the left and right bearings, respectively. The values $B_{1\mathrm{L}}^*$ and $B_{1\mathrm{R}}^*$ are given by

$$B_{1\mathrm{L}}^* = \frac{l_2 + l_3}{l_1 + l_2 + l_3} m_1^* r_1 \omega^2, \quad B_{1\mathrm{R}}^* = \frac{l_1}{l_1 + l_2 + l_3} m_1^* r_1 \omega^2 = \alpha_1 B_{1\mathrm{L}}^* \tag{4.6}$$

  where $\alpha_1 = l_1/(l_2 + l_3)$.
- **Step 2:** Suppose that mass m_2^* at the angle θ_2^* (at radius r_2) in correction plane II produces the force $-\boldsymbol{R}_2$. Then, forces with magnitudes $B_{2\mathrm{L}}^*$ and $B_{2\mathrm{R}}^*$ work in the direction θ_2^* on the left and right sides of the rotor, respectively:

$$B_{2\mathrm{L}}^* = \frac{l_3}{l_1 + l_2 + l_3} m_2^* r_2 \omega^2, \quad B_{2\mathrm{L}}^* = \frac{l_1 + l_2}{l_1 + l_2 + l_3} m_2^* r_2 \omega^2 = \alpha_2 B_{2\mathrm{L}}^* \tag{4.7}$$

  where $\alpha_2 = (l_1 + l_2)/l_3$.
- **Step 3:** Let the forces detected at both ends be $\boldsymbol{B}_\mathrm{L}$ and $\boldsymbol{B}_\mathrm{R}$. Then, the relationships $\boldsymbol{B}_\mathrm{R} = B_{1\mathrm{R}}^* + B_{2\mathrm{R}}^*$ and $\boldsymbol{B}_\mathrm{L} = B_{1\mathrm{L}}^* + B_{2\mathrm{L}}^*$ hold. If we represent their magnitudes by B_L and B_R and their angular positions by θ_L and θ_R respectively, the following relationships hold among the components in the x'-and y'-directions:

$$\left.\begin{aligned} B_{1\mathrm{L}}^* \cos\theta_1^* + B_{2\mathrm{L}}^* \cos\theta_2^* &= B_\mathrm{L} \cos\theta_\mathrm{L} \\ B_{1\mathrm{L}}^* \sin\theta_1^* + B_{2\mathrm{L}}^* \sin\theta_2^* &= B_\mathrm{L} \sin\theta_\mathrm{L} \\ \alpha_1 B_{1\mathrm{L}}^* \cos\theta_1^* + \alpha_2 B_{2\mathrm{L}}^* \cos\theta_2^* &= B_\mathrm{R} \cos\theta_\mathrm{R} \\ \alpha_1 B_{1\mathrm{L}}^* \sin\theta_1^* + \alpha_2 B_{2\mathrm{L}}^* \sin\theta_2^* &= B_\mathrm{R} \sin\theta_\mathrm{R} \end{aligned}\right\} \tag{4.8}$$

- **Step 4:** By solving Eq. (4.8) for $B_{1\mathrm{L}}^*, \theta_1^*, B_{2\mathrm{L}}^*$, and θ_2^*, we can determine the four unknown quantities related to the unbalance. By multiplying α_1 and α_2 to $B_{1\mathrm{L}}^*$ and $B_{2\mathrm{L}}^*$, we can obtain $B_{1\mathrm{R}}^*$ and $B_{2\mathrm{R}}^*$, respectively. Since we could obtain the unbalance m_1^* in correction plane I from Eq. (4.6) and m_2^* in plane II from Eq. (4.7), we can eliminate unbalance by attaching the same amounts of mass $m_1 (= m_1^*)$ and $m_2 (= m_2^*)$ in the opposite directions and at the same radii, respectively.

Principles of Soft-Bearing Balancing Machines Figure 4.8 illustrates a model of a soft-type balancing machine. We choose correction planes I and II at both sides of the cylinder and measurement planes a and b at the bearing locations. It is assumed that amplitudes A_a and A_b and phase angles γ_a and γ_b with a reference direction on planes a and b can be measured. The balancing procedure is as follows (Thearle, 1932; Macduff and Curreri, 1958; Rao, 1991):

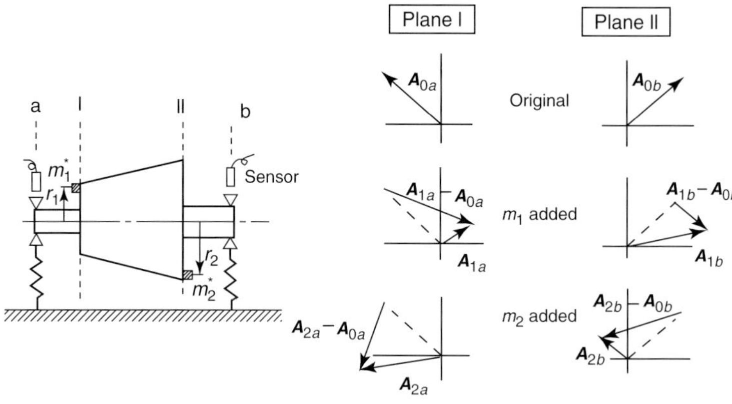

Figure 4.8 Vibrations at bearings of a soft-bearing balancing machine.

- **Step 1 (The first run):** In the original condition with unknown unbalance, we measure amplitudes and phase angles in planes a and b. Let them be A_{0a}, A_{0b} and γ_{0a}, γ_{0b}. The vectors $\boldsymbol{A}_{0a}$ and $\boldsymbol{A}_{0b}$ corresponding to these values are drawn in Figure 4.7.
- **Step 2 (The second run):** A trial mass m_1 is placed at radius r_1 and angle θ_1 in correction plane I. As seen from Eqs. (4.6) and (4.7), the effect of this change is determined by the product of the mass m_1 and radius r_1. Therefore, we introduce a parameter $m_1\boldsymbol{r}_1 = U_1$. This quantity is explained in detail in the next section. The unit of U_1 is given by the product of the trial units of mass and length. The measured amplitudes A_{1a} and A_{1b} and phase angles γ_{1a} and γ_{1b} are represented by the vectors $\boldsymbol{A}_{1a}$ and $\boldsymbol{A}_{1b}$, respectively, and shown in Figure 4.8.

By superimposing complex planes on planes a and b, we represent **A** utilizing a complex number denoting $\overline{A} = a + bi$. The bar over the symbol represents a complex number. This complex number is also represented by the polar form $\overline{A} = r(\cos\theta + i\sin\theta) = r\angle\theta$, where r (modulus) represents the distance between the point $\overline{A}$ and the origin and θ (argument) represents the angle between the direction of r and the positive direction of the real axis. In this polar expression, the multiplication of $\overline{A}_1 = r_1\angle\theta_1$ and $\overline{A}_2 = r_2\angle\theta_2$ is expressed by $\overline{A}_1\overline{A}_2 = (r_1r_2)\angle(\theta_1+\theta_2)$ and the division is represented by $\overline{A}_1/\overline{A}_2 = (r_1/r_2)\angle(\theta_1-\theta_2)$. The usage of these rules makes way for relatively simple calculations.

Since the difference $\boldsymbol{A}_{1b} - \boldsymbol{A}_{0b}$ occurs owing to a trial mass m_1 placed in correction plane I, that is by U_1, we can represent this relationship as follows:

$$\left.\begin{aligned} \overline{A}_{1a} - \overline{A}_{0a} &= \overline{\alpha}_{1a}\overline{U}_1 \\ \overline{A}_{1b} - \overline{A}_{0b} &= \overline{\alpha}_{1b}\overline{U}_1 \end{aligned}\right\} \tag{4.9}$$

where $\overline{\alpha}_{1a}$ and $\overline{\alpha}_{1b}$ are called the *influence coefficients*. These relationships represent the effect of an unbalance on the amplitudes and the phase angles. These influence coefficients are complex numbers. If there is no damping, the influence coefficients become real numbers.

- **Step 3 (The third run):** After removing the trial mass m_1, we place a trial mass m_2 at radius r_2 and angle θ_2 in correction plane II. Then, by measuring the response, we obtain the amplitude A_{2a} and A_{2b} and the phase angle γ_{2a} and γ_{2b}. These results are given by vectors $\boldsymbol{A}_{2a}$ and $\boldsymbol{A}_{2b}$ in Figure 4.8. Using the quantity $m_2\boldsymbol{r}_2 = \boldsymbol{U}_2$, we can obtain the following relationships:

$$\left.\begin{aligned} \overline{A}_{2a} - \overline{A}_{0a} &= \overline{\alpha}_{2a}\overline{U}_2 \\ \overline{A}_{2b} - \overline{A}_{0b} &= \overline{\alpha}_{2b}\overline{U}_2 \end{aligned}\right\} \tag{4.10}$$

From Eqs. (4.9) and (4.10), we can determine the coefficients $\overline{\alpha}_{1a}, \overline{\alpha}_{1b}, \overline{\alpha}_{2a}$, and $\overline{\alpha}_{2b}$.

- **Step 4:** We try to eliminate the original unbalance by placing correction weights in correction planes I and II. Let us take the unknown correction weights to be $\overline{U}_1^*$ and $\overline{U}_2^*$. The vibrations caused by these correction weights in planes a and b are obtained by multiplying the influence coefficients obtained above. Since we eliminate the original vibration by these counter vibrations, the following relationships must hold:

$$\left.\begin{aligned} \overline{A}_{0a} + \overline{\alpha}_{1a}\overline{U}_1^* + \overline{\alpha}_{2a}\overline{U}_2^* &= 0 \\ \overline{A}_{0b} + \overline{\alpha}_{1b}\overline{U}_1^* + \overline{\alpha}_{2b}\overline{U}_2^* &= 0 \end{aligned}\right\} \tag{4.11}$$

From this expression, we can determine $\overline{U}_1^*$ and $\overline{U}_2^*$, which give the magnitude and the direction of the correction weights.

4.3.3
Field Balancing

The shape of a rotor changes over time owing to various causes, such as the release of internal residual stress, wear, and corrosion. Dismantling the system for balancing is often not a suitable solution in terms of time and cost. In addition, we cannot perform balancing with large steam turbines and generators using balancing machines. In such cases, we must balance such installed rotors *in situ*. This type of balancing is called *field balancing*.

The principle of two-plane balancing is also used in field balancing. In fact, Thearle (1956) developed this procedure expressly for the purpose of field balancing. In the case of a soft-bearing balancing machine, balancing is performed at speeds far from the resonance. Therefore, a correction mass in plane I or II produces change in almost the same direction (or in an opposite direction) in planes a and b. Since the damping is generally large with regard to machines in the field, the balancing is often performed near the resonance to produce larger vibration. Therefore, in field balancing, the changes in planes a and b differ as shown in the following example. Concerning the influence coefficients $\overline{\alpha}_{1a}, \overline{\alpha}_{1b}, \overline{\alpha}_{2a}$, and $\overline{\alpha}_{2b}$, they are complex numbers in field balancing as shown in the following example; however, it should be noted that these coefficients are real numbers in the case of soft-bearing balancing machines.

Exercise 4.1: We obtained the following data in a balancing test. Determine the correction masses, considering the following conditions.

First trial: No trial weight, $A_{0a} = 3.5\,\mu\text{m}$, $\gamma_{0a} = 107^\circ$, $A_{0b} = 4.1\,\mu\text{m}$, $\gamma_{0b} = 61^\circ$

Second trial: Trial weight $m_1 = 15\text{g}$ was placed at radius $r_1 = 10\text{cm}$ and at angular position $\theta_1 = 0^\circ$. The observed vibrations were $A_{1a} = 3.3\,\mu\text{m}$, $\gamma_{1a} = 44^\circ$ and $A_{1b} = 3.1\,\mu\text{m}$, $\gamma_{1b} = 48.5^\circ$.

Third trial: Trial weight $m_2 = 15\text{g}$ was placed at radius $r_2 = 10\text{cm}$ and at angular position $\theta_2 = 0^\circ$. The observed vibrations were $A_{2a} = 2.0\,\mu\text{m}$, $\gamma_{2a} = 84^\circ$, $A_{2b} = 5.4\,\mu\text{m}$, $\gamma_{2b} = 41.5^\circ$

Answer: The given data are expressed by complex numbers as follows:

$$\left.\begin{array}{l} \overline{A}_{0a} = -1.023 + 3.347i,\ \overline{A}_{0b} = 1.988 + 3.586i,\ \overline{A}_{1a} = 2.374 + 2.293i \\ \overline{A}_{1b} = 2.054 + 2.322i,\ \overline{A}_{2a} = 0.209 + 1.989i,\ \overline{A}_{2b} = 4.045 + 3.578i \end{array}\right\} \quad (1)$$

And, trial weights are expressed by

$$\overline{U}_1 = 150, \quad \overline{U}_2 = 150 \quad (2)$$

Then, Eq. (4.9) becomes

$$\left.\begin{array}{l} 3.397 - 1.054i = \overline{\alpha}_{1a}(150) \\ 0.066 - 1.264i = \overline{\alpha}_{1b}(150) \end{array}\right\} \quad (3)$$

Solving these equations, we obtain

$$\overline{\alpha}_{1a} = 0.0225 - 0.00703i, \quad \overline{\alpha}_{1b} = 0.00044 - 0.00843i \quad (4)$$

Similarly, Eq. (4.10) becomes

$$\left.\begin{array}{l} 1.232 - 1.358i = \overline{\alpha}_{2a}(150) \\ 2.057 - 0.008i = \overline{\alpha}_{2b}(150) \end{array}\right\} \quad (5)$$

Solving these equations, we obtain

$$\overline{\alpha}_{2a} = 0.00822 - 0.00905i, \quad \overline{\alpha}_{2b} = 0.01371 - 0.00001i \quad (6)$$

From Eq. (4.11), the correction weights satisfy the following expressions:

$$\left.\begin{array}{l} (-1.023 + 3.347i) + (0.0225 - 0.00703i)\overline{U}_1^* + (0.00822 - 0.00905i)\overline{U}_2^* = 0 \\ (1.988 + 3.586i) + (0.00044 - 0.00843i)\overline{U}_1^* + (0.01371 - 0.00001i)\overline{U}_2^* = 0 \end{array}\right\} \quad (7)$$

Solving these equations, we obtain

$$\left.\begin{array}{l} \overline{U}_1^* = \dfrac{(62.79 - 34.41i) \times 1000}{381.08 - 23.33i} = \dfrac{71598\angle(-28.6^\circ)}{381.8\angle(-3.5^\circ)} = 187.5\angle 335^\circ \\ \overline{U}_2^* = \dfrac{(-42.17 - 56.61i) \times 1000}{381.08 - 23.33i} = \dfrac{70595\angle 233.2^\circ}{381.8\angle(-3.5^\circ)} = 184.9\angle 237^\circ \end{array}\right\} \quad (8)$$

Therefore, we can balance the rotor by placing correction weights 188 g cm at angle 335° in correction plane I, and 185 g cm at angle 237° in plane II.

4.3.4
Various Expressions of Unbalance

The above-mentioned description is an explanation from the viewpoint of dynamics. Since the same terminologies are used with different meanings in industry (Miwa and Shimomura, 1976), these meanings are explained in this section.

When a static unbalance exists, a centrifugal force $m\mathbf{e}\omega^2$ works. This unbalance force is eliminated if mass m_1, which satisfies the relationship $m\mathbf{e}\omega^2 + m_1\mathbf{a}\omega^2 = 0$, is added at radius a in the same plane at the center of gravity G. From this condition we know that the product $m\mathbf{e}$ or $m_1\mathbf{a}$ is more important than the eccentricity $\mathbf{e}$ itself. Therefore, the quantity

$$\mathbf{U} = m\mathbf{e}[\text{ g mm}] \tag{4.12}$$

is called an *unbalance vector* and its magnitude $U = me$ is called the *magnitude of unbalance*. These quantities are sometimes called simply *unbalance*. Equation (4.12) is also represented by $\mathbf{e} = \mathbf{U}/M[\text{mm}]$, which means unbalance per unit mass. Therefore, the eccentricity $\mathbf{e}$ is also called a *specific unbalance*.

Next, we consider a general case where an eccentricity e and a skew angle τ coexist. Three different types of expressions are explained below.

4.3.4.1 Resultant Unbalance *U* and Resultant Unbalance Moment *V*

In Figure 4.9, the coordinate axis z has its origin at the supporting point on the left-hand side. A thinly sliced disk with thickness dz, which is perpendicular to the rotor axis, is considered. Let the mass of this slice be $dm = \mu(z)dz$, where $\mu(z)$ is a line density of the rotor, and the eccentricity of the center of gravity be $\mathbf{e}(z)$. Then, its unbalance is represented by $d\mathbf{U}(z) = \mathbf{e}(z)dM = \mathbf{e}(z)\mu(z)dz$. The summation of such an unbalance, called a *resultant unbalance*, is given by

$$\mathbf{U} = \int_a^{a+h} d\mathbf{U}(z) = \int_a^{a+h} \mathbf{e}(z)\mu(z)dz \tag{4.13}$$

where a is the distance between the origin O and the rotor and h is the length of the rotor. Since this resultant unbalance $\mathbf{U}$ can be detected without spinning the

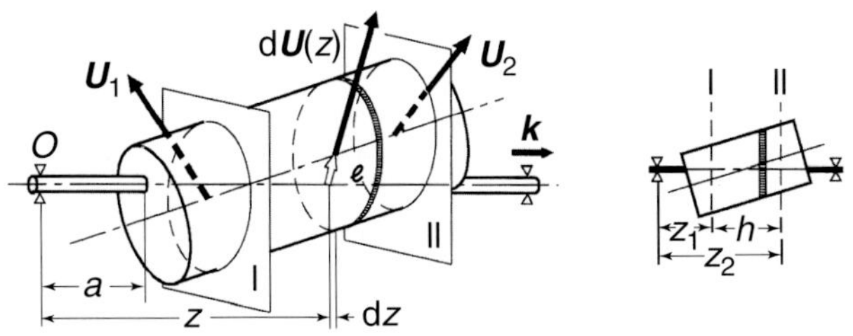

Figure 4.9 Equivalent concentrated unbalance U_1 and U_2.

rotor, it is also called *static unbalance*. Multiplying this by ω^2, we get the *resultant unbalance force* of the centrifugal force:

$$\boldsymbol{F} = \omega^2 \boldsymbol{U} \tag{4.14}$$

The quantity

$$\boldsymbol{V} = \int_a^{a+h} (z\boldsymbol{k}) \times d\boldsymbol{U}(z) = \int_a^{a+h} (z\boldsymbol{k}) \times \mathbf{e}(z)\mu(z)dz \tag{4.15}$$

is called a *resultant unbalance moment* concerning point O, where $\boldsymbol{k}$ is a unit vector in the direction of the bearing centerline and $\times$ is the outer product of vectors. Multiplying this by ω^2, we get the moment $\boldsymbol{N}$ produced by the centrifugal forces of all the elements:

$$\boldsymbol{N} = \omega^2 \boldsymbol{V} \tag{4.16}$$

This moment is called a *resultant moment of unbalance force*. Like this, we can represent the unbalance of a rigid rotor by using the resultant unbalance $\boldsymbol{U}$ and the resultant unbalance moment $\boldsymbol{V}$.

Exercise 4.2: Figure 1a is a rotor with constant eccentricity e_0 in the same angular position in the half of the rotor. The line density $\mu(z)$ is constant. Determine the resultant unbalance $\boldsymbol{U}$ and the resultant unbalance moment $\boldsymbol{V}$.

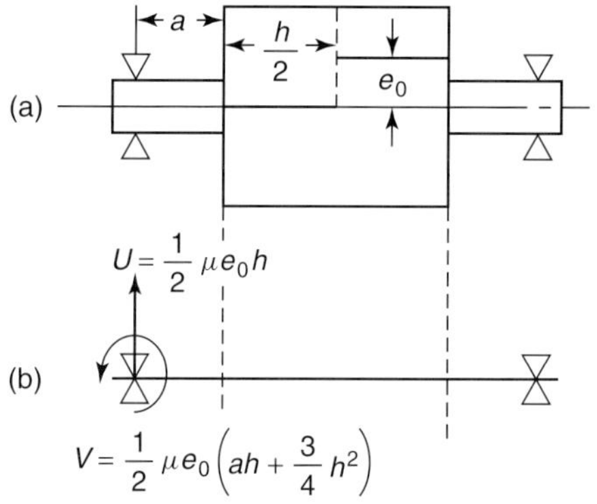

Figure 1 Resultant unbalance $\overline{U}$ and the resultant unbalance moment $\overline{V}$.

Answer: The magnitude of the resultant unbalance is obtained from Eq. (4.13) as

$$U = \int_a^{a+h} e(z)\mu(z)dz = \mu e_0 \int_{a+h/2}^{a+h} dz = \frac{1}{2}\mu e_0 h \tag{1}$$

and the resultant unbalance moment about point O is given by Eq. (4.15) as

$$V = \int_a^{a+h} ze(z)\mu(z)dz = \mu e_0 \int_{a+h/2}^{a+h} zdz = \frac{1}{2}\mu e_0 \left(ah + \frac{3}{4}h^2\right) \tag{2}$$

This vector $\boldsymbol{V}$ is perpendicular to the plane in which e_0 exists. Since Eq. (2) contains a, we know that its magnitude depends on the position of the origin O. From Eq. (4.16), $\boldsymbol{N}$ is obtained by multiplying ω^2 with $\boldsymbol{V}$. The result is shown in Figure 1b.

4.3.4.2 Dynamic Unbalance (U_1, U_2)

In Figure 4.9, $\boldsymbol{U}$ and $\boldsymbol{V}$ given by Eqs. (4.13) and (4.15), respectively, are replaced by the concentrated mass $\boldsymbol{U}_1$ and $\boldsymbol{U}_2$ in correction planes I and II, respectively. For this replacement, the following relationship must hold:

$$\left.\begin{aligned} &\boldsymbol{U}_1 + \boldsymbol{U}_2 = \boldsymbol{U} \\ &(z_1\boldsymbol{k}) \times \boldsymbol{U}_1 + (z_2\boldsymbol{k}) \times \boldsymbol{U}_2 = \boldsymbol{V} \end{aligned}\right\} \tag{4.17}$$

where z_1 and z_2 are the positions of the correction planes. The balancing is attained if we add $\boldsymbol{U}_{\text{I}}(= -\boldsymbol{U}_1)$ and $\boldsymbol{U}_{\text{II}}(= -\boldsymbol{U}_2)$ which cancel $\boldsymbol{U}_1$ and $\boldsymbol{U}_2$, respectively, in correction planes I and II. This set $[\boldsymbol{U}_1, \boldsymbol{U}_2]$ is called the *dynamic unbalance* represented at positions z_1 and z_2. We can also represent the unbalance by this set. Equation (4.17) shows that the static unbalance $\boldsymbol{U}$ is included in the dynamic unbalance. This corresponds to the fact that a static unbalance can also be detected by spinning (or dynamically) the rotor.

Exercise 4.3: Determine the dynamic unbalance of a rotor shown in Figure 1a.

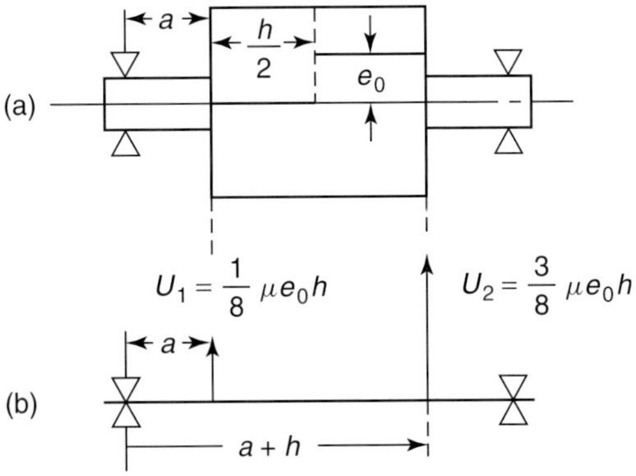

Figure 1 Dynamic unbalance $(\overline{U}_1, \overline{U}_2)$.

Answer: If we select correction planes I and II at both ends of the rotor, we get the following relationships from Eqs. (1) and (2) in Exercise 4.2 and Eq. (4.17).

$$\left.\begin{aligned} U_1 + U_2 &= \frac{1}{2}\mu e_0 h \\ aU_1 + (a+h)U_2 &= \frac{1}{2}\mu e_0 \left(ah + \frac{3}{4}h^2\right) \end{aligned}\right\} \tag{1}$$

Then, we obtain

$$U_1 = \frac{1}{8}\mu e_0 h, \quad U_2 = \frac{3}{8}\mu e_0 h \tag{2}$$

The result is shown in Figure 1b.

4.3.4.3 Static Unbalance $\boldsymbol{U}$ and Couple Unbalance $[\boldsymbol{U}_c, -\boldsymbol{U}_c]$

Now, we replace the dynamic unbalance $[\boldsymbol{U}_1, \boldsymbol{U}_2]$ by the static unbalance $\boldsymbol{U}$ and a couple whose forces are located in correction planes I and II as follows:

1) Suppose that a force $\boldsymbol{U}$ that is equal to the static unbalance and a force $-\boldsymbol{U}$ are added at position z_3 as shown in Figure 4.10a. Since $\boldsymbol{U} - \boldsymbol{U} = \boldsymbol{0}$ holds, balancing as a whole does not change because of this addition.
2) Since $-\boldsymbol{U} = -\boldsymbol{U}_1 - \boldsymbol{U}_2$ holds, the "summation of the dynamic unbalance $[\boldsymbol{U}_1, \boldsymbol{U}_2]$ at z_1 and z_2 and a given unbalance $-\boldsymbol{U}$ at z_3" is equivalent to the two couples which are "the set of unbalance $[\boldsymbol{U}_1, -\boldsymbol{U}_1]$ at z_1 and z_3 and the set $[-\boldsymbol{U}_2, \boldsymbol{U}_2]$ at z_3 and z_2" as shown in Figure 4.10b.
3) We remember that the effect of a couple is the same for any position of a rigid rotor. The couple $[\boldsymbol{U}_1, -\boldsymbol{U}_1]$ at z_1 and z_3 can be replaced by $[\boldsymbol{U}_{1c}, -\boldsymbol{U}_{1c}]$ at z_1 and z_2 under the condition that $(z_3 - z_1)\boldsymbol{U}_1 = (z_2 - z_1)\boldsymbol{U}_{1c}$. Similarly, the couple $[-\boldsymbol{U}_2, \boldsymbol{U}_2]$ at z_3 and z_2 can be replaced by $[-\boldsymbol{U}_{2c}, \boldsymbol{U}_{2c}]$ at z_1 and z_2 as shown in Figure 4.10c.
4) Using $-\boldsymbol{U}_{1c} + \boldsymbol{U}_{2c} = \boldsymbol{U}_c$, the summation of couples $[\boldsymbol{U}_{1c}, -\boldsymbol{U}_{1c}]$ and $[-\boldsymbol{U}_{2c}, \boldsymbol{U}_{2c}]$ is replaced by "an equivalent couple $[\boldsymbol{U}_c, -\boldsymbol{U}_c]$ at correction planes I and II" as shown in Figure 4.10d. This set of unbalance $[\boldsymbol{U}_c, -\boldsymbol{U}_c]$ is called a *couple unbalance*.

   From such replacements, we know that the unbalance of a rigid rotor can be represented by the static unbalance U and the couple unbalance $[\boldsymbol{U}_c, -\boldsymbol{U}_c]$.

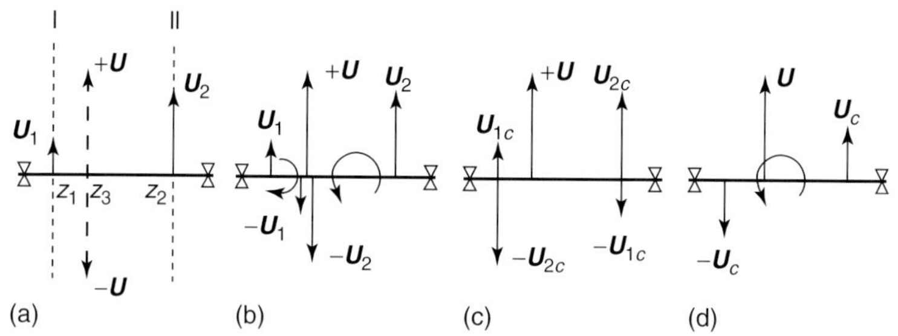

Figure 4.10 Transformation from a dynamic unbalance to a set of a static unbalance and a couple unbalance.

Exercise 4.4: Determine the static unbalance and the couple unbalance of a rotor shown in Figure 1a.

Answer: From Exercise 4.3, the magnitude of the static unbalance is

$$U = U_1 + U_2 = \frac{\mu e_0 h}{8} + \frac{3\mu e_0 h}{8} = \frac{\mu e_0 h}{2} \quad (1)$$

Now, we consider two unbalances $\boldsymbol{U}$ and $-\boldsymbol{U}$, whose magnitude is $\mu e_0 h/2$ and pointing toward the upper and lower directions, respectively, at the position $z_3 = a + b$ (Figure 4.10a). We replace $-\boldsymbol{U}$ and the dynamic unbalance

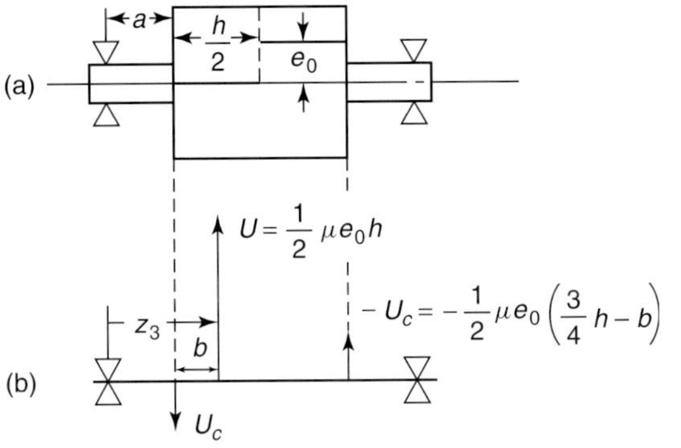

Figure 1 Static unbalance $\overline{U}$ and the couple unbalance $[\overline{U}_c, \overline{U}_c]$.

$[U_1, U_2]$ by (Figure 4.10b) the set $[U_1', U_2']$ in correction planes I and II (Figure 4.10d). Then the following relationships hold between them:

$$\left.\begin{aligned} U_1' + U_2' &= U_1 + U_2 - U = 0 \\ aU_1' + (a+h)U_2' &= aU_1 + (a+h)U_2 - (a+b)U \end{aligned}\right\} \quad (2)$$

Substituting Eq. (1) in Exercise 4.2 and Eq. (2) in Exercise 4.3 into this expression, we get

$$U_1' = -U_2' = -\frac{1}{2}\mu e_0 \left(\frac{3}{4}h - b\right) (\equiv U_c) \quad (3)$$

Namely, the magnitudes of $\overline{U}_1'$ and $\overline{U}_2'$ are equal and they constitute the couple unbalance.

The static unbalance U given by Eq. (1) and the couple unbalance $[U_c, -U_c]$ given by Eq. (3) are illustrated in Figure 1b.

Figure 4.11 illustrates three types of unbalance. Static unbalance shown in Figure 4.11a is a condition of unbalance for which the principal axis of inertia is displaced only in parallel to the shaft axis. Couple unbalance shown in Figure 4.11b is a condition of unbalance for which the principal axis intersects the shaft axis at the center of gravity. Dynamic unbalance shown in Figure 4.11c is a condition of

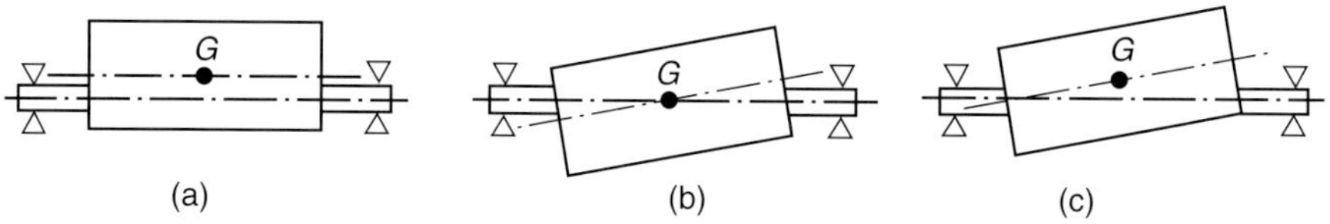

Figure 4.11 Rotors with various unbalances: (a) static unbalance, (b) couple unbalance, and (c) dynamic unbalance.

unbalance for which the principal axis is not parallel to and does not intersect the shaft axis at the center of gravity.

4.3.5 Balance Quality Grade of a Rigid Rotor

Ideally, a machine should be balanced to the level with no unbalance at all. However, in practice, we must take into account the time and cost necessary to balance the rotor. Therefore, it is appropriate to vary the permissibility of unbalance depending on the kind of rotating machinery. The International Organization for Standardization (ISO) published a standard ISO 1940/1 "Balance Quality Requirements of Rigid Rotors" and gave a guideline of the quality of balancing. This guideline shows what parameter is used to express the quality of balance and how to determine the balancing acceptance limit for a given rotor.

4.3.5.1 Balance Quality Grade

It is natural to consider that the influence of an unbalance on vibrations and stresses of a machine will increase as the eccentricity and the rotational speed become large. On the basis of many experiences, ISO 1940 established *balance quality grades* that permit a classification of the balance quality for typical machinery as shown in Table 4.1. The balance quality grades G are designated according to the magnitude of the product

$$e_{per}\omega \tag{4.18}$$

where e_{per} [mm] is a permissible eccentricity and ω [rad s^{-1}] is the maximum service angular speed. If the magnitude is equal to 6.3, the balance quality grade is designated G6.3 with prefix G. The balance quality grades are classified into 11 groups. The ratio of the successive number is 2.5 ($= 1/0.4 = 2.5/1 \approx 6.3/2.5 \approx \cdots$). This ratio was empirically determined. Table 4.1 gives recommended balance quality grade for various types of representative rotating machines.

Figure 4.12 shows the relationship between the maximum service speed n_{max} (rpm) and the *permissible residual specific unbalance* $U_{per}/M = e_{per}$ (g·mm kg^{-1}) (or permissible eccentricity e_{per} (μm)) where U_{per} is a *permissible residual unbalance.* When the maximum service speed is given, the permissible residual unbalance can be determined by this diagram. The lines are separated from each other by a factor of 2.5. However, sometimes grades between these lines are used. Each machine

Table 4.1 Guidance for Balance Quality Grades for Rotors in a Constant (Rigid) State.

Machinery types: general examples		Balance quality grade	Magnitude $e_{per}\omega$ (mm s^{-1})
Crankshaft drives for large slow marine diesel engines (piston speed below 9 s^{-1}), inherently unbalanced		G4000	4000
Crankshaft drives for large slow marine diesel engines (piston speed below 9 s^{-1}), inherently balanced		G1600	1600
Crankshaft drives, inherently unbalanced, elastically mounted		G630	630
Crankshaft drives, inherently unbalanced, rigidly mounted		G250	250
Complete reciprocating engines for cars, trucks, and locomotives		G100	100
Cars: wheels, wheel rims, wheel sets, drive shafts		G40	40
Crankshaft drives, inherently balanced, elastically mounted			
Agricultural machinery		G16	16
Crankshaft drives, inherently balanced, rigidly mounted			
Crushing machines			
Drive shafts (cardan shafts, propeller shafts)			
Aircraft gas turbines		G6.3	6.3
Centrifuge(separators, decanters)			
Electric motors and generators (of at least 80 mm shaft height), of maximum rated speeds up to 950 r min^{-1}			
Electric motors of shaft heights smaller than 80 mm			
Fans	Gears		
Machinery, general	Machine-tools		
Paper machines	Process plant machines		
Pumps	Turbo-chargers		
Water turbines			
Compressors	Computer drives	G2.5	2.5
Electric motors and generators (of at least 80 mm shaft height), of maximum rated speeds above 950 r min^{-1}			
Gas turbines and steam turbines			
Machine-tool drives	Textile machines		
Audio and video drives	Grinding machine drives	G1	1
Gyroscopes			
Spindles and drives of high-precision systems		G0.4	0.4

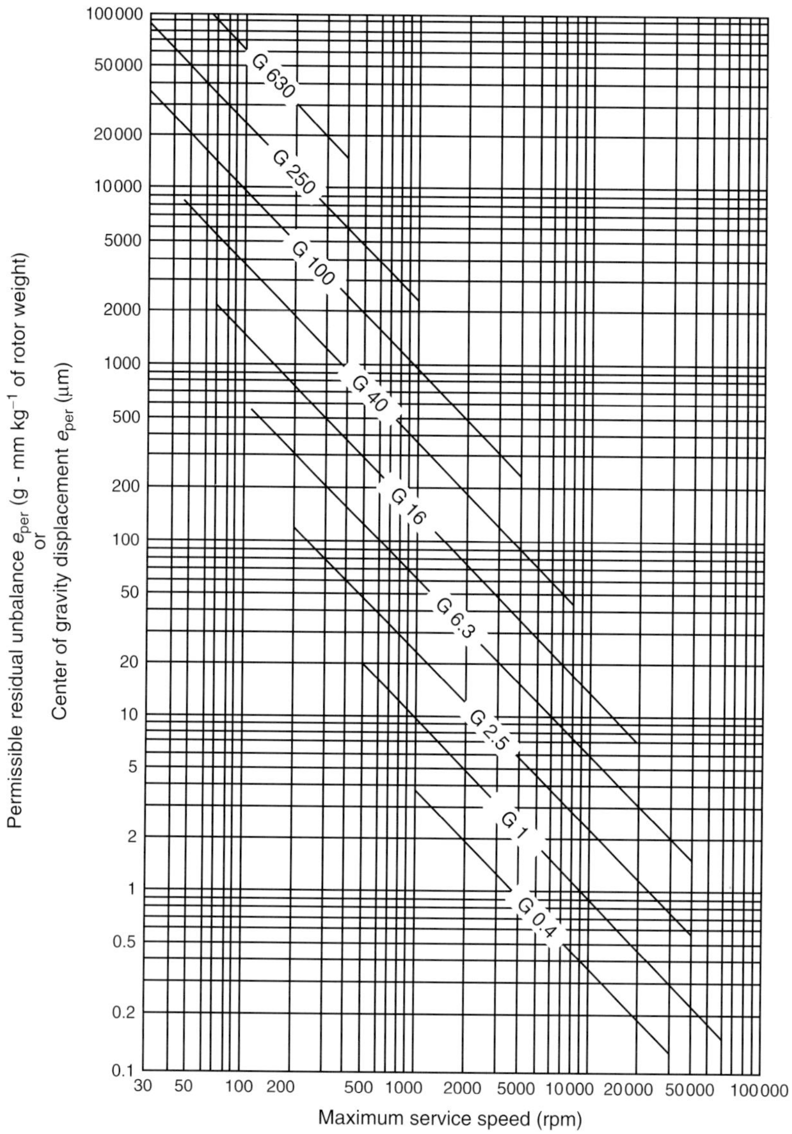

Figure 4.12 Maximum permissible residual unbalance.

has an appropriate service speed range and therefore the lines in Figure 4.12 are drawn in the finite range. We should not use the rotor outside these lines.

4.3.5.2 How to Use the Standards

We can determine a permissible residual unbalance using these standards in the following steps:

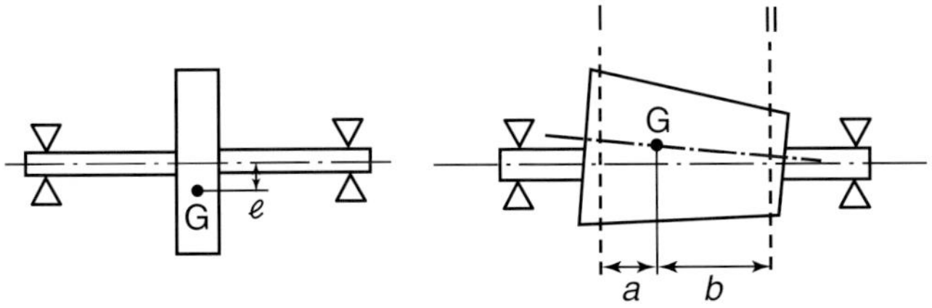

Figure 4.13 Examples for permissible residual unbalance: (a) single-plane balancing and (b) two-plane balancing.

- **Step 1:** Find the rotor type that most nearly describes the one to be balanced. Select a balance quality grade G from Table 4.1 on the basis of the rotor type.
- **Step 2:** Determine the maximum service speed ω_{max}. Use Figure 4.12 to determine the permissible residual specific unbalance e_{per}.
- **Step 3:** Determine the permissible residual unbalance U_{per} by multiplying the permissible residual specific unbalance e_{per} and the rotor mass m.
- **Step 4:** Allocate U_{per} to correction planes I and II.

Step 4 needs some explanation. The unbalance U_{per} obtained in Step 3 is the total permissible residual unbalance. In the case of a single-plane balancing shown in Figure 4.13a, all of U_{per} applies to that correction plane. In the case of a two-plane balancing shown in Figure 4.13b, U_{per} must be allocated to each correction planes on the basis of rotor configurations and dimensions. The following is one example. For other cases, refer to ISO. Suppose that the rotor shown in Figure 4.13b satisfies the following three conditions: (i) correction planes are between bearings; (ii) distance h between two correction planes is greater than one-third of distance l between bearings; (iii) the center of gravity exists in the middle part when the bearing distance is divided equally among three. Then, the permissible residual unbalance U_{perI} and U_{perII} in correction planes I and II, respectively, are given as follows:

$$U_{perI} = U_{per}\left(\frac{b}{a+b}\right), \quad U_{perII} = U_{per}\left(\frac{a}{a+b}\right) \tag{4.19}$$

Exercise 4.5: Figure 4.13b shows a turbine rotor. Let the mass of the rotor be $m = 3000\,\text{kg}$, maximum operation speed be $n_{max} = 3600\,\text{rpm}$, and the distances be $a = 500$ and $b = 1000\,\text{mm}$. Determine the permissible residual unbalance U_{perI} and U_{perII} at correction planes I and II.

Answer:

- **Step 1:** From Table 4.1, we know that the balance quality grade of turbine is G2.5.
- **Step 2:** $\omega_{max} = 2\pi n_{max}/60 = 377\,\text{rad s}^{-1}$. Then, $e_{per} = G/\omega_{max} = 2.5/377$ mm $= 6.6\,\mu\text{m}$.

- **Step 3:** $U_{\text{per}} = me_{\text{per}} = (3000 \times 10^3\ \text{g}) \times (6.6 \times 10^{-3}\ \text{mm}) = 1.98 \times 10^4\ \text{g} \cdot \text{mm}$
- **Step 4:** $U_{\text{perI}} = (1.98 \times 10^4\ \text{g} \cdot \text{mm}) \times 1000/(500 + 1000) = 13.2\text{g} \cdot \text{mm}$, $U_{\text{perII}} = (1.98 \times 10^4\ \text{g} \cdot \text{mm}) \times 500/(500 + 1000) = 6.6\text{g} \cdot \text{mm}$.

4.4 Balancing of a Flexible Rotor

4.4.1 Effect of the Elastic Deformation of a Rotor

The balancing technique of rigid rotors mentioned in the previous sections is applicable only when the operation range is far below the first critical speed and the rotor does not bend. We cannot use it when the rotor deforms in high-speed range. For example, Figure 4.14a shows the same rotor as that shown in Figure 4.2a. It is assumed that the rotor is almost balanced and the relationships $F_{\text{I}} + F_{\text{II}} = F$ and $aF_{\text{I}} = bF_{\text{II}}$ hold among centrifugal forces. When the rotational speed approaches the critical speed of the first mode, the shaft deforms as shown in Figure 4.14b. Then the position of the center of gravity of the rotor and those of the balancing weights change. As a result, the centrifugal forces working at these points change. This means that the balancing condition in Figure 4.14a, which was attained by putting masses m_1 and m_2, is broken. We see from this example that, since the deflection curves change depending on the rotational speed, it is impossible to balance a rotor not to cause vibrations in all its operation range. In other words, a flexible rotor can be balanced only at one rotational speed.

In the case of a continuous rotor, the eccentricity $e(s)$ changes spatially as a function of the position s as shown in Figure 3.9. Suppose that we try to balance a rotor by installing a set of correction masses m_1 and m_2 in correction planes I and II in Figure 4.15. Balancing is attained if the following conditions are satisfied (Bishop and Gladwell, 1959). First, the summation of the centrifugal forces must be zero, that is,

$$\left.\begin{aligned} &\int_0^l A\rho e_\xi(s)\omega^2 ds + m_1 a_{1\xi}\omega^2 + m_2 a_{2\xi}\omega^2 = 0 \\ &\int_0^l A\rho e_\eta(s)\omega^2 ds + m_1 a_{1\eta}\omega^2 + m_2 a_{2\eta}\omega^2 = 0 \end{aligned}\right\} \tag{4.20}$$

Second, in order to prevent the transmission of the moment from the rotor to the bearings, the summation of the moment due to centrifugal forces must be zero, that is,

$$\left.\begin{aligned} &\int_0^l A\rho e_\xi(s)\omega^2 s ds + m_1 a_{1\xi}\omega^2 s_1 + m_2 a_{2\xi}\omega^2 s_2 = 0 \\ &\int_0^l A\rho e_\eta(s)\omega^2 s ds + m_1 a_{1\eta}\omega^2 s_1 + m_2 a_{2\eta}\omega^2 s_2 = 0 \end{aligned}\right\} \tag{4.21}$$

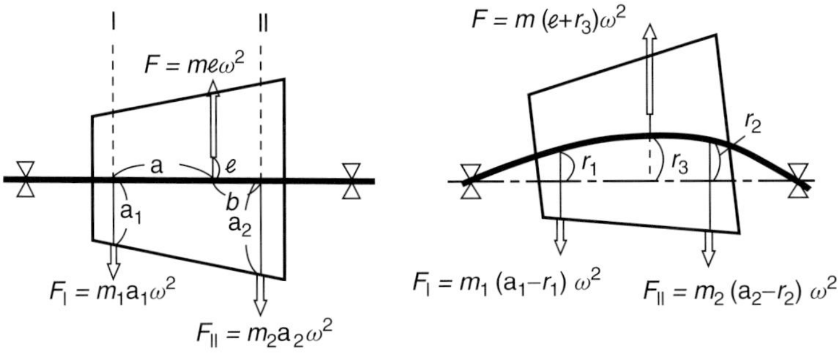

Figure 4.14 Decrease of balancing effect due to deformation: (a) far below critical speeds and (b) near a critical speed.

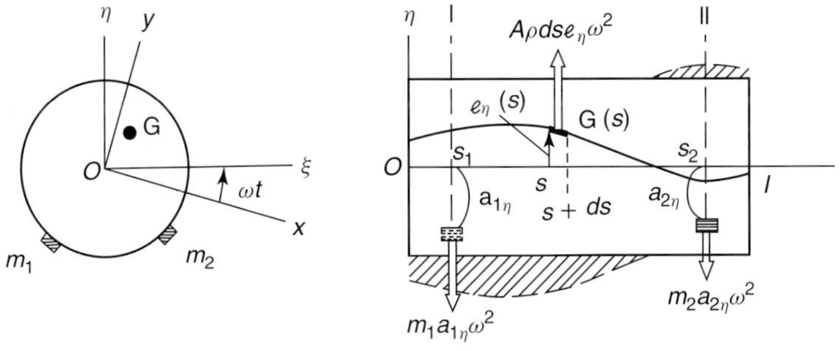

Figure 4.15 Balancing a flexible rotor considering it as a rigid rotor.

must hold simultaneously. However, as mentioned above, even if these conditions are satisfied by attaching m_1 and m_2 at a certain rotational speed, these conditions do not hold when the rotor deflects at a different rotational speed. In Figure 4.15, the rotor is balanced for all rotational speeds if an infinite number of correction weights having the same form of distribution as the unbalance, but in the opposite directions, are attached so as to cancel the unbalances as shown by the dashed line. However, such a balancing procedure is practically impossible. Instead, various types of practical balancing techniques for flexible rotors have been proposed. Here, two representative balancing methods called *modal balancing method* and *influence coefficient method* are explained. The balancing of a rigid rotor is also called *low-speed balancing* and that of a flexible rotor is called *high-speed balancing*.

4.4.2 Modal Balancing Method

As mentioned earlier, the deflection curve of a continuous rotor can be expressed by the superposition of characteristic modes. The amplitudes of these modes are determined according to the responses to the modal components that are obtained by expanding the eccentricity $e(s)$ by the modal functions $\varphi_n(s)$. On the basis of

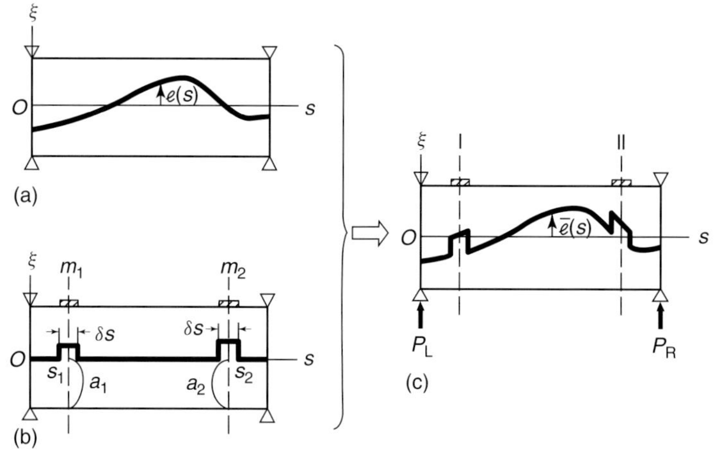

Figure 4.16 Composition of eccentricity: (a) initial eccentricity, (b) eccentricity made by correction weight, and (c) eccentricity after balancing.

this principle, the unbalance components are eliminated step by step from the lower mode *in the modal balancing method.* There are two types of modal balancing: *N-plane modal balancing,* in which the amplitudes of 1 to N modes are diminished by using N correction planes, and the other is *(N + 2)-plane modal balancing,* in which forces transmitted to the bearings are also diminished by using two additional correction planes.

4.4.2.1 N-Plane Modal Balancing

The following explanation follows the results of Bishop and Gladwell (1959). For simplicity, it is supposed that the eccentricity exists in one plane, as shown in Figure 4.16a, and we take the ξ-axis in this plane. In the general case shown in Figure 3.9, we can apply the results obtained here to each component in the $O\xi$-plane and $O\eta$-plane. Considering the operation range, we adopt the assumption that it is practically enough to balance up to the second mode. We prepare two correction planes at $s = s_1$ and s_2. The initial eccentricity shown in Figure 4.16a is expanded as

$$e(s) = \sum_{n=1}^{\infty} e_n \varphi_n(s) \tag{4.22}$$

where

$$e_n = \frac{1}{Z}\int_0^l e(s)\varphi_n(s)ds, \quad Z = \int_0^l \varphi_n(s)^2 ds \tag{4.23}$$

Similarly, the eccentricity made by the correction masses m_1 and m_2 shown in Figure 4.16b is expanded by eigenfunctions. For this purpose, we first calculate the mass per unit length m_1' and m_2' at the part where the correction masses are added. They are given by $m_1' = m_1/\delta s$ and $m_2' = m_2/\delta s$, where δs is the length of the correction masses. The eccentricity of this part is given by

$$e'(s) = \frac{m_1{}'a_1}{\rho A}, \quad e'(s) = \frac{m_2{}'a_2}{\rho A} \tag{4.24}$$

where $\rho A(s)$ is the mass of the rotor per unit length, a_1 and a_2 are the radii of the shaft at the position at which they are attached, and $\rho A >> m_1{}', m_2{}'$ holds. When function $e'(s)$ is expanded in a series in a way similar to Eq. (4.22), the coefficient of the nth mode is given by

$$\begin{aligned} e_n{}' &= \frac{m_1{}'a_1}{\rho AZ}\int_{s_1-\delta s/2}^{s_1+\delta s/2}\varphi_n(s)ds + \frac{m_2{}'a_2}{\rho AZ}\int_{s_2-\delta s/2}^{s_2-\delta s/2}\varphi_n(s)ds \\ &\rightarrow \frac{1}{\rho AZ}\{m_1a_1\varphi_n(s_1) + m_2a_2\varphi_n(s_2)\} \end{aligned} \tag{4.25}$$

The last term is the limit when $\delta s \rightarrow 0$.

From Eqs. (4.23) and (4.24), we know that the eccentricity after balancing $\bar{e}(s)$ shown in Figure 4.16 is given by

$$\bar{e}(s) = \sum_{n=1}^{\infty} \bar{e}_n\varphi_n(s) \tag{4.26}$$

where

$$\bar{e}_n = e_n + \frac{m_1a_1}{\rho AZ}\varphi_n(s_1) + \frac{m_2a_2}{\rho AZ}\varphi_n(s_2) \tag{4.27}$$

In modal balancing, the eccentricities $\bar{e}_1$ and $\bar{e}_2$ are diminished by adding m_1 and m_2. The necessary masses m_1 and m_2 are determined in the following procedure. First, we make sure that there is no vibration by adding mass m_{11} in correction plane I in the first mode. To attain this, the eccentricity of the first mode $\bar{e}_1$ must be zero; that is, the condition

$$e_1 + \frac{m_{11}a_1}{\rho AZ}\varphi_1(s_1) = 0 \tag{4.28}$$

must hold. We decide the plane in which the first modal eccentricity $\bar{e}_1$ exists by operating the rotor near the first critical speed and measuring the bending by an appropriate means. Then, we add the correction mass m_{11}, which is determined by Eq. (4.28). In practical operation, since it is difficult to determine e_1 theoretically, we determine it by trial and error.

After accomplishing the balancing of the first mode, we proceed to eliminate the second modal eccentricity $\bar{e}_2$ by adding correction masses m_{12} and m_{22} in correction planes I and II. In this case, the following two conditions must hold:

$$\left.\begin{aligned} 0 + \frac{m_{12}a_1}{\rho AZ}\varphi_1(s_1) + \frac{m_{22}a_2}{\rho AZ}\varphi_1(s_2) = 0 \quad \text{(i)} \\ e_2 + \frac{m_{12}a_1}{\rho AZ}\varphi_2(s_1) + \frac{m_{22}a_2}{\rho AZ}\varphi_2(s_2) = 0 \quad \text{(ii)} \end{aligned}\right\} \tag{4.29}$$

The first condition is that the balancing of the first mode is not lost by adding correction masses m_{12} and m_{22}. The second condition is required to balance the second mode. It is theoretically possible to determine m_{12} and m_{22} from these two

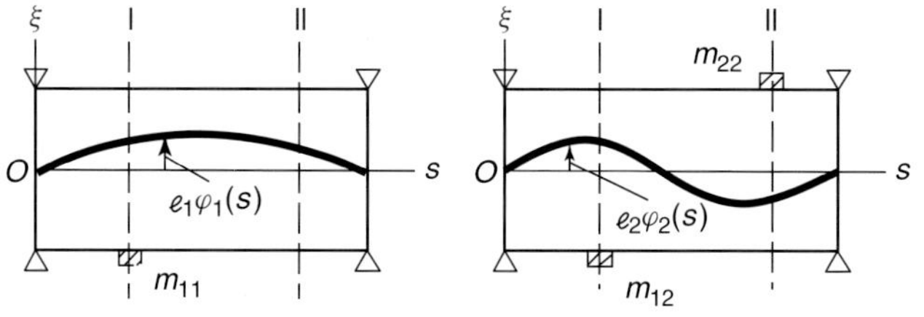

Figure 4.17 Balancing in each mode: (a) elimination of the first mode component of eccentricity by m_{11}; (b) elimination of the second mode component of eccentricity by m_{12} and m_{22}.

conditions if the coefficient determinant is not zero. However, since the quantity e_2 is not known practically, we determine the correction masses again by trial and error. That is, at a rotational speed near the second critical speed, we add correction masses, keeping the ratio m_{12}/m_{22} constant until the vibration of the second mode diminishes. In such a way, the correction masses shown in Figure 4.17 are determined. Then, we can diminish the resonance of the first and second modes by attaching $m_{11} + m_{22}$ in correction plane I and m_{22} in plane II. This is the outline of the modal balancing method proposed by Bishop and Gladwell (1959).

4.4.2.2 (N + 2)-Plane Modal Balancing

The forces transmitted to the bearings in Figure 4.16 are obtained as follows. Let the forces transmitted to the left and right bearings be $P_{\rm L}$ and $P_{\rm R}$, respectively. They are determined from the following two conditions:

$$P_{\rm L} + P_{\rm R} = \int_0^l A\rho\bar{e}(s)\omega^2 ds, \quad lP_{\rm R} = \int_0^l A\rho\bar{e}(s)\omega^2 s ds \tag{4.30}$$

In the N-plane modal balancing, we diminished modal eccentricity up to the Nth mode by using N correction planes. In the $(N+2)$-plane modal balancing, we also diminish the forces transmitted to the bearing by using $N+2$ correction planes. If the bearing pedestals are rigid and the rotor is comparatively light, we can use N-plane modal balancing. However, if the rotor is heavy and the deflections of the supports are not negligible, we must use $(N+2)$-plane modal balancing. The conditions to make $P_{\rm L} = P_{\rm R} = 0$ are obtained from Eq. (4.30) and they agree with the conditions of Eqs. (4.20) and (4.21) for a rigid rotor.

4.4.3
Influence Coefficient Method

In a system with large damping or in a system where some rotors are connected to each other such as in a steam turbine generator system, the characteristic modes do not appear clearly. In such a case, the *influence coefficient method* is used. In this method, m correction planes and n measurement points along the shaft are used, and the relationships between the magnitude of the correction mass (input) and the magnitude of vibration (output) are investigated. Its procedure is as follows.

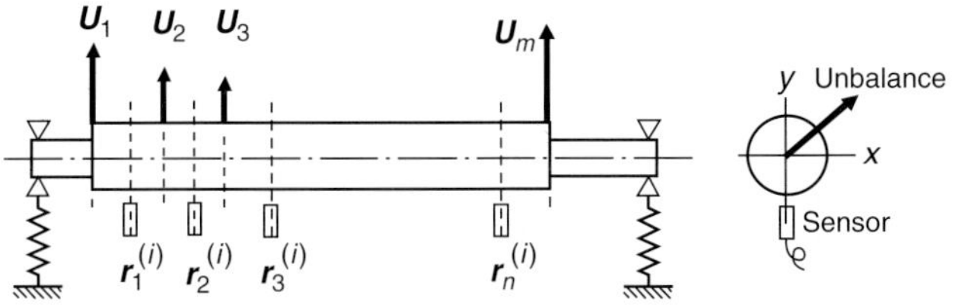

Figure 4.18 Correction planes and measurement points.

First, in Figure 4.18, the distributed unbalance $e(s)$ is replaced by the concentrated unbalance $\mathbf{e}_1$, $\mathbf{e}_2$, ... , and $\mathbf{e}_m$ approximately. Then, using the expression of Eq. (4.12), we represent these unbalances by $\boldsymbol{U}_1, \boldsymbol{U}_2, \ldots, \boldsymbol{U}_m$ where $\boldsymbol{U}_i = m_i\mathbf{e}_i$. As explained in Section 4.3.2, we use complex numbers $\overline{U}_1, \overline{U}_2, \ldots, \overline{U}_m$ to represent these vector quantities having both magnitude and phase. Suppose that the rotor is operated at k-different rotational speed $\omega_i (i = 1, \ldots, k)$ and the deflections $\bar{r}_1^{(i)}, \ldots, \bar{r}_n^{(i)} (i = 1, \ldots, k)$ at n points are measured. We assume the following linear relationships:

$$\left.\begin{aligned}
\bar{r}_1{}^{(1)} &= \bar{a}_{11}{}^{(1)}\overline{U}_1 + \cdots + \bar{a}_{1m}{}^{(1)}\overline{U}_m \\
&\vdots \\
\bar{r}_n{}^{(1)} &= \bar{a}_{n1}{}^{(1)}\overline{U}_1 + \cdots + \bar{a}_{nm}{}^{(1)}\overline{U}_m \\
&\vdots \\
\bar{r}_1{}^{(k)} &= \bar{a}_{11}{}^{(k)}\overline{U}_1 + \cdots + \bar{a}_{1m}{}^{(k)}\overline{U}_m \\
&\vdots \\
\bar{r}_n{}^{(k)} &= \bar{a}_{n1}{}^{(k)}\overline{U}_1 + \cdots + \bar{a}_{nm}{}^{(k)}\overline{U}_m
\end{aligned}\right\} \tag{4.31}$$

The coefficients $\bar{a}_{ij}{}^{(k)}$ are called *influence coefficients*. This expression is represented by matrices as follows:

$$[\bar{r}] = [\bar{a}]\left[\overline{U}\right] \tag{4.32}$$

The matrix $[\bar{a}]$ has $(k \times n)$ rows and m columns. Balancing is attained if we can make $\bar{r}_1, \ldots, \bar{r}_n$ zero by attaching masses m_i $(i = 1, \ldots, m)$ in m correction planes. Suppose that the influence coefficients $[\bar{a}]$ are known. If this matrix is square, that is, $k \times n = m$ holds, the inverse matrix $[\bar{a}]^{-1}$ exists and we can determine the initial unbalance $[\boldsymbol{U}]$ by

$$\left[\overline{U}\right] = [\bar{a}]^{-1}[\bar{r}] \tag{4.33}$$

Then, we can balance the rotor by attaching the correction masses $\left[-\overline{U}\right]$.

We can determine the influence coefficients, which are supposed to be known above, as follows: We attach known trial masses in the correction planes and measure deflections at the above-mentioned rotational speeds $\omega_i (i = 1, \ldots, k)$. Then, we calculate the differences between these deflections and the responses for the initial unbalance. The relationship between this difference and the trial mass gives the influence coefficients. For example, $\bar{r}_1{}^{(s)\prime}, \ldots, \bar{r}_n{}^{(s)\prime}$ be the deflections obtained at the rotational speed ω_s when we add a known mass m_0 at radius $\bar{b}_j$ in the jth correction plane. The following relationship holds between the change in response and the added mass:

$$\left.\begin{aligned} \bar{r}_1{}^{(s)\prime} - \bar{r}_1{}^{(s)} &= \bar{a}_{1j}{}^{(s)} m_0 \bar{b}_j \\ &\vdots \\ \bar{r}_n{}^{(s)\prime} - \bar{r}_n{}^{(s)} &= \bar{a}_{nj}{}^{(s)} m_0 \bar{b}_j \end{aligned}\right\} \tag{4.34}$$

We can determine $\bar{a}_{1j}{}^{(s)}, \ldots, \bar{a}_{nj}{}^{(s)}$ from this expression. By repeating such operations, we can determine all the coefficients of $[a]$.

References

Bishop, R.E.D. and Gladwell, G.M.L. (1959) The vibration and balancing of an unbalanced flexible rotor. *J. Mech. Eng. Sci.*, **1**(l), 66–77.

Miwa, S. and Shimomura, G. (1976) *Balancing of Rotating Machinery*, Corona Publishing Co., Tokyo (in Japanese).

Macduff, J.N. and Curreri, J.R. (1958) *Vibration Control*, McGraw-Hill, New York.

Rao, J.S. (1991) *Rotor Dynamics*, 2nd edn, John Wiley & Sons, Inc., New York.

Thearle, E.L. (1932) A new type of dynamic-balancing machine. *Trans. ASME, J. Appl. Mech.*, **54**(12), 131–141.

Thearle, W.T. (1956) Dynamic balancing of rotating machinery in the field. *Trans. ASME, J. Appl. Mech.*, **56**, 745–753.

5
Vibrations of an Asymmetrical Shaft and an Asymmetrical Rotor

5.1
General Considerations

Some types of rotating machinery have asymmetry in shaft stiffness or in moment of inertia. For example, a rotor of the two-pole alternating current generator shown in Figure 5.1 has slots for coils, and therefore the bending stiffness of rotor differs depending on the direction. In the case of a propeller, its moment of inertia has a directional difference. Such an *asymmetrical rotor system* with a rotating asymmetry has a unique vibration characteristic different from that of a *symmetrical rotor system*, which is composed of a shaft with a circular cross section and a disk. For example, unstable vibration appears in the vicinity of the major critical speed and, in addition, the shape of the resonance curves of a steady-state vibration on both sides of this unstable zone change depending on the angular position of the unbalance. Furthermore, an asymmetrical shaft has a resonance at the rotational speed at half the major critical speed when it is supported horizontally. Since the equations of motion governing these asymmetrical systems have coefficients that change with time, its mathematical treatment is much more difficult than that for a symmetrical system.

In this chapter, two basic rotor models, composed of a rotor and a massless elastic shaft, are studied. One is a system composed of a disk and a massless elastic shaft with a directional difference in stiffness. The other is a system composed of an elastic shaft with a circular cross section and a rotor with difference in moment of inertia. The former is called an *asymmetrical shaft system* and the latter an asymmetrical rotor system. In this chapter, vibration phenomena in such systems are explained. For simplicity, systems with 2 DOF are used.

For further study, equation of motions of a 4 DOF system where the deflection motion and the inclination motion couple with each other are given in Appendix D. Concerning asymmetrical rotors with distributed masses, refer to Dimentberg (1961) or Tondl (1965).

Linear and Nonlinear Rotordynamics: A Modern Treatment with Applications, Second Edition.
Yukio Ishida and Toshio Yamamoto.

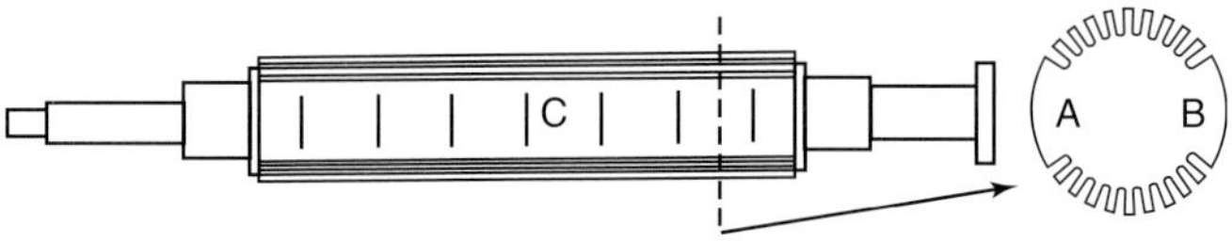

Figure 5.1 Two-pole generator rotor.

5.2 Asymmetrical Shaft with a Disk at Midspan

We study a deflection motion of a rotor mounted at the midspan of an asymmetrical shaft. The analytical model is given by replacing the shaft of the system in Figure 2.4 by an asymmetrical shaft.

5.2.1 Equations of Motion

A *principal axis of area of cross section* is defined as an axis whose product of area of section vanishes. Regardless of the shape of the cross section, the principal axes always make a right angle; therefore, the directions of the maximum and minimum stiffnesses also make a right angle. From a dynamic point of view, every type of cross section is equivalent to a rectangular shape that is considered in this section. Figure 5.2 shows a whirling rotor that rotates at the rotational speed ω. The system O-xy is a coordinate system fixed in space, and O-$x'y'$ is a rotating coordinate system fixed to the asymmetrical shaft and rotates with an angular speed ω. The x'-axis coincides with the direction of smaller bending stiffness as the y'-axis does with that of larger bending stiffness. The time $t = 0$ is selected at the instance when the x'-axis coincides with the x-axis. Then, the relationship $\angle x'\text{O}x = \omega t$ holds. Let the spring constants in the directions of the x'- and y'-axes be $k - \Delta k$ and $k + \Delta k$, respectively. Then, the restoring forces F'_x

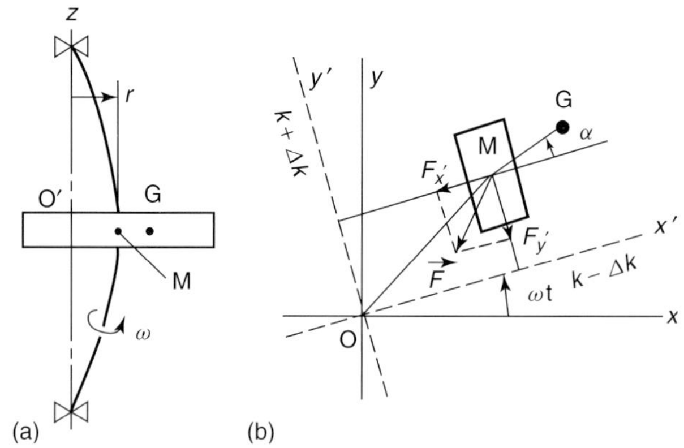

Figure 5.2 Asymmetrical shaft system: (a) physical model, (b) coordinates.

and F'_y in these directions are expressed as

$$\left.\begin{aligned} F'_x &= -(k - \Delta k)x' \\ F'_y &= -(k + \Delta k)y' \end{aligned}\right\} \tag{5.1}$$

The following relationships hold between the coordinates and forces, respectively.

$$\left.\begin{aligned} x' &= x\cos\omega t + y\sin\omega t \\ y' &= -x\sin\omega t + y\cos\omega t \end{aligned}\right\} \tag{5.2}$$

$$\left.\begin{aligned} F_x &= F'_x\cos\omega t - F'_y\sin\omega t \\ F_y &= F'_x\sin\omega t + F'_y\cos\omega t \end{aligned}\right\} \tag{5.3}$$

Substituting Eqs. (5.1) and (5.2) into Eq. (5.3) gives

$$\left.\begin{aligned} F_x &= -\left\{kx - \Delta k(x\cos 2\omega t + y\sin 2\omega t)\right\} \\ F_y &= -\left\{ky - \Delta k(x\sin 2\omega t - y\cos 2\omega t)\right\} \end{aligned}\right\} \tag{5.4}$$

The right-hand sides of Eq. (2.1) are replaced by these expressions, and x_G and y_G on its left-hand side are expressed by x and y. After adding damping forces, we get the following equations of motion:

$$\left.\begin{aligned} m\ddot{x} + c\dot{x} + kx - \Delta k(x\cos 2\omega t + y\sin 2\omega t) &= me\omega^2\cos(\omega t + \alpha) \\ m\ddot{y} + c\dot{y} + ky - \Delta k(x\sin 2\omega t - y\cos 2\omega t) &= me\omega^2\sin(\omega t + \alpha) \end{aligned}\right\} \tag{5.5}$$

where α is an initial phase angle of the center of gravity of the rotor. For convenience, we use the notations $c_0 = c/m$, $p_0 = \sqrt{k/m}$, and $\Delta = \Delta k/m$. Then, we have

$$\left.\begin{aligned} \ddot{x} + c_0\dot{x} + p_0^2 x - \Delta(x\cos 2\omega l + y\sin 2\omega t) &= e\omega^2\cos(\omega t + \alpha) \\ \ddot{y} + c_0\dot{y} + p_0^2 y - \Delta(x\sin 2\omega t - y\cos 2\omega t) &= e\omega^2\sin(\omega t + \alpha) \end{aligned}\right\} \tag{5.6}$$

As shown in Eqs. (5.5) and (5.6), the motions in the x- and y-directions couple with each other through asymmetries Δk and Δ even in the deflection motion. In addition, since the coefficients of x and y are functions of time, this system is classified into the category of a parametrically excited system with an external force. As these equations are linear, the general solutions are expressed as the summation of the solution of the homogeneous equations, obtained by putting the right-hand side equal to zero, and the special solution of the nonhomogeneous equations. The former solution corresponds to free vibration, while the latter does to forced vibration. In the following sections, we discuss each of these solutions.

5.2.2
Free Vibrations and Natural Frequency Diagrams

We calculate the natural angular frequency p of a free vibration in a system with no damping force and no unbalance force. From Eq. (5.6), the equations of motion of such a system are given by

$$\left.\begin{aligned} \ddot{x} + p_0^2 x - \Delta(x\cos 2\omega t + y\sin 2\omega t) &= 0 \\ \ddot{y} + p_0^2 y - \Delta(x\sin 2\omega t - y\cos 2\omega t) &= 0 \end{aligned}\right\} \tag{5.7}$$

In Section 2.4.2, in order to obtain natural frequencies, we assumed a circular motion of $x = A\cos(pt + \beta)$ and $y = A\sin(pt + \beta)$ as shown by Eq. (2.22). In the case of an asymmetrical shaft, if we substitute this expression into the equations above, a component with frequency $2\omega - p$ appears. Therefore, we must take this derived component into account in the assumed solution. However, as the component with the frequency $2\omega - p$ in the assumed solution derives only the component with the frequency p because of the special form of these parametric terms, the solution is closed and the two terms p and $2\omega - p$ adequately express the exact solution. From such a consideration, we assume the solution in the following form

$$\left.\begin{aligned} x &= A\cos(pt + \beta) + B\cos\{(2\omega - p)t - \beta\} \\ y &= A\sin(pt + \beta) + B\sin\{(2\omega - p)t - \beta\} \end{aligned}\right\} \tag{5.8}$$

where β is a phase angle. Substituting this into Eq. (5.7) and equating the coefficients of the same frequency term on the right- and left-hand sides, we can obtain the frequency equation that determines p.

Here, we permit complex numbers as root p and investigate the characteristics of the solution. For this purpose, the coordinate plane O-xy shown in Figure 5.2 is overlapped on the complex plane and the complex number $z = x + iy$ is defined.

Equation (5.7) are expressed by complex numbers as follows:

$$\ddot{z} + p_0^2 z - \Delta\bar{z}e^{i2\omega t} = 0 \tag{5.9}$$

where $\bar{z} = x - iy$ is a complex conjugate. Similarly, the symbol "–" over other notations also indicates complex conjugates. We assume the solution in the form $z = Ae^{i(pt+\beta)} = ae^{ipt}$, where $a = Ae^{i\beta}$ is a complex number. When this solution is substituted into Eq. (5.9), the third term becomes

$$\begin{aligned} \Delta\bar{z}e^{i2\omega t} &= \Delta\overline{ae^{ipt}}e^{i2\omega t} = \Delta\bar{a}\overline{e^{ipt}}e^{i2\omega t} \\ &= \Delta Ae^{-i\beta}e^{-i\bar{p}t}e^{i2\omega t} = \Delta Ae^{-i\beta}e^{i(2\omega-\bar{p})t} \end{aligned} \tag{5.10}$$

This suggests that we must include the term with frequency $2\omega - \bar{p}$ in the assumed solution. As this component $2\omega - \bar{p}$ derives the component $2\omega - \overline{(2\omega - \bar{p})} = p$, which has the same frequency as the original frequency p, the solution is closed. Therefore, we can assume the solution in the form

$$r = ae^{ipt} + be^{i(2\omega-\bar{p})t} \tag{5.11}$$

where the coefficients a and b and the frequency p may become complex. If p is real, this expression coincides with Eq. (5.8).

Substituting Eq. (5.11) into Eq. (5.9) and equating the coefficients of the terms e^{ipt} and $e^{i(2\omega-\bar{p})t}$ in both sides, respectively, we get

$$\left.\begin{aligned} &\left(p_0^2 - p^2\right)a - \Delta\bar{b} = 0 \\ &\left\{p_0^2 - (2\omega - \bar{p})^2\right\}b - \Delta\bar{a} = 0 \end{aligned}\right\} \tag{5.12}$$

Furthermore, taking the complex conjugates of both sides gives

$$\left.\begin{aligned} &\left(p_0^2 - \bar{p}^2\right)a - \Delta b = 0 \\ &-\Delta a + \left\{p_0^2 - (2\omega - p)^2\right\}\bar{b} = 0 \end{aligned}\right\} \tag{5.13}$$

In order to derive the frequency equation, we eliminate the amplitudes. Equating the expressions for a/b in Eqs. (5.12) and (5.13), we get

$$\left(p_0^2 - p^2\right)\left\{p_0^2 - (2\omega - p)^2\right\} - \Delta^2 = 0 \tag{5.14}$$

Similarly, equating the expressions for $\overline{a}/b$, we get

$$\left(p_0^2 - \overline{p}^2\right)\left\{p_0^2 - (2\omega - \overline{p})^2\right\} - \Delta^2 = 0 \tag{5.15}$$

If p is real, these two expressions are the same. If it is complex, we know that a pair composed of complex p and $\overline{p}$ is given as the roots of the same equations. Therefore, it is enough to use Eq. (5.14) as a frequency equation. From Eq. (5.14), we get

$$(p-\omega)^4 - 2\left(\omega^2 + p_0^2\right)(p-\omega)^2 + \left\{\omega^2 - \left(p_0^2 + \Delta\right)\right\}\left\{\omega^2 - \left(p_0^2 - \Delta\right)\right\} = 0 \tag{5.16}$$

This equation is a quadratic equation in $(p-\omega)^2$, and we can easily obtain the roots $p_1, \cdots, p_4$ as follows:

$$p_1 = \omega + \nu, p_2 = \omega - \nu, \quad p_3 = \omega + \mu, p_4 = \omega - \mu \tag{5.17}$$

where

$$\left.\begin{array}{l} \nu = \sqrt{\omega^2 + p_0^2 + \sqrt{4p_0^2\omega^2 + \Delta^2}} \\ \mu = \sqrt{\omega^2 + p_0^2 - \sqrt{4p_0^2\omega^2 + \Delta^2}} \end{array}\right\} \tag{5.18}$$

Figure 5.3 shows natural frequencies $p_1, \cdots, p_4$ as a function of the rotational speed ω. As ν is real, p_1 and p_2 are real for any value of ω. On the contrary, p_3 and p_4 are not real for some range of ω. Let the rotational speeds where μ vanishes be ω_{c1} and ω_{c2}. From Eq. (5.18), we get

$$\omega_{c1} = \sqrt{p_0^2 + \Delta} = \sqrt{\frac{k + \Delta k}{m}}, \qquad \omega_{c2} = \sqrt{p_0^2 - \Delta} = \sqrt{\frac{k - \Delta k}{m}} \tag{5.19}$$

The rotational speeds ω_{c1} and ω_{c2} correspond to the natural frequencies in the directions of higher and lower stiffnesses. The natural frequencies p_3 and p_4 are

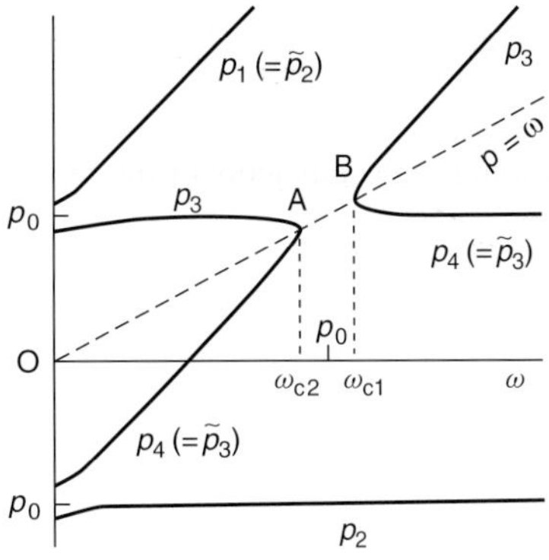

Figure 5.3 Natural frequency diagram of an asymmetrical shaft.

real only in the ranges $\omega > \omega_{c1}$ and $\omega < \omega_{c2}$. From Eq. (5.17), we see that the relationships

$$p_1 = 2\omega - p_2, \qquad p_4 = 2\omega - p_3 \tag{5.20}$$

hold. When we assumed the solution of Eq. (5.8), we guessed from the terms in the equations of motion that components with frequencies p and $2\omega - p$ coexist. In fact, this equation shows that p_1 and p_2 and p_3 and p_4 correspond to such a coexisting couple, respectively. For convenience, we define new notations $2\omega - p \equiv \tilde{p}$ and use, for example, $\tilde{p}_2$ and $\tilde{p}_3$ instead of p_1 and p_4 to show explicitly that they have relationships to p_2 and p_3, respectively.

The solution for a free vibration is given by the summation of expressions that are obtained by substituting $p_1, \cdots, p_4$ into Eq. (5.11). However, since these frequencies $p_1, \cdots, p_4$ have the relationships mentioned above and the amplitudes and frequencies may take complex values, it is difficult to understand its concrete form from that expression. Therefore, we rearrange this solution and discuss it in further detail

5.2.2.1 Solutions in the Ranges $\omega > \omega_{c1}$ and $\omega < \omega_{c2}$

Substituting $p_1, \cdots p_4$ into Eq. (5.11), we have

$$\begin{aligned} z &= (a_1 + b_2)e^{i(\omega+\nu)t} + (a_2 + b_1)e^{i(\omega-\nu)t} \\ &+ (a_3 + b_4)e^{i(\omega+\mu)t} + (a_4 + b_3)e^{i(\omega-\mu)t} \end{aligned} \tag{5.21}$$

If we put $a_n = A_n e^{i\alpha_n}$ and $b_n = B_n e^{i\beta_n}$ $(n = 1, \cdots, 4)$, we have the following relationships from Eqs. (5.12) and (5.13).

$$\left.\begin{aligned} \frac{A_1}{B_1}e^{i(\alpha_1+\beta_1)} &= \frac{B_2}{A_2}e^{-i(\alpha_2+\beta_2)} = \kappa_1 \\ \frac{A_3}{B_3}e^{i(\alpha_3+\beta_3)} &= \frac{B_4}{A_4}e^{-i(\alpha_4+\beta_4)} = \kappa_2 \end{aligned}\right\} \tag{5.22}$$

where κ_1 and κ_2 are real and are given by

$$\kappa_1 = \frac{\Delta}{p_0^2 - (\omega+\nu)^2}, \qquad \kappa_2 = \frac{\Delta}{p_0^2 - (\omega+\mu)^2} \tag{5.23}$$

Substituting the relationship among real quantities obtained from Eq. (5.22), such as $A_1 = \kappa_1 B_1$ and $\alpha_1 = -\beta_1$, into Eq. (5.21) and using the transformations

$$\left.\begin{aligned} A_1 e^{i\alpha_1} + \kappa_1 A_2 e^{-i\alpha_2} &= P_1 e^{i\theta_1} \\ A_3 e^{i\alpha_3} + \kappa_2 A_4 e^{-i\alpha_4} &= P_2 e^{i\theta_2} \end{aligned}\right\} \tag{5.24}$$

Finally, we obtain the solution in the following form

$$\begin{aligned} z &= P_1 e^{i\theta_1} e^{i(\omega+\nu)t} + \frac{1}{\kappa_1} P_1 e^{-i\theta_1} e^{i(\omega-\nu)t} \\ &+ P_2 e^{i\theta_2} e^{i(\omega+\mu)t} + \frac{1}{\kappa_2} P_2 e^{-i\theta_2} e^{i(\omega-\mu)t} \end{aligned} \tag{5.25}$$

This can be separated into components in the x- and y-directions as follows:

$$\left.\begin{aligned}
x &= P_1\cos\{(\omega+\nu)t+\theta_1\} + \frac{1}{\kappa_1}P_1\cos\{(\omega-\nu)t-\theta_1\}\\
&\quad + P_2\cos\{(\omega+\mu)t+\theta_2\} + \frac{1}{\kappa_2}P_2\cos\{(\omega-\mu)t-\theta_2\}\\
y &= P_1\sin\{(\omega+\nu)t+\theta_1\} + \frac{1}{\kappa_1}P_1\sin\{(\omega-\nu)t-\theta_1\}\\
&\quad + P_2\sin\{(\omega+\mu)t+\theta_2\} + \frac{1}{\kappa_2}P_2\sin\{(\omega-\mu)t-\theta_2\}
\end{aligned}\right\} \tag{5.26}$$

This result shows that rotor motion is composed of four circular motions with constant amplitudes and, therefore, is finite.

5.2.2.2 Solutions in the Range $\omega_{c1} > \omega > \omega_{c2}$

We can use the same procedure as that used in Section 5.2.2.1. That is, the unknown quantities $a_1, \cdots, b_4$ in Eq. (5.21) are lessened by substituting the relationships given by Eqs. (5.12) and (5.13). However, since this calculation is relatively cumbersome, we obtain the solution in a different way, as described below.

Since μ in Eq. (5.18) becomes imaginary, we put

$$\left.\begin{aligned}
\mu &= in\\
n &= \sqrt{\sqrt{rp_0^2\omega^2+\Delta^2}-(\omega^2+p_0^2)}
\end{aligned}\right\} \tag{5.27}$$

where n is a real. Since we already know the frequencies as given by Eq. (5.17), we use the following expression that includes only four amplitudes $c_1, \cdots, c_4$ instead of eight unknown amplitudes $a_1, \cdots, b_4$.

$$z = c_1 e^{i(\omega+\nu)t} + c_2 e^{i(\omega-\nu)t} + c_3 e^{(+n+i\omega)t} + c_4 e^{(-n+i\omega)t} \tag{5.28}$$

where $c_1, \cdots, c_4$ are generally complex numbers. Substituting this assumed solution into Eq. (5.9) and comparing the coefficients, we have

$$\left.\begin{aligned}
&\text{(i)}\quad \frac{c_1}{\bar{c}_2} = \frac{\Delta}{p_0^2-(\omega+\nu)^2}, \qquad \text{(ii)}\quad \frac{c_1}{\bar{c}_2} = \frac{p_0^2-(\omega-\nu)^2}{\Delta}\\
&\text{(iii)}\quad \frac{c_3}{\bar{c}_3} = \frac{\Delta}{p_0^2+(n+i\omega)^2}, \qquad \text{(iv)}\quad \frac{c_4}{\bar{c}_4} = \frac{\Delta}{p_0^2+(-n+i\omega)^2}
\end{aligned}\right\} \tag{5.29}$$

If we substitute $p_1 = \omega + \nu$ or $p_2 = \omega - \nu$ into Eq. (5.14), we see that the right-hand sides of (i) and (ii) in Eq. (5.29) are the same and take the real value k_1 given in Eq. (5.23). The absolute values on the right-hand sides of (iii) and (iv) are clearly 1 because they are the ratios of complex conjugates. Representing c_k in the forms $c_k = C_k e^{i\gamma_k}(k = 1, \cdots, 4)$ and substituting them into Eq. (5.29), we have

$$C_1 = k_1 C_2, \quad \gamma_2 = -\gamma_1, \quad \gamma_3 = -\gamma_4 = \varphi \tag{5.30}$$

where

$$\varphi = \frac{1}{2}\tan^{-1}\frac{-2\omega n}{p_0^2-\omega^2+n^2} \tag{5.31}$$

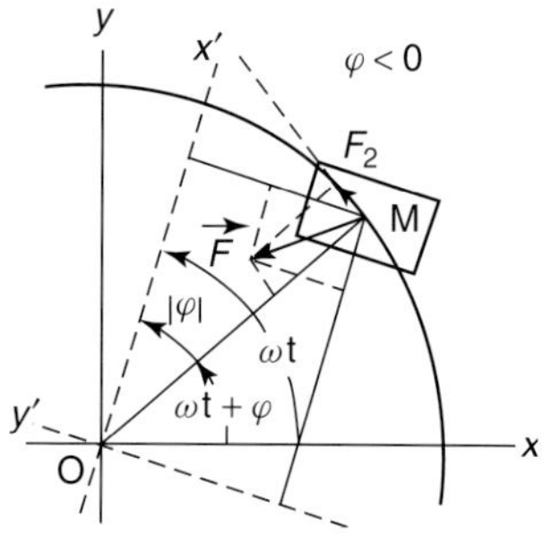

Figure 5.4 Occurrence of unstable vibration ($\omega_{c1} > \omega > \omega_{c2}$).

In (iii) in Eq. (5.29), since the imaginary part of $\overline{c_3}/c_3 = e^{i(-2\gamma_3)} = (p_0^2 - \omega^2 + n^2 + 2n\omega i)/\Delta$ is positive, $0 < -2\gamma_3 < \pi$, that is, $0 > \varphi > -\pi/2$ holds. From the relationship of Eqs. (5.30), (5.28) become

$$z = C_1 e^{i\gamma_1} e^{i(\omega+\nu)t} + \frac{C_1}{k_1} e^{-i\gamma_1} e^{i(\omega-\nu)t} + C_3 e^{+nt} e^{i(\omega t+\varphi)} + C_4 e^{-nt} e^{i(\omega t-\varphi)} \tag{5.32}$$

The third term increases exponentially with time, and the fourth term decreases. In practical systems with inevitable damping, the first and second terms also damp out. Therefore, the third term,

$$\left.\begin{aligned} x &= C_3 e^{+nt} \cos(\omega t + \varphi) \\ y &= C_3 e^{+nt} \sin(\omega t + \varphi) \end{aligned}\right\} \tag{5.33}$$

becomes more predominant with time. This shows that an unstable vibration with frequency ω appears in this range called an *unstable range*.

Here, we discuss the mechanism of how this unstable vibration occurs. In the unstable range $\omega_{c1} > \omega > \omega_{c2}$, the phase angle φ takes the value $(-\pi/2) < \varphi < 0$ as mentioned above. Therefore, as shown in Figure 5.4, the asymmetrical shaft deflects in the direction that differs from the x'-direction (the direction with smaller stiffness) by φ. In the case of a symmetrical shaft having equal stiffness k in every direction, the restoring force is directed toward the origin O. However, in the case of the asymmetrical shaft treated here, the restoring force deviates from the direction to the origin because spring constants multiplied to the deflections x' and y' are different. As a result, a tangential component F_2 of the restoring force F appears. Owing to the work of this component, energy is charged into the system and the whirling motion grows.

Outside the unstable range, the whirling motion passes through a trajectory of a very complex shape given by Eq. (5.26). In this case, the time averaging of the work is zero and the energy is not charged.

5.2.3
Synchronous Whirl in the Vicinity of the Major Critical Speed

Equations of motion with an unbalance are given by Eq. (5.6). The unbalance force causes a whirling motion with the frequency ω. By the same process mentioned in Section 5.2.2, this component may produce other frequency components. However, this component $z = Pe^{i\omega t}$ produces the same frequency component as follows:

$$\Delta \bar{z} e^{i2\omega t} = \Delta p e^{-i\omega t} e^{i2\omega t} = \Delta p e^{i\omega t} \tag{5.34}$$

With such a consideration, we can assume a solution in the form

$$\left.\begin{aligned} x &= p\cos(\omega t + \beta) = P_1 \cos \omega t - P_2 \sin \omega t \\ y &= P\sin(\omega t + \beta) = p_1 \sin \omega t - P_2 \cos \omega t \end{aligned}\right\} \tag{5.35}$$

where $P_1 = P\cos\beta$ and $P_2 = P\sin\beta$. First, we assume that P and β are constant. Substituting this into Eq. (5.6) and comparing the coefficients of $\cos\omega t$ and $\sin\omega t$ on both sides, we get

$$\left.\begin{aligned} -c_0\omega P_2 + \left(p_0^2 - \Delta - \omega^2\right) P_1 &= e\omega^2 \cos\alpha \\ +c_0\omega P_1 + \left(p_0^2 + \Delta - \omega^2\right) P_2 &= e\omega^2 \sin\alpha \end{aligned}\right\} \tag{5.36}$$

We can obtain the steady-state solutions $P_1 = P_{10}$ and $P_2 = p_{20}$ from this equation. For simplicity, we discuss the case with no damping in the following. From Eq. (5.36) with $c_0 = 0$, we get

$$P_{10} = \frac{e\omega^2 \cos\alpha}{\omega_{c2}^2 - \omega^2}, \quad P_{20} = \frac{e\omega^2 \sin\alpha}{\omega_{c1}^2 - \omega^2} \tag{5.37}$$

where ω_{c1} and ω_{c2} are given by Eq. (5.19). The amplitude is obtained by $P_0 = \sqrt{P_{10}^2 + P_{20}^2}$.

In an asymmetrical shaft, the stability problem of the steady-state solution is very important in understanding the behavior of the rotor. The stability is defined as follows: In practical machineries, the rotor always deviates slightly from the steady-state solution due to inevitable disturbance. If the rotor returns to the position corresponding to the steady-state solution, the system is said to be stable. If the rotor leaves it, the system is said to be unstable. Only the stable solution can be observed in practice.

To investigate the stability, we must assume that the amplitudes P_1 and P_2 in Eq. (5.35) are functions of time. Substituting Eq. (5.35) into Eq. (5.6) and comparing the coefficients of $\cos\omega t$ and $\sin\omega t$ on both sides, respectively, we get

$$\left.\begin{aligned} \ddot{P}_1 - 2\omega\dot{P}_2 + \left(p_0^2 - \Delta - \omega^2\right) P_1 &= e\omega^2 \cos\alpha \\ \ddot{P}_2 + 2\omega\dot{P}_1 + \left(p_0^2 + \Delta - \omega^2\right) P_2 &= e\omega^2 \sin\alpha \end{aligned}\right\} \tag{5.38}$$

(If we assume that P_1 and P_2 are slowly varying functions of time as is shown in Section 6.4.1, the calculation becomes easier than that shown here.) Considering deviations ξ_1 and ξ_2, we have

$$P_1 = P_{10} + \xi_1, \; P_2 = P_{20} + \xi_2 \tag{5.39}$$

and when we substitute them into Eq. (5.38), we get

$$\left.\begin{aligned} \ddot{\xi}_1 - 2\omega\dot{\xi}_2 + \left(\omega_{c2}^2 - \omega^2\right)\xi_1 &= 0 \\ \ddot{\xi}_2 + 2\omega\dot{\xi}_1 + \left(\omega_{c1}^2 - \omega^2\right)\xi_2 &= 0 \end{aligned}\right\} \tag{5.40}$$

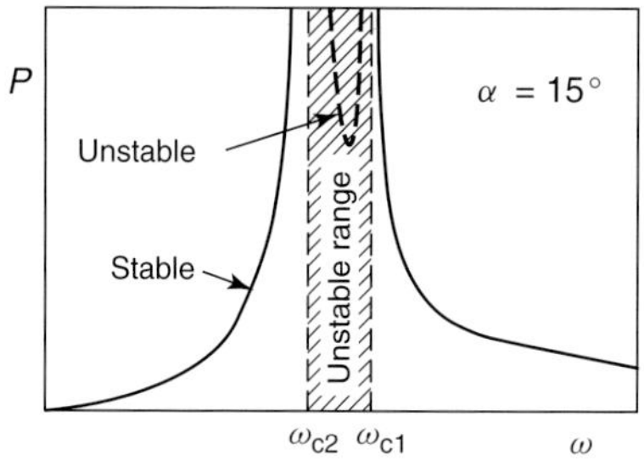

Figure 5.5 Resonance curves in the vicinity of the major critical speed.

We assume the solution in the form

$$\xi_1(t) = Ae^{st},\ \xi_2(t) = Be^{st} \tag{5.41}$$

and substitute this into Eq. (5.40). In order to have a nontrivial solution for A and B, the determinant of the coefficients of A and B must be zero. This determinant is called a *characteristic determinant*. Then, the following expression must be satisfied.

$$s^4 + \left(\omega_{c1}^2 + \omega_{c2}^2 + 2\omega^2\right)s^2 + \left(\omega_{c1}^2 - \omega^2\right)\left(\omega_{c2}^2 - \omega^2\right) = 0 \tag{5.42}$$

From this characteristic equation, we know that the root s has a positive real part when $\omega_{c2} < \omega < \omega_{c1}$. That is, in the range AB in Figure 5.3, the small deviations $\xi_1(t)$ and $\xi_2(t)$ increase with time and the steady-state solution becomes unstable.

Figure 5.5 shows resonance curves obtained from Eq. (5.37) in the vicinity of the major critical speed. The shape of this resonance curve changes depending on the direction α of the unbalance. As an example, the case of $\alpha = 15^\circ$ is shown here. A solid line and a dashed line mean stable and unstable solutions, respectively. The amplitude becomes infinite at the rotational speeds corresponding to the natural frequency ω_{c1} in the direction of the higher stiffness and ω_{c2} in that of the lower stiffness, and the solution becomes unstable between these two rotational speeds. As stated above, in an asymmetrical shaft system, both free and forced vibrations become unstable vibrations with frequency ω in the rotational range $\omega_{c2} < \omega < \omega_{c1}$.

5.3
Inclination Motion of an Asymmetrical Rotor Mounted on a Symmetrical Shaft

An asymmetrical rotor system has characteristics similar to those of an asymmetrical shaft system. As shown in Chapter 2 and Appendix A, a rotor has three principal axes of the moment of inertia. Let one of them coincide with the shaft center-line. We represent the moment of inertia about it as I_p and the other two moments of inertia as I_1 and I_2. A rotor of $I_1 = I_2$ is called a *symmetrical rotor* and that of $I_1 \neq I_2$ called an *asymmetrical rotor*.

5.3.1
Equations of Motion

We study an inclination vibration of an asymmetrical rotor mounted at the midspan of an elastic shaft. Unlike the case of Section 2.4, where the equations of motion were derived from the angular momentum change, we derive its equations of motion using Lagrange's equation (Yamamoto and Ota, 1969). When the system becomes complicated, this analytical method is easier than that shown in Section 2.4. We consider an asymmetrical rotor of the type shown in Figure 5.6. Let the three axes of the rectangular coordinate system M-$X_2Y_2Z_2$ coincide with the principal axes of the moment of inertia. Plane B defined by the X_2- and Y_2-axes is the same as the disk plane in Figure 2.16. The coordinate system M-$X_1Y_1Z_1$ in Figure 2.16, whose position is determined according to the position of the dynamic unbalance, and the system M-$X_2Y_2Z_2$ make angle η as shown in Figure 5.6.

Figure 5.7 shows a rotor in whirling motion. The shaft inclines by θ in a direction that differs from the x-axis by φ. We represent equations of motion by the projections of the inclination angle of the shaft $\theta_x = \theta\cos\varphi$ and $\theta_y = \theta\sin\varphi$.

First, we derive the kinetic energy T. Let the angular velocity components in the MX_2-, MY_2-, and MZ_2-directions be ω_{X2}, ω_{Y2}, and ω_{Z2}, respectively. Then, the kinetic energy is represented by

$$T = \frac{1}{2}\left(I_p\omega_{Z2}^2 + I_1\omega_{Y2}^2 + I_2\omega_{X2}^2\right) \tag{5.43}$$

Let the Eulerian angles representing the coordinate system M-$X_2Y_2Z_2$ be θ_2, φ_2, and ψ_2, as shown in Figure 2.22. Similar to Eq. (2.63), we have

$$\left.\begin{aligned}\omega_{X_2} &= \dot{\theta}_2\sin\phi_2 - \dot{\varphi}_2\sin\theta_2\cos\phi_2\\ \omega_{Y_2} &= \dot{\theta}_2\cos\phi_2 + \dot{\varphi}_2\sin\theta_2\sin\phi_2\\ \omega_{Z_2} &= \dot{\varphi}_2\cos\theta_2 + \dot{\phi}_2\end{aligned}\right\} \tag{5.44}$$

We substitute this into Eq. (5.43). Representing $I_1 = I + \Delta I, I_2 = I - \Delta I$, and using the notation

$$\Theta_2 = \varphi_2 + \phi_2, \quad \theta_{x2} = \theta_2\cos\varphi_2, \quad \theta_{y2} = \theta_2\sin\varphi_2 \tag{5.45}$$

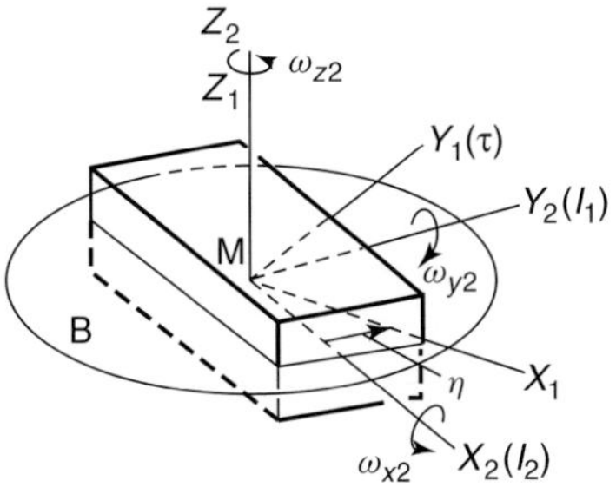

Figure 5.6 Principal axes of moment of inertia and coordinate systems.

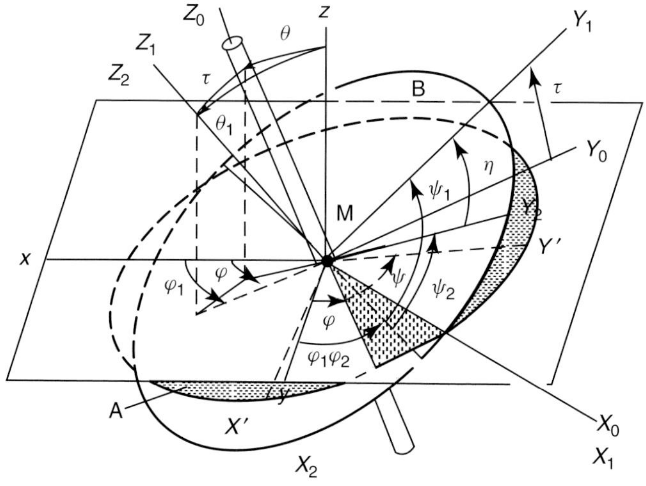

Figure 5.7 Skew angle τ and coordinate system.

To an accuracy of the second order of a small quantity, Eq. (5.43) becomes

$$
\begin{aligned}
T \approx \frac{1}{2}\Big[& I_p \left\{\dot{\Theta}_2^2 - \dot{\Theta}_2 \left(\theta_{x_2}\dot{\theta}_{y_2} - \dot{\theta}_{x_2}\theta_{y_2}\right)\right\} + I\left(\dot{\theta}_{x_2}^2 + \dot{\theta}_{y_2}^2\right) \\
& + \Delta I\left\{\left(\dot{\theta}_{x_2}^2 - \dot{\theta}_{y_2}^2\right)\cos 2\Theta_2 + 2\dot{\theta}_{x2}\dot{\theta}_{y2}\sin 2\Theta_2\right\}\Big]
\end{aligned}
\tag{5.46}
$$

In deriving this expression, we used the approximations $\sin\theta_2 \approx \theta_2$ and $\cos\theta_2 \approx 1 - \theta_2^2/2$, which hold under the assumption that θ_2 is small and the following relations:

$$
\left.
\begin{aligned}
& \dot{\theta}_{y_2}\theta_{x_2} - \dot{\theta}_{x_2}\theta_{y_2} = \theta_2^2\dot{\varphi}_2 \\
& \dot{\theta}_{x_2}^2 + \dot{\theta}_{y_2}^2 = \dot{\theta}_2^2 + \theta_2^2\dot{\varphi}_2^2 \\
& 2\dot{\theta}_{x_2}\dot{\theta}_{y_2} = \left(\dot{\theta}_2^2 - \theta_2^2\dot{\varphi}_2^2\right)\sin 2\varphi_2 + 2\theta_2\dot{\theta}_2\dot{\varphi}_2\cos 2\varphi_2 \\
& \dot{\theta}_{x_2}^2 - \dot{\theta}_{y_2}^2 = \left(\dot{\theta}_2^2 - \theta_2^2\dot{\varphi}_2^2\right)\cos 2\varphi_2 - 2\theta_2\dot{\theta}_2\dot{\varphi}_2\sin 2\varphi_2
\end{aligned}
\right\}
\tag{5.47}
$$

Next, we transform the expression from M-$X_2Y_2Z_2$ to M-$X_1Y_1Z_1$ with Eulerian angles $\theta_1, \varphi_1, \psi_1$ whose Y_1-axis is taken in the direction of the unbalance. From Figure 5.6, the transformations are given by

$$
\left.
\begin{aligned}
& \theta_2 = \theta_1 \quad \left(\theta_{x_2} = \theta_{x_1},\ \theta_{y_2} = \theta_{y_1}\right) \\
& \varphi_2 = \varphi_1,\ \ \phi_2 + \eta = \phi_1
\end{aligned}
\right\}
\tag{5.48}
$$

Using new notation $\Theta_1 = \varphi_1 + \phi_1$, we have $\Theta_2 + \eta = \Theta_1$. Then, Eq. (5.46) becomes as follows after transformation.

$$T \approx \frac{1}{2}\Big[I_p \left\{ \dot{\Theta}_1^2 - \dot{\Theta}_1 \left(\theta_{x_1}\dot{\theta}_{y_1} - \dot{\theta}_{x_1}\theta_{y_1} \right) \right\} + I \left(\dot{\theta}_{x_1}^2 + \dot{\theta}_{y_1}^2 \right) \\ + \Delta I \left\{ \left(\dot{\theta}_{x_1}^2 - \dot{\theta}_{y_1}^2 \right) \cos 2\left(\Theta_1 - \eta\right) + 2\dot{\theta}_{x_1}\dot{\theta}_{y_1} \sin 2\left(\Theta_1 - \eta\right) \right\} \Big] \quad (5.49)$$

Here, we define the coordinate systems M-$X_0Y_0Z_0$ and M-$X'Y'Z_0$ on the plane A perpendicular to the shaft. The former coincides with the system M-$X_1Y_1Z_1$, as the latter does with the system M-$X_2Y_2Z_2$ when $\tau = 0$; that is, plane B, including the rotor, coincides with plane A. Let the angle between the MY'-axis and the crossline of plane A and the horizontal plane be ψ. Then, the system M-$X'Y'Z_0$ is represented by θ, φ, ψ, and the system M-$X_0Y_0Z_0$ is represented by $\theta, \varphi, \psi + \eta (\equiv \psi_0)$.

Next, we transform the expression from M-$X_1Y_1Z_1$ to M-$X_0Y_0Z_0$. To derive the relationships between the Euler angles $\theta_1, \varphi_1, \psi_1$ representing the former and θ, φ, ψ_0 representing the latter, we utilize the relationships among direction cosines obtained in Section 2.5.1. The relationship between M-$X_1Y_1Z_1$ and O-xyz is given in Table 5.1a and that between M-$X_0Y_0Z_0$ and O-xyz is given in Table 5.1b. In

Table 5.1 Direction Cosines Representing Relationships between Coordinates.

(a)	X_1	Y_1	Z_1
x	$l_1' = \cos\Theta_1 - (\theta_1^2/2)\cos\varphi_1\cos\psi_1$	$l_2' = -\sin\Theta_1 + (\theta_1^2/2)\cos\varphi_1\sin\psi_1$	$l_3' = \theta_1\cos\varphi_1 \equiv \theta_{x_1}$
y	$m_1' = \sin\Theta_1 - (\theta_1^2/2)\sin\varphi_1\cos\psi_1$	$m_2' = \cos\Theta_1 + (\theta_1^2/2)\sin\varphi_1\sin\psi_1$	$m_3' = \theta_1\sin\varphi_1 \equiv \theta_{y_1}$
z	$n_1' = -\theta_1\cos\psi_1$	$n_2' = \theta_1\sin\psi_1$	$n_3 = \cos\theta_1 = 1 - \theta^2/2$

(b)	X_0	Y_0	Z_0
x	$l_1 = \cos\Theta_0 - (\theta^2/2)\cos\varphi\cos\psi_0$	$l_2 = -\sin\Theta_0 + (\theta^2/2)\cos\varphi\sin\psi_0$	$l_3 = \theta\cos\varphi \equiv \theta_x$
y	$m_1 = \sin\Theta_0 - (\theta^2/2)\sin\varphi\cos\psi_0$	$m_2 = \cos\Theta_0 + (\theta^2/2)\sin\varphi\sin\psi_0$	$m_3 = \theta\sin\varphi = \theta_y$
z	$n_1 = -\theta\cos\psi_0$	$n_2 = \theta\sin\psi_0$	$n_3 = \cos\theta = 1 - \theta^2/2$

(c)	X_0	Y_0	Z_0
X_1	1	0	0
Y_1	0	$\cos\tau = 1 - \tau^2/2$	$\sin\tau = \tau$
Z_1	0	$-\sin\tau = -\tau$	$\cos\tau = 1 - \tau^2/2$

deriving these relationships, the angles θ and θ_1 are assumed to be small and approximations such as $\sin\theta \approx \theta$ and $\cos\theta \approx 1 - \theta^2/2$ are used. In addition, we used new notations $\Theta = \varphi + \psi$ and $\Theta_0 = \varphi + \psi_0 = \varphi + \psi + \eta$. The relationship between M-$X_1Y_1Z_1$ and M-$X_0Y_0Z_0$ is obtained easily and shown in Table 5.1c. From Table 5.1b,c, we can get the relationship between M-$X_1Y_1Z_1$ and O-xyz. By comparing this relationship and Table 5.1a, we obtain

$$\left.\begin{aligned}
&l_1' = l_1,\ l_2' = \left(1 - \frac{\tau^2}{2}\right) l_2 + \tau l_3,\ l_3' = -\tau l_2 + \left(1 - \frac{\tau^2}{2}\right) l_3 \\
&m_1' = m_1,\ m_2' = \left(1 - \frac{\tau^2}{2}\right) m_2 + \tau m_3,\ m_3' - \tau m_2 + \left(1 - \frac{\tau^2}{2}\right) m_3 \\
&n_1' = n_1,\ n_2' = \left(1 - \frac{\tau^2}{2}\right) n_2 + \tau n_3,\ n_3' = -\tau n_2 + \left(1 - \frac{\tau^2}{2}\right) n_3
\end{aligned}\right\} \tag{5.50}$$

These relationships are expressed more concretely by

$$\left.\begin{aligned}
&\cos\Theta_1 - \frac{\theta_1^2}{2}\cos\varphi_1\cos\psi_1 = \cos\Theta_0 - \frac{\theta^2}{2}\cos\varphi\cos\psi_0 \\
&\sin\Theta_1 - \frac{\theta_1^2}{2}\cos\varphi_1\sin\psi_1 = \sin\Theta_0 - \frac{\theta^2}{2}\cos\varphi\sin\psi_0 - \tau\theta_x - \frac{\tau^2}{2}\sin\Theta_0 \\
&\theta_{x_1} = \tau\sin\Theta_0 + \theta_x \\
&\sin\Theta_1 - \frac{\theta_1^2}{2}\sin\varphi_1\cos\psi_1 = \sin\Theta_0 - \frac{\theta^2}{2}\sin\varphi\cos\psi_0 \\
&\cos\Theta_1 + \frac{\theta_1^2}{2}\sin\varphi_1\sin\psi_1 = \cos\Theta_0 + \frac{\theta^2}{2}\sin\varphi\sin\psi_0 + \tau\theta_y - \frac{\tau^2}{2}\cos\Theta_0 \\
&\theta_{y_1} = -\tau\cos\Theta_0 + \theta_y \\
&\theta_1\cos\psi_1 = \theta\cos\psi_0 \\
&\theta_1\sin\psi_1 = \theta\sin\psi_0 + \tau \\
&\frac{\theta_1^2}{2} = \tau\theta\sin\psi_0 + \frac{\theta^2}{2} + \frac{\tau^2}{2}
\end{aligned}\right\} \tag{5.51}$$

From the first, fifth, and ninth expressions or the second, fourth, and ninth expressions, we have

$$\left.\begin{aligned}
\cos\Theta_1 &= \cos\Theta_0\left[1 + \frac{\tau}{2}\left(\theta_x\sin\Theta_0 - \theta_y\cos\Theta_0\right)\right] + \frac{\tau\theta_y}{2} \\
\sin\Theta_1 &= \sin\Theta_0\left[1 + \frac{\tau}{2}\left(\theta_x\sin\Theta_0 - \theta_y\cos\Theta_0\right)\right] - \frac{\tau\theta_x}{2}
\end{aligned}\right\} \tag{5.52}$$

where approximations such as $1/(2-\theta_1^2/2) \approx (1/2+\theta_1^2/8)$ are used. Differentiating the second expression by time and substituting the first expression into its $\cos\Theta_1$, we obtain

$$\dot{\Theta}_1 = \dot{\Theta} - \frac{\tau}{2}\left(\dot{\theta}_x \cos\Theta_0 + \dot{\theta}_y \sin\Theta_0\right) + \frac{\tau\dot{\Theta}}{2}\left(\theta_x \sin\Theta_0 - \theta_y \cos\Theta_0\right) \tag{5.53}$$

In the derivation of this expression, we used the approximations $1/\cos\Theta_1 = 1/(\cos\Theta_0 + \Delta) \approx 1/\cos\Theta_0 - \Delta/\cos^2\Theta_0$ under the assumption that Δ is small. Substituting Eq. (5.53) and the third and sixth expressions in Eq. (5.51) into Eq. (5.49), we can finally express the kinetic energy T by the inclination angles θ_x and θ_y and the rotation angle Θ to an accuracy of the second order of small quantities as follows:

$$\begin{aligned}
T &= \frac{1}{2}I_p\Big\{\dot{\Theta}^2 - \tau\dot{\Theta}\left(\dot{\theta}_x\cos\Theta_0 + \dot{\theta}_y\sin\Theta_0\right) \\
&\quad + \tau\dot{\Theta}^2\left(\theta_x\sin\Theta_0 - \theta_y\cos\Theta_0\right)\Big\} \\
&\quad - \frac{1}{2}I_p\dot{\Theta}\Big\{(\tau\sin\Theta_0 + \theta_x)\left(\tau\dot{\Theta}\sin\Theta_0 + \dot{\theta}_y\right) \\
&\quad - \left(\tau\dot{\Theta}\sin\Theta_0 + \dot{\theta}_x\right)(-\tau\cos\Theta_0 + \theta_x)\Big\} \\
&\quad + \frac{1}{2}I\Big\{\left(\tau\dot{\Theta}\cos\Theta_0 + \dot{\theta}_x\right)^2 + \left(\tau\dot{\Theta}\sin\Theta_0 + \dot{\theta}_y\right)^2\Big\} \\
&\quad + \frac{1}{2}\Delta I\Big[\Big\{\left(\tau\dot{\Theta}\cos\Theta_0 + \dot{\theta}_x\right)^2 - \left(\tau\dot{\Theta}\sin\Theta_0 + \dot{\theta}_y\right)^2\Big\}\cos 2(\Theta_0 - \eta) \\
&\quad + 2\left(\tau\dot{\Theta}\cos\Theta_0 + \dot{\theta}_x\right)\left(\tau\dot{\Theta}\sin\Theta_0 + \dot{\theta}_y)\right)\sin 2(\Theta - \eta)\Big] \\
&= \frac{1}{2}I_p\Big[\left(1-\tau^2\right)\dot{\Theta}^2 + \dot{\Theta}\left(\dot{\theta}_x\theta_y - \theta_x\dot{\theta}_y\right) \\
&\quad - 2\tau\dot{\Theta}\left\{\dot{\theta}_x\cos(\Theta+\eta) + \dot{\theta}_y\sin(\Theta+\eta)\right\}\Big] \\
&\quad + \frac{1}{2}I\Big[\tau^2\dot{\Theta}^2 + \dot{\theta}_x^2 + \dot{\theta}_y^2 + 2\tau\dot{\Theta}\left\{\dot{\theta}_x\cos(\Theta+\eta)\right. \\
&\quad \left. + \dot{\theta}_y\sin(\Theta+\eta)\right\}\Big] \\
&\quad + \frac{1}{2}\Delta I\Big[\tau^2\dot{\Theta}^2\cos 2\eta + 2\tau\dot{\Theta}\left\{\dot{\theta}_x\cos(\Theta-\eta) + \dot{\theta}_y\sin(\Theta-\eta)\right\} \\
&\quad + \left(\dot{\theta}_x^2 - \dot{\theta}_y^2\right)\cos 2\Theta + 2\dot{\theta}_x\dot{\theta}_y\sin 2\Theta\Big]
\end{aligned} \tag{5.54}$$

The relationship between $\Theta_0 = \Theta + \eta$ and Θ_1 is given by Eq. (5.52). However, in the derivation of Eq. (5.54), we could use the relationship $\Theta_1 \approx \Theta_0 = \Theta + \eta$ of $O(\varepsilon^0)$ for the terms with $\theta_{x_1}\dot{\theta}_{y_1} - \dot{\theta}_{x_1}\theta_{y_1}, \dot{\theta}_{x_1}^2 - \dot{\theta}_{y_1}^2, 2\dot{\theta}_x\dot{\theta}_y$, and so on, which are of $O(\varepsilon^2)$.

Suppose that the shaft is rotating with a constant rotational speed ω. Let $t = 0$ when the Y_2-axis corresponding to the principal axis with I_1 passes through the zy-plane. As $\dot{\Theta}$ and Θ are represented by

$$\dot{\Theta} = \dot{\varphi} + \dot{\psi} = \omega, \qquad \Theta = \omega t - \frac{\pi}{2} \tag{5.55}$$

the kinetic energy becomes

$$\begin{aligned} T = & \frac{1}{2} I_\mathrm{p}\Big[\left(1-\tau^2\right)\omega^2 + \omega\left(\dot{\theta}_x\dot{\theta}_y - \theta_x\dot{\theta}_y\right) \\ & -2\omega\tau\left\{\dot{\theta}_x \sin(\omega t+\eta) - \dot{\theta}_y \cos(\omega t+\eta)\right\}\Big] \\ & +\frac{1}{2} I\Big[\tau^2\omega^2 + \dot{\theta}_x^2 + \dot{\theta}_y^2 \\ & +2\omega\tau\left\{\dot{\theta}_x \sin(\omega t+\eta) - \dot{\theta}_y \cos(\omega t+\eta)\right\}\Big] \\ & +\frac{1}{2}\Delta I\Big[2\omega\tau\left\{\dot{\theta}_x \sin(\omega t-\eta) - \dot{\theta}_y \cos(\omega t-\eta)\right\} \\ & -\left(\dot{\theta}_x^2 - \dot{\theta}_y^2\right)\cos 2\omega t - 2\dot{\theta}_x\dot{\theta}_y \sin 2\omega t + \tau^2\omega^2\cos 2\eta\Big] \end{aligned} \tag{5.56}$$

Next, we discuss the potential energy and the dissipation function. The potential energy V is represented by

$$V = \frac{1}{2}\delta\left(\theta_x^2 + \theta_y^2\right) \tag{5.57}$$

where δ is the spring constant. The dissipation function F is represented by

$$F = \frac{1}{2}c\left(\dot{\theta}_x^2 + \dot{\theta}_y^2\right) \tag{5.58}$$

where c is the damping coefficient. We substitute these results into Lagrange's equation,

$$\frac{d}{dt}\left(\frac{\partial T}{\partial \dot{q}_s}\right) - \frac{\partial T}{\partial q_s} + \frac{\partial V}{\partial q_s} + \frac{\partial F}{\partial \dot{q}_s} = Q_s \ (s = 1, 2) \tag{5.59}$$

where q_s are generalized coordinates and Q_s are non conservative generalized forces.

Substituting Eqs. (5.56–5.58) into Eq. (5.59), we arrive at

$$\left.\begin{aligned} & I\ddot{\theta}_x + I_\mathrm{p}\omega\dot{\theta}_y + c\dot{\theta}_x + \delta\theta_x - \Delta I\frac{\mathrm{d}}{\mathrm{d}t}\left(\dot{\theta}_x \cos 2\omega t + \dot{\theta}_y \sin 2\omega t\right) \\ & = \tau\omega^2\left\{\left(I_\mathrm{p} - I\right)\cos(\omega t+\eta) - \Delta I\cos(\omega t-\eta)\right\} \\ & I\ddot{\theta}_y - I_\mathrm{p}\omega\dot{\theta}_x + c\dot{\theta}_y + \delta\theta_y - \Delta I\frac{\mathrm{d}}{\mathrm{d}t}\left(\dot{\theta}_x \sin 2\omega t - \dot{\theta}_y \cos 2\omega t\right) \\ & = \tau\omega^2\left\{\left(I_\mathrm{p} - I\right)\sin(\omega t+\eta) - \Delta I\sin(\omega t-\eta)\right\} \end{aligned}\right\} \tag{5.60}$$

5.3.2
Free Vibrations and a Natural Frequency Diagram

Equation (5.60) can be solved in the same way as that of an asymmetrical shaft mentioned earlier. In this section, we obtain a natural angular frequency diagram corresponding to Figure 5.3. Considering only real natural angular frequencies p, we assume the solution corresponding to Eq. (5.8) in the form

$$\left.\begin{aligned} \theta_x &= A\cos(pt+\beta) + B\cos(\tilde{p}t-\beta) \\ \theta_y &= A\sin(pt+\beta) + B\sin(\tilde{p}t-\beta) \end{aligned}\right\} \tag{5.61}$$

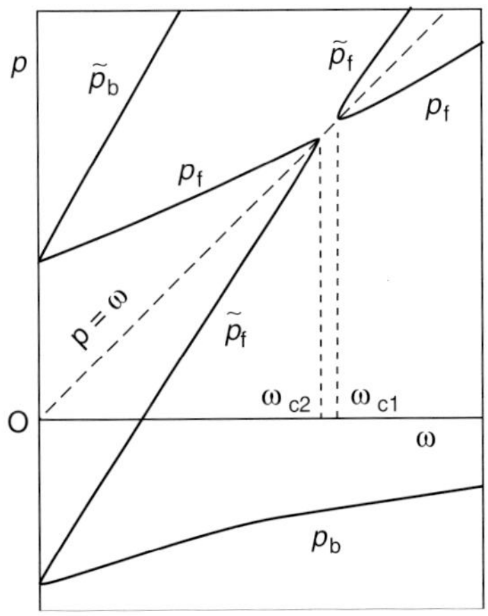

Figure 5.8 Frequency diagram of an asymmetrical rotor.

where $\tilde{p} = 2\omega - p$. Substituting this solution into Eq. (5.60) with $c = 0$ and $\tau = 0$ and comparing the coefficients, we have

$$\left.\begin{array}{l}\left(\delta + I_p\omega p - Ip^2\right) A - \Delta I p\tilde{p} B = 0 \\ -\Delta I p\tilde{p} A + \left(\delta + I_p\omega\tilde{p} - I\tilde{p}^2\right) B = 0\end{array}\right\} \tag{5.62}$$

Putting the characteristic determinant to be zero, we get the following frequency equation:

$$\left(\delta + I_p\omega p - Ip^2\right)\left(\delta + I_p\omega\tilde{p} - I\tilde{p}^2\right) - (\Delta I p\tilde{p})^2 = 0 \tag{5.63}$$

This is a quartic equation on p and has four roots at most. Figure 5.8 shows natural angular frequencies as a function of the rotational speed ω. The range $\omega_{c1} > \omega > \omega_{c2}$ is an unstable range where unstable vibration occurs. The boundary of this range is given by putting $p = \tilde{p} = \omega$ in this expression. They are given by

$$\omega_{c1} = \sqrt{\frac{\delta}{(I - \Delta I) - I_p}}, \qquad \omega_{c2} = \sqrt{\frac{\delta}{(I + \Delta I) - I_p}} \tag{5.64}$$

This unstable range exists when $I_2 = I - \Delta I > I_p$, which means that it exists in the case of a cylinder-like rotor

5.3.3
Synchronous Whirl in the Vicinity of the Major Critical Speed

We consider the case with no damping in Eq. (5.60). Since we can solve in the same manner as shown in Section 5.2.3, we show only the result of analysis.

Steady-state solution has the form

$$\left.\begin{array}{l}\theta_x = P_0\cos(\omega t + \beta) = P_{10}\cos\omega t - P_{20}\sin\omega t \\ \theta_y = P_0\sin(\omega t + \beta) = P_{10}\sin\omega t + P_{20}\cos\omega t\end{array}\right\} \tag{5.65}$$

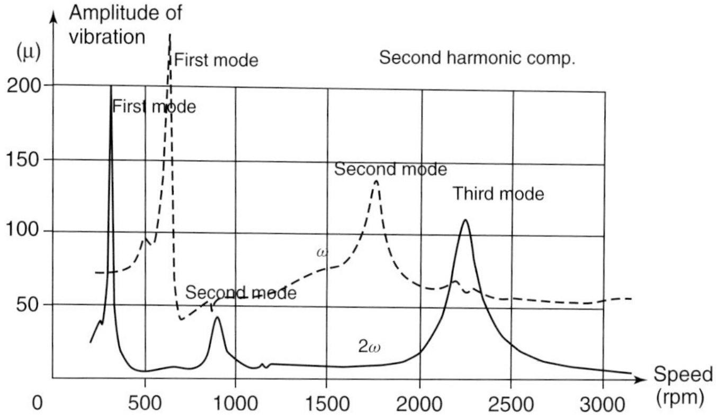

Figure 5.9 Response of a 600 MW generator (Courtesy of Prof. Bachschmid.)

where

$$\left.\begin{aligned} P_{10} &= -\frac{\tau\omega^2\left(i_\mathrm{p}-1-\Delta\right)\cos\eta}{(\delta/I)+\left(i_\mathrm{p}-1-\Delta\right)\omega^2} \\ P_{20} &= -\frac{\tau\omega^2\left(i_\mathrm{p}-1+\Delta\right)\sin\eta}{(\delta/I)+\left(i_\mathrm{p}-1+\Delta\right)\omega^2} \end{aligned}\right\} \tag{5.66}$$

The notations $\Delta I/I = \Delta$ and $i_\mathrm{p} = I_\mathrm{p}/I$ are used here. Resonance curves have the same shape as the case shown in Figure 5.5. The amplitude becomes infinite at the rotational speeds ω_{c1} and ω_{c2}, which make the denominators of Eq. (5.66) zero. The solution becomes unstable in the range $\omega_{c1} > \omega > \omega_{c2}$. The resonance curves depend on the magnitude and direction of the unbalance.

5.4 Double-Frequency Vibrations of an Asymmetrical Horizontal Shaft

Another important phenomenon observed in a horizontal asymmetrical shaft system is the resonance that occurs at half the major critical speed (Soderberg, 1932, Laffoon and Rose, 1940). As this vibration has the frequency of twice the rotational speed, it is sometimes called a *double-frequency vibration* and its resonance speed is called a *secondary critical speed*. Figure 5.9 shows a response of a 600 MW two-pole horizontal turbogenerator (Bachschmid and Diana, 1980). The broken line represents a synchronous (one-per-revolution) vibration. Resonances occurred in the vicinities of 628 rpm (the first mode) and 1740 rpm (the second mode). The solid line represents a twice-per-revolution vibration. Resonances occurred in the vicinities of 316 rpm (the first mode) and 895 rpm (the second mode), which are about half of the critical speeds shown by the broken lines. In addition, the resonance of twice-per-revolution vibration corresponding to the third mode appeared at 2226 rpm.

When an asymmetrical shaft is supported horizontally, the interplay between the asymmetry in stiffness and the gravity causes this resonance. For example,

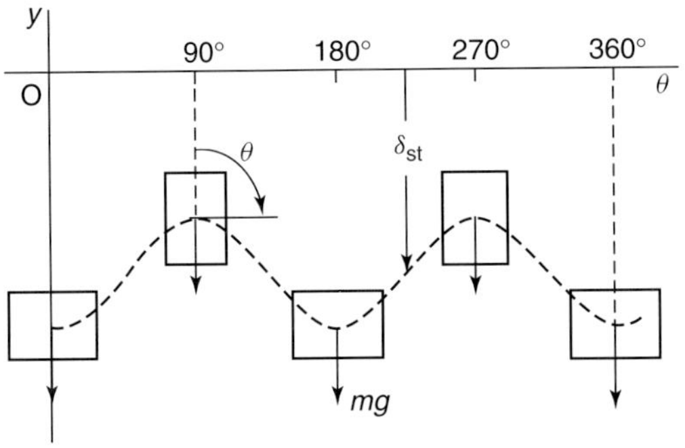

Figure 5.10 Variation of static deflection during rotation.

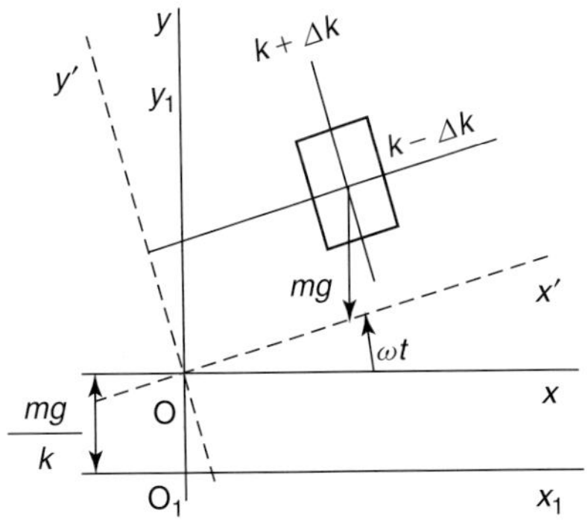

Figure 5.11 Horizontal asymmetrical shaft and coordinate systems.

suppose that an asymmetrical shaft rotates quasistatically as shown in Figure 5.10. The static deflection becomes large when the direction of weak stiffness coincides with the direction of gravity, and it becomes small when the direction of strong stiffness coincides with that of gravity. This up-and-down motion occurs twice during one rotation of the shaft and, therefore, may cause the double-frequency vibration.

Now, we analyze a simple case of a rotor mounted at the midspan of a shaft and with no unbalance and no damping. The coordinate systems are shown in Figure 5.11. The origin O is taken at the bearing centerline. The gravitational force mg works downward (the $-y$-direction). The nondimensionalized equations of motion are obtained by adding the gravitational force to Eq. (5.7):

$$\left.\begin{aligned} \ddot{x} + p_0^2 x - \Delta(x\cos 2\omega t + y\sin 2\omega t) &= 0 \\ \ddot{y} + p_0^2 y - \Delta(x\sin 2\omega t - y\cos 2\omega t) &= -g \end{aligned}\right\} \tag{5.67}$$

where g is the acceleration of gravity. The coordinate system O-x_1y_1 is parallel to O-xy and separated from it downward by $g/p_0^2 (= mg/k)$, which is the average value of the static deflection. Substituting the transformation

$$x = x_1, \quad y = y_1 - \frac{g}{p_0^2} \tag{5.68}$$

into Eq. (5.67), we arrive at the following equations of motion:

$$\left.\begin{aligned}\ddot{x}_1 + p_0^2 x_1 - \Delta(x_1 \cos 2\omega t + y_1 \sin 2\omega t) &= \Delta \frac{g}{p_0^2} \cos\left(2\omega t + \frac{\pi}{2}\right)\\ \ddot{y}_1 + p_0^2 y_1 - \Delta(x_1 \sin 2\omega t - y_1 \cos 2\omega t) &= \Delta \frac{g}{p_0^2} \sin\left(2\omega t + \frac{\pi}{2}\right)\end{aligned}\right\} \tag{5.69}$$

From this expression, we see that a force with magnitude $\Delta g/p_0^2$ and frequency 2ω of a forward whirling mode works substantially on the system. We can solve these equations of motion in the manner shown in Section 5.2.3. Considering the frequency 2ω of the external force and the derivations of vibration components through the third terms on the left-hand sides, we assume the solution as follows:

$$\left.\begin{aligned}x_1 &= R\cos(2\omega t + \theta) + A_x\\ y_1 &= R\sin(2\omega t + \theta) + A_y\end{aligned}\right\} \tag{5.70}$$

Substituting this into Eq. (5.69) and equating the coefficients concerning $\cos(2\omega t + \theta)$ and $\sin(2\omega t + \theta)$ in the right- and left-hand sides, respectively, we get algebraic equations on R, θ, A_x, and A_y. Substituting these values into Eq. (5.70) and

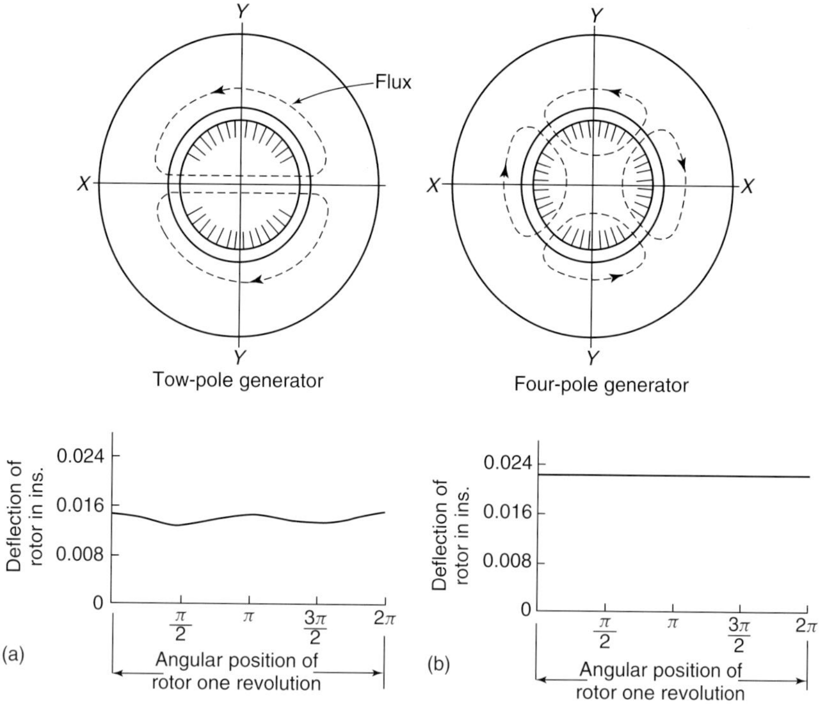

Figure 5.12 Vertical deflections of (a) two-pole and (b) four-pole generator rotors (Laffoon and Rose, (1940), Courtesy of ASME.)

transforming the coordinates into the original ones, x and y, we obtain

$$\left.\begin{aligned} x &= \frac{g\Delta}{\left(p_0^2 - 4\omega^2\right)p_0^2 - \Delta^2}\cos\left(2\omega t + \frac{\pi}{2}\right) \\ y &= \frac{g\Delta}{\left(p_0^2 - 4\omega^2\right)p_0^2 - \Delta^2}\sin\left(2\omega t + \frac{\pi}{2}\right) - \frac{g\left(p_0^2 - 4\omega^2\right)}{\left(p_0^2 - 4\omega^2\right)p_0^2 - \Delta^2} \end{aligned}\right\} \tag{5.71}$$

Since denominators of these expressions diminish at the rotational speed

$$\omega = \frac{1}{2}\sqrt{p_0^2 - \frac{\Delta^2}{p_0^2}} = \frac{1}{2}p_0\sqrt{1 - \left(\frac{\Delta k}{k}\right)^2} \tag{5.72}$$

we know that resonance of a forward whirling motion occurs with frequency 2ω. If the value $\Delta k/k$ representing the directional difference in stiffness is comparatively small, this resonance speed is about half of the major critical speed. This expression also teaches us that a backward whirling motion with frequency 2ω does not occur. It is clear that we cannot decrease this double-frequency vibration by balancing the rotor because, as mentioned earlier, the cause of this vibration is the interplay of asymmetry Δ and gravity. To diminish this vibration, we must decrease the difference in stiffness. For that purpose, the following two means are used (Bishop and Parkinson, 1965). The first is to cut dummy slots in the longitudinal direction at positions A and B in Figure 5.1. Although steel bars are inserted into these slots to protect the decrease in magnetic flux, we can still make the difference small. The second is to make slots at position C, which are at right angles to the centerline of the rotor, as shown in Figure 5.1.

Topic: Two-Pole and Four-Pole Turbine-Generator Rotor

Laffoon and Rose (1940) investigated the double-frequency vibration of a 3600 rpm generator rotor in detail. In his paper, he showed vertical deflection of a rotor, which is shown in Figure 5.12. The deflection changes twice during the rotor rotates 2π in a two-pole generator but does not in a four-pole generator rotor.

References

Bachschmid, N. and Diana, G. (1980) Reduction of twice per revolution vibration levels due to weight effect in large turbogenerators. *IMechE, Proceedings of the 2nd International Conference on Vibrations in Rotating Machinery*, 203–208.

Bishop, R.E.D. and Parkinson, A.G. (1965) Second order vibration of flexible shafts. *Philos. Trans. R. Soc. Lond., Ser. A*, **259** (1095), 1–31.

Dimentberg, F.M. (1961) *Flexural Vibrations of Rotating Shafts*, Butterworths, London.

Laffoon, C.M. and Rose, B.A. (1940) Special problems of two-pole turbine generators. *Trans. AIEE*, **59**, 30–56.

Soderberg, R.C. (1932) On the subcritical speeds of the rotating shaft. *Trans. ASME*, **54**, 45–52.

Tondl, A. (1965) *Some Problems of Rotor Dynamics*, Czechoslovak Academy of Sciences, Prague.

Yamamoto, T. and Ota, H. (1969) On the vibrations of a shaft carrying an unsymmetrical rotor. *Mem. Fac. Eng., Nagoya Univ.*, **21** (1), 8–12.

6 Nonlinear Vibrations

6.1 General Considerations

Until now, we studied linear systems where restoring force terms, damping terms, and inertia terms are represented by the first-order functions of deflection, velocity, and acceleration. However, such equations of motion are approximate expressions under the assumption that the deflections are small. When the deflection becomes large, phenomena due to nonlinearity may occur. In addition, some mechanical elements constituting a rotor system make various types of nonlinearity. For example, clearance in a ball bearing, oil film in a journal bearing, clearance in a squeeze-film damper bearing, and magnetic force between the rotor and the stator in a motor produce nonlinearity in restoring forces. It is also known that friction between a shaft and a mounted element creates nonlinearity in damping forces. In systems with these nonlinearities, interesting phenomena such as jump phenomena, subharmonic resonances, combination resonances, chaotic vibrations, and limit cycles appear.

The analyses of nonlinear phenomena are far more difficult compared to those of linear phenomena. In addition, since a rotor executes a whirling motion because of gyroscopic moment, analytical methods used in the analyses of rectilinear systems sometimes cannot be applied directly to rotor systems.

In this chapter, various types of nonlinear phenomena are introduced, and analytical methods applicable to these are explained. And in this chapter, we treat only the nonlinearities in restoring forces. Phenomena due to the nonlinearities in damping forces are explained in Chapter 7.[1]

6.2 Causes and Expressions of Nonlinear Spring Characteristics: Weak Nonlinearity

In rotating machineries, nonlinear spring characteristics appear because of the various causes mentioned above. When the restoring force is expressed as a

1) More detailed explanations for nonlinear phenomena introduced in this Chapter and Chapter 7 are given in the book by Guran *et al.* (1999).

Linear and Nonlinear Rotordynamics: A Modern Treatment with Applications, Second Edition.
Yukio Ishida and Toshio Yamamoto.

function of the deflection, the nonlinearity is often classified into two types. It is called *weak nonlinearity* when the deviation from the linear relationship called *Hooke's law* is small, and is called *strong nonlinearity* when it deviates appreciably from it. In the former case, since its characteristics can be approximated by a power series whose terms have small coefficients, the theoretical treatment is comparatively easy. However, the theoretical treatment of the latter system is difficult, in general. In this section, the representation of weak nonlinear spring characteristics is explained.

Figure 6.1 shows two cases of spring characteristics that are obtained when the lower end of an elastic shaft is supported by two different types of bearings. The upper end is supported by a double-row self-aligning ball bearing in both cases. Figure 6.1b is the case where the lower end is supported by a double-row self-aligning ball bearing. As the inner surface of the outer ring forms a part of a sphere, the inner ring can turn freely and the supporting condition is a simple support. In this case, the spring characteristics of the shaft become linear. Figure 6.1c is the case where the lower end is supported by a single-row deep-groove ball bearing (Yamamoto, Ishida, and Kawasumi, 1977). Because the balls roll in the grooves carved in both the inner and outer rings, the inner ring cannot incline

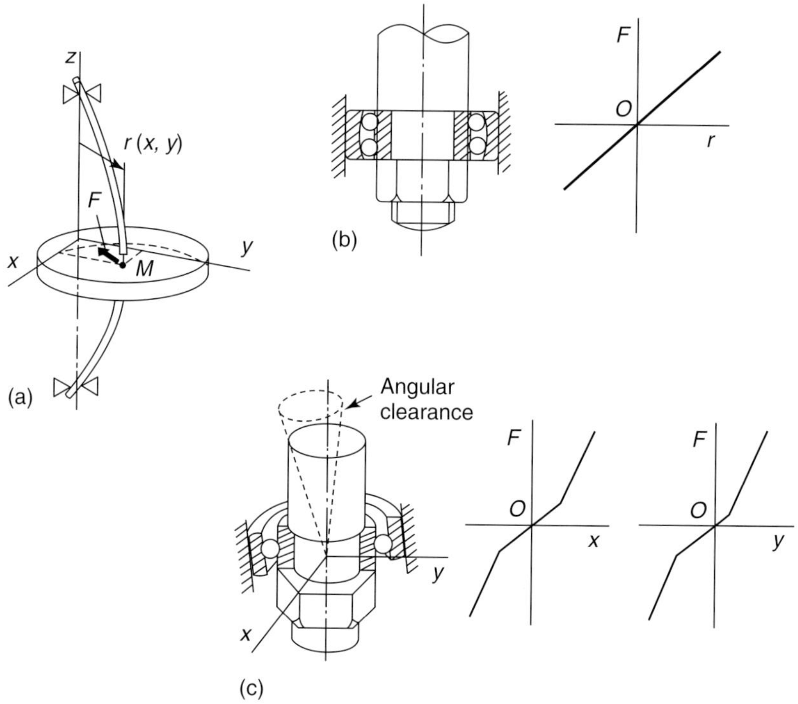

Figure 6.1 Clearance in bearings and spring characteristics: (a) rotor system, (b) double-row self-aligning ball bearing (linear), and (c) single-row deep-groove ball bearing (nonlinear).

relative to the outer ring and therefore the supporting condition is a fixed support. However, owing to small *clearances* among the inner ring, balls, and the outer ring, the inner ring can incline a little and the supporting condition becomes a simple support within this clearance. In bearing engineering, the term *radial clearance* and *axial clearance* are commonly used to express the accuracy of the bearings. But, as it is most convenient to express clearance by the angle in the treatment of restoring forces, the term *angular clearance* is used in this book. This angular clearance exists in every direction and can be expressed by a cone as shown in Figure 6.1c. The supporting condition is a simple support when the centerline of the shaft is located in this cone and becomes fixed when it is out of the cone. As a result, the restoring force has nonlinear characteristics of piecewise linear type, as shown in Figure 6.1c. In a practical setup, the transition from a simple support to a fixed support occurs gradually because clearances around each ball disappear one by one as the inclination of the shaft increases. Therefore, the practical transition is comparatively smooth and the spring characteristics can be approximated by a power series of low order. That is, the setup shown in Figure 6.1 generally has a weak nonlinearity.

Now, we consider a case where the lower end is supported by a single-row deep-groove ball bearing. If the bearing inclines a little from the *bearing centerline* connecting the centers of the upper and lower bearings because of misalignment in the assembly, the centerline in the state of static equilibrium shifts from the center of the cone, and the restoring forces differ depending on the direction. For example, suppose that the outer ring inclines in the yz-plane as shown in Figure 6.1b, then the spring characteristics in the x-direction is symmetrical, but that in the y-direction is asymmetrical. We know from such an example that the nonlinear spring characteristics of a rotor system must be considered two dimensionally.

To make it easy to treat nonlinear spring characteristics in theoretical analyses, we express these two-dimensional spring characteristics approximately by the power series of the shaft deflections x and y (Yamamoto and Ishida, 1977). Here, the terms higher than the third power are neglected in the restoring forces. Then, the corresponding potential energy is represented by terms up to the fourth order as follows:

$$\begin{aligned} V = {} & k_{20}x^2 + k_{11}xy + k_{02}y^2 + \varepsilon_{30}x^3 + \varepsilon_{21}x^2y + \varepsilon_{12}xy^2 + \varepsilon_{03}y^3 \\ & + \beta_{40}x^4 + \beta_{31}x^3y + \beta_{22}x^2y^2 + \beta_{13}xy^3 + \beta_{04}y^4 \end{aligned} \tag{6.1}$$

where k_{ij} are spring constants, and ε_{ij} and β_{ij} are the coefficients of asymmetrical and symmetrical nonlinear terms, respectively. This is the most general expression of the series up to the fourth power.

In this section, to proceed with the approximation analysis considering the magnitudes of various parameters, we use nondimensional expressions for equations of motion. Suppose that all the parameters are nondimensionalized by an appropriate mean. In addition, let k_{11} be zero by rotating the direction of the coordinates. Then,

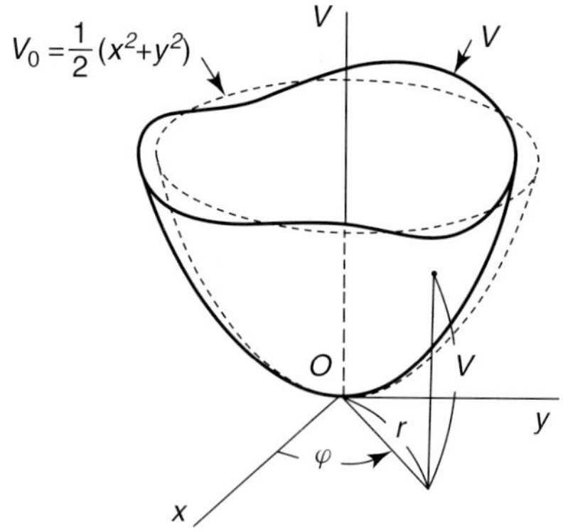

Figure 6.2 Two-dimensional distribution of a potential energy.

Eq. (6.1) becomes

$$\begin{aligned} V = &\frac{1}{2}(x^2+y^2) + \varepsilon_{30}x^3 + \varepsilon_{21}x^2y + \varepsilon_{12}xy^2 + \varepsilon_{03}y^3 \\ &+ \beta_{40}x^4 + \beta_{31}x^3y + \beta_{22}x^2y^2 + \beta_{13}xy^3 + \beta_{04}y^4 \end{aligned} \tag{6.2}$$

The two-dimensional distribution of this potential energy illustrated in Figure 6.2 deviates irregularly from the paraboloid $V = (x^2+y^2)/2(\equiv V_0)$. The restoring forces F_x and F_y are obtained by partially differentiating Eq. (6.2) by x and y, respectively, as follows:

$$F_x = -\frac{\partial V}{\partial x} = -(x+N_x), \quad F_y = -\frac{\partial V}{\partial y} = -(y+N_y) \tag{6.3}$$

where the nonlinear terms N_x and N_y are given by

$$\begin{aligned} N_x &= 3\varepsilon_{30}x^2 + 2\varepsilon_{21}xy + \varepsilon_{12}y^2 + 4\beta_{40}x^3 + 3\beta_{31}x^2y + 2\beta_{22}xy^2 + \beta_{13}y^3 \\ N_y &= \varepsilon_{21}x^2 + 2\varepsilon_{12}xy + 3\varepsilon_{03}y^2 + \beta_{31}x^3 + 2\beta_{22}x^2y + 3\beta_{13}xy^2 + 4\beta_{04}y^3 \end{aligned} \tag{6.4}$$

Suppose that the perfect assembly of the setup shown in Figure 6.1b is attained. Then, the bearing centerline is located at the center of the cone and the nonlinear spring characteristics are the same for every direction, that is, isotropic. Therefore, when the rotor whirls with a constant radius, the restoring force is constant and the character of nonlinearity does not apparently appear in this motion. In other words, the character of nonlinearity does not appear in the circular motion but does only in the radial motion. From such a consideration, we find that the polar coordinate expression is more proper than the rectangular coordinate expression. Therefore, using the transformation

$$x = r\cos\varphi, \quad y = r\sin\varphi \tag{6.5}$$

we express Eq. (6.2) by the polar coordinates (r, φ) as follows:

$$\begin{aligned}
V &= \frac{1}{2}r^2 + (\varepsilon_c^{(1)} \cos\varphi + \varepsilon_s^{(1)} \sin\varphi + \varepsilon_c^{(3)} \cos 3\varphi + \varepsilon_s^{(3)} \sin 3\varphi)r^3 \\
&\quad + (\beta^{(0)} + \beta_c^{(2)} \cos 2\varphi + \beta_s^{(2)} \sin 2\varphi + \beta_c^{(4)} \cos 4\varphi + \beta_s^{(4)} \sin 4\varphi)r^4 \\
&= \frac{1}{2}r^2 + \left\{\varepsilon^{(1)} \cos(\varphi - \varphi_1) + \varepsilon^{(3)} \cos 3(\varphi - \varphi_3)\right\} r^3 \\
&\quad + \left\{\beta^{(0)} + \beta^{(2)} \cos 2(\varphi - \varphi_2) + \beta^{(4)} \cos 4(\varphi - \varphi_4)\right\} r^4
\end{aligned} \tag{6.6}$$

The following relationships hold between the coefficients of Eq. (6.2) and those of Eq. (6.6):

$$\begin{aligned}
&\varepsilon_c^{(1)} = \frac{1}{4}(3\varepsilon_{30} + \varepsilon_{12}), \quad \varepsilon_s^{(1)} = \frac{1}{4}(\varepsilon_{21} + 3\varepsilon_{03}), \quad \varepsilon_c^{(3)} = \frac{1}{4}(\varepsilon_{30} - \varepsilon_{12}), \\
&\varepsilon_s^{(3)} = \frac{1}{4}(\varepsilon_{21} - \varepsilon_{03}), \quad \beta^{(0)} = \frac{1}{8}(3\beta_{40} + \beta_{22} + 3\beta_{04}), \\
&\beta_c^{(2)} = \frac{1}{2}(\beta_{40} - \beta_{04}), \quad \beta_s^{(2)} = \frac{1}{4}(\beta_{31} + \beta_{13}), \\
&\beta_c^{(4)} = \frac{1}{8}(\beta_{40} + \beta_{22} + \beta_{04}), \quad \beta_s^{(4)} = \frac{1}{8}(\beta_{31} - \beta_{13}), \\
&\varepsilon^{(n)} = \sqrt{\varepsilon_c^{(n)2} + \varepsilon_s^{(n)2}}, \quad \varphi_n = \tan^{-1}\left(\frac{\varepsilon_s^{(n)}}{\varepsilon_c^{(n)}}\right) \quad (n = 1, 3) \\
&\beta^{(n)} = \sqrt{\beta_c^{(n)2} + \beta_s^{(n)2}}, \quad \varphi_n = \tan^{-1}\left(\frac{\beta_s^{(n)}}{\beta_c^{(n)}}\right) \quad (n = 2, 4)
\end{aligned} \tag{6.7}$$

From Eq. (6.6), we find that the potential energy can bc decomposed into the components having a regular shape of distribution. For example, Figure 6.3 shows the potential energy distribution and the variation of its magnitude at the position $r =$ constant, in the case of $V = (1/2)r^2 + (\varepsilon_c^{(1)} \cos\varphi + \varepsilon_s^{(1)} \sin\varphi)r^3$, where only the first three terms of Eq. (6.6) exist. In this case, the magnitude of the potential energy changes once, while the direction angle φ changes from 0 to 2π. Figure 6.4 shows equipotential lines, that is, the cross sections of the curved surface $V(r, \varphi)$ cut by the horizontal plane $V =$ constant. Each figure shows the case where only one of $\varepsilon^{(1)}, \varepsilon^{(3)}, \beta^{(0)}, \beta^{(2)}$, and $\beta^{(4)}$ exists. We see that the notations $\varepsilon^{(n)}$ and $\beta^{(n)}$ are

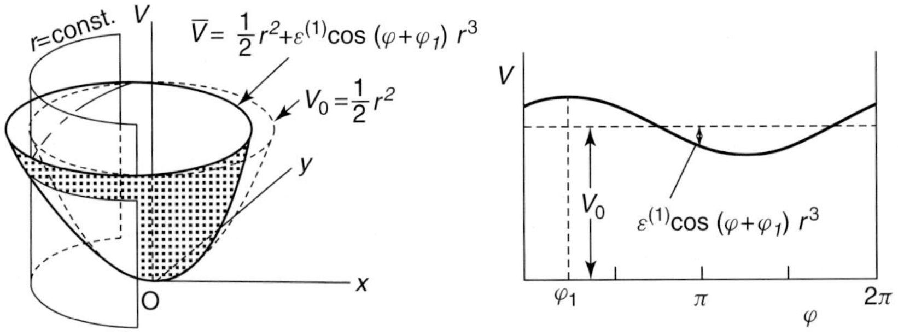

Figure 6.3 Potential energy of a system with only the component $\varepsilon^{(1)}$.

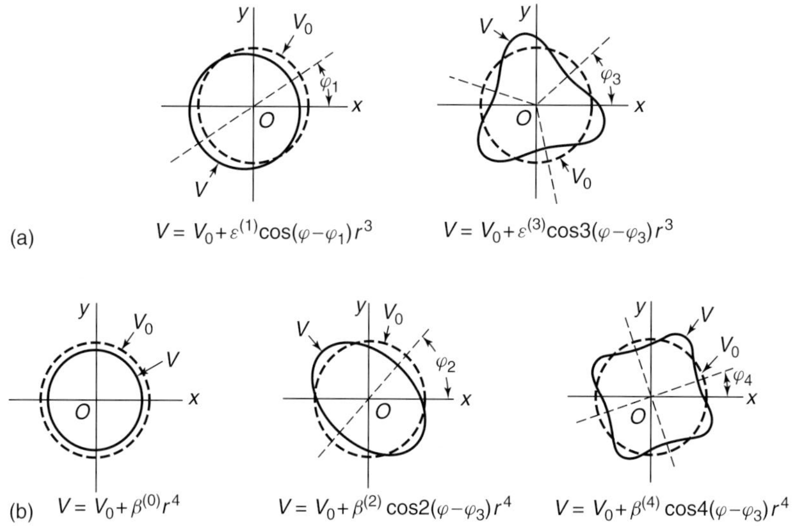

Figure 6.4 Equipotential line: (a) cases with only one asymmetrical nonlinear component and (b) cases with only one symmetrical nonlinear component.

the coefficients of nonlinear terms that represent a nonlinear component whose potential energy changes n times, while the direction angle φ changes from 0 to 2π with constant r. This nonlinear component is represented by the notation$N(n)$.

Let us assume that the rotor system is assembled with no misalignment. Then the spring characteristics are isotropic and the potential energy V corresponding to Eq. (6.2) is represented by $V = (1/2)r^2 + \beta^{(0)}r^4 + \gamma^{(0)}r^6 + \cdots$, that is, a power series with only even power terms of deflection r. The magnitudes of these coefficients generally have the relationship $\beta^{(0)} \gg \gamma^{(0)} \gg \cdots$ When the equilibrium position shifts by a in the x-direction owing to a misalignment, the spring characteristics become asymmetric and the corresponding potential energy is obtained by replacing x in $r^2 = x^2 + y^2$ by $x + a$. Since terms of third power or less are induced from the term $\beta^{(0)}r^4$, terms of fifth power or less do from the term $\gamma^{(0)}r^6$, and so on, and other kinds of component appear. For example, $\beta^{(0)}r^4$ becomes $\beta^{(0)}\left\{(x+a)^2 + y^2\right\}^2 =$ (terms of second power or less) $+4\beta^{(0)}a(x^3 + xy^2) + \beta^{(0)}(x^2 + y^2)^2$. Therefore, the coefficients in Eq. (6.2) are given by $\varepsilon_{30} = \varepsilon_{12} = 4\beta^{(0)}a$ and $\varepsilon_{21} = \varepsilon_{03} = 0$, and then we get $\varepsilon_c^{(1)} = 4\beta^{(0)}a$, $\varepsilon_s^{(1)} = \varepsilon_c^{(3)} = \varepsilon_s^{(3)} = 0$ from Eq. (6.7). Namely, the component $N(1)$ appears from $\beta^{(0)}r^4$. In a similar way, we know that the components $N(1)$, $N(2)$, and $N(3)$ appear from $\gamma^{(0)}r^6$.

In practical machineries, components derived from $\beta^{(0)}$ are more likely to appear than those derived from $\gamma^{(0)}$, and components $N(0)$ and $N(1)$ with simple shapes of distribution are more likely to appear than the components $N(3)$ and $N(4)$ with comparatively complex shapes of distribution. Namely, nonlinear resonances due to $N(0)$ may appear with appreciable intensity when the assembly is good, and those due to $N(1)$ may appear when there is a misalignment. Later, we explain that the characteristics of nonlinear resonances in rotor systems can be clarified easily if we pay attention to these nonlinear components expressed by polar coordinates.

6.3 Expressions of Equations of Motion Using Physical and Normal Coordinates

Equation (2.48) represents the most simple rotor system with gyroscopic moment. In this section, we investigate nonlinear resonance using this 2 DOF (degrees of freedom) equation of motion with the addition of nonlinear terms. The results concerning a 4 DOF rotor system where a disk is mounted at an arbitrary position are summarized in Appendix E.

Introducing a representative angle τ_0 that has a magnitude of the same order as the amplitude of vibration, we define the following dimensionless quantities:

$$\left.\begin{array}{l} \theta_x^* = \dfrac{\theta_x}{\tau_0}, \quad \theta_y^* = \dfrac{\theta_y}{\tau_0}, \quad \tau^* = \dfrac{\tau}{\tau_0}, \quad i_p = \dfrac{I_p}{I}, \quad c^* = \dfrac{c}{\sqrt{\delta\, I}} \\ t^* = t\sqrt{\delta/I}, \quad \omega^* = \dfrac{\omega}{\sqrt{\delta/I}}, \quad \varepsilon_{ij}^* = \dfrac{\varepsilon_{ij}\tau_0}{\delta}, \quad \beta_{ij}^* = \dfrac{\beta_{ij}\tau_0^2}{\delta} \end{array}\right\} \tag{6.8}$$

Using these parameters, equations of motion are expressed as follows:

$$\left.\begin{array}{l} \ddot{\theta}_x + i_p\omega\dot{\theta}_y + c\dot{\theta}_x + \theta_x + N_{\theta x} = F\cos\omega t \\ \ddot{\theta}_y - i_p\omega\dot{\theta}_x + c\dot{\theta}_y + \theta_y + N_{\theta y} = F\sin\omega t \end{array}\right\} \tag{6.9}$$

where $F = (1 - i_p)\tau\omega^2$ and nonlinear terms $N_{\theta x}$ and $N_{\theta y}$ are obtained by replacing x and y in N_x and N_y of Eq. (6.4) by θ_x and θ_y. The asterisks representing dimensionless parameters are eliminated in these equations and in subsequent sections.

The frequency equation derived from Eq. (6.9) is

$$G(p) = 1 + i_p\omega p - p^2 = 0 \tag{6.10}$$

The two roots $p_f(>0)$ and $p_b(<0)$ of this equation have the following relationships:

$$p_f + p_b = i_p\omega, \quad p_f p_b = -1 \tag{6.11}$$

There are two representative methods to analyze nonlinear vibrations in rotor systems. In one method the solution is obtained directly from Eq. (6.9) written in the physical coordinates (θ_x, θ_y), and in the other method the solution is obtained after the equations of motion are transformed into those written in the normal coordinates (X_1, X_2); both methods are explained in this chapter.

First, we derive the equations of motion in a normal coordinate expression from Eq. (6.9). In a linear system, by selecting the coordinates properly, we can transform a set of coupled equations of motion into a set of uncoupled independent equations of motion. Each equation in these uncoupled equations governs the motion of a different mode. Using such coordinates, called *normal coordinates*, we can analyze phenomena by solving the minimum number of necessary equations, even in a system with many DOF. Here, we recall the following result of analytical dynamics: In general multi-DOF rectilinear system, the transformation matrix $[a_{ij}]$ leading from equations of motion in terms of physical coordinates$\{x_i\}$ to equations of motion in terms of normal coordinates $\{X_i\}$ in the form $\{x_i\} = [a_{ij}]\{X_i\}$ can be obtained by composing it of modal vectors. That is, the column elements of the transformation matrix $[a_{ij}]$ represent the ratios of the components of modal vectors (Meirovitch, 1975). However, in the case of rotor systems with a gyroscopic moment, we cannot use such a conventional method. Instead, the transformation matrix can be obtained in the following way (Yamamoto, 1956).

The rotor system given by Eq. (6.9) has a forward whirling mode with the natural frequency p_{f} and a backward whirling mode with the natural frequency p_{b}. Suppose that the rotor whirls with a forward whirling modep_{f}, then the orbit is represented by $\theta_x = R\cos p_{\mathrm{f}}t$ and $\theta_y = R\sin p_{\mathrm{f}}t$. If we can select the coordinate X_1 that satisfies the relationships $\theta_x = X_1$ and $\theta_y = -\dot{X}_1/p_{\mathrm{f}}$, this coordinate is a normal coordinate, which, by itself, can describe the forward whirling mode. We can select a similar coordinate X_2 for a backward whirling mode. From such a consideration, we know that the transformation from the physical coordinates (θ_x, θ_y) to the normal coordinate (X_1, X_2) is given by

$$\theta_x = X_1 + X_2, \quad \theta_y = -\frac{\dot{X}_1}{p_{\mathrm{f}}} - \frac{\dot{X}_2}{p_{\mathrm{b}}} \tag{6.12}$$

In the equations of motion (Eq. (6.9)), the damping terms and the nonlinear terms have magnitudes of the same order as that of the small parameter ε. This order is expressed by the symbol $O(\varepsilon)$. Transferring these small terms to the right-hand side of Eq. (6.9) and substituting the transformation (Eq. (6.12)) onto the left-hand side, we obtain the following expression after some rearrangement:

$$\left.\begin{aligned} \frac{1}{p_{\mathrm{f}}^2}(\ddot{X}_1 + p_{\mathrm{f}}^2 X_1) + \frac{1}{p_{\mathrm{b}}^2}(\ddot{X}_2 + p_{\mathrm{b}}^2 X_2) &= -c\dot{\theta}_x - N_{\theta x} + F\cos\omega t \\ \frac{1}{p_{\mathrm{f}}}(\dddot{X}_1 + p_{\mathrm{f}}^2 \dot{X}_1) + \frac{1}{p_{\mathrm{b}}}(\dddot{X}_2 + p_{\mathrm{b}}^2 \dot{X}_2) &= c\dot{\theta}_y + N_{\theta y} - F\sin\omega t \end{aligned}\right\} \tag{6.13}$$

After integrating the second equation, we solve Eq. (6.13) for $(\ddot{X}_1 + p_{\mathrm{f}}^2 X_1)$ and $(\ddot{X}_2 + p_{\mathrm{b}}^2 X_2)$, rearrange the obtained expressions considering Eq. (6.11), and finally arrive at the following equations of motion expressed in the normal coordinates:

$$\left.\begin{aligned} \ddot{X}_1 + p_{\mathrm{f}}^2 X_1 &= \frac{p_{\mathrm{f}}^2}{1 + p_{\mathrm{f}}^2}\left\{\left(\frac{p_{\mathrm{f}}}{\omega} + 1\right) F\cos\omega t + c(-\dot{\theta}_x + p_{\mathrm{f}}\theta_y) - N_{\theta x} + p_{\mathrm{f}}\int N_{\theta y}\mathrm{d}t\right\} \\ \ddot{X}_2 + p_{\mathrm{b}}^2 X_2 &= \frac{p_{\mathrm{b}}^2}{1 + p_{\mathrm{b}}^2}\left\{\left(\frac{p_{\mathrm{b}}}{\omega} + 1\right) F\cos\omega t + c(-\dot{\theta}_x + p_{\mathrm{b}}\theta_y) - N_{\theta x} + p_{\mathrm{b}}\int N_{\theta y}\mathrm{d}t\right\} \end{aligned}\right\} \tag{6.14}$$

Although these two expressions are coupled through small damping terms and nonlinear terms, they are independent of the accuracy of $O(\varepsilon^0)$.

We can solve most of the phenomena by using both Eq. (6.9) in the physical coordinate expression and Eq. (6.14) in the normal coordinate expression. However, some types of nonlinear phenomena can be solved only by one of them. For example, in an asymmetrical shaft system and an asymmetrical rotor system, the equations in the θ_x- and θ_y-directions couple with each other through an asymmetry as seen from Eqs. (5.5) and (5.60). Therefore, when the asymmetry in shaft stiffness or in moment of inertia of the rotor is not small, we cannot transform the equations of motion into two independent equations to an accuracy of $O(\varepsilon^0)$ using the procedure mentioned above. On the contrary, when we investigate nonstationary oscillations by the asymptotic method proposed by Mitropol'skii, we must use Eq. (6.14) in the normal coordinate expression (Evan-Iwanowski, 1976).

6.4
Various Types of Nonlinear Resonances

When a rotor has nonlinear spring characteristics, the shape of the resonance curves of a harmonic resonance at the major critical speed changes, and, in addition, various types of nonlinear resonances such as subharmonic resonance, combination resonance, superharmonic resonance, and combination tone may occur when natural frequencies and rotational speed have specific relationships. These resonance rotational speeds are also called *critical speeds*.

Figure 6.5 shows the relationships between the frequencies of the external forces and the frequency components of the vibration concerning the following four cases: (i) In the vicinity of the rotational speed where the rotational speed ω becomes an integral multiple of a natural frequency p_i, that is, when $\omega \approx \pm mp_i$ holds, a subharmonic resonance with the frequency $\omega_i = \pm\omega/m(\approx p_i)$ occurs. (ii) In the opposite situation, that is, when $m\omega(\approx p_i)$ holds, a superharmonic resonance with the frequency $\omega_i = \pm m\omega(\approx p_i)$ occurs. (iii) In a multi-DOF system with natural frequencies $p_i, p_j, \ldots$, when the relationship $\omega = \pm mp_i \pm np_j \pm \cdots$ holds, a combination resonance with frequency components $\omega_i(\approx p_i), \omega_j(\approx p_j), \ldots$, occurs. (iv) In a system with multiple external forces with frequencies $\omega, \Omega_i, \ldots$, a combination tone with the frequency $\omega(= \pm m\omega \pm n\Omega_i \pm \cdots \approx p_i)$ occurs. Furthermore super-subharmonic resonance, supercombination resonance, and subcombination tone

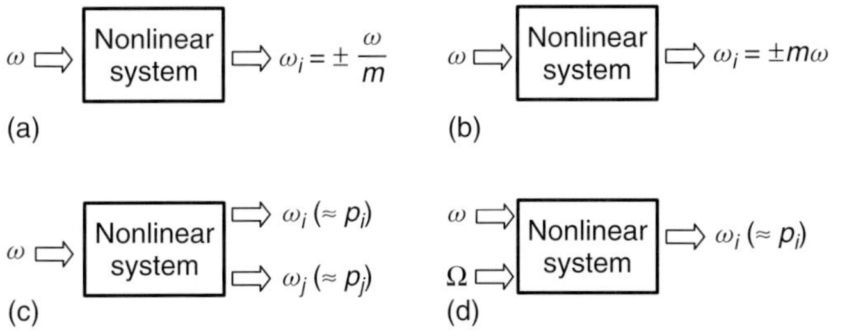

Figure 6.5 Types of nonlinear resonances and relationships between frequencies: (a) subharmonic resonance; (b) superharmonic resonance; (c) combination resonance; and (d) combination tones.

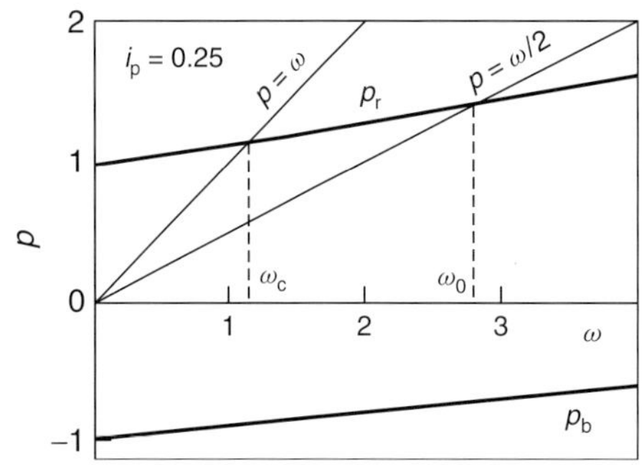

Figure 6.6 Critical speeds of harmonic resonance [p_f] and subharmonic resonance [$2p_f$].

that have dual characteristics may also occur. In this section, analytical procedures and characteristics of harmonic, subharmonic, and combination resonances are explained.

6.4.1 Harmonic Resonance

In the vicinity of the intersection of the curve p_f and the straight line $p = \omega$ in Figure 6.6, where the rotational speed ω becomes almost equal to the natural frequency p_f, a forward whirling motion with the frequency ω (called a *synchronous whirl*) occurs intensely. This critical speed $\omega_c = 1/\sqrt{1 - i_p}$ is obtained by putting $p = \omega$ in the frequency equation (Eq. (6.10)). To distinguish resonances, we use the notation [*], which means that its critical speed is given by $* = \omega$. For example, the harmonic resonance, which occurs at $p_f = \omega$, is denoted by the notation [p_f].

A harmonic resonance also occurs in linear systems. However, in the case of nonlinear systems, the shape of resonance curves changes and jump phenomena may occur. In this section, we derive resonance curves of this harmonic resonance under the assumption that the external force F due to unbalance is small.

6.4.1.1 Solution by the Harmonic Balance Method

We solve Eq. (6.9) written in physical coordinates by the harmonic balance method. In this method, under the assumption that a periodic solution exists, we represent the solution in Fourier series. We keep the terms that are considered to appear with appreciable magnitude as the first approximation, and substitute them into the equations of motion. Comparing the coefficients of the terms with the frequency ω on the right- and left-hand sides, we obtain coupled equations. In ordinary rectilinear systems, these equations give the amplitudes and phases. However, in rotor systems, since the number of such equations obtained from the equations of motion in the θ_x- and θ_y-directions is larger than the number of unknown quantities, if we apply the method in a way similar to the method mentioned above no solution is obtained. For example, let us assume the solution in the form of Eq. (6.15) and consider that this solution is accurate in $O(\varepsilon)$. If we substitute it into

Eq. (6.9) and compare the coefficients of $\cos(\omega t + \beta)$ and $\sin(\omega t + \beta)$, we obtain a total of four equations. Since there are two unknown quantities P and β, no solution satisfies the four equations simultaneously. Therefore, we revise the method in the following way, which takes into account the character of rotor systems.

First, we can assume a solution of $O(\varepsilon^0)$ in the vicinity of the major critical speed ω_c in the following form:

$$\left.\begin{aligned} \theta_x &= P\cos(\omega t + \beta) \\ \theta_y &= P\sin(\omega t + \beta) \end{aligned}\right\} \tag{6.15}$$

where the magnitudes of P and β are of $O(\varepsilon^0)$. Substituting this into the equation of motion and equating the coefficients, we obtain the same expressions from the equations of motion in the θ_x- and θ_y-directions to an accuracy of $O(\varepsilon)$. Since F is assumed to have a magnitude of $O(\varepsilon)$, it does not remain in these expressions. In this case, we get only a trivial solution $P = 0$, which gives no resonance. Thus, we must assume a solution to the accuracy of $O(\varepsilon)$ in which small deviations in the orbit are considered.

When the solution in the form of Eq. (6.15), representing a circular motion with the angular velocity ω, is substituted into Eq. (6.9), small components with various frequencies are derived through nonlinear terms. Namely, the component of $O(\varepsilon^0)$ with the frequency $+\omega$ produces components of $O(\varepsilon)$ with constant value and with $\pm\omega, \pm 2\omega, \pm 3\omega, \ldots$, where + represents a forward whirl and – represents a backward one. To analyze to an accuracy of $O(\varepsilon)$, we must include these higher order components in the assumed solution from the beginning. As we pay attention only to the frequency ω, we assume the solution, including only the terms with the frequencies $\pm\omega$, as follows:

$$\left.\begin{aligned} \theta_x &= P\cos(\omega t + \beta) + \varepsilon\left\{a_1\cos(\omega t + \delta_1) + b_1\cos(-\omega t + \delta_2)\right\} \\ \theta_y &= P\sin(\omega t + \beta) + \varepsilon\left\{a_1\sin(\omega t + \delta_1) + b_1\sin(-\omega t + \delta_2)\right\} \end{aligned}\right\} \tag{6.16}$$

The second and third terms represent the derived components of forward and backward whirling modes. To show explicitly that a term has the magnitude of $O(\varepsilon)$, the small parameter ε is used in the expressions. Since the expression in Eq. (6.16) is inconvenient due to the existence of many types of phases with the same frequency, we use the following expression, which is equivalent to Eq. (6.16):

$$\left.\begin{aligned} \theta_x &= P\cos(\omega t + \beta) + \varepsilon\left\{a\cos(\omega t + \beta) + b\sin(\omega t + \beta)\right\} \\ \theta_y &= P\sin(\omega t + \beta) + \varepsilon\left\{a'\sin(\omega t + \beta) + b'\cos(\omega t + \beta)\right\} \end{aligned}\right\} \tag{6.17}$$

Four unknown equations $\varepsilon a_1, \varepsilon b_1, \delta_1$, and δ_2 in Eq. (6.16) are replaced equivalently with four unknown quantities $\varepsilon a, \varepsilon b, \varepsilon a'$, and $\varepsilon b'$ in Eq. (6.17). We substitute Eq. (6.17) into Eq. (6.9) and equate the coefficients of $\cos(\omega t + \beta)$ and $\sin(\omega t + \beta)$ on the right- and left-hand sides to an accuracy of $O(\varepsilon)$. In this derivation, the amplitudes P, a, b, a', and b' and the phase angle β are assumed to be functions of slowly varying time, and this assumption is represented concretely by setting $P = O(\varepsilon^0)$, $\dot{P} = O(\varepsilon)$, and $\ddot{P} = O(\varepsilon^2)$ in the analytical process. From the first and second equations of Eq. (6.9), we get

$$\left.\begin{aligned}
&(i_p-2)\omega\dot{P}-c\omega P+(1-\omega^2)\varepsilon b-i_p\omega^2\varepsilon b'+N_{\theta xs}=F\sin\beta && \text{(i)}\\
&(i_p-2)\omega P\dot{\beta}+G(\omega)P+(1-\omega^2)\varepsilon a+i_p\omega^2\varepsilon a'+N_{\theta xc}=F\cos\beta && \text{(ii)}\\
&(i_p-2)\omega P\dot{\beta}+G(\omega)P+i_p\omega^2\varepsilon a+(1-\omega^2)\varepsilon a'+N_{\theta ys}=F\cos\beta && \text{(iii)}\\
&-(i_p-2)\omega\dot{P}+c\omega P-i_p\omega^2\varepsilon b+(1-\omega^2)\varepsilon b'+N_{\theta yc}=-F\sin\beta && \text{(iv)}
\end{aligned}\right\} \tag{6.18}$$

where the notations $N_{\theta xc}$, $N_{\theta xs}$, $N_{\theta yc}$, and $N_{\theta ys}$ are the coefficients of $N_{\theta x}=N_{\theta xc}\cos(\omega t+\beta)+N_{\theta xs}\sin(\omega t+\beta)+\cdots$ and $N_{\theta y}=N_{\theta yc}\cos(\omega t+\beta)+N_{\theta ys}\sin(\omega t+\beta)+\cdots$, which are obtained when the first terms of Eq. (6.17) are substituted into $N_{\theta x}$ and $N_{\theta y}$ in Eq. (6.9). In such an analysis, we noted that the coefficients of nonlinear terms are small quantities of $O(\varepsilon)$. Finally, we obtain

$$\left.\begin{aligned}
&N_{\theta xc}=\left(3\beta_{40}+\frac{\beta_{22}}{2}\right)P^3,\quad N_{\theta xs}=\frac{3}{4}(\beta_{31}+\beta_{13})P^3\\
&N_{\theta yc}=\frac{3}{4}(\beta_{31}+\beta_{13})P^3,\quad N_{\theta ys}=\left(\frac{\beta_{22}}{2}+3\beta_{04}\right)P^3
\end{aligned}\right\} \tag{6.19}$$

The magnitude of $G(\omega)$ given by Eq. (6.10) is $O(\varepsilon)$ in the vicinity of the major critical speed ω_c. Although the four equations in Eq. (6.18) have six unknown quantities $P,\beta,\varepsilon a,\varepsilon b,\varepsilon a'$, and $\varepsilon b'$, the combinations (ii) + (iii) and (i) − (iv) create the following two equations, including only amplitude P and phase angle β if we neglect the terms $G(\omega)\varepsilon(a+a')$ and $G(\omega)\varepsilon(b+b')$ of $O(\varepsilon^2)$:

$$\left.\begin{aligned}
&2(i_p-2)\omega P\dot{\beta}+2G(\omega)P+(N_{\theta xc}+N_{\theta ys})=2F\cos\beta\\
&2(i_p-2)\omega\dot{P}-2c\omega P+(N_{\theta xs}-N_{\theta yc})=2F\sin\beta
\end{aligned}\right\} \tag{6.20}$$

Substituting Eq. (6.19) into Eq. (6.20) and transferring the coefficients to those in the polar coordinate expressions using Eq. (6.7), we obtain

$$\left.\begin{aligned}
&(i_p-2)\omega P\dot{\beta}+G(\omega)P+4\beta^{(0)}P^3=F\cos\beta\\
&(i_p-2)\omega\dot{P}-c\omega P=F\sin\beta
\end{aligned}\right\} \tag{6.21}$$

In this expression, only $\beta^{(0)}$among the coefficients in Eq. (6.7) is included. This means that to an accuracy of $O(\varepsilon)$, only the isotropic component $N(0)$ among components $N(0),\ldots,N(4)$ influences this harmonic resonance.

Setting $\dot{P}=0$ and $\dot{\beta}=0$ in Eq. (6.21) and rearranging the equations, we obtain the following expressions, which give the steady-state solutions $P=P_0$ and $\beta=\beta_0$:

$$\left.\begin{aligned}
&\left[\left\{G(\omega)+4\beta^{(0)}P_0^2\right\}^2+(c\omega)^2\right]P_0^2=F^2\\
&\beta_0=-\tan^{-1}\left(\frac{c\omega}{G(\omega)+4\beta^{(0)}P^2}\right)
\end{aligned}\right\} \tag{6.22}$$

Backbone curves representing the inclination of the resonance curves are obtained by setting $c=0$ and $F=0$ in Eq. (6.22). This expression for a backbone curve also gives the relationship between the amplitude and the natural frequency in a free whirling motion.

From a practical viewpoint, not only the existence of the solution P_0 but also the stability of the solution is an important factor. To investigate the stability, we consider small variations $\xi(t)$ and $\eta(t)$ of $O(\varepsilon)$ as follows:

$$P = P_0 + \xi, \quad \beta = \beta_0 + \eta \tag{6.23}$$

If $\xi(t)$ and $\eta(t)$ decrease with time, the solution is stable; otherwise, the solution is unstable. Substituting Eq. (6.23) into Eq. (6.21) and linearizing the variations $\xi(t)$ and $\eta(t)$ by neglecting the small terms of $O(\varepsilon^3)$ and less, we have

$$\left.\begin{aligned} \dot{\eta} &= a^*\xi + b^*\eta \\ \dot{\xi} &= c^*\xi + d^*\eta \end{aligned}\right\} \tag{6.24}$$

where

$$\left.\begin{aligned} a^* &= \frac{G + 12\beta^{(0)}P_0^2}{(2 - i_p)\omega P_0}, \quad b^* = \frac{F\sin\beta_0}{(2 - i_p)\omega P_0} = \frac{-c}{2 - i_p} \\ c^* &= \frac{-c}{2 - i_p}, \quad d^* = \frac{-F\cos\beta_0}{(2 - i_p)\omega} = \frac{-(G + 4\beta^{(0)}P_0^2)P_0}{(2 - i_p)\omega} \end{aligned}\right\} \tag{6.25}$$

Assuming the solutions of Eq. (6.24) to be $\xi(t) = Ae^{st}$ and $\eta(t) = Be^{st}$ and substituting them into Eq. (6.24), we obtain algebraic equations on A and B. The equations have a nontrivial solution if the determinant composed of the coefficients is zero. This condition gives the following characteristic equation:

$$s^2 - (b^* + c^*)s - (a^*d^* - b^*c^*) = 0 \tag{6.26}$$

From the Routh–Hurwitz criteria, we know that the real parts of the roots s are positive when the conditions

$$\left.\begin{aligned} &-(b^* + c^*) > 0 \qquad &\text{(i)} \\ &-(a^*d^* - b^*c^*) > 0 \qquad &\text{(ii)} \end{aligned}\right\} \tag{6.27}$$

hold. As condition (i) always holds when $i_p < 1$ (the case that the major critical speed exists), we have the following stability condition from condition (ii):

$$\left\{G(\omega) + 12\beta^{(0)}P_0^2\right\}\left\{G(\omega) + 4\beta^{(0)}P_0^2\right\} + c^2\omega^2 > 0 \tag{6.28}$$

From such an analysis, we obtain the resonance curve shown in Figure 6.7. In this figure, resonance curves are illustrated for various values of $\beta^{(0)}$ relating to isotropic component $N(0)$. The resonance curve is a hard spring type when $\beta^{(0)} > 0$ and a soft spring type when $\beta^{(0)} < 0$. In these cases, when the rotor is accelerated or decelerated slowly, the amplitude jumps from the resonance curve of large amplitude to that of small amplitude. In the case of $\beta^{(0)} = 0$, the shape of the resonance curve is the same as that of a linear system even if other nonlinear components exist. The solid and dashed lines represent stable and unstable solutions, respectively. Points with a vertical tangent are the boundaries of the stable and unstable solutions. This can be proved by showing that the expression obtained by differentiating Eq. (6.22) by P_0 and putting $\partial\omega/\partial P_0 = 0$ coincides with the stability condition of Eq. (6.28) (Yamamoto and Hayashi, 1963).

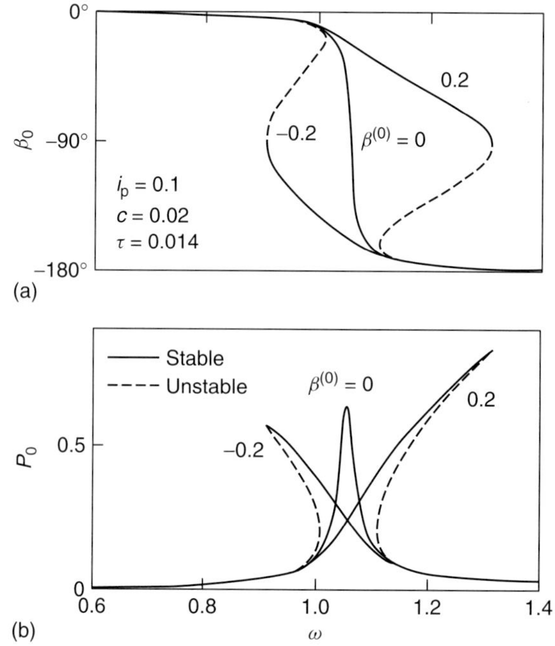

Figure 6.7 Resonance curves at the major critical speed: (a) phase angle and (b) amplitude.

6.4.1.2 Solution Using Normal Coordinates

Here, we obtain the expression for resonance curves using normal coordinates. If the equation of motion is expressed in the form of Eq. (6.14), various quantitative analytical methods developed in the field of nonlinear dynamics can be used. For example, we can analyze using Duffing's iteration method as follows (Stoker, 1976).

Since the mode X_1 relating to the natural frequency p_f resonates at the major critical speed ω_c, we use the first equation of Eq. (6.14). As the frequency of the oscillation is the same as the excitation frequency ω, we rearrange it in the following form:

$$\ddot{X}_1 + \omega^2 X_1 = (\omega^2 - p_f^2)X_1 + \frac{p_f^2}{1+p_f^2}\left\{\left(\frac{p_f}{\omega}+1\right) F\cos\omega t + c(-\dot{\theta}_x + p_f\theta_y) - N_{\theta x} + p_f\int N_{\theta y}\mathrm{d}t\right\} \tag{6.29}$$

In the vicinity of ω_c, the magnitude of the detuning $\sigma = \omega^2 - p_f^2$ is $O(\varepsilon)$. In addition, the magnitudes of the force F, the damping coefficient c, and nonlinear terms are assumed to be $O(\varepsilon)$.

First, we adopt the following solution of $\ddot{X}_{10} + \omega^2 X_{10} = 0$ as the first approximation:

$$X_{10} = P\cos(\omega t + \beta) \tag{6.30}$$

The corresponding approximate solution is $X_{20} = 0$.

Next, we substitute these solutions into the right-hand side of Eq. (6.29) and look for the second approximation X_{11} from $\ddot{X}_{11} + \omega^2 X_{11} = f(X_{10}, X_{20})$. As the magnitudes of all the terms on the right-hand side of Eq. (6.29) are $O(\varepsilon)$, solutions X_{10} and X_{20} are adequate substitutes. Therefore, we can substitute the following approximate solution into Eq. (6.29):

$$\left.\begin{aligned} &X_{10} = P\cos(\omega t + \beta), \quad X_{20} = 0, \\ &\theta_x = X_{10} + X_{20} = X_{10} = P\cos(\omega t + \beta) \\ &\theta_y = -\frac{\dot{X}_{10}}{p_f} - \frac{\dot{X}_{20}}{p_b} = -\frac{\dot{X}_{10}}{p_f} \approx P\sin(\omega t + \beta) \end{aligned}\right\} \tag{6.31}$$

To an accuracy of $O(\varepsilon^0)$, the approximations $p_f \approx \omega_c = 1/\sqrt{1 - i_p}$, $p_f^2/(1 + p_f^2) \approx 1/(2 - i_p)$, and $p_f/\omega \approx 1$ hold. Considering these approximations, we obtain

$$\begin{aligned} \ddot{X}_{11} + \omega^2 X_{11} &= (\omega^2 - p_f^2)P\cos(\omega t + \beta) + \frac{1}{2 - i_p}\left\{2F\cos\omega t\right. \\ &\quad \left. + c\frac{2p}{\sqrt{1 - i_p}}\sin(\omega t + \beta) - P^3(3\beta_{04} + \beta_{22} + 3\beta_{40})\cos(\omega t + \beta)\right\} \\ &\quad + [\text{terms except } \omega] \\ &= \left\{(\omega^2 - p_f^2)P + \frac{2F}{2 - i_p}\cos\beta - \frac{8\beta^{(0)}P^3}{2 - i_p}\right\}\cos(\omega t + \beta) \\ &\quad + \frac{2}{2 - i_p}\left(F\sin\beta + c\frac{P}{\sqrt{1 - i_p}}\right)\sin(\omega t + \beta) \\ &\quad + [\text{terms except } \omega] \end{aligned} \tag{6.32}$$

If terms $\cos(\omega t + \beta)$ and $\sin(\omega t + \beta)$ exist on the right-hand side, *secular terms* in the form $t\cos(\omega t + \beta)$, which increase with time, will appeal in the solution X_{11}. Therefore, for the existence of periodic solutions, the coefficients of these secular terms must be zero:

$$\left.\begin{aligned} &-\left(1 - \frac{i_p}{2}\right)(\omega^2 - p_f^2)P + 4\beta^{(0)}P^3 = F\cos\beta \\ &-\frac{cP}{\sqrt{1 - i_p}} = F\sin\beta \end{aligned}\right\} \tag{6.33}$$

Eliminating the phase angle β gives the following expression for resonance curves:

$$\left[\left\{-\left(1 - \frac{i_p}{2}\right)(\omega^2 - p_f^2) + 4\beta^{(0)}P^2\right\}^2 + \frac{c^2}{1 - i_p}\right]P^2 = F^2 \tag{6.34}$$

This expression is substantially the same as Eq. (6.22) obtained previously. In fact, as p_f is a root of the characteristic equation (i.e., the relationship $1 + i_p\omega p_f - p_f^2 = 0$

holds), the following holds to an accuracy of $O(\varepsilon)$:

$$\left.\begin{aligned} G(\omega) &= 1 + i_p\omega^2 - \omega^2 = (p_f^2 - i_p\omega p_f) - i_p\omega^2 - \omega^2 \\ &= (\omega - p_f)(i_p\omega - p_f - \omega) \approx -(\omega - p_f)2\omega(1 - i_p/2) \\ &\approx -(1 - i_p/2)(\omega^2 - p_f^2) \\ \omega \approx \omega_c &= 1\Big/\sqrt{1 - i_p} \end{aligned}\right\} \tag{6.35}$$

Then, we know that Eq. (6.34) coincides with Eq. (6.22).

6.4.2
Subharmonic Resonance of Order 1/2 of a Forward Whirling Mode

If the system has asymmetrical nonlinear characteristics expressed by second power terms of coordinates, a subharmonic resonance of order 1/2 occurs when the relationship $2p_f = \omega$ or $-2p_b = \omega$ holds. In this section we discuss the subharmonic resonance of order 1/2 of a forward whirling mode, which occurs in the vicinity of the rotational speed ω_0 where the relationship

$$2p_f = \omega \tag{6.36}$$

holds between the natural frequency p_f and the rotational speed ω (Figure 5.6). Following the notation mentioned previously, this resonance is represented by the symbol $[2p_f]$.

Figure 6.8 shows time histories and an orbit obtained by numerical simulation. The dashed lines show components of subharmonic vibrations. As the time history of the dashed line in the x-direction is advanced more than that in the y-direction by a quarter period, we know that this vibration is a forward whirling motion.

Unlike the case of the harmonic resonance, the magnitude of force F is assumed to be $O(\varepsilon^0)$ in the analysis of this resonance. Although this assumption is adopted for the sake of analytical convenience, a comparatively large unbalance is necessary for the occurrence of a subharmonic resonance, in practice. In the vicinity of this critical speed ω_0, a subharmonic component with the frequency $(1/2)\omega$ occurs in

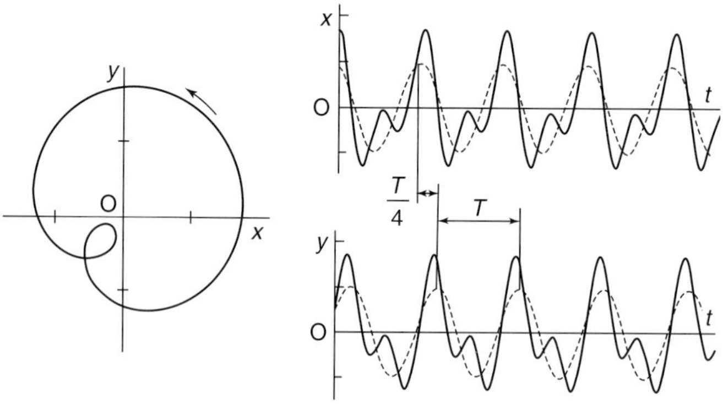

Figure 6.8 Time histories and an orbit of a subharmonic resonance $[2p_f]$.

addition to the harmonic component that always exists. Therefore, we assume the solution to an accuracy of $O(\varepsilon)$ as follows:

$$\left.\begin{aligned}\theta_x &= R\cos\theta_f + P\cos(\omega t+\beta) + \varepsilon(a\cos\theta_f + b\sin\theta_f)\\ \theta_y &= R\sin\theta_f + P\sin(\omega t+\beta) + \varepsilon(a'\sin\theta_f + b'\cos\theta_f)\end{aligned}\right\} \tag{6.37}$$

where $\theta_f = (1/2)\omega t + \delta$. Substituting Eq. (6.37) into Eq. (6.9), we compare the coefficients of $\cos\theta_f$ and $\sin\theta_f$. From a procedure similar to that executed in the process from Eqs. (6.18)–(6.21), we finally obtain the following expressions:

$$\left.\begin{aligned}(1-i_p)\omega\dot{R} &= -c\left(\frac{\omega}{2}\right)R + 2\left\{\varepsilon_c^{(1)}\sin(2\delta-\beta) - \varepsilon_s^{(1)}\cos(2\delta-\beta)\right\}RP\\ (1-i_p)\omega R\dot{\delta} &= G\left(\frac{\omega}{2}\right)R + 2\left\{\varepsilon_c^{(1)}\cos(2\delta-\beta) + \varepsilon_s^{(1)}\sin(2\delta-\beta)\right\}RP\\ &\quad + 4\beta^{(0)}(R^2+2P^2)R\end{aligned}\right\} \tag{6.38}$$

Since this resonance speed is well separated from the major critical speed, we can use the following solution obtained to an accuracy of $O(\varepsilon^0)$ for the harmonic component in Eq. (6.38):

$$P = -\frac{F}{G(\omega)}, \quad \beta = -\pi \tag{6.39}$$

Steady-state solutions $R = R_0$ and $\delta = \delta_0$ are obtained by putting $\dot{R} = 0$ and $\dot{\delta} = 0$ in Eq. (6.38). By eliminating the phase angle δ_0 from these equations, we get

$$\left.\begin{aligned}&R_0 = 0\\ &\left\{G\left(\frac{\omega}{2}\right) + 4\beta^{(0)}(R_0^2+2P^2)\right\}^2 + \left(\frac{c\omega}{2}\right)^2 = 4\varepsilon^{(1)2}P^2\end{aligned}\right\} \tag{6.40}$$

where $\varepsilon^{(1)2} = \varepsilon_c^{(1)2} + \varepsilon_s^{(1)2}$. Thus, both a trivial solution and a solution with a finite amplitude coexist in this subharmonic resonance of order 1/2 of a forward whirling mode.

Next, we investigate the stability of the steady-state solutions R_0 and δ_0. The stability of the solution with a finite amplitude can be investigated in a manner similar to the case of the major critical speed mentioned before. As a result, we get the following stability condition:

$$\beta^{(0)}\left\{G\left(\frac{\omega}{2}\right) + 4\beta^{(0)}(R_0^2+2P^2)\right\} > 0 \tag{6.41}$$

On the contrary, in the case of a trivial solution, we cannot use Eq. (6.38) because the phase angle δ_0 is not defined. Instead, we transform the variables into u and v by the transformations

$$u = R\cos\delta, \quad v = R\sin\delta \tag{6.42}$$

and express Eq. (6.38) by these new variables. The new variables u and v are adopted because the values $(u, v) = (0, 0)$ are determined even if $(R, \delta) = (0, \delta_0)$ are not determined. From a similar analysis, we get the following stability condition for $(u_0, v_0) = (0, 0)$:

$$\left\{G\left(\frac{\omega}{2}\right) + 8\beta^{(0)}P^2\right\}^2 + \frac{1}{4}c^2\omega^2 - 4\varepsilon^{(1)2}P^2 > 0 \tag{6.43}$$

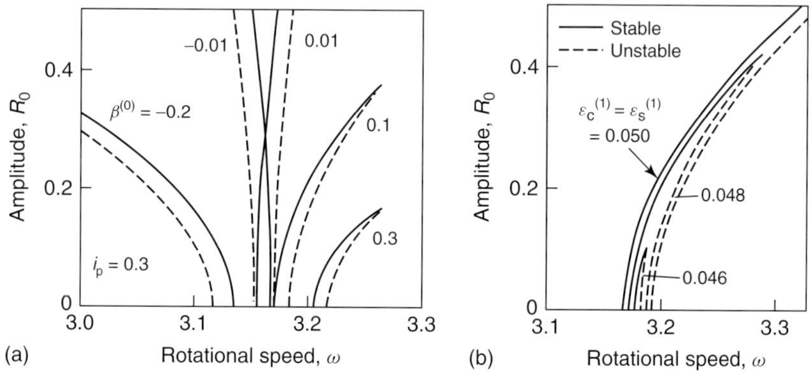

Figure 6.9 Influence of nonlinear components on subharmonic resonance [$2p_f$]: (a) influence of $N(0)$ and (b) influence of $N(1)$.

Since the second equation in Eq. (6.40) coincides with the left-hand side of this equation when $R_0 \to 0$, we know that the intersection points of the curve $R_0 \neq 0$ and the ω-axis are the boundaries of the stable and unstable branches of the curve $R_0 = 0$. Since only the coefficients $\beta^{(0)}$ and $\varepsilon^{(1)}$ are contained in Eq. (6.40), we know that only the nonlinear components $N(0)$ and $N(1)$ have influence on this subharmonic resonance of order 1/2 of a forward whirling mode. Figure 6.9 shows influences of these nonlinear components on the resonance curves. Figure 6.9a shows that $N(0)$ has influence on the inclination of the resonance curves, and Figure 6.9b shows that $N(1)$ has influence on the magnitude of the subharmonic resonance.

6.4.3 Subharmonic Resonance of Order 1/3 of a Forward Whirling Mode

Nonlinear spring characteristics are classified into asymmetrical nonlinearity and symmetrical nonlinearity. Subharmonic resonance of order 1/2 occurs because of the former. Two of the representative resonances due to the latter are subharmonic resonances of order 1/3. For example, a subharmonic resonance of order 1/3 of a forward whirling mode and that of a backward whirling mode may occur. In this section, we show the result of an analysis of subharmonic resonance of order 1/3 of a forward whirling mode [$3p_f$] that occurs in the vicinity of the rotational speed ω_0 where the relationship

$$3p_f = \omega \tag{6.44}$$

holds (Yamamoto and Ishida, 1977).

If we put the phase as $\theta_f = (1/3)\omega t + \delta$, the solution has the same form as that of Eq. (6.37). The expression for the resonance curve corresponding to Eq. (6.40) is given by

$$\left.\begin{aligned} &R_0 = 0 \\ &\left\{G\left(\frac{\omega}{3}\right) + 4\beta^{(0)}(R_0^2 + 2P^2)\right\}^2 + \left(\frac{c\omega}{3}\right)^2 = 9\beta^{(2)2}P^2R_0^2 \end{aligned}\right\} \tag{6.45}$$

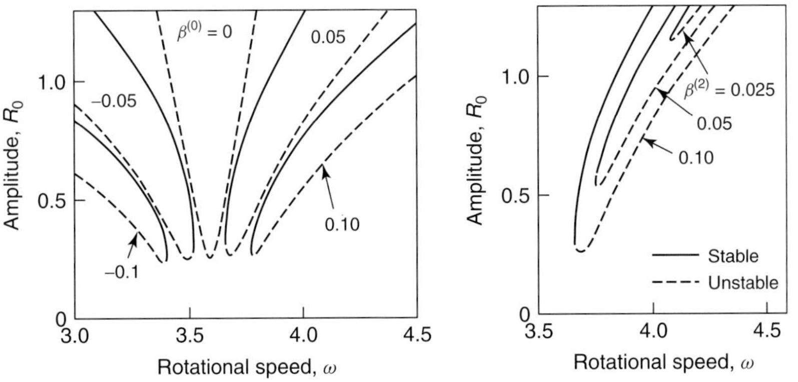

Figure 6.10 Influence of nonlinear components on subharmonic resonance [$2p_f$]: (a) influence of $N(0)$ and (b) influence of $N(2)$.

where $\beta^{(2)2} = \beta_c^{(2)2} + \beta_s^{(2)2}$. Similar to the case of the subharmonic resonance of order 1/2, this has a trivial solution and a solution with finite amplitude. Nonlinear components $N(0)$ and $N(2)$ have an influence on this resonance.

The stability of the steady-state solutions R_0 and δ_0 are as follows. The trivial solution is stable for all the rotational speeds. The solution with finite amplitude is stable when the following condition is satisfied.

$$\beta^{(0)}\left\{G\left(\frac{\omega}{3}\right) - 4\beta^{(0)}(R_0^2 - 2P^2)\right\}\left\{G\left(\frac{\omega}{3}\right) + 4\beta^{(0)}(R_0^2 + 2P^2)\right\} + \left(\frac{c\omega}{3}\right)^2 < 0 \quad (6.46)$$

The influence of nonlinear components on resonance curves is shown in Figure 6.10. Similar to the case of subharmonic resonance [$2p_f$] nonlinear component $N(0)$ has an influence on the inclination of resonance curves. However, nonlinear component $N(2)$ has an influence on the magnitude of resonance. This resonance does not appear if $N(2)$ does not exist.

Unlike the case of subharmonic resonance [$2p_f$], here the curve with finite amplitude is separated from the curve for a trivial solution. Therefore, the rotor can be operated in the vicinity of this resonance speed without the occurrence of subharmonic resonance if a disturbance large enough to jump to the resonance curve of a nontrivial solution does not work.

6.4.4 Combination Resonance

In a multi-DOF, in the vicinity of the rotational speed where the relationship $\pm lp_i \pm mp_i \pm \cdots = n\omega$ (l, m, and n are integers) holds among natural frequencies $p_i, p_j, \ldots$ and the rotational speed ω, several frequency components may appear simultaneously in addition to the harmonic component (Yamamoto, 1957b, 1957c). This resonance is called a *combination resonance* or a *summed-and-differential harmonic resonance*.

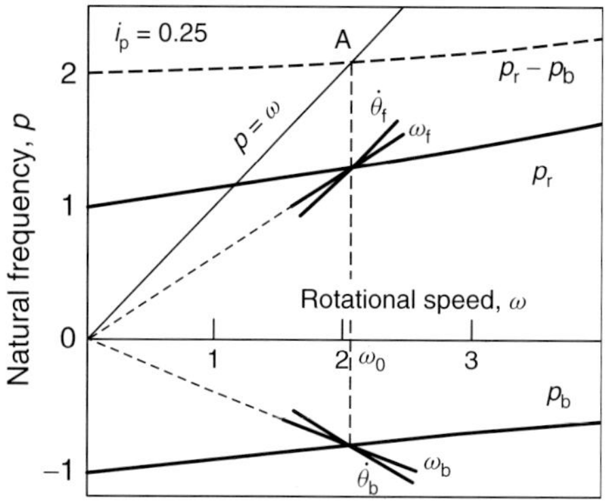

Figure 6.11 Critical speed of combination resonance $[p_f - p_b]$.

For example, when a rotor has asymmetrical nonlinearity represented by quadratic terms of coordinates, components whose frequencies are almost equal to p_f and p_b occur in the vicinity of the rotational speed ω_0 as shown in Figure 6.11, where the relationship $p_f - p_b = \omega$ holds. The solution to an accuracy of $O(\varepsilon^0)$ is represented as follows:

$$\left.\begin{aligned}\theta_x &= R_f \cos\theta_f + R_b \cos\theta_b + P\cos(\omega t + \beta)\\ \theta_y &= R_f \sin\theta_f + R_b \sin\theta_b + P\sin(\omega t + \beta)\end{aligned}\right\} \tag{6.47}$$

In this combination resonance $[p_f - p_b]$, the relationship

$$\dot{\theta}_f + \dot{\theta}_b = \omega \tag{6.48}$$

holds among the frequencies of the components. Experimental results show that the frequencies $\dot{\theta}_f$ and $\dot{\theta}_b$ are almost proportional to the rotational speed ω. Therefore, we can adopt the following expressions as the frequencies in the accuracy of $O(\varepsilon^0)$:

$$\dot{\theta}_f = \frac{p_{f0}}{\omega_0}\omega(\equiv\omega_f), \quad \dot{\theta}_b = \frac{p_{b0}}{\omega_0}\omega(\equiv\omega_b) \tag{6.49}$$

where p_{f0} and p_{b0} are the values of p_f and p_b at $\omega = \omega_0$, respectively. Then the frequencies in the accuracy of $O(\varepsilon^0)$ are represented by

$$\dot{\theta}_f = \omega_f + \dot{\delta}_f, \quad \dot{\theta}_b = \omega_b + \dot{\delta}_b \tag{6.50}$$

where $\dot{\delta}_f$ and $\dot{\delta}_b$ have a magnitude of $O(\varepsilon)$. Equation (6.48) gives

$$\dot{\delta}_f - \dot{\delta}_b = 0 \tag{6.51}$$

Therefore, we know that each of δ_f and δ_b is not constant, but the difference $\psi \equiv \delta_f - \delta_b$ is constant in the steady-state vibration of combination resonance.

Similar to Eq. (6.37), we assume a solution to an accuracy of $O(\varepsilon)$ and substitute it into the equations of motion. Applying the harmonic balance method, we obtain the trivial solution $R_{f0} = R_{b0} = 0$ and the following expressions for the solutions with finite amplitude:

$$\left.\begin{aligned}
\dot{R}_{\mathrm{f}} &= \frac{1}{2\omega_{\mathrm{f}} - i_{\mathrm{p}}\omega}(-c\omega_{\mathrm{f}}R_{\mathrm{f}} - 2\varepsilon^{(1)}R_{\mathrm{b}}P\sin\psi) \\
\dot{R}_{\mathrm{b}} &= \frac{1}{2\omega_{\mathrm{b}} - i_{\mathrm{p}}\omega}(-c\omega_{\mathrm{b}}R_{\mathrm{b}} + 2\varepsilon^{(1)}R_{\mathrm{f}}P\sin\psi) \\
\dot{\psi} &= \frac{1}{2\omega_{\mathrm{f}} - i_{\mathrm{p}}\omega}\left\{G_{\mathrm{f}} + 4\beta^{(0)}(R_{\mathrm{f}}^2 + 2R_{\mathrm{b}}^2 + 2P^2) - 2\varepsilon^{(1)}\left(\frac{R_{\mathrm{b}}}{R_{\mathrm{f}}}\right)P\cos\psi\right\} \\
&\quad - \frac{1}{2\omega_{\mathrm{b}} - i_{\mathrm{p}}\omega}\left\{G_{\mathrm{b}} + 4\beta^{(0)}(2R_{\mathrm{f}}^2 + R_{\mathrm{b}}^2 + 2P^2) - 2\varepsilon^{(1)}\left(\frac{R_{\mathrm{f}}}{R_{\mathrm{b}}}\right)P\cos\psi\right\}
\end{aligned}\right\} \tag{6.52}$$

For the convenience of representation, we rotate the coordinate axis so as to make angle φ_1 relating to $N(1)$ equal to zero (refer to Figure 6.4). Then $\varepsilon_c^{(1)}$ becomes $\varepsilon^{(1)}$ and $\varepsilon_s^{(1)}$ becomes 0. From this expression we obtain the steady-state solution $R_{\mathrm{f}} = R_{\mathrm{f}0}$, $R_{\mathrm{b}} = R_{\mathrm{b}0}$, and $\psi = \psi_0$. The stability of this solution is determined using the same procedure as is done in the case of a harmonic resonance mentioned previously. As this expression contains the coefficients $\beta^{(0)}$ and $\varepsilon^{(1)}$, we know that only the nonlinear components $N(0)$ and $N(1)$ have influence on this combination resonance $[p_{\mathrm{f}} - p_{\mathrm{b}}]$. The component $N(0)$ determines the inclination of resonance curves, and the component $N(1)$ determines the magnitude of the resonance. The resonance curves for R_{f} and R_{b} obtained from Eq. (6.52) have the same shape as those in Figure 6.9.

From the first and second equations in Eq. (6.52), we obtain the following expression for the steady-state solution:

$$\frac{R_{\mathrm{f}0}^2}{R_{\mathrm{b}0}^2} = -\frac{\omega_{\mathrm{b}}}{\omega_{\mathrm{f}}} \tag{6.53}$$

This expression is valid because, as seen from Eq. (6.49), $\omega_{\mathrm{f}} > 0$ and $\omega_{\mathrm{b}} < 0$ hold. However, from some types of combination resonance, the expression corresponding to Eq. (6.53) does not hold. For example, in the case of the combination resonance $[p_{\mathrm{f}} + p_{\mathrm{b}}]$, which may appear in the vicinity of the rotational speed where $p_{\mathrm{f}} + p_{\mathrm{b}} = \omega$ holds, the following expression is obtained in the analysis:

$$\frac{R_{\mathrm{f}0}^2}{R_{\mathrm{b}0}^2} = \frac{\omega_{\mathrm{b}}}{\omega_{\mathrm{f}}} \tag{6.54}$$

This relationship is clearly invalid. Therefore, we know that this combination resonance $[p_{\mathrm{f}} + p_{\mathrm{b}}]$ does not occur. From such an analysis, we can prove that "the combination resonance of the additive type of the absolute values occurs but that of the differential type does not occur" (Yamamoto, 1961). For example, the resonances $[p_{\mathrm{f}} - p_{\mathrm{b}}]$ and $[2p_{\mathrm{f}} - p_{\mathrm{b}}]$ occur because $p_{\mathrm{f}} - p_{\mathrm{b}} = |p_{\mathrm{f}}| + |p_{\mathrm{b}}| = \omega$ and $2p_{\mathrm{f}} - p_{\mathrm{b}} = 2|p_{\mathrm{f}}| + |p_{\mathrm{b}}| = \omega$ hold. However, the resonances $[p_{\mathrm{f}} + p_{\mathrm{b}}]$ and $[2p_{\mathrm{f}} + p_{\mathrm{b}}]$ do not occur because $p_{\mathrm{f}} + p_{\mathrm{b}} = |p_{\mathrm{f}}| - |p_{\mathrm{b}}| = \omega$ and $2p_{\mathrm{f}} + p_{\mathrm{b}} = 2|p_{\mathrm{f}}| - |p_{\mathrm{b}}| = \omega$ hold.

6.4.5
Summary of Nonlinear Resonances

Analysis and characteristics have been explained for several nonlinear resonances. The results of various types of subharmonic and combination resonances that are expected to occur in a 2 DOF rotor system with the quadratic and cubic nonlinearities are summarized in Table 6.1 (Yamamoto and Ishida, 1975, 1977). We know the following characteristics from this summary:

1) The isotropic nonlinear component $N(0)$ and one of the components from $N(0)$ to $N(4)$ have influence on each nonlinear resonance.
2) The isotropic nonlinear component $N(0)$ determines the inclination of the resonance curves. The resonance curves become a hard spring type when the coefficient $\beta^{(0)}$ is positive and a soft spring type when it is negative.
3) One of the nonlinear components from $N(0)$ to $N(4)$ has influence on each nonlinear resonance and determines the intensity, that is, the magnitude of the amplitude.
4) The subharmonic resonance $[2p_{\mathrm{f}}]$ and the combination resonance $[p_{\mathrm{f}} - p_{\mathrm{b}}]$ occur when the asymmetrical nonlinear component $N(1)$ exists.
5) The subharmonic resonance $[-2p_{\mathrm{b}}]$ occurs when the asymmetrical nonlinear component $N(3)$ exists.
6) The combination resonance $[2p_{\mathrm{f}} - p_{\mathrm{b}}]$ occurs when the symmetrical nonlinear component $N(0)$ exists.
7) The subharmonic resonance $[3p_{\mathrm{f}}]$ and the combination resonance $[p_{\mathrm{f}} - 2p_{\mathrm{b}}]$ occur when the symmetrical component $N(2)$ exists.
8) The subharmonic resonance $[-3p_{\mathrm{b}}]$ occurs when the symmetrical component $N(4)$ exists.

In practical rotating machinery, it is considered that the nonlinear components with a simple shape of distribution appear more intensely than those with a complex shape of distribution. For example, when the system is well assembled, the bearing centerline is located at the center of the clearance and the symmetrical component $N(0)$ appears predominantly. On the contrary, when there is a misalignment in the assembly, the asymmetrical component $N(1)$ appears predominantly (Yamamoto, Ishida, and Kawasumi, 1977). Therefore, from such a consideration and the result mentioned above, we know that the combination resonance $[2p_{\mathrm{f}} - p_{\mathrm{b}}]$ influenced by $N(0)$ is most likely to appear when the rotor is well assembled and the subharmonic resonance $[2p_{\mathrm{f}}]$ and the combination resonance $[p_{\mathrm{f}} - p_{\mathrm{b}}]$ influenced by $N(1)$ are most likely to appear when the rotor system has some misalignment.

Nonlinear resonances in a 4 DOF rotor system, where a deflection motion and an inclination motion couple with each other, can be analyzed in a similar manner (Yamamoto and Ishida, 1975). The details of analysis are omitted here. However, we can surmise the result of a 4 DOF system from that of a 2 DOF system as follows: Let a 4 DOF rotor system have natural frequencies $p_1, \ldots, p_4$. If we note that p_{f} and p_{b} in the 2 DOF system correspond to p_1, p_2 and p_3, p_4 in the 4 DOF system, respectively, we can guess the nonlinear components that influence

Table 6.1 Summary of theoretical analysis and experiments.

Cause and classification	Whirling direction	Theoretical result (2 DOF)			Experimental result (4 DOF)			
		Types of resonance	Influential component		Types of resonance	Bad assembly	Good assembly	
			Inclination	Intensity			Example 1	Example 2
Harmonic resonance	F	p_f	$N(0)$		p_2	Occurs also in linear systems		
Symmetrical nonlinearity								
Subharmonic resonance	F	$3p_f$	$N(0)$	$N(2)$	$3p_2$	×	×	Large
	B	$-3p_b$	$N(0)$	$N(4)$	$-3p_3, -3p_4$	×	×	×
	F + F		$N(0)$	$N(2)$	No resonance			
	2F + B	$2p_f - p_b$	$N(0)$	$N(0)$	$2p_2 - p_3, 2p_2 - p_4$	×	Large	Large
Combination resonance	F + 2B	$p_f - 2p_b$	$N(0)$	$N(2)$	$p_2 - 2p_3, p_2 - 2p_4$	×	×	Large
	B + B		$N(0)$	$N(4)$	$-2p_3 - p_4, -p_3 - 2p_4$	×	×	×
	F + F + ˜B		$N(0)$	$N(0)$	No resonance			
	F + B + ˜B		$N(0)$	$N(2)$	$p_2 - p_3 - p_4$	×	×	Large
Asymmetrical nonlinearity								
Subharmonic resonance	F	$2p_f$	$N(0)$	$N(1)$	$2p_2$	Large	Large	Large
	B	$-2p_b$	$N(0)$	$N(3)$	$-2p_3, -2p_4$	Small	×	×
	F + F		$N(0)$	$N(1)$				
Combination resonance	F + B	$p_f - p_b$	$N(0)$	$N(1)$	$p_2 - p_3, p_2 - p_4$	Large	Large	Large
	B + B		$N(0)$	$N(3)$	$-p_3 - p_4$	Small	×	×

The symbol × represent the case that the resonance was not observed in experiment.

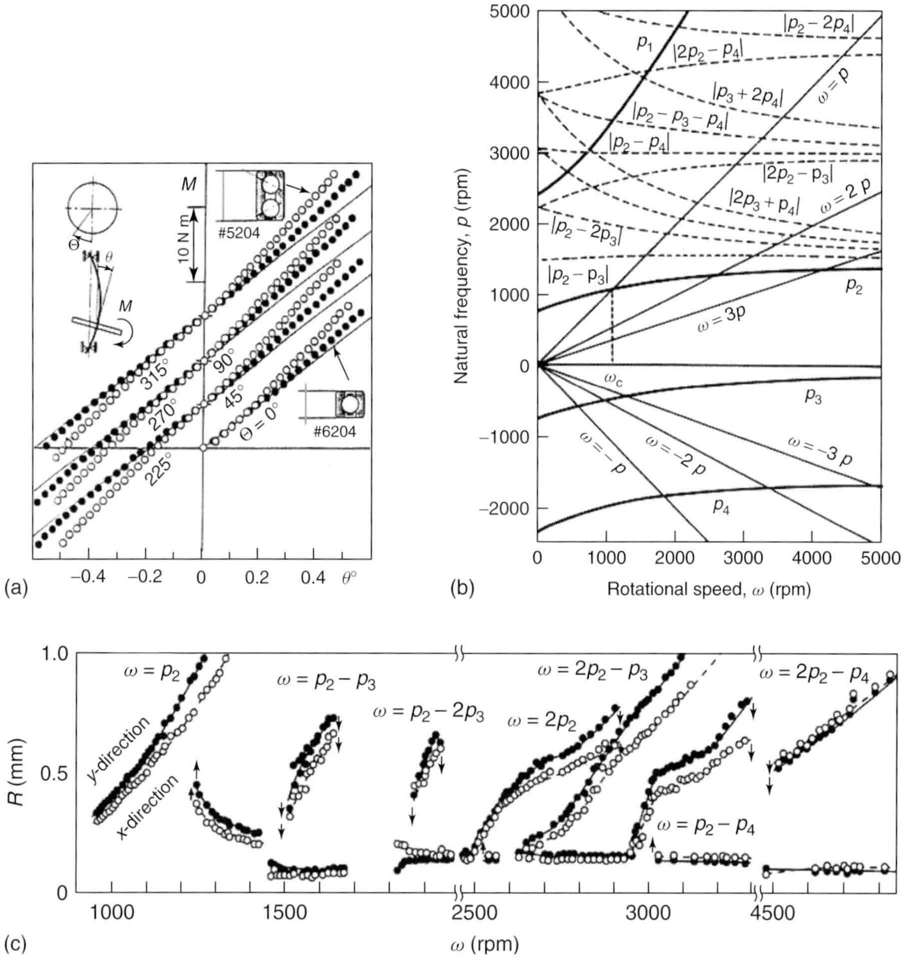

Figure 6.12 Experimental results: (a) spring characteristics, (b) natural frequency diagram, and (c) resonance curve.

nonlinear resonance in the 4 DOF rotor system. For example, by replacing p_2 by p_f and p_3 by p_b, we know that the combination resonance $[p_2 - 2p_3]$ is influenced by the same nonlinear component as the combination resonance $[p_f - 2p_b]$.

Experimental results are shown in Figure 6.12. In the experimental setup shown in Figure 6.12a, the upper and lower ends of the elastic shaft S are supported by a double-row self-aligning ball bearing (No.1200) and a single-row deep-groove ball bearing (No.6204), respectively. A disk R (diameter 481.3 mm, thickness 5.55 mm) is mounted on an elastic shaft S (length 700 mm, diameter 12 mm, $a : b = 1 : 4$). Nonlinearity is caused by the clearance of the single-row deep-groove ball bearing. Figure 6.12a shows relationships between the moment made by a weight hung at the disk edge and the inclination angle at various angles Θ. The characteristic is a hard spring type at every direction; however, its intensity changes depending on the direction. Although this nonlinearity is made by clearance, these characteristic curves are not a piecewise linear type but are comparatively smooth because clearances around the balls disappear one by one as the deflection increases. For comparison, characteristics of a system where the shaft is supported by a double-row angular contact ball bearing (No. 5204) at the lower end are also shown. They show stronger nonlinearity because the thickness of this bearing is larger than that of a single-row deep-groove ball bearing. The corresponding natural frequency diagram is shown in Figure 6.12b, and the corresponding resonance curves are shown in Figure 6.12c. The resonance curve of the harmonic resonance in the vicinity of about 1150 rpm becomes a hard spring type with a jump phenomenon and a hysteresis phenomenon. In addition, various kinds of subharmonic and combination resonance occur.

The experimental results executed by this and similar setups are summarized in Table 6.1. From the comparison between these theoretical and experimental results, it becomes clear that resonances influenced by a symmetrical component $N(0)$ occurred intensely when the assembly was good, and those influenced by an asymmetrical component $N(1)$ did so when the assembly was not so good. Thus, we can predict the occurrence of nonlinear resonances to some extent if we pay attention to the related nonlinear components expressed by polar coordinates.

We note the following regularity concerning the nonlinear components that determines the intensity, that is, the occurrence of the vibration shown in Table 6.1. For example, the component $N(1)$ influences the subharmonic resonance $[2p_{\mathrm{f}}]$ that occurs at $2p_{\mathrm{f}} = \omega$ as $N(2)$ does the combination resonance $[p_{\mathrm{f}} - 2p_{\mathrm{b}}]$ that occurs at $p_{\mathrm{f}} - 2p_{\mathrm{b}} = \omega$. If we arrange these relationships to the forms $2p_{\mathrm{f}} - \omega = 0$ and $p_{\mathrm{f}} - 2p_{\mathrm{b}} - \omega = 0$ and calculate the absolute values of the summations of the coefficients, the values $2 - 1 = 1$ and $1 - 2 - 1 = -2$ obtained from the coefficients give the values of k in $N(k)$. As this rule was found empirically, we should execute the analysis mentioned above to obtain an exact conclusion. However, it is recommended that this rule be used to check the possibility of occurrence before executing laborious calculation.

6.5 Nonlinear Resonances in a System with Radial Clearance: Strong Nonlinearity

The system discussed above is a system with weak nonlinearity and its restoring force can be approximated by a power series with low-order terms. However, some machine elements cause strong nonlinearity. For example, since ball or roller bearings have little damping effect, aircraft gas turbine engines generally adopt *squeeze-film damper bearings* for some of their bearings. A simplified model is illustrated in Figure 6.13b. Damper oil is supplied between the bearing holder A and the casing B. Holder A is supported by a weak spring S and is kept at the center of casing B. (Other types of squeeze-film damper bearings have no such spring S.) When this shaft vibrates, element A moves in relation to element B, and

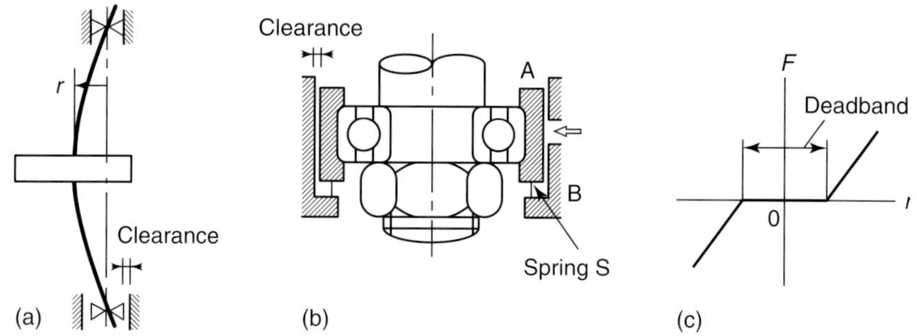

Figure 6.13 Rotor model with a squeeze-film damper bearing. (a) rotor system, (b) squeeze-film damper bearing, and (c) spring characteristics.

therefore the oil dampens the vibration. When this type of bearing is used in the rotor system shown in Figure 6.13a, the relationship between the rotor deflection r and the restoring force F shows strong nonlinearity with a *deadband*, as shown in Figure 6.13c. This figure shows a case where symmetrical nonlinearity appears.

Ehrich observed various kinds of nonlinear resonances in aircraft gas turbine engines. For example, he obtained a subharmonic resonance shown in Figure 6.14. Figure 6.14a is a frequency diagram that shows the frequencies of the predominant vibrations for the rotational speed, and Figure 6.14b shows the amplitude of the rotational speed. Owing to the damping effect of a squeeze-film damper bearing,

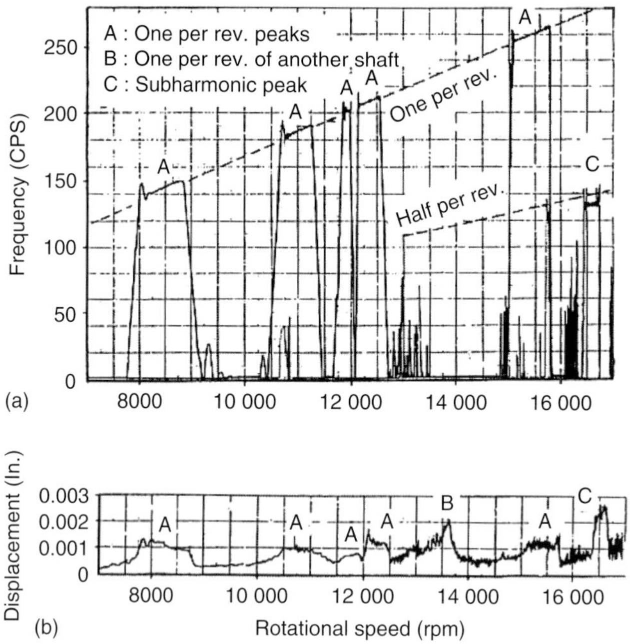

Figure 6.14 Experimental results of an aircraft gas turbine engine. (Courtesy of Dr. Ehrich.)

resonances are suppressed and the peaks are not sharp. The shaft of this turbine is dual and composed of a high-speed turbine and a low-speed turbine, which rotate at different speeds. Therefore, two unbalance forces with different angular velocities work. The peaks indicated by A are harmonic resonances responding to one of them. Since a practical engine is complicated and has many vibration modes, several harmonic resonances occur. The peak indicated by B is a harmonic resonance in response to the other unbalance force. The peak denoted by C is a subharmonic resonance of order 1/2 that has a frequency of half the rotational speed. The harmonic resonance with the same mode as this subharmonic resonance occurs at about 8300 rpm, which is half of the resonance speed of the subharmonic resonance.

Ehrich reported various types of nonlinear resonances, such as subharmonic resonances up to the order 1/9 (Ehrich, 1988) and chaotic vibration (Ehrich, 1991), which he observed in practical gas turbine engines. Since the nonlinearity in such systems is strong, it is not appropriate to approximate their spring characteristics by a power series as mentioned above. In this section, the outline of the phenomena is explained by using the results obtained by Ehrich.

6.5.1
Equations of Motion

Figure 6.15 shows an analytical model, horizontally supported by three bearings. Both ends of the shaft are supported simply by double-row self-aligning ball bearings and a squeeze-film damper bearing is used between them. When the deflection r_1 from the bearing centerline is small, the stiffness of the shaft is determined by the shaft (spring constant k_s), which is supported simply at both ends only. When the deflection becomes large and the radial clearance disappears, the squeeze-film damper bearing (spring constant k_b) also supports the shaft and the total shaft stiffness increases. Consequently, the shaft has nonlinear spring characteristics of a piecewise linear type. Let the deflection of the rotor when the clearance diminishes be R_0. Then, the spring constant is $k_1 = k_s$when $r_1 \leq R_0$ and $k_2 = k_s + k_b$ when $r_1 \geq R_0$. Owing to the gravitational force mg, the static equilibrium position moves

Figure 6.15 Rotor model with radial clearance (asymmetrical case).

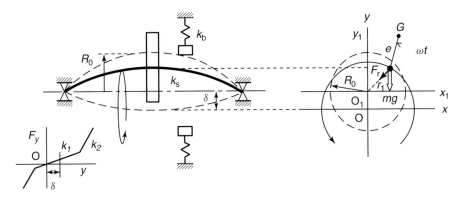

downward by δ, and asymmetrical nonlinear characteristics appear in the rotor. This static deflection δ is determined by the following relationship:

$$\left.\begin{aligned} mg &= k_1\delta && \text{when } \delta \le R_0 \\ mg &= k_1 R_0 + k_2(\delta - R_0) && \text{when } \delta \ge R_0 \end{aligned}\right\} \tag{6.55}$$

As shown in Figure 6.15, the rectangular coordinate system O$-xy$, which has its origin at the equilibrium position of the nonrotating shaft, is prepared. Let the rotor position be (x, y) and the position of the center of gravity of the disk with the eccentricity e be (x_G, y_G). In addition, a rectangular coordinate system $O_1-x_1y_1$, which has its origin at the bearing centerline, is also prepared. Let the rotor deflection on this coordinate system be (x_1, y_1). The nonlinear restoring force works in the direction of the origin O_1 and its magnitude F_r is represented by

$$\left.\begin{aligned} F_r &= k_1 r_1 && \text{when } r_1 \le R_0 \\ F_r &= k_1 R_0 + k_2(r_1 - R_0) && \text{when } r_1 \ge R_0 \end{aligned}\right\} \tag{6.56}$$

where $r_1 = \sqrt{x_1^2 + y_1^2}$. Let the mass of the rotor be m and the damping coefficient be c. The equations of motion are expressed as follows:

$$\left.\begin{aligned} m\ddot{x}_G &= -c\dot{x}_1 - F_r\frac{x_1}{r_1} \\ m\ddot{y}_G &= -c\dot{y}_1 - F_r\frac{y_1}{r_1} - mg \end{aligned}\right\} \tag{6.57}$$

Substituting the relationships $x_G = x + e\cos\omega t = x_1 + e\cos\omega t$ and $y_G = y + e\sin\omega t = y_1 + \delta + e\sin\omega t$ into this equation, we get

$$\left.\begin{aligned} m\ddot{x}_1 + c\dot{x}_1 + F_r\frac{x_1}{\sqrt{x_1^2 + y_1^2}} &= me\omega^2\cos\omega t \\ m\ddot{y}_1 + c\dot{y}_1 + F_r\frac{y_1}{\sqrt{x_1^2 + y_1^2}} &= me\omega^2\sin\omega t - mg \end{aligned}\right\} \tag{6.58}$$

This equation is solved numerically by the Runge–Kutta method. The results are explained below.

6.5.2 Harmonic Resonance and Subharmonic Resonances

The strength of the nonlinearity of this system is given by the ratio of the stiffnesses $\beta = k_1/k_2$. Figure 6.16 shows the response curve of a steady-state vibration when the nonlinearity is strong and the damping is small. The figure shows a variation of the maximum deflection. Since the disk is mounted at the midspan of the shaft, there is no coupling by the gyroscopic terms. It is coupled only by the nonlinear terms. Therefore, the trajectory is not circular, and the motion in the x- and y-directions change independently. Various types of subharmonic resonances

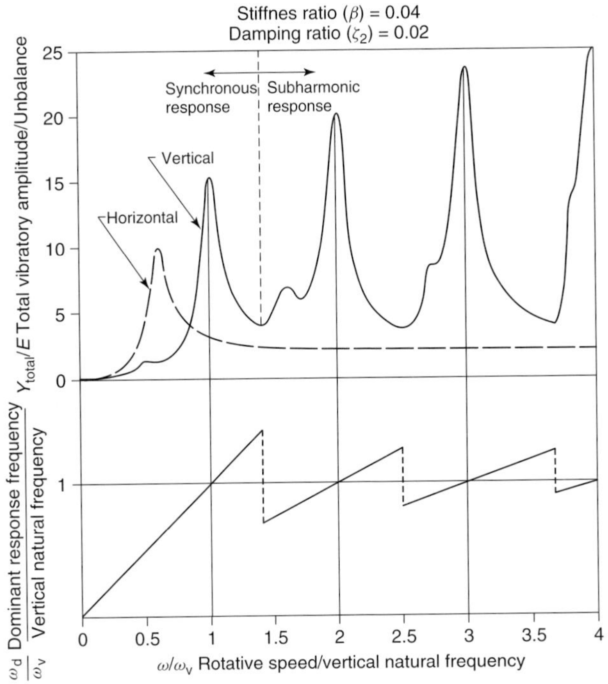

Figure 6.16 Amplitudes and frequencies of the harmonic and subharmonic resonances (Ehrich, 1988, courtesy of ASME).

occur at the rotational speeds that are integer times the major critical speed where a harmonic resonance (a synchronous oscillation) occurs. As shown in the lower figure, the frequencies of these resonances are close to a natural frequency of the system and have a value of $1/n$ (n: an integer) of the rotational speed.

If the system is vertical and the alignment is good, the static equilibrium position is at the center of the clearance (i.e., $\delta = 0$). In this case, the spring characteristics become those of isotropic symmetrical nonlinearity. Then, the subharmonic resonances due to symmetrical nonlinearity, such as the subharmonic resonance of order 1/3, occur. Ehrich obtained theoretically a resonance curve at the major critical speed, which is shown in Figure 6.17. The two vertical dashed lines represent critical speeds of the corresponding linear system where the spring constant is supposed to be k_1 or k_2. Figure 6.17a is a case with small clearance, and Figure 6.17b is that with large clearance. In the latter, the resonance curve is a hard spring type and jump phenomena appear. In the speed range where three values of amplitude exist, the middle one is unstable.[2)]

2) One of the authors obtained resonance curves of the harmonic resonance in a vertical shaft system with a deadband ($mg = 0, k_1 = 0$) in a different way. Refer to Yamamoto (1955) or Childs (1993).

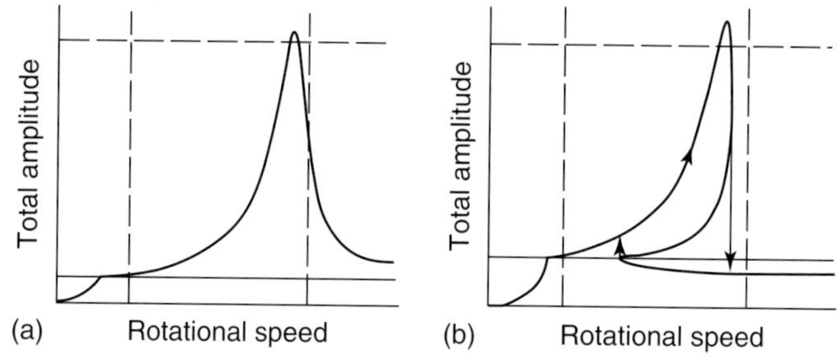

Figure 6.17 Harmonic resonance in a system with symmetric nonlinearity: (a) small clearance and (b) large clearance (Ehrich and O'Connor, 1967; Ehrich, 1992, courtesy of ASME).

6.5.3 Chaotic Vibrations

Ehrich (1991) found seemingly random vibrations in the transition zones between subharmonic resonances, which are shown in Figure 6.16. Figure 6.18 shows details in the transition zone between the subharmonic resonances of orders 1/2 and 1/3 in a system with smaller damping and larger nonlinearity than those shown in Figure 6.16. A zone with an irregular change in amplitude appears.

Figure 6.19 shows the local maximum value of the deflection in the trajectory, that is, the distance between the Poincaré point (filled circle) and the origin,

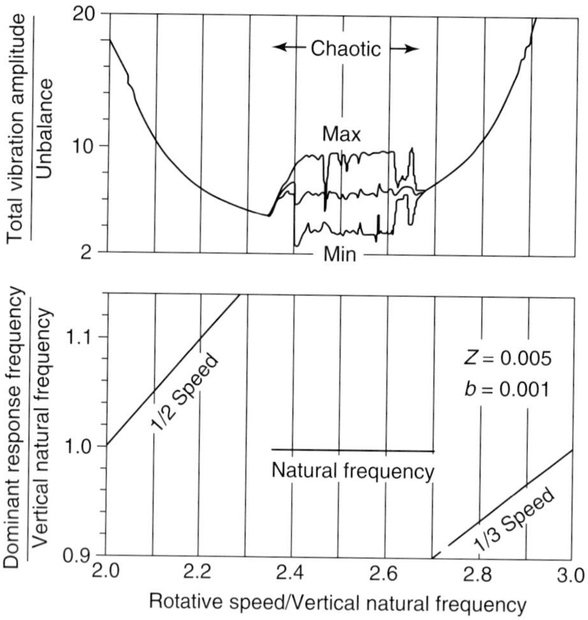

Figure 6.18 Computed amplitude and frequency response in the transition zone between second- and third-order subharmonic resonances (Ehrich, 1991, courtesy of ASME).

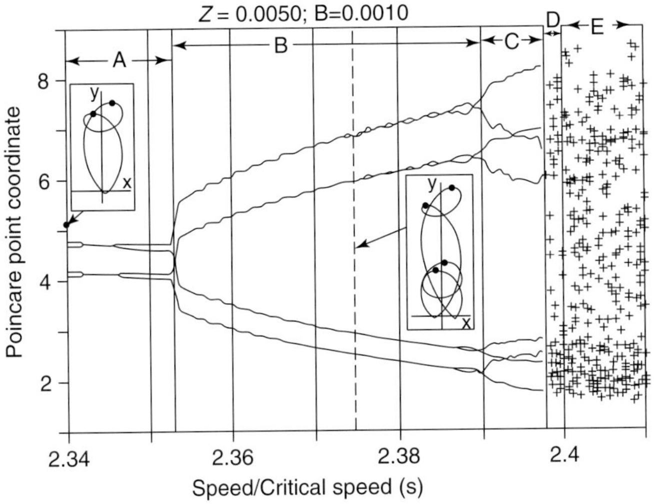

Figure 6.19 Bifurcation map of transition from second-order subharmonic resonance to chaotic motion (Ehrich, 1991, courtesy of ASME).

for the rotational speed on the left-half of the irregular zone in Figure 6.18. As the rotational speed increases, the following phenomena occur. A subharmonic resonance of order 1/2 occurs in region A. Its orbit has two extreme values. When the rotational speed enters into region B, the orbit duplicates (this phenomenon is called *bifurcation*) and has four extreme values. Such a bifurcation is repeated, and, finally, a nonperiodic whirling motion appears in region E. If the rotational speed is increased further, the change in phenomena occurs inversely and converges into a subharmonic resonance of order 1/3. The phenomenon in region E is a *chaotic vibration*, and the bifurcations before and after these chaotic vibrations are called *period doubling*. The ratio of the intervals between the bifurcations is constant and the value is

$$\frac{B}{C} \approx \frac{0.037}{0.008} \approx 4.625 \tag{6.59}$$

In the theory of chaotic vibrations, it is known that a system experiences such a period doubling process before the occurrence of chaotic vibrations, and the adjacent parameter ranges have a constant ratio $\delta = 4.6692016\ldots$ This ratio is called the *Feigenbaum number* (Szemplinska-Stupnica *et al.*, 1988). A close value is also obtained in this case.

6.6 Nonlinear Resonances of a Continuous Rotor

In the preceding sections, we considered nonlinear spring characteristics due to clearances in mechanical elements. We experience this kind of nonlinearity most often in practical rotating machineries. However, other types of nonlinearity exist. For example, two kinds of continuous rotor models are shown in Figure 6.20. If a bearing is allowed to move in the direction of the shaft's centerline as shown in Figure 6.20a, the restoring force of the shaft has linear spring characteristics. On the contrary, if the movement of the bearing is locked as shown in Figure 6.20b, the shaft elongates when it deflects and acquires nonlinear spring characteristics of hard spring type. This type of nonlinearity is called *geometrical nonlinearity*. This effect becomes more apparent when the shaft deflection becomes large. In this section, the analysis of nonlinear resonances of a continuous rotor with such geometrical nonlinearity is explained (Ishida, Nagasaka, and Lee, 1996b).

6.6.1
Representations of Nonlinear Spring Characteristics and Equations of Motion

In Section 3.2, in the derivation of equations of motion for a linear continuous rotor shown in Figure 6.20a, we got Eq. (3.5) for the transverse motion of a sliced element. In the case of a nonlinear continuous rotor shown in Figure 6.20b, since the movements of the bearings at both ends are restricted, the tension T due to shaft elongation must be added. Considering this effect, we derive the equations of motion as follows.

Figure 6.21 corresponds to Figures 3.2 and 3.3. It is assumed that a sliced element with length ds deflects during a whirling motion. The geometrical center M shifts from position $(0, 0, s_0)$ to (x, y, s_s), and the deflections along the coordinate axes are represented by u, v, and w, respectively. At the same time, point P shifts from position (a, b, s_0) to (x_P, y_P, s_P), and this deflection is represented by u_P, v_P, and w_P. From Figure 6.21, we have

$$u_P = u, \quad v_P = v, \quad w_P = w - a\frac{\partial u}{\partial s} - b\frac{\partial v}{\partial s} \tag{6.60}$$

Similarly, point $Q(a,\ b,\ s_0 + ds)$, separated from point P by ds, shifts by u_Q, v_Q, and w_Q, which are given approximately by

$$u_Q = u_P + \frac{\partial u_P}{\partial s}ds, \quad v_Q = v_P + \frac{\partial v_P}{\partial s}ds, \quad w_Q = w_P + \frac{\partial w_P}{\partial s}ds \tag{6.61}$$

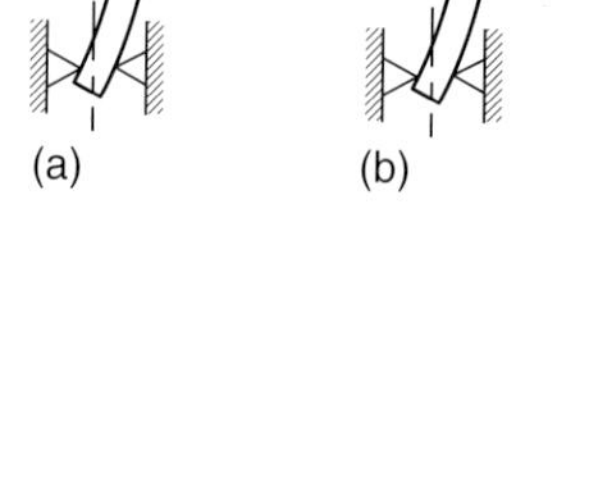

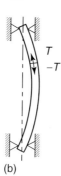

Figure 6.20 Geometrical nonlinearity: (a) linear system and (b) nonlinear system.

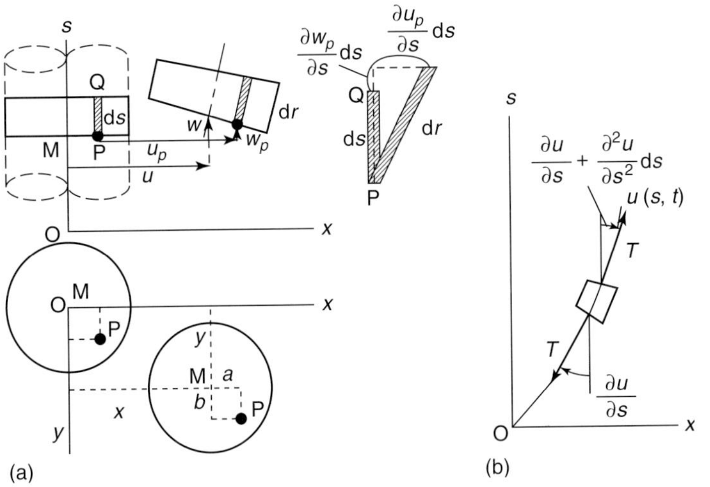

Figure 6.21 Sliced element with tension T: (a) sliced element and (b) deflection curve.

Therefore, after deformation, the length dr of the sliced element at point P becomes

$$\begin{aligned} dr &= \sqrt{\left(ds + \frac{\partial w_P}{\partial s}ds\right)^2 + \left(\frac{\partial u_P}{\partial s}ds\right)^2 + \left(\frac{\partial v_P}{\partial s}ds\right)^2} \\ &\approx ds\left\{1 + \frac{\partial w_P}{\partial s} + \frac{1}{2}\left(\frac{\partial u_P}{\partial s}\right)^2 + \frac{1}{2}\left(\frac{\partial v_P}{\partial s}\right)^2\right\} \end{aligned} \tag{6.62}$$

where the condition $\partial w_P/\partial s << \partial u_P/\partial s \approx \partial w_P/\partial s$ and the approximations $\sqrt{1+\Delta} \approx 1 + \Delta/2$ (for $\Delta \ll 1$) are used. From Eqs. (6.60) and (6.62), the strain $\varepsilon(a, b)$ in the s-direction is expressed by

$$\varepsilon = \frac{dr - ds}{ds} = \frac{\partial w}{\partial s} - a\frac{\partial^2 u}{\partial s^2} - b\frac{\partial^2 v}{\partial s^2} + \frac{1}{2}\left\{\left(\frac{\partial u}{\partial s}\right)^2 + \left(\frac{\partial v}{\partial s}\right)^2\right\} \tag{6.63}$$

By deriving the stress at point P using Hook's law $\sigma = E\varepsilon$ and integrating it into the entire cross section with area A, we can get the expression for tension T (an axial force) as follows:

$$T = \int_A \sigma dA = EA\frac{\partial w}{\partial s} + \frac{1}{2}EA\left[\left(\frac{\partial u}{\partial s}\right)^2 + \left(\frac{\partial v}{\partial s}\right)^2\right] \tag{6.64}$$

Integrating this expression along the s-direction considering that the tension T is constant in that direction, we obtain

$$w(s) = \frac{1}{EA}Ts - \frac{1}{2}\int_0^s \left[\left(\frac{\partial u}{\partial s}\right)^2 + \left(\frac{\partial v}{\partial s}\right)^2\right] ds + C \tag{6.65}$$

Applying the condition that the deflection is zero at $s = 0$, namely $w(0) = 0$, we get the constant $C = 0$. From the condition $w(l) = 0$, we get the expression for T:

$$T = \frac{EA}{2l} \int_0^l \left[\left(\frac{\partial u}{\partial s} \right)^2 + \left(\frac{\partial v}{\partial s} \right)^2 \right] ds \tag{6.66}$$

As shown in Figure 6.21b, this tension T works in the tangential direction of the shaft deflection curve given by $u(s, t)$ and $v(s, t)$. For example, in the xs-plane, since this deflection curve has the gradient $\partial u/\partial s$ at the lower position of the element and $\partial u/\partial s + (\partial^2 u/\partial s^2)ds$ at the upper position, the summation of these two tensions does not cancel each other and the component of tension T in the x-direction appears. Therefore, we must add $+(\partial^2 u/\partial s^2)Tds$ and $+(\partial^2 v/\partial s^2)Tds$ to the right-hand sides of the first and second equations of Eq. (3.4) for the deflection motion. Consequently, adding the terms

$$\left.\begin{array}{l} \text{(First equation)} \quad -\dfrac{EA}{2l}\dfrac{\partial^2 u}{\partial s^2} \displaystyle\int_0^l \left[\left(\frac{\partial u}{\partial s} \right)^2 + \left(\frac{\partial v}{\partial s} \right)^2 \right] ds \\ \text{(Second equation)} \quad -\dfrac{EA}{2l}\dfrac{\partial^2 v}{\partial s^2} \displaystyle\int_0^l \left[\left(\frac{\partial u}{\partial s} \right)^2 + \left(\frac{\partial v}{\partial s} \right)^2 \right] ds \end{array}\right\} \tag{6.67}$$

to the left-hand sides of Eq. (3.10), we can obtain the equations of motion.

We introduce a representative quantity e_0 which is of the same order as the rotor unbalance and define the following dimensionless quantities:

$$\left.\begin{array}{l} x^* = \dfrac{u}{e_0}, \quad y^* = \dfrac{v}{e_0}, \quad s^* = \dfrac{s}{l}, \quad e_\xi^* = \dfrac{e_\xi}{e_0}, \quad e_\eta^* = \dfrac{e_\eta}{e_0} \\ t^* = \left(\dfrac{\pi}{l} \right)^2 \left(\dfrac{EA}{\rho A} \right)^{1/2} t (\equiv \gamma t), \quad \omega^* = \dfrac{\omega}{\gamma} \\ \varkappa = \dfrac{\pi^2 R^2}{4l^2}, \quad c^* = \dfrac{cl^4 \gamma}{\pi^4 EI}, \quad \alpha = \dfrac{Ae_0^2}{2I} \end{array}\right\} \tag{6.68}$$

The equations of motion expressed by these quantities are as follows:

$$\left.\begin{array}{l} \dfrac{1}{\pi^4}\dfrac{\partial^4 u}{\partial s^4} + \dfrac{\partial^2 u}{\partial t^2} - \dfrac{\kappa}{\pi^2}\left[\dfrac{\partial^4 u}{\partial s^2 \partial t^2} + 2\omega \dfrac{\partial^3 v}{\partial s^2 \partial t} \right] + c\dfrac{\partial u}{\partial t} \\ \qquad - \dfrac{a}{\pi^4}\dfrac{\partial^2 u}{\partial s^2} \displaystyle\int_0^1 \left[\left(\frac{\partial u}{\partial s} \right)^2 + \left(\frac{\partial v}{\partial s} \right)^2 \right] ds = \omega^2 (e_\xi \cos \omega t - e_\eta \sin \omega t) \\ \dfrac{1}{\pi^4}\dfrac{\partial^4 u}{\partial s^4} + \dfrac{\partial^2 v}{\partial t^2} - \dfrac{\kappa}{\pi^2}\left[\dfrac{\partial^4 v}{\partial s^2 \partial t^2} - 2\omega \dfrac{\partial^3 u}{\partial s^2 \partial t} \right] + c\dfrac{\partial v}{\partial t} \\ \qquad - \dfrac{a}{\pi^4}\dfrac{\partial^2 v}{\partial s^2} \displaystyle\int_0^1 \left[\left(\frac{\partial u}{\partial s} \right)^2 + \left(\frac{\partial v}{\partial s} \right)^2 \right] ds = \omega^2 (e_\xi \sin \omega t + e_\eta \cos \omega t) \end{array}\right\} \tag{6.69}$$

where the asterisks have been eliminated for simplicity in the representation.

These expressions are also represented by a complex number $z = u + iv$ and its complex conjugate $\bar{z}$ as follows:

$$\frac{1}{\pi^4}\frac{\partial^4 z}{\partial s^4} + \frac{\partial^2 z}{\partial t^2} - \frac{\kappa}{\pi^2}\left[\frac{\partial^4 z}{\partial s^2 \partial t^2} - 2i\omega\frac{\partial^3 z}{\partial s^2 \partial t}\right] + c\frac{\partial v}{\partial t}$$
$$-\frac{a}{\pi^4}\frac{\partial^2 z}{\partial s^2}\int_0^1\left(\frac{\partial z}{\partial s}\right)\left(\frac{\partial \bar{z}}{\partial s}\right)\mathrm{d}s = \omega^2(e_\xi + ie_\eta)\mathrm{e}^{iwt} \tag{6.70}$$

In the following, we solve this equation under the condition that the shaft is supported simply at both ends. This boundary condition is given by putting $l = 1$ in Eq. (3.14).

6.6.2 Transformation to Ordinary Differential Equations

In this section we transform the equations of motion (Eq. (6.69)) to ordinary differential equations by utilizing eigenfunctions of Eq. (3.26). As a result, we can decompose the original equations to modal equations that correspond to Eq. (6.9) for a concentrated mass system. However, these modal equations are not yet divided into forward and backward modes. After this transformation, we can solve by using, among others, the harmonic balance method or normal coordinates as shown in Section 6.4.1.

First, similar to Eq. (3.39), we expand the solution by the characteristic functions given by

$$\overline{\varphi}_n(s) = \sin n\pi s \tag{6.71}$$

where s/l in Eq. (3.37) is replaced by the dimensionless quantity s:

$$u(s,t) = \sum_{n=1}^{\infty} u_n(t)\overline{\varphi}_n(s), \quad v(s,t) = \sum_{n=1}^{\infty} v_n(t)\overline{\varphi}_n(s) \tag{6.72}$$

Similarly, we expand the unbalances as

$$e_\xi(s) = \sum_{n=1}^{\infty} e_{\xi n}\overline{\Phi}_n(s), \quad e_\eta(s) = \sum_{n=1}^{\infty} e_{\eta n}\overline{\Phi}_n(s) \tag{6.73}$$

Substituting these expressions into Eq. (6.70) and utilizing the orthogonality of modal shapes given by Eq. (3.38), we got nonlinear ordinary differential equations. This procedure is the same as the *Galerkin procedure* used in continuous systems. In this analysis, we must calculate the following term:

$$\int_0^1\left[\left(\frac{\partial u}{\partial s}\right)^2 + \left(\frac{\partial v}{\partial s}\right)^2\right]ds = \int_0^1\left[\left(\sum_{j=1}^{\infty} u_j j\pi\cos j\pi s\right)^2 + \left(\sum_{j=1}^{\infty} v_j j\pi\cos j\pi s\right)^2\right]ds \tag{6.74}$$

Here, we utilize the fact that the integration of the product term of $\cos i\pi s \cos j\pi s$ is zero when $i \neq j$ and 1/2 when $i = j$. After some calculation we arrive at the following expressions:

$$\left.\begin{aligned}&(1+\kappa n^2)\ddot{u}_n + 2x\omega n^2\dot{v}_n + c\dot{u}_n + n^4 u_n + \alpha n^2 u_n \sum_{j=1}^{\infty}\frac{j^2}{2}(u_j^2+v_j^2)\\&\quad = \omega^2(e_{\xi n}\cos\omega t - e_{\eta n}\sin\omega t)\\&(1+\kappa n^2)\ddot{v}_n - 2x\omega n^2\dot{u}_n + c\dot{v}_n + n^4 v_n + \alpha n^2 v_n \sum_{j=1}^{\infty}\frac{j^2}{2}(u_j^2+v_j^2)\\&\quad = \omega^2(e_{\xi n}\sin\omega t + e_{\eta n}\cos\omega t) \qquad (n=1,2,\ldots)\end{aligned}\right\} \tag{6.75}$$

6.6.3
Harmonic Resonance

A natural frequency diagram of a continuous rotor is given by Figure 3.7 or 3.8. We derive a resonance curve at the major critical speed where the rotational speed coincides with a forward natural frequency p_{fn}. Similar to Eq. (6.17), we assume the solution to an accuracy of $O(\varepsilon)$ as follows:

$$\left.\begin{aligned}u_n &= P\cos(\omega t+\beta) + \varepsilon\left\{a\cos(\omega t+\beta) + b\sin(\omega t+\beta)\right\}\\v_n &= P\sin(\omega t+\beta) + \varepsilon\left\{a'\sin(\omega t+\beta) + b'\cos(\omega t+\beta)\right\}\end{aligned}\right\} \tag{6.76}$$

Substituting this into Eq. (6.75), we use the harmonic balance method. It is assumed that the amplitude P and phase angle β are functions that change slowly with time. Equating the coefficients of $\cos(\omega t+\beta)$ and $\sin(\omega t+\beta)$ on both sides to an accuracy of $O(\varepsilon)$, we arrive at

$$\left.\begin{aligned}2\omega P\dot{\beta} &= G_n(\omega)P + \frac{1}{2}\alpha n^4 P^3 - \omega^2(e_{\xi n}\cos\beta + e_{\eta n}\sin\beta)\\2\omega\dot{P} &= -c\omega P - \omega^2(e_{\xi n}\sin\beta - e_{\eta n}\cos\beta)\end{aligned}\right\} \tag{6.77}$$

where $G_n(\omega) = n^4 + \kappa\omega^2 n^2 - \omega^2$ and has a magnitude of $O(\varepsilon)$ in the vicinity of the major critical speed under consideration. The analysis hereafter is the same as that mentioned in Section 6.4.1. The resonance curve is given by

$$\left\{G_n(\omega)P_0 + \frac{1}{2}\alpha n^4 P_0^3\right\}^2 + (c\omega P_0)^2 = \omega^4(e_{\xi n}^2 + e_{\eta n}^2) \tag{6.78}$$

Computed resonance curves are shown in Figure 6.22, where the two cases for different shaft diameters are plotted. Solid lines represent stable solutions and dashed lines represent unstable solutions. Because the nonlinearity is caused by the stiffening effect of rotor elongation, these resonance curves become a hard spring type. In this figure, we cannot compare the inclinations of the resonance curves directly because these inclinations change through contraction or extension when the resonance speed is shifted to 1.0 by nondimensionalyzing the parameters. However, we ascertained that the inclination becomes stronger as the shaft becomes more slender by drawing the resonance curves using parameters with dimensions.

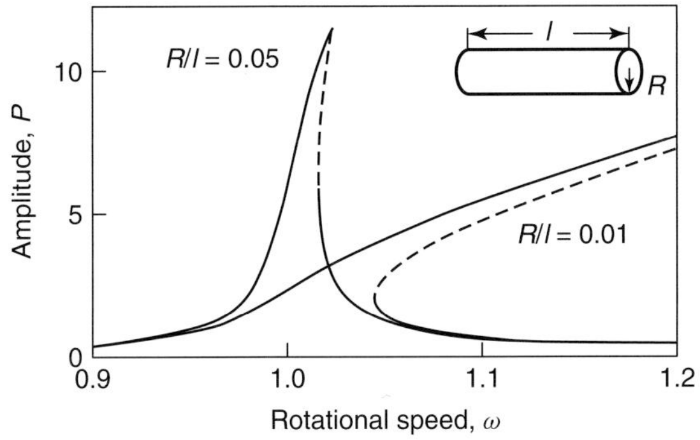

Figure 6.22 Harmonic resonance.

6.6.4 Summary of Nonlinear Resonances

In a system with nonlinear spring characteristics, various types of nonlinear resonances, such subharmonic resonances and combination resonances, are expected to occur. However, from the similar analysis of these resonances, it becomes clear that almost all of them have only the trivial solution (Ishida, Nagasaka, and Lee, 1996b). The results are summarized in Table 6.2. Since the nonlinearity expressed by the integration of Eq. (6.70) is isotropic, that is, has no directional difference,

Table 6.2 Summary of Nonlinear Resonances in a Continuous Rotor System.

Type of resonance		Linear	Nonlinear
Main resonance	p_{fn}	Occurs	Occurs
	$2p_{fn}$		×
Subharmonic resonance	$-2p_{bn}$		×
	$3p_{fn}$		×
	$-3p_{bn}$		×
	$p_{fm} - p_{bn}$		×
	$p_{fm} + p_{fn}$		×
	$-p_{bm} - p_{bn}$		×
	$2p_{fm} + p_{fn}$		×
Combination resonance	$2p_{fm} - p_{bn}$		Occurs when $m = n$
	$p_{fm} - 2p_{bn}$		×
	$-2p_{bm} - p_{bn}$		×
	$p_{fl} + p_{fm} + p_{fn}$		×
	$p_{fl} + p_{fm} - p_{bn}$		Occurs when $l = n$ or $m = n$
	$p_{fl} - p_{bm} - p_{bn}$		×
	$-p_{bl} - p_{bm} - p_{bn}$		×

[a]The symbol × represents a case in which only the steady-state solution with zero-amplitude exists; in other words, this type of oscillation does not occur.

it corresponds to the disk–shaft system with only the component $N(0)$. Therefore, we can imagine the results given in Table 6.2 from the results given in Table 6.1.

The above-mentioned conclusion holds in a vertical rotor system. In a horizontal rotor system with gravity or a rotor system with a lateral force due to some cause, the static equilibrium position of the shaft centerline shifts from the centerline of the bearing and asymmetrical nonlinearity appears. As a result, nonlinear resonances, such as subharmonic resonance of order 1/2 (Ishida, Nagasaka and Lee, 1996b) and combination resonance $[p_{fm} - p_{bn}]$ (Nagasaka Ishida, Lee and Kojima 1997), occur. In addition, in a slender continuous rotor with asymmetrical nonlinear spring characteristics, the effect of internal resonance must be considered. This is because, as is discussed in the next section, internal resonance phenomena appear in a system where absolute values of forward and backward natural frequencies are equal to each other, and an asymmetrical nonlinearity connecting these natural frequencies exist. The horizontal continuous rotor mentioned here satisfies these conditions.

6.7 Internal Resonance Phenomenon

In a multi-DOF system with nonlinear spring characteristics, unique and complex phenomena occur when a simple relationship such as $p_i = 2p_j$ or $p_i + p_j = p_k$ holds among natural frequencies (Nayfeh and Balachandran, 1989). This phenomenon, called the *internal resonance phenomenon*, occurs because of the energy transfer through nonlinearity between natural frequencies. In this section, internal resonance phenomena in a rotor system are explained.

6.7.1 Examples of the Internal Resonance Phenomenon

Figure 6.23a is a natural frequency diagram of a 4 DOF rotor system where a disk is mounted at a lower part of an elastic shaft. Since the natural frequencies change as a function of the rotational speed due to the gyroscopic moment, the relationship of internal resonance holds at various rotational speeds. For example, the relationship $p_2 : p_3 = 2 : (-1)$ holds at the rotational speed ω_P and $p_2 : p_3 = 3 : (-1)$ holds at ω_Q. If the critical speed of harmonic resonance A or that of subharmonic resonance B comes close to or coincides with P or Q, a condition of internal resonance is satisfied in the resonance region, and, as a result, the characteristics of the resonance may differ from that of a single resonance.

Figure 6.23b shows natural frequencies of a lateral vibration of the Jeffcott rotor. In this case, the absolute values of forward and backward natural frequencies are always equal and the relationship $p_f : p_b = 1 : (-1)$ holds at any rotational speed. Therefore, the effect of internal resonance may appear in any type of resonance. The Jeffcott rotor is a representative rotor model that is widely used in the analysis of rotor vibration. When we use this model for nonlinear analysis, we must pay

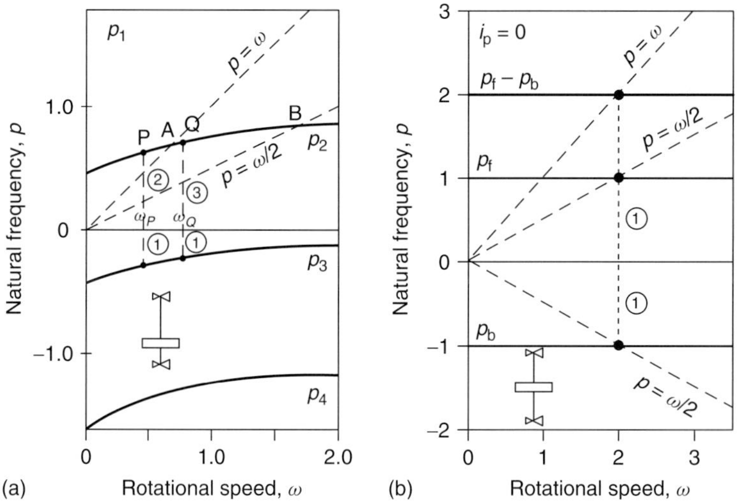

Figure 6.23 Examples of internal resonance in rotor systems: (a) 4 DOF rotor system and (b) Jeffcott rotor.

attention to the effect of this internal resonance. In some cases, this model gives a completely different result from that of one with a single resonance.

If a continuous rotor treated in Chapter 3 is slender, the gyroscopic moment is small and therefore the natural frequencies of forward and backward whirling mode have the relationship $p_{fn} : p_{bn} = 1 : (-1)$, similar to the Jeffcott rotor. In addition, as shown in Figure 3.10, they have the relationship of, for example, $p_{f1} : p_{f2} = 1 : 4$, between the first and the second modes.

6.7.2
Subharmonic Resonance of Order 1/2

Internal resonance phenomena in the Jeffcott rotor is explained taking a forward subharmonic resonance of order 1/2 as an example (Ishida and Inoue, 2004). In the lateral vibrations of the Jeffcott rotor with the natural frequency diagram shown in Figure 6.23b, critical speeds of a forward subharmonic resonance of order 1/2 [$2p_f$], a backward subharmonic resonance of order l/2 [$-2p_b$], and a combination resonance [$p_f - p_b$] coincide precisely. However, to investigate the transition from a single resonance (refer to Section 6.4) to a resonance with internal resonance, we use the 2 DOF inclination model shown in Figure 2.11. Its equations of motion are given by Eq. (6.9), and a natural frequency diagram is given by Figure 6.24a. These three critical speeds are separated from each other in this figure, but they come close and finally coincide as $i_p \to 0$. Putting $i_p = 0$ in Eq. (6.9), we obtain the equations of motion of the Jeffcott rotor, and its natural frequency diagram is given by Figure 6.23b.

So far, we discussed single nonlinear resonances and obtained the following results:

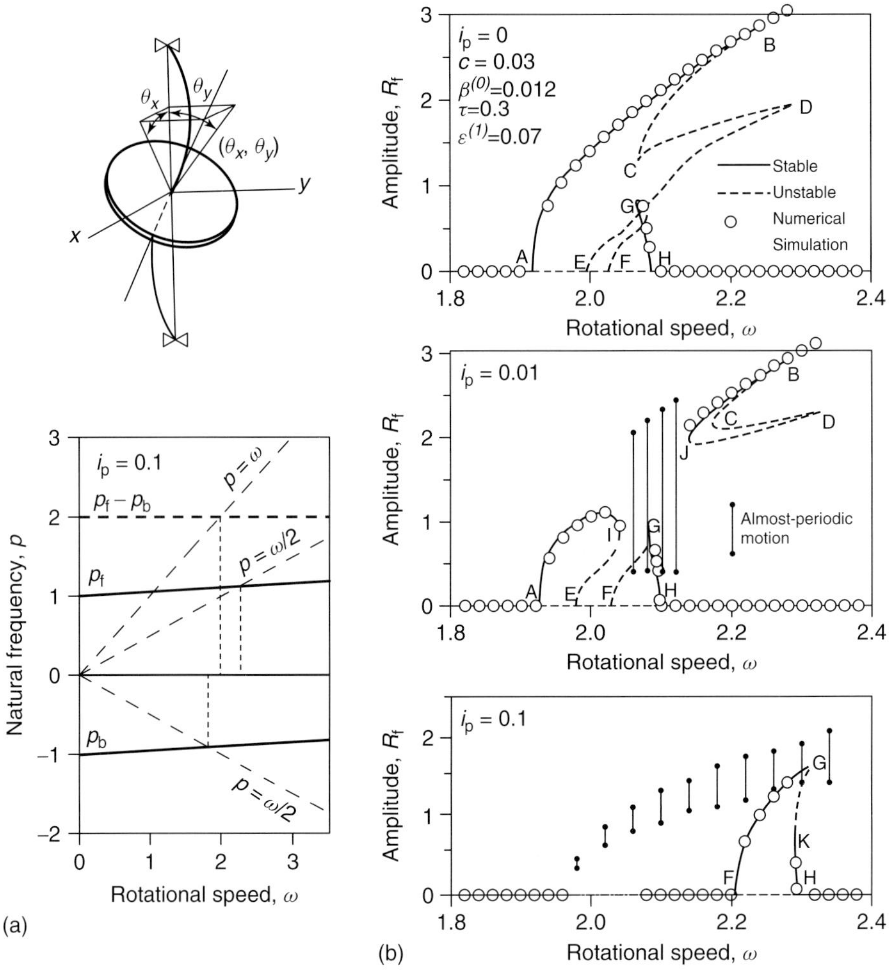

Figure 6.24 Change in the internal resonance phenomenon by the discrepancy of critical speeds: (a) natural frequency diagram and (b) change of resonance curves with parameter i_p.

1) Subharmonic resonance $[2p_f]$ and combination resonance $[p_f - p_b]$ are influenced by $N(0)$ and $N(1)$.
2) Subharmonic resonance $[-2p_b]$ is influenced by $N(0)$ and $N(3)$.
3) Component $N(0)$ determines the inclination of the resonance curves.
4) Components $N(1)$ and $N(3)$ have influence on the magnitude of the resonance.

In the following analysis, we assume that the rotor system has only components $N(0)$ and $N(1)$ since component $N(3)$ is not as large in practical machineries. In such a case, when the resonance occurs as a single resonance, subharmonic resonance $[2p_f]$ and combination resonance $[p_f - p_b]$ occur, but subharmonic resonance $[-2p_b]$ does not occur. In this section, we note especially the transition from the state where three resonances $[2p_f]$, $[p_f - p_b]$, and $[-2p_b]$ coincide exactly with each other to the state where they separate from each other and appear as a single resonance. Therefore, we assume the solution in the following form, which can also express the case where the components with the frequencies $\omega_f = +(1/2)\omega(\approx p_f)$ and $\omega_b = -(1/2)\omega(\approx p_b)$ occur simultaneously under the influence of internal resonance:

$$\left.\begin{aligned}\theta_x &= R_f \cos(\omega_f t + \delta_f) + R_b \cos(\omega_b t + \delta_b) + P\cos(\omega t + \beta) + A_x + \varepsilon(\cdots)\\ \theta_y &= R_f \sin(\omega_f t + \delta_f) + R_b \sin(\omega_b t + \delta_b) + P\sin(\omega t + \beta) + A_y + \varepsilon(\cdots)\end{aligned}\right\} \tag{6.79}$$

In this expression, terms of order $O(\varepsilon)$ that are necessary in the analysis are not shown, for simplicity. Using the procedure explained in Section 6.4.2, we can obtain the following result.

The analytical result is shown in Figure 6.24b. Depending on the value of i_p, it changes as follows : (i) The case $i_p = 0$ corresponds to the Jeffcott rotor. Unlike the curves in Figure 6.9, which have no influence of internal resonance, the resonance curves are composed of the large branch ABCDE on the lower speed side and the small branch FGH on the higher speed side. Branches AB and GH are stable. (ii) The figure for $i_p = 0.01$ shows a case where the critical speeds are slightly different from each other. Branch AB observed for $i_p = 0$ split into branches AI and JB, and no steady-state solution exists between them. Almost periodic motions appear in this zone in numerical simulation. Figure 6.25 shows time histories and spectra of a steady-state vibration and an almost periodic motion for the same parameter values. In addition to the harmonic component, a constant component, as well as

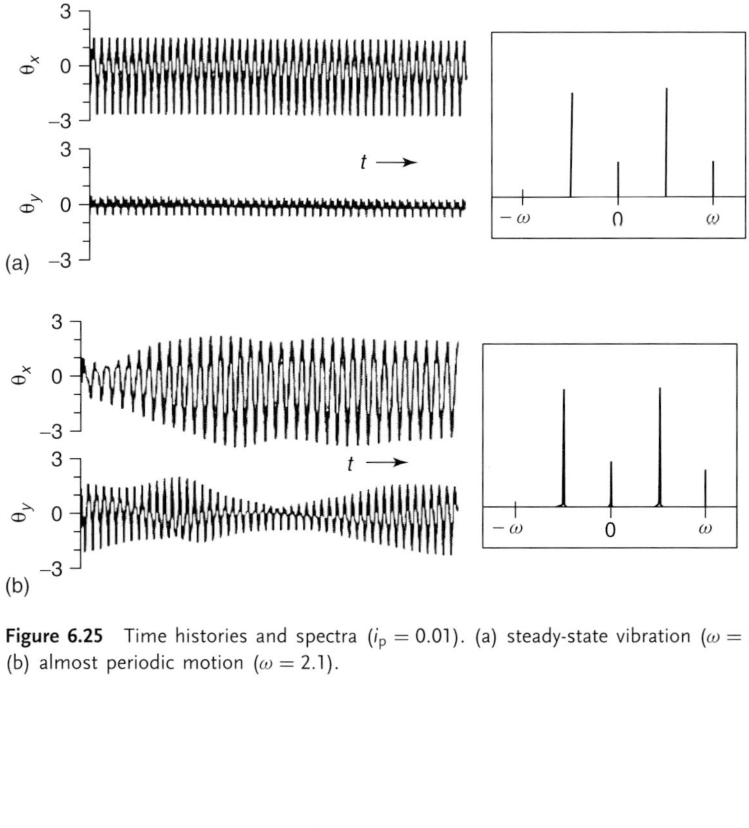

Figure 6.25 Time histories and spectra ($i_p = 0.01$). (a) steady-state vibration ($\omega = 2.0$) and (b) almost periodic motion ($\omega = 2.1$).

forward and backward subharmonic components, appears in the spectra. In the case of an almost periodic motion, these spectra have some width. The occurrence of almost periodic motions is a typical characteristic of a system with an internal resonance. (iii) The figure for $i_p = 0.1$ shows a case where the critical speeds are separated to some extent. The branches BCDJ and AIE observed in the case of $i_p = 0.01$ disappear and the zone of almost periodic motion expands. (iv) If the critical speeds separate further, as shown in Figure 6.9, the influence of internal resonance disappears. As a result, branch FGKH converges to a resonance curve of a single subharmonic resonance and the almost periodic motion converges to a resonance curve of a single combination resonance $[p_f - p_b]$. Thus, the internal resonance has a remarkable effect on the vibration characteristics.

6.7.3
Chaotic Vibrations in the Vicinity of the Major Critical Speed

In Section 6.5.3, it was explained that chaotic vibrations generally occur in rotor systems with strong nonlinearity. However, chaotic vibrations may appear in a system with weak nonlinearity if a condition of internal resonance holds (Inoue, Ishida, and Kondo, 2006).

Let us consider phenomena in the vicinity of the major critical speed in a rotor system governed by Eq. (6.9) in the case of small i_p. In this rotational speed range, critical speeds of a harmonic resonance $[p_f]$, a harmonic resonance of a backward whirling mode $[-p_b]$, and a combination resonance (a super-summed-and-differential harmonic resonance) $[(p_f - p_b)/2]$ are close to each other. We assume that an isotropic symmetrical nonlinear component $N(0)$ and an asymmetrical component $N(1)$ exist here. By analysis that considers the internal resonance, we can obtain the resonance curves shown in Figure 6.26a. The solid lines represent stable solutions, and the dashed and dash-dotted lines represent unstable solutions. The dashed line indicates instability due to saddle-node bifurcation, and the dash-dotted line indicates that due to Hopf bifurcation (Nayfeh and Balachandran, 1995). The shape of the resonance curves is similar to that with no internal resonance shown in Figure 6.7. However, unlike Figure 6.7, the large-amplitude part (dash-dotted line) of the resonance curve of the lower speed side is unstable.

Steady-state solutions and solutions with varying amplitudes are also depicted in the figures. As the rotational speed increases, this solution with varying amplitude changes from an almost periodic motion to a chaotic motion. Figure 6.26b shows a time history, spectra, and a Poincaré map of chaotic motion obtained at point A. A *Poincaré map* is a diagram depicted by sampling points on a trajectory at every interval of the external force, and is used to investigate the periodicity of a nonlinear phenomenon (Szemplinska-Stupnica, Ioose and Moon 1988). This figure represents the projection state of the variables $(\theta_x, \dot{\theta}_x, \theta_y, \dot{\theta}_y)$ in the four-dimensional phase space onto the two-dimensional $\theta_x\theta_y$-plane. We see in these figures that the amplitude variation is irregular, spectra distribute continuously, and points in the Poincaré map do not return to the same location after an integer multiple of the period of rotor rotation. All of these characteristics are typical of chaotic motions.

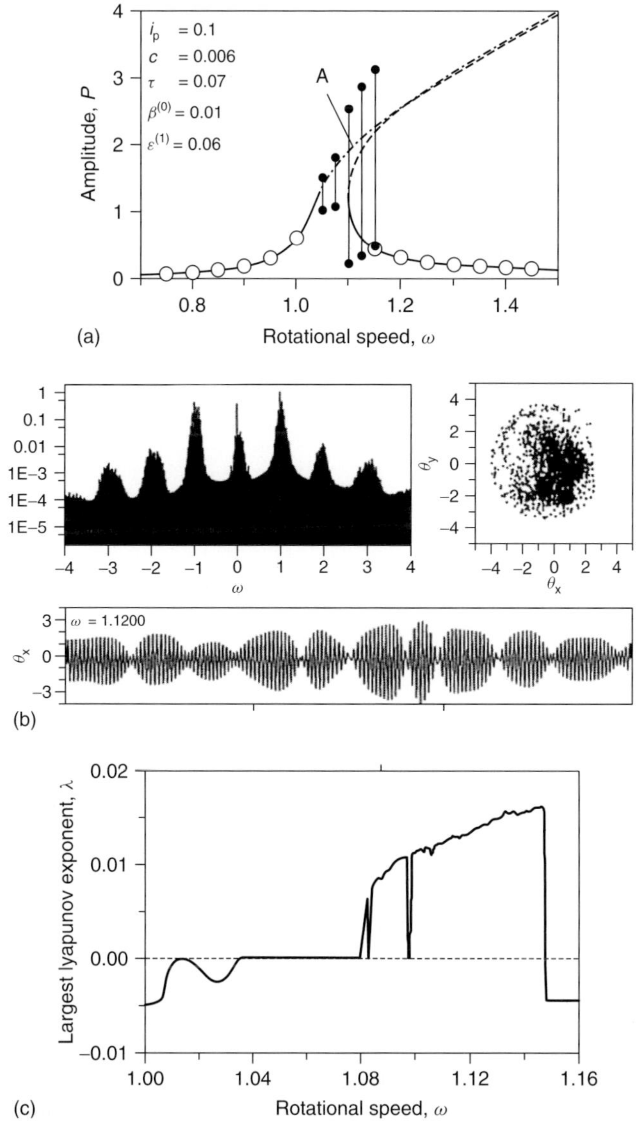

Figure 6.26 Vibrations effected by an internal resonance at the major critical speed: (a) resonance curve, (b) chaotic motion at point A ($\omega = 1.12$), and (c) maximum Lyapunov exponent.

To ascertain further that this is a chaotic motion, the *maximum Lyapunov exponent* is calculated. The result is shown in Figure 6.26c. The Lyapunov exponent λ is defined as follows. Let us consider two trajectories at distance d_0 at $t = 0$ in phase space. We assume that the two points on these trajectories have a distance d expressed approximately by $d(t) = d_0 e^{\lambda t}$. The λ in this expression gives the Lyapunov exponent.

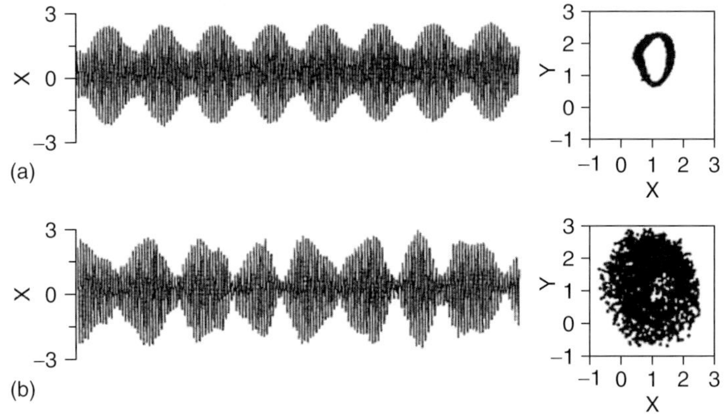

Figure 6.27 Experimental results: (a) almost periodic motion (889 rpm) and (b) chaotic vibration (894 rpm).

If the Lyapunov exponent has a positive value, these trajectories diverge rapidly. This means that the solution depends sensitively on the initial condition, and this sensitive dependence on the initial condition is a characteristic of chaotic vibration (Szemplinska-Stupnica, Ioose and Moon 1988). Figure 6.26c shows the maximum value of Lyapunov exponents, which was obtained by the method proposed by Wolf, Swift, Swinney and Vastano (1985). The maximum Lyapunov exponent above $\omega = 1.08$ is positive, and this means the occurrence of chaotic vibrations.

Experimental results are shown in Figure 6.27. The setup is the same type as that shown in Figure 6.12, but the dimensions are different. The shaft is 700 mm ($a = 125$ mm, $b = 575$ mm) in length and 12 mm in diameter, and the disk is 300 mm in diameter and 14 mm in thickness. The lower end of the shaft is supported by a single-row deep-groove ball bearing (No.6204) and the upper end is supported by a double-row self-aligning ball bearing (No.1200). This result was obtained when the center of the upper bearing shifted from the centerline of the lower bearing by about 4 mm.

References

Childs, D. (1993) *Turbomachinery Rotordynamics*, John Wiley & Sons, Inc., New York.

Ehrich, F.F. (1988) High order subharmonic response of high speed rotors in bearing clearance. *Trans. ASME, J. Vib. Acoust. Stress Reliab. Des.*, **113** (1), 50–56.

Ehrich, F.F. (1991) Some observations of chaotic vibration phenomena in high-speed rotordynamics. *Trans. ASME, J. Vib. Acoust. Stress Rehab. Des.*, **113** (1), 50–56.

Ehrich, F.F. (1992) *Handbook of Rotordynamics*, McGraw-Hill, New York.

Ehrich, F.F. and O'Connor, J.J. (1967) Stator whirl with rotors in bearing clearance. *Trans. ASME, J. Eng. Ind.*, **89** (3), 381–390.

Evan-Iwanowski, R.M. (1976) *Resonance Oscillations in Mechanical Systems*, Elsevier, New York.

Guran, A., Bajaj, A., Ishida, Y., Parkins, N., D'Eleuterio, G., and Pieme, C. (1999)

Stability of Gyroscopic Systems, World Scientific, River Edge, NJ, pp. 103–191.

Ishida, Y. and Inoue, T. (2004) Internal resonance of the Jeffcott rotor (critical speed of twice the major critical speed). *Trans. ASME.J. Vib. Acoust.*, **126** (4), 31–36.

Inoue, T., Ishida, Y., and Kondo, K. (2006) Chaotic vibration and internal resonance speed. *Trans. ASME J. Vib. Acoust. Stress Reliab. Des.*, **128** (2), 156–169.

Ishida, Y., Nagasaka, I., and Lee, S. (1996b) Forced oscillations of a continuous rotor with geometric nonlinearity (subharmonic resonance of order 1/2 and intemal resonances). *Trans. JSME, Ser.C*, **62** (596), 10–15. (in Japanese).

Meirovitch, L. (1975) *Elements of Vibration Analysis*, McGraw-Hill, New York.

Nagasaka, I., Ishida, Y., Lee, S., and Kojima, S. (1997) Forced oscillations of a horizontal rotor with geometric nonlinearity (combination and super-combination resonances and internal resonance). *Trans. JSME, Ser. C*, **63** (611), 2206–2213.

Nayfeh, A.H. and Balachandran, B. (1989) Modal interactions in dynamical and structural systems. *Appl. Mech. Rev.*, **42** (11, Pt. 2), 175–201.

Nayfeh, A.H. and Balachandran, B. (1995) *Applied Nonlinear Dynamics*, John Wiley & Sons, Inc.

Stoker, J.J. (1976) *Nonlinear Vibrations in Mechanical and Electric Systems*, Interscience, New York.

Szemplinska-Stupnica, W., Ioose, G., and Moon, F.C. (1988) *Chaotic Motions in Nonlinear Dynamical Systems*, CISM Course and Lectures, Vol. **298**, Springer-Verlag, New York, pp. 1–50.

Wolf, A., Swift, B., Swinney, H., and Vastano, J. (1985) Determining Lyapunov exponents from a time series. *Physica*, **16D** (3), 285–317.

Yamamoto, T. (1955) On the critical speed of a shaft of sub-harmonic oscillation. *Trans. JSME*, **21** (111), 853–858. (in Japanese).

Yamamoto, T. (1956) On sub-harmonic oscillations and on vibrations of peculiar modes in non-linear systems having multiple degrees of freedom. *Trans. JSME*, **22** (123), 868–875. (in Japanese).

Yamamoto, T. (1957b) On the vibrations of a rotating shaft, Chapter V: on subharmonic oscillations and on "summed and differential harmonic oscillations" in non-linear systems having multiple degrees of freedom. *Mem. Fac. Eng., Nagoya Univ.*, **9** (1), 53–71.

Yamamoto, T. (1957c) On the vibrations of a rotating shaft, Chapter VII: on the critical speeds of shaft appearing at lower rotating speeds. *Mem. Fac. Eng., Nagoya Univ.*, **9** (1), 79–86.

Yamamoto, T. (1961) On sub-harmonic and summed and differential harmonic oscillations of rotating shaft. *Bull. JSME*, **4** (13), 51–58.

Yamamoto, T. and Hayashi, S. (1963) On the response curves and the stability of "summed and differential harmonic" oscillations. *Bull. JSME*, **6** (23), 420–429.

Yamamoto, T. and Ishida, Y. (1975) Theoretical discussions of a rotating shaft with nonlinear spring characteristics. *Trans. JSME*, **4J** (345), 1374–11384.

Yamamoto, T. and Ishida, Y. (1977) Theoretical discussions of a rotating shaft with nonlinear spring characteristics. *Ing. -Arch.*, **46** (2), 125–135.

Yamamoto, T., Ishida, Y., and Kawasumi, J. (1977) Oscillations of a rotating shaft with symmetrical nonlinear spring characteristics. *Bull. JSME*, **18** (123), 965–975.

7
Self-Excited Vibrations due to Internal Damping

7.1
General Considerations

Friction is inevitable when a rotor rotates. This friction generally dampens free vibrations and suppresses the amplitude of forced vibrations. However, under some special conditions, this friction also makes the system unstable and causes self-excited vibrations.

Friction generates between two parts moving relative to each other. In rotor systems, friction is classified into two categories, based on where it works, because its effects differ qualitatively. The first category is friction that works between a moving part and a stationary part, such as a rotating disk and the surrounding air. This type of friction, called *external friction* or *external damping*, always stabilizes a system. The second category is friction that works between two rotating parts. This type of friction, called *internal friction* or *internal damping*, stabilizes the system in the speed range lower than the major critical speed but destabilizes it in the postcritical range. Since this friction causes self-excited vibrations, understanding it is essential for the safe operation of rotor systems. These internal frictions are further classified into *hysteretic damping*, which is due to friction in the shaft material, and *structural damping*, which is due to the sliding between the shaft and mounted elements such as bearings and gears.

In this chapter, internal frictions and self-excited vibrations are explained. Self-excited vibrations due to other causes, such as dry friction generated when a rotor touches a casing, collision between a bearing outer ring and a housing, and some fluid effects, are explained in Chapter 9.

7.2
Friction in Rotor Systems and Its Expressions

In this section, the characteristics and expressions of representative friction in rotor systems are explained. Because precise quantitative experiments on friction are generally difficult, most of the theories proposed for internal friction are qualitative. Some of these theories are explained below. However, it is difficult to determine which is most appropriate.

Linear and Nonlinear Rotordynamics: A Modern Treatment with Applications, Second Edition.
Yukio Ishida and Toshio Yamamoto.

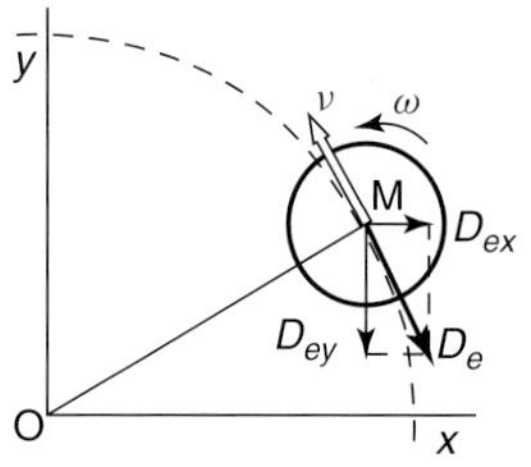

Figure 7.1 External damping (viscous damping).

7.2.1
External Damping

As shown in Figure 7.1, external damping works on the rotor when it moves in fluid, for example, air or steam. This type of damping is often represented as viscous damping because of its simplicity in theoretical treatment. Namely, the magnitude is proportional to the velocity in the static coordinate system and the direction is opposite to the velocity. It is represented as follows:

$$\left.\begin{aligned} D_{ex} &= -c\dot{x} \\ D_{ey} &= -c\dot{y} \end{aligned}\right\} \tag{7.1}$$

where c is a damping coefficient.

7.2.2
Hysteretic Internal Damping

When a rotor whirls with an angular velocity different from the rotational speed ω, the shaft material extends and contracts periodically. This hysteretic damping is due to the friction between particles or fibers in the shaft material.

When the rotor deforms, damping forces as well as elastic forces appear in the material. The *Kelvin–Voigt model* is often used as a representative model of a viscoelastic material. In this model, the following relationship holds among stress σ, strain ε, and strain velocity $\dot{\varepsilon}$ (Gasch and Pfutzner, 1975)

$$\sigma = E(\varepsilon + K\dot{\varepsilon}) \tag{7.2}$$

where E is Young's modulus and K is a coefficient relating to damping. Here, we derive the relationship between stress σ and strain ε. It is assumed that viscoelastic material extends and contracts periodically. Substituting $\varepsilon(t) = a\sin\omega t$ into this expression gives

$$\sigma = E\varepsilon + EKa\omega\cos\omega t = E\varepsilon + EK\omega\sqrt{a^2 - \varepsilon^2} \tag{7.3}$$

Then, we get

$$(\sigma - E\varepsilon)^2 + (EK\omega)^2\varepsilon^2 = (EK\omega)^2 a^2 \tag{7.4}$$

This relationship between stress and strain is represented by an elliptic hysteresis loop as shown in Figure 7.2. (This figure is enlarged in the ordinate direction. In

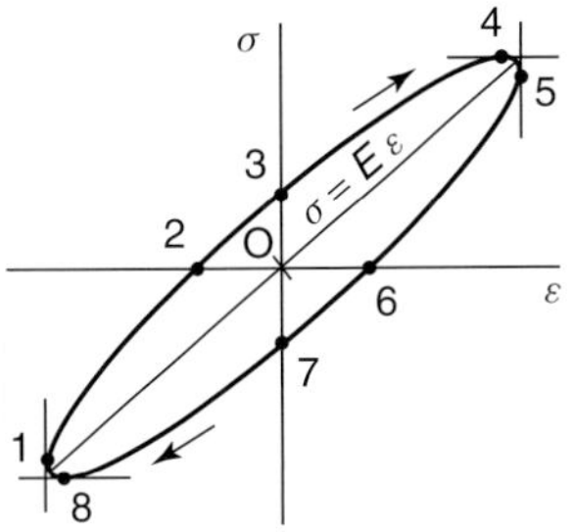

Figure 7.2 Hysteresis due to internal damping of Eq. (7.2).

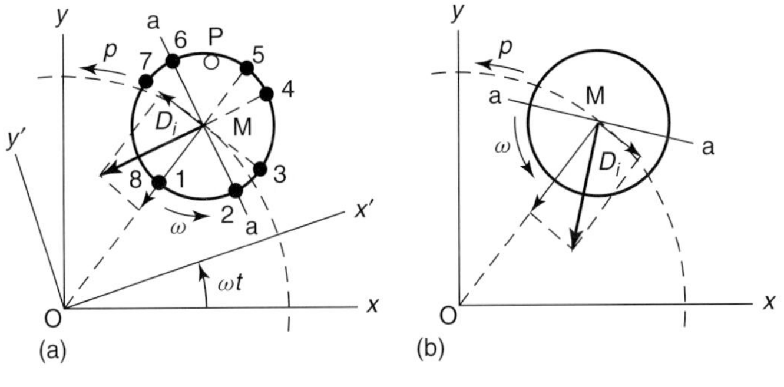

Figure 7.3 Destabilizing and damping forces due to hysteretic loop. (a) Postcritical range ($\omega > p$) and (b) precritical range ($\omega < p$).

real material such as steel, this loop is very narrow.) In the following discussion, two explanations for the destabilizing force are introduced.

Timoshenko (1955) and Den Hartog (1956) explained how the self-excited or damping forces tangent to the whirling orbit occur when such hysteretic material is used in rotors as follows: Figure 7.3 shows a shaft cross section of an elastic rotor that rotates with an angular velocity ω and whirls with an angular velocity p. It is assumed that the rotor is perfectly balanced and therefore no unbalance force works. Figure 7.3a shows the rotor in the postcritical range ($\omega > p$). The fiber at point P changes its position in the order 1, 2, …, extending or contracting with angular velocity $\omega - p$. The strain becomes maximum at points 1 and 5 and becomes zero at points 3 and 7. Points where the stress becomes maximum (points 8 and 4) or zero (points 2 and 6) are located a little before these points. The restoring force works in the direction perpendicular to the stress neutral line a-a. Therefore, the direction of the restoring force deviates from point O on the bearing centerline and the component D_i tangential to the orbit appears. This component appears as a destabilizing force that charges energy into the rotor. On the contrary, in the precritical range ($\omega < p$), the fiber at point P moves in the opposite direction and the tangential component works as a damping force as shown in Figure 7.3b.

Kimball (1925, 1932) and Kimball and Lovell (1926, 1927) gave another explanation about the occurrence of this destabilizing force. Figure 7.4 corresponds to Figure 7.3a. If internal damping does not exist, only an elastic force works. Compression and tension are generated in the half-sections on the left and right

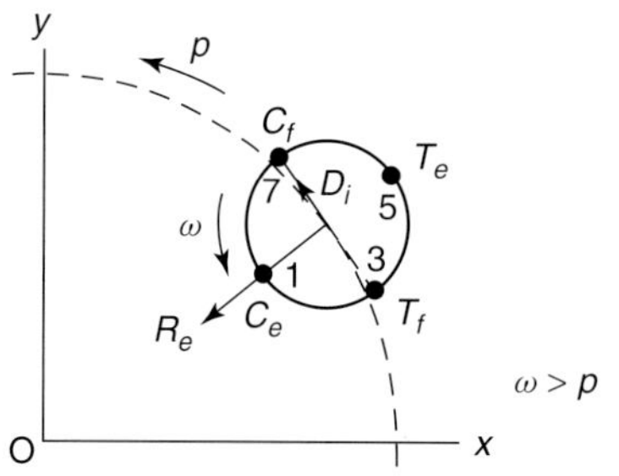

Figure 7.4 Occurrence of a destabilizing force (theory 2, Kimball, 1932).

sides of the line 7 to 3, respectively. The former state is denoted by C_e and the latter by T_e. As a result, an elastic restoring force R_e works form the tension side to the compression side, that is, in the direction from point 5 to point l. Next, the case where a force works because of internal damping is considered. When $\omega > p$, the fiber at point P contracts while moving from point 5 to point 1 via point 7. Since the internal friction works to resist this motion, a small compression works in the upper half of this cross section. This state is denoted by C_f. On the contrary, in the lower half, where the fiber extends while moving from points 1 to 5 via point 3, a small tension works. This state is denoted by T_f. If we compare these small friction forces to the above-mentioned elastic force, we know that force D_i works from the tension side to the compression side, that is, from point 3 to point 7. This component appears in the same direction as the whirling motion and works as a destabilizing force. When $\omega < p$, the fiber moves in the clockwise direction, which is inverse to the case above; therefore, we know that the component works as a resisting force.

The practical shape of the hysteresis loop differs from that of Figure 7.2. Many researchers have investigated the characteristics, but their results vary. Perhaps, this is because the results of experiments are easily affected by loading conditions, previous stress and strain conditions of the test piece, and differences in the types of material.

To take this internal damping into the equations of motion, we must represent it in mathematical form. In the following discussion, characteristics of various types of hysteresis loops and their mathematical expressions are explained. Since the internal damping force is related to the velocity relative to the rotating shaft, we consider the coordinates $O\text{-}x'y'$ rotating with the angular velocity ω. In addition, for simplicity of representation, we occasionally use the complex variables $z = x + iy$ and $z' = x' + iy'$ that were defined in Chapter 5,

1) We consider the case of the hysteresis loop in Figure 7.2, which is based on the viscous damping force of Eq. (7.2). As explained in Figure 7.3, this internal damping force is proportional to strain velocity, which is determined by the difference in whirling angular velocity p and the rotational speed ω. Therefore, we discuss this internal damping in the rotating coordinate system $O\text{-}x'y'$. Consequently, internal damping force $D'_i = D'_{ix} + iD'_{iy}$ in the rotating coordinate system is given by

$$D'_i = -h\dot{z}' \tag{7.5}$$

where $h > 0$.

The expression derived from Eq. (7.2), which is based on the assumption of viscous characteristics of material, is often used in the analysis because such a linear expression is easy to deal with. However, although this expression is suitable for understanding roughly the fundamental characteristics of internal damping, we cannot explain the phenomena well enough quantitatively.

The area of the hysteresis loop corresponds to energy that enters or dissipates during one periodic change of the stress. We can see from the Eq. (7.4) for the elliptic loop that the area is proportional to the square of the amplitude a and also depends on the frequency ω. This expression, which shows that the dissipative energy depends on the frequency ω, contradicts the results of many hysteresis loop experiments.

2) To obtain an expression that is correct quantitatively, we must first investigate the characteristics of the hysteresis loop experimentally and then derive an expression for the internal damping force based on it. The results of this experimental research on various types of material are given by Robertson (1935) and Tondl (1965), although their conclusions are not exactly the same. The most widely accepted result is given by Kimball and Lovell (1926, 1927), who determined the magnitude of internal damping by measuring the difference in the equilibrium positions of rotating and nonrotating horizontal rotors, that is, a nonrotating rotor or a rotor with no internal damping shifts downward because of gravity and stands still when it is balanced. When a rotor with internal damping rotates, the rotor shifts a little in the horizontal direction, tracing a circular path due to the tangential component in Figure 7.3a, and then balances itself. From the angle φ between these two equilibrium positions, the ratio of the internal damping force and the gravitational force can be determined. Experimental results show that the angle φ is constant in a wide range of rotational speeds. From such experiments, they derived the following characteristic: "Internal damping force, that is, a hysteresis loop, is independent of the rotational speed in the wide speed range, and its area is proportional to the square of the maximum deflection." Dimentberg (1961) also adopted this conclusion. Owing to this experimental result, the following expressions of the internal damping force are proposed.

Some researchers divided Eq. (7.5) by $(p-\omega)$ to eliminate the effect of the frequency to some extent and obtained the following expression (Tondl, 1965):

$$D_i' = -h\frac{\dot{z}'}{|p-\omega|} \tag{7.6}$$

However, this expression still depends on the frequency ω.

Dimentberg considered that hysteresis is made by adding or subtracting a constant force to an elastic force, depending on the sign of the strain velocity, as shown in Figure 7.5. He obtained

$$D_t' = -hA\frac{\dot{z}'}{|\dot{z}'|} \tag{7.7}$$

where A is the maximum deflection of a periodic oscillation. In the case of Figure 7.5, ε_{max} corresponds to A. In the case of a whirling motion of a rotor, we may adopt the maximum deflection $|z'|_{\max}$ as A. For example, when a rotor is whirling with a constant amplitude r, we adopt $A = r = \sqrt{x'^2 + y'^2}$.

Baker (1933) derived the expression for the internal damping force considering energy input or output during one cycle on the hysteresis loop in the rotating coordinate. Timoshenko (1955) introduced Baker's theory with some revision. In the following discussion, this revised theory is introduced. In Figure 7.3, which shows a rotor rotating with an angular velocity ω, let us to say a force F (= kr, k: spring constant) is applied to cancel the elastic restoring force in the radial direction and a force Q to cancel the internal damping force in the tangential direction and make the rotor stay in balance at the position shown in the figure. When the rotor rotates once, point P in the elastic shaft elongates and contracts once. Since the energy E dissipated during this one cycle is proportional to the square of the deflection OM $= |z| = r$, we can express it as

$$E = 2\pi h r^2 \tag{7.8}$$

where h is a constant. To keep the angular velocity ω constant, we must supply the corresponding amount of energy from outside the system. This energy is produced by the torque applied to the rotor. When point P rotates once, the work done by this torque is $M \cdot (2\pi)$. Therefore, we get the relationship

$$M \cdot (2\pi) = 2\pi h r^2 \tag{7.9}$$

The torque M balances the moment Q due to the couple made by the applied force Q and the reaction force $-Q$ at the bearing. Substituting $M = Qr$ into Eq. (7.9), we have

$$Q = hr \tag{7.10}$$

The force Q, that is, added virtually to stop the whirling motion, corresponds to the magnitude of the internal damping force D_i. This expression was originally derived by Baker. If we multiply a unit vector $\dot{z}'/|\dot{z}'|$ by Eq. (7.10), we obtain an expression that can represent both magnitude and direction. The result is the same as Eq. (7.7).

3) Some researchers obtained experimental results where the area of the hysteresis loop depends on the maximum amplitude but is not proportional to the square of it. Therefore, Tondl (1965) assumed the following: "Internal damping force works in the same direction as the relative velocity. Its magnitude is constant during one cycle and is a function of the maximum deflection A." Under this assumption, he generalized Eq. (7.7) and proposed the following expression:

$$D'_t = -h(A)\frac{\dot{z}'}{|\dot{z}'|} \tag{7.11}$$

This expression gives a hysteresis loop of the same shape as Figure 7.5, but its width differs from that of Figure 7.5 because $h(A)$ and hA do not have the same value. It is proposed in this expression that when the hysteresis loop occurs around the constant strain ε_0, the summation of the amplitude

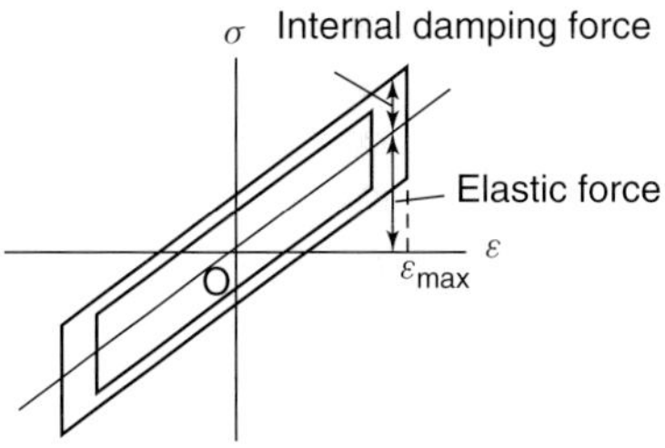

Figure 7.5 Hysteresis corresponding to Eq. (7.7).

of a periodic vibration and a constant deflection should be adopted as the maximum amplitude $A = \max|z|$. Namely, $A = \text{amp}\,|z|$ in Eq. (7.7) and $A = \text{amp}\,|z| + \varepsilon_0$ in Eq. (7.11).

7.2.3 Structural Internal Damping

In rotating machinery, mechanical elements such as disks and bearings are mounted on an elastic shaft. They are often shrink fitted on the shaft as shown in Figure 7.6. It is known that friction generated between a shaft and a hub produces internal damping force in a manner similar to the case of hysteretic damping. From the experiments on hysteretic damping mentioned above, Kimball and Lovell (1926, 1927) obtained a value of about 0.001–0.002 for the ratio of the internal damping force and elastic force in ordinary metals. Since these values are very small, the effect of internal damping diminishes if external damping exists. On the contrary, the structural damping treated here often becomes large enough to cause self-excited vibrations and is therefore of practical importance.

Friction between rotating parts produces a destabilizing force in the following way. Let us consider the case when the deflection is comparatively large and the static friction cannot prevent the sliding between the shaft and the rotor. When a deflected elastic shaft whirls with an angular velocity that differs from the rotational speed, the fibers of the shaft elongate or contract in the hub. A fiber on the shaft surface contracts while it moves from the outside to the inside of the orbit circle and elongates while it moves from the inside to the outside. Then, dry friction works from the hub to the shaft surface to prevent this motion. Since this friction

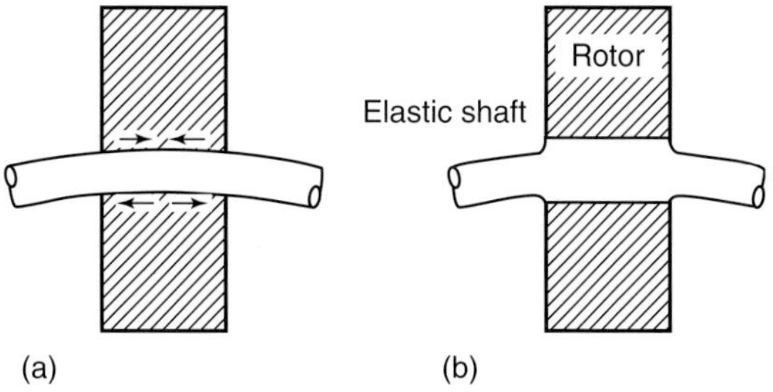

Figure 7.6 Fitting between a disk and a shaft. (a) Sliding and (b) countermeasure.

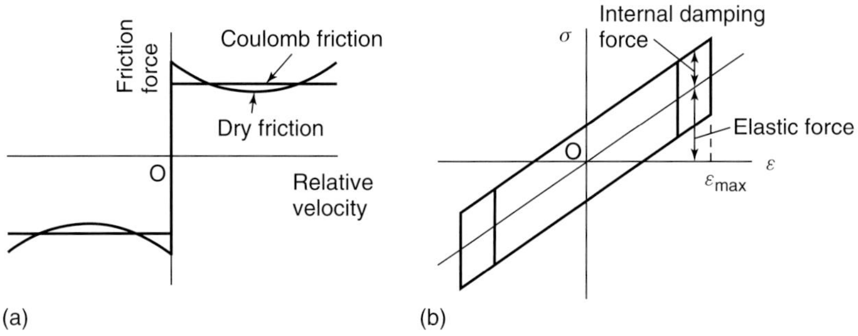

Figure 7.7 Dry friction and Coulomb friction. (a) Approximation of friction and (b) hysteresis loop of Coulomb friction.

works in the same way as Kimball's explanation mentioned above, we know that a self-excited vibration occurs. It is apparent that we can suppress this self-excited vibration by preventing sliding between the disk and the shaft. In Figure 7.6, case (b) is preferable to case (a).

To find a mathematical expression, we review the characteristics of dry friction. Friction on the contact surfaces works in a direction opposite to the relative velocity and has a magnitude proportional to the normal stress. When the normal stress is constant, the dry friction changes as a function of the relative velocity, as shown in Figure 7.7a. In a mathematical treatment, this characteristic curve is often approximated to have a constant magnitude. This simplified model is called *Coulomb damping*. The hysteresis loop for this Coulomb damping is shown in Figure 7.7b. In this case, the width of the hysteresis loop is independent of the amplitude of the stress variation. The energy dissipated in one cycle is proportional to amplitude $\varepsilon_{\max}$.

Tondl (1965) expressed the internal damping force as follows:

$$D_i' = -h\left(|\dot{z}'|\right)\frac{\dot{z}'}{|\dot{z}'|} \tag{7.12}$$

where h is a function of the magnitude of relative velocity $|\dot{z}'|$. The force made by Coulomb damping is expressed by

$$D_i' = -h\frac{\dot{z}'}{|\dot{z}'|} \tag{7.13}$$

where h is constant.

7.3 Self-Excited Vibrations due to Hysteretic Damping

Let us consider vibrations under the action of hysteretic damping. For simplicity, we investigate the deflection motion of a 2 DOF system. The equations of motion are obtained by adding external damping $c\dot{x}, c\dot{y}$ and internal damping D_{ix}, D_{iy} to

Eq. (2.4) as follows:

$$\left.\begin{aligned}\ddot{x} + c\dot{x} + p^2x &= F\cos\omega t + D_{ix}\\ \ddot{y} + c\dot{y} + p^2y &= F\sin\omega t + D_{iy}\end{aligned}\right\} \qquad (7.14)$$

where $p = \sqrt{k/m}$ and $F = e\omega^2$. For an analysis of self-excited vibrations, use of the complex variable $z = x + iy$ often makes the analysis easy. The equation of motion in a complex variable expression is

$$\ddot{z} + c\dot{z} + p^2z = Fe^{iwt} + D_i \qquad (7.15)$$

where $D_i = D_{ix} + iD_{iy}$. The various expressions for internal damping forces mentioned above are substituted into D_i. The following transformations to the quantities in the rotating coordinate system are used if necessary:

$$z = z'e^{iwt},\ D_i = D_i'e^{iwt} \qquad (7.16)$$

7.3.1 System with Linear Internal Damping Force

Here we analyze the case of a linear damping force, expressed by Eq. (7.5). From Eq. (7.16), we get

$$D_i = D_i'e^{iwt} = -h\dot{z}e^{iwt} = -h(\dot{z} - iwz) \qquad (7.17)$$

Substituting this into Eq. (7.15), we have the equation of motion. The real number expression for it is

$$\left.\begin{aligned}\ddot{x} + c\dot{x} + h(\dot{x} + \omega y) + p^2x &= F\cos\omega t\\ \ddot{y} + c\dot{y} + h(\dot{y} - \omega x) + p^2y &= F\sin\omega t\end{aligned}\right\} \qquad (7.18)$$

The solution is given by the summation of a forced solution with frequency ω and a free solution, which is a solution of the homogeneous equations.

We assume the forced solution as

$$\left.\begin{aligned}x &= P\cos(\omega t + \beta)\\ y &= P\sin(\omega t + \beta)\end{aligned}\right\} \qquad (7.19)$$

Substituting Eq. (7.18) and comparing the terms with frequency ω, we have

$$\left.\begin{aligned}P &= \frac{F}{\sqrt{\left(p^2 - \omega^2\right)^2 + c^2\omega^2}},\\ \beta &= \tan^{-1}\frac{-c\omega}{p^2 - \omega^2}\end{aligned}\right. \qquad (7.20)$$

This solution is the same as the case of no internal damping. This is understandable because the fibers in the shaft with the angular velocity ω never change their length; therefore, internal damping does not work.

In the analysis of free vibration, we use complex numbers for convenience. The equations of motion are obtained by putting $F = 0$ in Eq. (7.18) as follows:

$$\ddot{z} + c\dot{z} + h(\dot{z} - i\omega z) + p^2z = 0 \qquad (7.21)$$

We assume the solution as

$$z = Ae^{\lambda t} \tag{7.22}$$

Substituting this into Eq. (7.21), we have the frequency equation:

$$\lambda^2 + (c+h)\lambda + p^2 - i\omega h = 0 \tag{7.23}$$

The two roots λ_1 and λ_2 are

$$\lambda_1, \lambda_2 = \frac{1}{2}\left\{-(c+h) \pm \sqrt{(c+h)^2 - 4\left(p^2 - i\omega h\right)}\right\} \tag{7.24}$$

Hence, Eq. (7.22) becomes

$$z = A_1 \exp(\lambda_1 t) + A_2 \exp(\lambda_2 t) \tag{7.25}$$

For convenience, we sometimes use this type of expression instead of Eq. (7.22) for an exponential function. We approximate Eq. (7.24) as follows. If we assume that the damping coefficients c and h are small in $O(\varepsilon)$ and use the approximation $\sqrt{1+\varepsilon} \approx 1 + \varepsilon/2$, we have

$$\left.\begin{aligned} \lambda_1 &= -\frac{1}{2}\left(c + h - \frac{\omega h}{p}\right) + ip \\ \lambda_2 &= -\frac{1}{2}\left(c + h + \frac{\omega h}{p}\right) - ip \end{aligned}\right\} \tag{7.26}$$

Therefore, Eq. (7.25) becomes

$$z = A_1 \exp\left[-\frac{1}{2}\left(c + h - \frac{\omega h}{p}\right)t\right] e^{ipt} + A_2 \exp\left[-\frac{1}{2}\left(c + h + \frac{\omega h}{p}\right)t\right] e^{-ipt} \tag{7.27}$$

Since the quantity in brackets in the second term is always negative, the second term dampens and diminishes. The quantity in brackets in the first term becomes positive when

$$-\left(c + h - \frac{\omega h}{p}\right) > 0 \quad \therefore \omega > \left(1 + \frac{c}{h}\right)p \tag{7.28}$$

When this condition is satisfied, a self-excited vibration with frequency p occurs.

Stability of the solution is easily determined by the Routh–Hurwitz criteria for equations with complex coefficients, which is explained in Appendix F. Since the expression $\lambda = i\mu$ is used in this appendix, we must transform Eq. (7.23) into an expression on μ. Then we get $\mu^2 - i(c+h)\mu - p^2 + i\omega h = 0$. From Appendix F, the imaginary part of μ is positive, that is, the real part of $\lambda = i\mu$ becomes negative when the following condition is satisfied:

$$\begin{aligned} &-\begin{vmatrix} 1 & 0 \\ 0 & -(c+h) \end{vmatrix} = c + h > 0 \quad \text{(i)} \\ &\begin{vmatrix} 1 & 0 & -p^2 & 0 \\ 0 & -(c+h) & h\omega & 0 \\ 0 & 1 & 0 & -p^2 \\ 0 & 0 & -(c+h) & h\omega \end{vmatrix} = \left\{p(c+h) - h\omega\right\}\left\{p(c+h) + h\omega\right\} > 0 \quad \text{(ii)} \end{aligned} \tag{7.29}$$

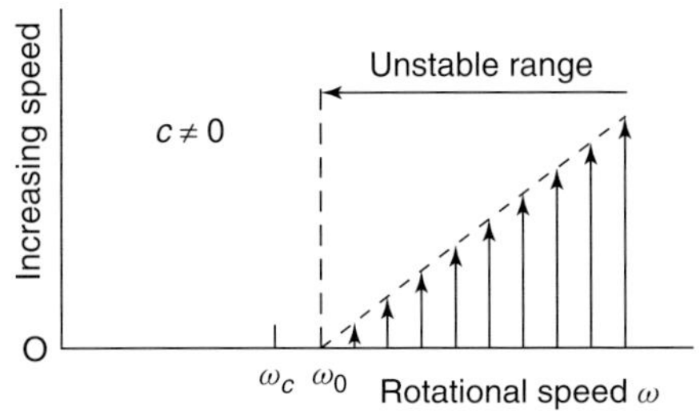

Figure 7.8 Self-excited vibration in a system with Eq. (7.5).

Since the external damping coefficient c and the internal damping coefficient h are positive, condition (i) always holds. Therefore, from condition (ii), we know that the system is stable when $p(c + h) - h\omega > 0$, which coincides with Eq. (7.28).

This result is illustrated in Figure 7.8. When the external damping does not exist ($c = 0$), the major critical speed ω_c is the boundary of stable and unstable speed ranges. When the external damping c exists, the onset speed of instability, $\omega_0 = (1 + c/h)p$, moves to the higher-speed side and self-excited vibrations occur in the range $\omega > \omega_0$.

As seen from Eq. (7.27), this self-excited vibration is a forward whirl with frequency p and its amplitude increases more rapidly as the rotational speed increases. In Figure 7.8, this increasing speed is denoted by the length of arrows.

7.3.2
System with Nonlinear Internal Damping Force

In this section, a system with an internal damping force whose magnitude is proportional to the deflection is explained. Let us assume that a rotor is executing a circular whirling motion with a constant amplitude c in the range higher than the major critical speed, that is, $\omega > \omega_c = p$. Substituting $z = re^{ip}$ and $A = r$ into Eq. (7.7) and considering the relationships $D_i' = D_i e^{-i\omega t}$ and $z' = ze^{-i\omega t} = re^{i(p-\omega)t}$, we get

$$D_i = -hr\frac{ir(p-\omega)e^{i(p-\omega)t}}{\left|ir(p-\omega)e^{i(p-\omega)t}\right|}e^{i\omega t} = ihz = h(-y + ix) \tag{7.30}$$

This force works in the tangential direction of the circular orbit as shown in Figure 7.9. In addition, an external damping force with magnitude $|c\dot{z}| = cpr$ also works. Therefore, when $|D_i| = hr > |D_e| = cpr$, that is, $h > cp$ holds, the entire speed range above the major critical speed becomes unstable and a self-excited vibration occurs. A self-excited vibration does not occur when $h < cp$.

Next, we investigate the growing speed using Baker's theory (1933). A forward whirling motion above the major critical speed ($\omega > \omega_c = p$) is governed by the following equation, which is obtained from Eq. (7.15) with $F = 0$ and Eq. (7.30).

$$\ddot{z} + c\dot{z} + p^2 z - ihz = 0 \tag{7.31}$$

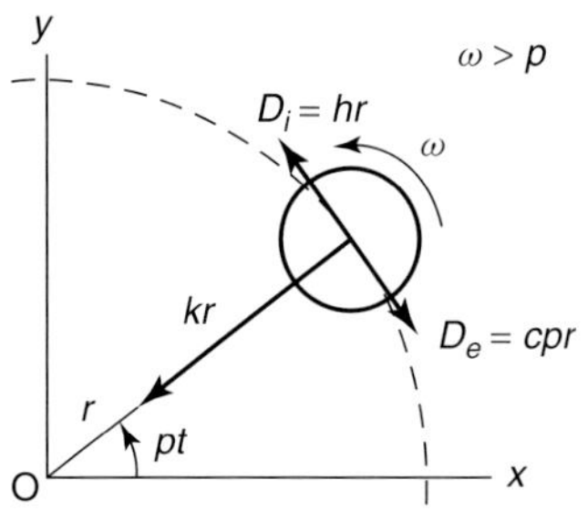

Figure 7.9 Internal and external damping forces (hysteretic damping).

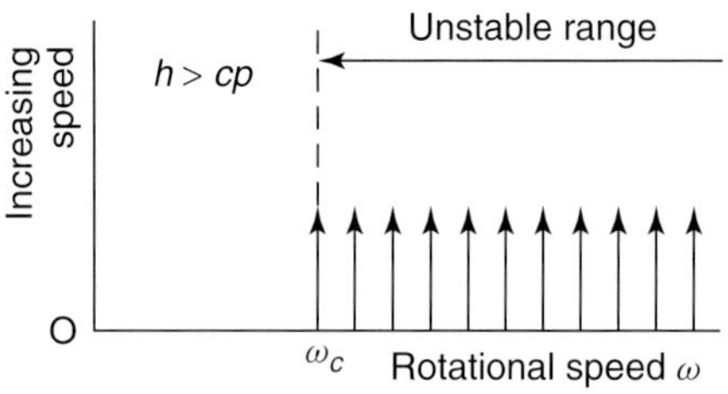

Figure 7.10 Self-excited oscillation in a system with Eq. (7.7).

Here, we assume a whirling motion whose amplitude does not increase so much after one cycle and therefore can use Eq. (7.30) approximately. Substituting a free solution in the form of Eq. (7.22), we have the following frequency equation:

$$\lambda^2 + c\lambda + p^2 - ih = 0 \tag{7.32}$$

This has two roots, given by

$$\lambda_1, \lambda_2 = \frac{1}{2}\left\{-c \pm \sqrt{c^2 - 4(p^2 - ih)}\right\} \approx -\frac{1}{2}(c \mp \frac{h}{p}) \pm ip \tag{7.33}$$

However, we considered only a forward whirl that corresponds to positive λ and thus only λ_1 is appropriate. Therefore, the self-excited vibration is given by

$$z = A_1 \exp\left[-\frac{1}{2}\left(c - \frac{h}{p}\right)t\right] e^{ipt} \tag{7.34}$$

We see from this expression that in the speed range above the major critical speed, the system is unstable when $-(c - h/p)/2 > 0$ (i.e., $h > cp$) and stable when $h < cp$. The increasing speed of the self-excited vibration is determined by the quantity in brackets. Since the natural frequency p is constant in this rotor system, the increasing speed is constant in the postcritical range.

The result is shown in Figure 7.10. Unlike Figure 7.8, the boundary of the unstable range does not change even if external damping exists and the increasing speed of the amplitude does not depend on the rotational speed. Figure 7.10, based on experimental results on the hysteresis loop, coincides with practical phenomena.

7.4 Self-Excited Vibrations due to Structural Damping

Vibration due to structural damping is explained in this section. Unlike Section 7.3, where a 2 DOF deflection model is used, a 2 DOF inclination model with gyroscopic moment is adopted here. The equations of motion are given by

$$\left.\begin{aligned}\ddot{\theta}_x + i_p\omega\dot{\theta}_y + c\dot{\theta}_x + \theta_x &= F\cos\omega t + D_{ix}\\ \ddot{\theta}_y - i_p\omega\dot{\theta}_x + c\dot{\theta}_y + \theta_y &= F\sin\omega t + D_{iy}\end{aligned}\right\} \tag{7.35}$$

where $F = (1 - i_p)\tau\omega^2$. We can represent them using a complex variable $z_\theta = \theta_x + i\theta_y$ as follows:

$$\ddot{z}_\theta - ii_p\omega\dot{z}_\theta + c\dot{z}_\theta + z_\vartheta = Fe^{iwt} + D_i \tag{7.36}$$

The term D_i is given by Eq. (7.13), representing an internal damping force due to Coulomb damping. This term is represented in a static coordinate system as follows:

$$D_i = D_i' e^{i\omega t} = -h\frac{z_\theta - i\omega z_\theta}{|z_\theta - i\omega z_\theta|} = -h\frac{(\dot{\theta}_x + \omega\theta_y) + i(\dot{\theta}_y - \omega\theta_x)}{\sqrt{(\dot{\theta}_x + \omega\theta_y)^2 + (\dot{\theta}_y - \omega\theta_x)^2}} \tag{7.37}$$

Since this term is nonlinear, analysis of Eqs. (7.35) and (7.36) is generally difficult. In Section 7.3.2, where we analyzed a system with nonlinear terms given by Eq. (7.7), the internal damping force was approximated by a linear expression on the assumption that the orbit is a circular whirling motion of a forward whirling mode. We could adopt this assumption in Section 7.3.2 because the solution is composed of only a self-excited vibration.

Since an unbalance force is considered in this section, the orbit has a complex shape composed of a self-excited vibration and a harmonic vibration. The analysis is performed without approximate replacement of the nonlinear internal damping force by other simple expressions. First, we investigate what types of phenomena occur by integrating Eq. (7.35) using the Runge–Kutta method. The parameter value $i_p = 0.3$ is adopted. Figure 7.11 shows frequencies of self-excited vibrations

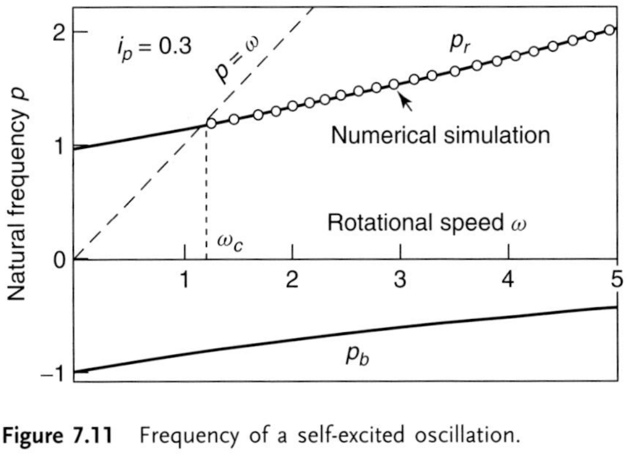

Figure 7.11 Frequency of a self-excited oscillation.

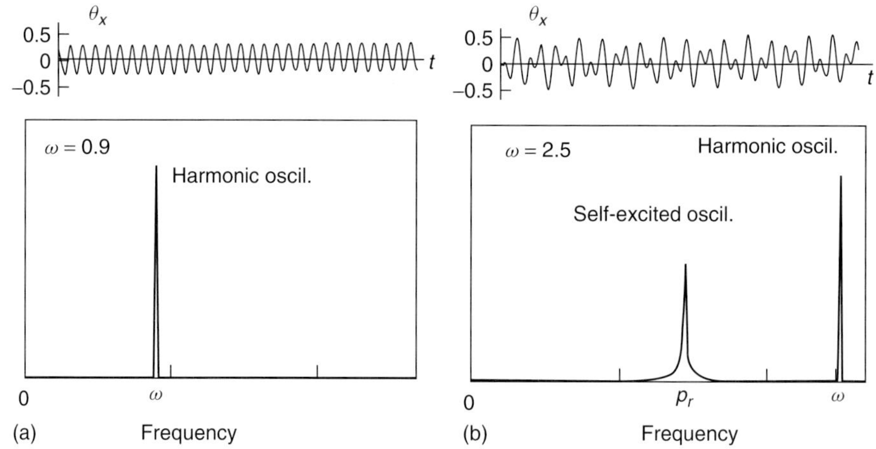

Figure 7.12 Time histories and their spectra. (a) Precritical range $\omega = 0.9$ and (b) postcritical range $\omega = 2.5$.

obtained numerically. A theoretical frequency diagram of a corresponding system but with no internal damping is shown in Figure 7.11 for comparison. Figure 7.12 shows time histories and their spectrums. In Figure 7.12a, which is a case in the precritical speed region, only a harmonic vibration occurs in response to an unbalance force. On the other hand, in Figure 7.12b, which is a case in the postcritical speed range, a self-excited vibration of a forward whirling mode with frequency p_f occurs in addition to a harmonic component. In this case, the self-excited vibration does not continue to increase but saturates and converges to a steady-state vibration with constant amplitude. This vibration, called a *self-sustained vibration*, forms a limit cycle in a phase plane.

Next, we derive response curves of the self-sustained vibration. From the results of numerical simulation, we can assume the solution as follows:

$$\begin{aligned} \theta_x &= R\cos\left(p_f t + \delta_f\right) + P\cos(\omega t + \beta) \\ \theta_y &= R\sin\left(p_f t + \delta_f\right) + P\sin(\omega t + \beta) \end{aligned} \tag{7.38}$$

We do not consider a trivial solution here since Eq. (7.37) becomes indeterminate for such a solution. We substitute this solution into Eq. (7.35) and compare the coefficients of the terms with frequencies p_f and ω on the right- and left-hand sides, respectively, to an accuracy of $O(\varepsilon)$. It is assumed that amplitudes R and P and phase angles δ_f and β change slowly with time. Concerning nonlinear terms, we can obtain the following simple expression if we note that $\omega > p_f$ and h is in $O(\varepsilon)$:

$$D_i = h\left\{i\cos\left(p_f t + \delta_f\right) - \sin\left(p_f t + \delta_f\right)\right\} \tag{7.39}$$

Therefore, the terms in Eq. (7.35) become

$$\begin{aligned} D_{ix} &= -h\sin\left(p_f t + \delta_f\right) \\ D_{iy} &= h\cos\left(p_f t + \delta_f\right) \end{aligned} \tag{7.40}$$

From a comparison of the terms with frequency p_f, we have

$$\begin{aligned}\left(2p_f - i_p\omega\right)\dot{R} &= -cp_f\omega R + h \\ \left(2p_f - i_p\omega\right)R\dot{\delta}_f &= 0\end{aligned} \tag{7.41}$$

where the relationship $G(p_f) = 0$ obtained from Eq. (6.10) is used in the derivation. From a comparison of the terms with frequency ω, we have

$$\begin{aligned}\left(2 - i_p\omega\right)\dot{P} &= -c\omega P - F\sin\beta \\ \left(2 - i_p\right)\omega P\dot{\beta} &= G(\omega)P - F\cos\beta\end{aligned} \tag{7.42}$$

where $G(\omega) = 1 + i_p\omega^2 - \omega^2$. Putting $\dot{R} = \dot{\delta}_f = 0$ in Eq. (7.41), we have the amplitude $R = R_0$ of the self-excited vibration in the postcritical speed range $(\omega > \omega_c)$:

$$R_0 = \frac{h}{cp_f} \tag{7.43}$$

Steady-state solutions of a harmonic component, $P = P_0$ and $\beta = \beta_0$, are obtained by putting $\dot{P} = \dot{\beta} = 0$ in Eq. (7.42) as follows:

$$P_0 = \frac{F}{\sqrt{G(\omega)^2 + c^2\omega^2}}, \quad \tan\beta_0 = \frac{-c\omega}{G(\omega)} \tag{7.44}$$

The result is the same as that of a system with no internal damping.

Concerning the solution below the major critical speed, we assume the solution in the same form as Eq. (7.38) and execute the same calculation. Corresponding to the first expression of Eq. (7.41), we have $(2p_f - i_p\omega)\dot{R} = -cp_f R - h$. Therefore, the steady-state solution is given by $cp_f R_0 + h = 0$. However, this expression gives no positive value for R_0; thus, no self-excited vibration occurs.

The stability of these steady-state solutions can be determined by the same procedure as shown in Chapter 6 for nonlinear resonances. The final result is illustrated in Figure 7.13 (Ishida and Yamamoto, 1993).

Unlike the case of internal material damping, a self-excited vibration occurs above the major critical speed, which converges to a certain amplitude. The magnitude

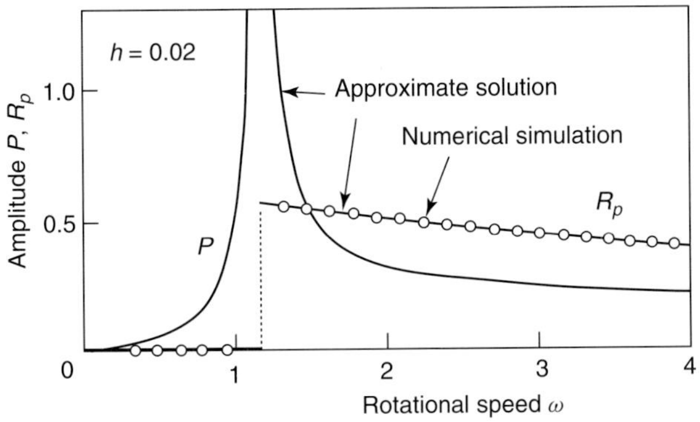

Figure 7.13 Response of a system with internal damping.

of this amplitude is almost constant above the major critical speed. However, it decreases gradually as the rotational speed increases because the natural frequency p_f in the denominator of Eq. (7.43) increases gradually owing to the gyroscopic moment. This result was ascertained by experiment (Ishida and Yamamoto, 1993).

References

Baker, J.G. (1933) Self-induced vibration. *Trans. ASME, J. Appl. Mech.*, **1** (l), 5–13.

Den Hartog, J.P. (1956) *Mechanical Vibration*, 4th edn, McGraw-Hill, New York.

Dimentberg, F.M. (1961) *Flexural Vibrations of Rotating Shafts*, Applied Science Publishers, London.

Gasch, R. and Pfutzner, H. (1975) *Rotordynamik: Eine Einfuhrung*, Springer-Verlag, Berlin.

Ishida, Y. and Yamamoto, T. (1993) Forced oscillations of a rotating shaft with nonlinear spring characteristics and internal damping (1/2-order subharmonic oscillations and entertainment). *Nonlinear Dyn.*, **4** (5), 413–43l.

Kimball, A.L. (1925) Internal friction as a cause of shaft whirling. *Philos. Mag. A*, **49**, 724–727.

Kimball, A.L. (1932) *Vibration Prevention in Engineering*, John Wiley & Sons, Ltd, London.

Kimball, A.L. and Lovell, D.E. (1926) Internal friction in solids. *Trans. ASME*, **48**, 479–500.

Kimball, A.L. and Lovell, D.E. (1927) Internal friction in solids. *Mech. Eng.*, **49** (5), 440–444.

Robertson, D. (1935) Hysteric influences on the whirling of rotors. *Proc. IME.*, **131**, 513–537.

Timoshenko, S. (1955) *Vibration Problems in Engineering*, 3rd edn, D. Van Nostrand, New York.

Tondl, A. (1965) *Some Problems of Rotor Dynamics*, Czechoslovak Academy of Sciences, Prague.

8
Nonstationary Vibrations during Passage through Critical Speeds

8.1
General Considerations

The rated speed of some rotating machines is above the major critical speed. For example, a steam turbine generator has a rated speed of about twice the first critical speed, and an aircraft gas turbine is operated in the speed range up to a speed above the second or third critical speed. In such rotating machines, the rotor must pass the critical speed with a small amplitude at startup or shutdown for safety.

Research on nonstationary vibrations during passage through critical speeds are classified into two categories based on whether or not they consider the relationship between a driving torque and an angular speed of a rotor. The first group investigated nonstationary dynamic characteristics on the assumption that the rotational speed changes with a constant angular acceleration. This assumption holds in a system where the driving torque is large enough not to be affected by variation of the dynamic load due to vibration, that is, in a system with a large driving torque and a small unbalance. The second group investigated them considering the mutual interaction between the vibration of the rotor and the energy source. In a system with a low driving torque compared to the unbalance, it may happen that the rotor cannot pass through the critical speed because the dynamic load increases when the amplitude becomes large at the resonance range. In such a case, the time variations of the driving torque and the rotational speed are not known beforehand; therefore, we must obtain them in the process of solving equations of motion. The representative research on this problem was done by Kononenko (1969).

It is difficult to obtain the nonstationary response analytically in most cases. Dimentberg (1961) obtained the analytical solution expressed by the Fresnel integrals when the Jeffcott rotor changes its speed with constant angular acceleration. But this method cannot be applied to other, more complex, rotor systems. The methods used most widely to tackle this type of problem are numerical methods, such as the Runge–Kutta method, the complex-FFT method utilizing signal processing theory (Ishida, Yasuda, and Murakami, 1997), the asymptotic method (Mitropol'skii, 1965), and the method of multiple scales (Nayfeh and Mook, 1979; Neal and Nayfeh, 1990). These methods can all be used with nonlinear systems.

Linear and Nonlinear Rotordynamics: A Modern Treatment with Applications, Second Edition.
Yukio Ishida and Toshio Yamamoto.

The complex-FFT method is also applicable not only to numerical simulation but also to analysis of experimental data.

In this chapter, nonstationary characteristics of a rotor passing through critical speeds are explained primarily using the numerical integration method. The asymptotic method is also explained. The complex-FFT method is discussed in Chapter 12.

8.2 Equations of Motion for Lateral Motion

Figure 8.1 shows a rotor system where a disk is mounted at the midpoint of an elastic shaft. This model is the same as that shown in Figure 2.4, but nonstationary motion is investigated in this section. First, we derive the equations of motion. The centerline of an elastic shaft passes through the geometrical center $M(x, y)$. The center of gravity $G(x_G, y_G)$ is separated by e from point M. The rotational angle of the rotor is denoted by the angle ψ, which is the angle between the line MG and the x-axis. Let the mass of the disk be m, the polar moment of inertia be I_p, the force working at the center of gravity be $\mathbf{F}(F_x, F_y)$, and the torque around the center of gravity be T_G. From Newton's second law, we have

$$m\ddot{x}_G = F_x, \quad m\ddot{y}_G = F_y, \quad I_p\ddot{\psi} = T_G \tag{8.1}$$

where the force $\mathbf{F}(F_x, F_y)$ includes an elastic force $\mathbf{F}_e(-kx, -ky)$ and an external damping force $\mathbf{D}(-c\dot{x}, -c\dot{y})$ and the torque T_G includes a driving torque $T_d(\dot{\psi})$ applied by the motor, a resisting torque $-T_r(\dot{\psi})$ due to the bearings and surrounding medium and a torque $T_e(= -L)$ due to an elastic restoring force $\mathbf{F}_e$. The torques T_d and T_r working around point M can be replaced by those around point G since the effect of a couple is equal at any point. The torque due to an external damping force $\mathbf{D}$ is ignored since it is very small compared to T_e. The torque $L = (kx)e\sin\psi - (ky)e\cos\psi$, which depends on the magnitude of vibration, is

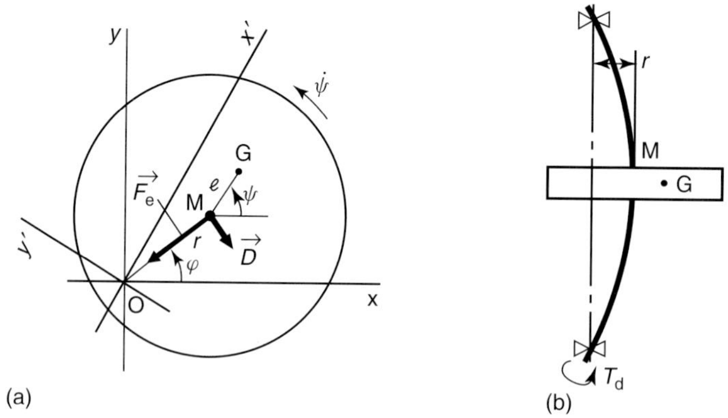

Figure 8.1 2 DOF deflection model: (a) physical model. (b) coordinates

called a *dynamic load torque*. Substituting these forces and torques on the right-hand side of Eq. (8.1) and transferring the coordinates by

$$\left.\begin{aligned} x_G &= x + e\cos\psi \\ y_G &= y + e\sin\psi \end{aligned}\right\} \tag{8.2}$$

we have

$$\left.\begin{aligned} m\ddot{x} + c\dot{x} + kx &= me\dot{\psi}^2\cos\psi + me\ddot{\psi}\sin\psi \\ m\ddot{y} + c\dot{y} + ky &= me\dot{\psi}^2\sin\psi - me\ddot{\psi}\cos\psi \\ I_p\ddot{\psi} + T_r &= T_d - ke(x\sin\psi - y\cos\psi) \end{aligned}\right\} \tag{8.3}$$

These equations are nonlinear equations in x, y, and ψ. Here, we define the following dimensionless quantities using $\delta_{st} = mg/k$ as a reference quantity. This quantity is a static deflection that occurs when it is assumed to be supported horizontally.

$$\left.\begin{aligned} &t' = t\sqrt{k/m}, \quad x' = x/\delta_{st}, \quad y' = y/\delta_{st} \\ &e' = e/\delta_{st}, \quad c' = c/\sqrt{mk}, \quad T_d' = mT_d/kI_p \\ &T_r' = mT_r/kI_p, \quad K = m\delta_{st}^2/I_p \end{aligned}\right\} \tag{8.4}$$

The equations of motion with these dimensionless quantities are expressed by

$$\left.\begin{aligned} \ddot{x} + c\dot{x} + x &= e\dot{\psi}^2\cos\psi + e\ddot{\psi}\sin\psi \\ \ddot{y} + c\dot{y} + y &= e\dot{\psi}^2\sin\psi - e\ddot{\psi}\cos\psi \\ \ddot{\psi} + T_r &= T_d - Ke(x\sin\psi - y\cos\psi) \end{aligned}\right\} \tag{8.5}$$

where the primes denoting dimensionless quantities are eliminated for simplicity. Using the transformations

$$x = r\cos\varphi, \quad y = r\sin\varphi \tag{8.6}$$

these equations are represented by the polar coordinates (r, φ) as follows:

$$\left.\begin{aligned} \ddot{r} - r\dot{\varphi}^2 + c\dot{r} + r &= e\dot{\psi}^2\cos(\psi - \varphi) + e\ddot{\psi}\sin(\psi - \varphi) \\ 2\dot{r}\dot{\varphi} + r\ddot{\varphi} + cr\dot{\varphi} &= e\dot{\psi}^2\sin(\psi - \varphi) - e\ddot{\psi}\cos(\psi - \varphi) \\ \ddot{\psi} + T_r &= T_d - Ker\sin(\psi - \varphi) \end{aligned}\right\} \tag{8.7}$$

8.3 Transition with Constant Acceleration

In this section, we consider nonstationary vibrations during passage through a critical speed with constant acceleration λ. The phenomena are investigated numerically by the Runge–Kutta method. In Figure 8.1, a rotor rotating with constant angular velocity ω_0 starts to accelerate at the instance when the unbalance

MG passes through the direction $\psi = \psi_0$. If time origin $t = 0$ is taken at the start of acceleration, the angle ψ is given by

$$\ddot{\psi} = \lambda, \quad \dot{\psi} = \lambda t + \omega_0, \quad \psi = \frac{1}{2}\lambda t^2 + \omega_0 t + \psi_0 \tag{8.8}$$

When the angle ψ is specified in this way, the first and second equations of Eq. (8.5),

$$\left.\begin{aligned} \ddot{x} + c\dot{x} + x &= e\dot{\psi}^2 \cos\psi + e\ddot{\psi}\sin\psi \\ \ddot{y} + c\dot{y} + y &= e\dot{\psi}^2 \sin\psi - e\ddot{\psi}\cos\psi \end{aligned}\right\} \tag{8.9}$$

and the first and second equations of Eq. (8.7)

$$\left.\begin{aligned} \ddot{r} - r\dot{\varphi}^2 + c\dot{r} + r &= e\dot{\psi}^2 \cos(\psi - \varphi) + e\ddot{\psi}\sin(\psi - \varphi) \\ 2\dot{r}\dot{\varphi} + r\ddot{\varphi} + cr\dot{\varphi} &= e\dot{\psi}^2 \sin(\psi - \varphi) - e\ddot{\psi}\cos(\psi - \varphi) \end{aligned}\right\} \tag{8.10}$$

can be solved independent of the third one.

Time histories obtained by integrating Eq. (8.10) numerically are shown in Figure 8.2. The deflection increases rapidly and becomes large when the rotor enters resonance range and then decreases, changing its amplitude comparatively slowly. This slow vibration is a beat phenomenon because of the coexistence of a free vibration excited at the resonance region and a forced response due to an unbalance. Figure 8.3a corresponds to an envelope of the time history in Figure 8.2. The unique variation in the angular velocity in Figure 8.3b was first shown by Yanabe and Tamura (1971).

We can understand this phenomenon if we observe the whirling motion in the rotating coordinate system O-$x'y'$ rotating with the same angular velocity $\dot{\psi}$ as the rotor as follows (Ishida, Ikeda, and Yamamoto, 1987): In Figure 8.1, the rotating coordinates x', y' and the polar coordinates r, φ have the following relationship

$$\left.\begin{aligned} x' &= r\cos(\varphi - \psi) = r\cos\varphi' \\ y' &= r\sin(\varphi - \psi) = r\sin\varphi' \end{aligned}\right\} \tag{8.11}$$

where $\varphi' = \varphi - \psi$. In Figure 8.4a, when the rotational speed $\dot{\psi}$ is constant ($\dot{\psi} = \omega$), a steady-state vibration is represented by a fixed point. If the rotational

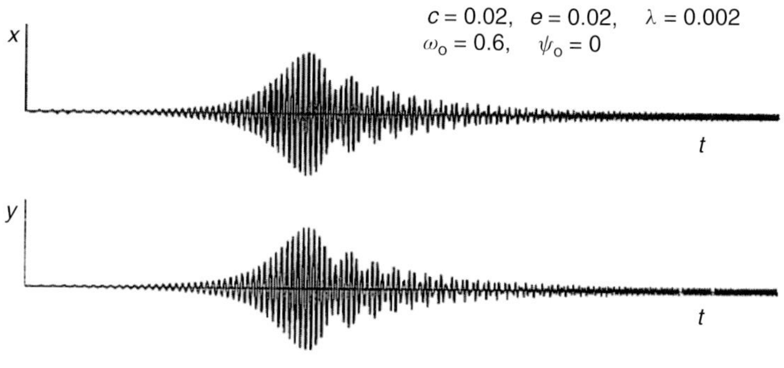

Figure 8.2 Time histories (constant acceleration).

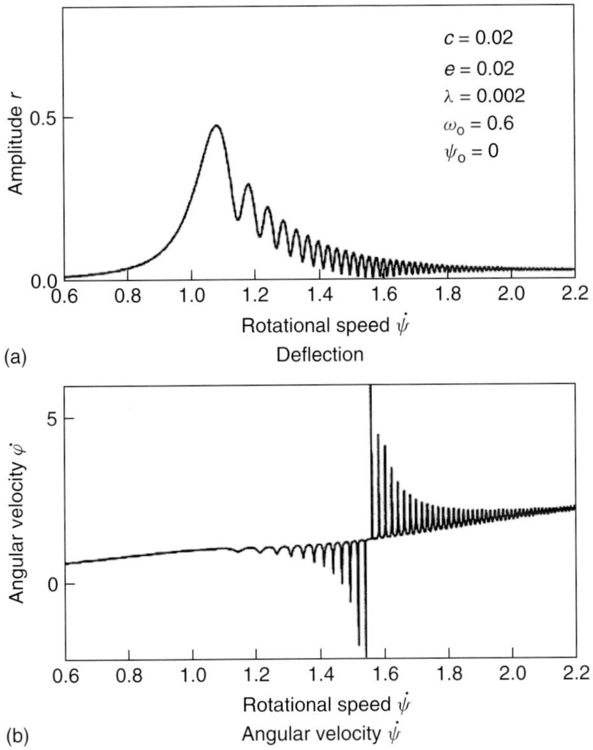

Figure 8.3 Nonstationary response during transition. (a) Deflection and (b) angular velocity $\dot{\psi}$.

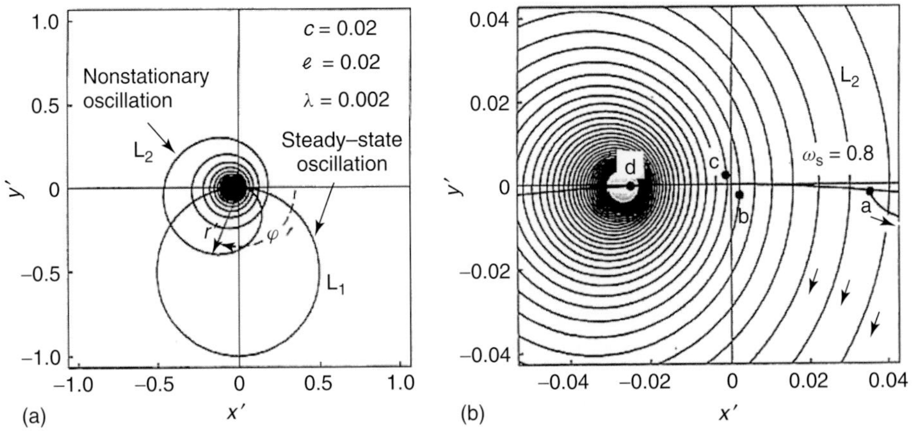

Figure 8.4 Orbit in the rotating coordinates (Ishida, Ikeda, and Yamamoto, 1987). (a) Stationary and nonstationary solutions and (b) enlarged orbit.

speed ω changes quasistatically, such a fixed point moves and traces the line L_1. If the rotational speed changes with some angular acceleration, the point representing the nonstationary vibration moves, pursuing the point corresponding to the steady-state solution on the line L_1. Figure 8.4b is an enlarged figure from the center area of Figure 8.4a. For example, when the rotor is accelerated from $\omega_s = 0.8$, the shaft center M leaves from point a, increases its deflection rapidly, traces a spiral orbit L_2, and approaches to a point corresponding to a steady-state solution on L_1. The final stage of the spiral line is not drawn; however, it will reach to point d if the acceleration is continued. The deflection r in Figure 8.3a is the distance between the center of the rotor and the origin O. Since the angular acceleration $\dot{\psi}$ increases linearly in this case of constant acceleration, a variation of the phase angle $\dot{\varphi}$ appears correspondingly in the angle $\dot{\varphi}'$. When the origin O is located inside the spiral orbit, the sign of $\dot{\varphi}'(<0)$ does not change, although its magnitude varies. The variation of $\dot{\varphi}'$ becomes larger as the rotor approaches the origin. When the origin moves from inside to outside due to the decreasing orbit, the angular velocity $\dot{\varphi}'$ changes abruptly from a large negative value (e.g., in the case of point b) to a large positive value (for point c), as shown in the vicinity of $\dot{\psi} = 1.55$ in Figure 8.3b. However, we should note that no specific change occurs in the whirling motion itself. Subsequently, the circular orbit leaves from the origin while decreasing its diameter, and therefore the variation of the angular velocity decreases.

Figure 8.5 shows an influence of the angular acceleration to the variation of the radius of the whirling orbit. The curve for $\lambda = 0$ corresponds to the resonance curve for steady-state vibrations. As the acceleration λ increases, the maximum radius r_{max} decreases.

When we design rotating machinery, it is sometimes necessary to estimate the maximum radius r_{max} during passage through the critical speed. Since it is difficult to obtain this value theoretically, the results of approximate analysis were given as a diagram (Lewis, 1932) and were also given as approximate expressions. For example, Yanabe and Tamura (1972) gave the following expression that gives the maximum radius r_{max} and the angular velocity $\dot{\psi}_{max}$ corresponding to this

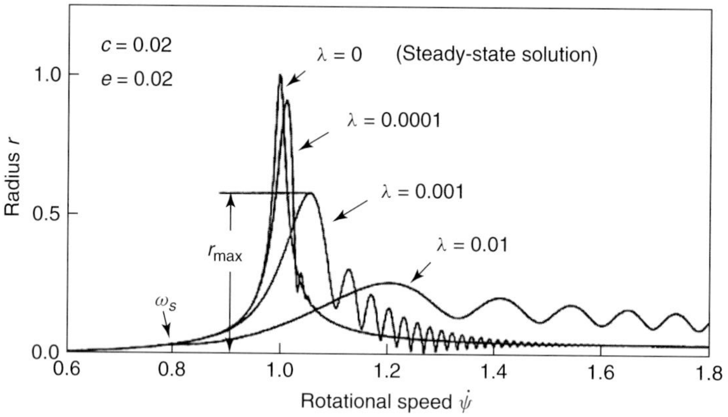

Figure 8.5 Influence of angular acceleration λ to radius change.

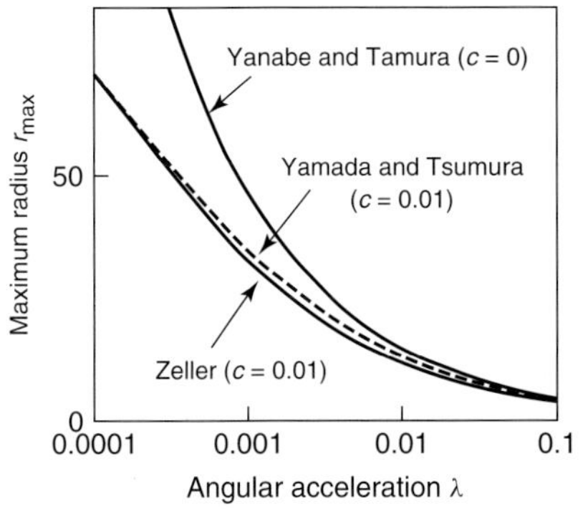

Figure 8.6 Comparison of various approximations for the maximum amplitude.

maximum amplitude in a case with no damping.

$$\left.\begin{aligned} \dot{\psi}_{\max} &\approx 1+\sqrt{\frac{3\pi}{2}\lambda} \approx 1+2.17\sqrt{\lambda} \\ r_{\max} &\approx \sqrt{\frac{\pi}{2\lambda}}\left[1+\frac{1}{\pi\sqrt{3}(1+\sqrt{3\pi\lambda/8})}\right] \approx 1.48\frac{1}{\sqrt{\lambda}} \end{aligned}\right\} \tag{8.12}$$

In the case of a system with damping, Yamada and Tsumura (1951) proposed the following approximate expression

$$r_{\max} \approx \frac{3.78}{\sqrt{2\pi\lambda}} \exp\left\{-1.16\left(\frac{1}{2\pi\lambda}\right)^{0.379} c^{0.7}\right\} \tag{8.13}$$

while Zeller (1949) did the following:

$$r_{\max} \approx \frac{1}{c\sqrt{1-(c/2)^2}}\left\{1-\exp\left(-c\frac{\pi}{\sqrt{2\lambda}}\right)\right\} \tag{8.14}$$

The results given by these approximate expressions are compared in Figure 8.6. For damping $c = 0.01$, the results obtained by Yamada and Tsumura and that by Zeller agree well.

8.4 Transition with Limited Driving Torque

8.4.1 Characteristics of Power Sources

In the preceding section, the driving source is not affected by the vibration of the rotor, and the rotational speed increases with a constant angular acceleration. Such a driving source is called an *ideal driving energy source*. However, some rotating machines do not have enough power; therefore, the driving source experiences reciprocal influence from the vibration of the rotor. In such a system, the vibratory subsystem and the driving subsystem interact mutually. In the analysis of a rotor system with a limited power supply, we must solve simultaneously a set of three

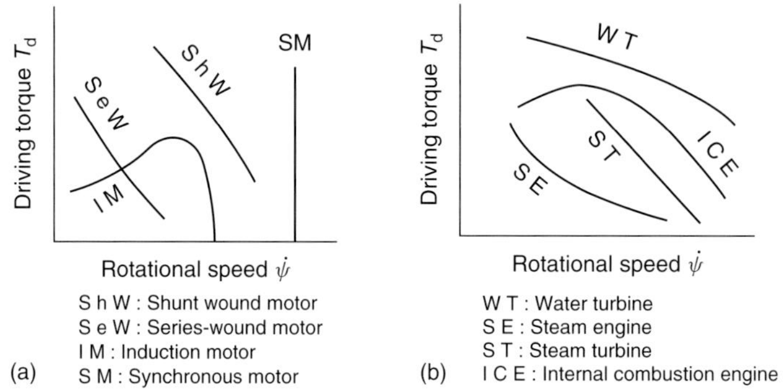

Figure 8.7 (a,b) Driving torque of various machinery (Kononenko, 1969).

equations in Eq. (8.5) or (8.7) considering the characteristic of a driving torque T_d and that of a resisting torque T_r. In general, a driving torque T_d and a resisting torque T_r are functions of angle ψ and angular velocity $\dot{\psi}$. However, in the following, we assume that they are functions of $\dot{\psi}$ only.

The characteristics of driving torque are generally measured with a constant angular velocity. If the angular acceleration is not so large, the result can be used in the study of nonstationary vibrations. The characteristics of driving torque differ depending on the driving source. For example, Figure 8.7, showing the characteristics of various types of driving energy sources, is a figure from the book by Kononenko (1969). Many driving sources have negative gradients, that is, the driving torque decreases as the rotational speed increases. In the analysis, we must represent this characteristic curve $T_d(\dot{\psi})$ by an appropriate mathematical expression.

We must also determine the resisting torque $T_r(\dot{\psi})$, which generally becomes large as the rotational speed increases. However, for simplicity, the resisting torque is neglected in the following analysis. (We may consider $T_d - T_r$ to be a substantial driving torque.)

Since the driving torque decreases as the velocity increases in many cases shown in Figure 8.7, we analyze the case

$$T_d(\dot{\psi}) = A - B\dot{\psi} \tag{8.15}$$

Substituting this equation into Eq. (8.7), we have

$$\left.\begin{aligned} &\ddot{r} - r\dot{\varphi}^2 + c\dot{r} + r = e\dot{\psi}^2\cos(\psi - \varphi) + e\ddot{\psi}\sin(\psi - \varphi) \\ &2\dot{r}\dot{\varphi} + r\ddot{\varphi} + cr\dot{\varphi} = e\dot{\psi}^2\sin(\psi - \varphi) - e\ddot{\psi}\cos(\psi - \varphi) \\ &\ddot{\psi} = A - B\dot{\psi} - Ker\sin(\psi - \varphi) \end{aligned}\right\} \tag{8.16}$$

8.4.2
Steady-State Vibration

Before analyzing nonstationary vibration, we consider a steady-state vibration. Let us assume that a driving torque cancels the other torques and a rotor executes a steady-state whirling motion with radius R_0 and phase angle $\beta_0(=\varphi-\psi)$ at the rotational speed ω_0. Substituting the solution

$$\left.\begin{aligned} x &= R_0\cos(\omega_0 t+\beta_0)\\ y &= R_0\sin(\omega_0 t+\beta_0)\\ \dot{\psi} &= \omega_0 \end{aligned}\right\} \tag{8.17}$$

namely, $r=R_0$, $\varphi=\omega_0 t+\beta_0$, and $\psi=\omega_0 t$ into Eq. (8.16), we have

$$\left.\begin{aligned} &-R_0{\omega_0}^2+R_0=e{\omega_0}^2\cos(-\beta_0)\\ &cR_0\omega_0=e{\omega_0}^2\sin(-\beta_0)\\ &A-B\omega_0-KeR_0\sin(-\beta_0)=0 \end{aligned}\right\} \tag{8.18}$$

From the first and second equations, we have the following expression for resonance curves:

$$\left\{(1-{\omega_0}^2)^2+c^2{\omega_0}^2\right\}{R_0}^2=e^2{\omega_0}^4 \tag{8.19}$$

Kononenko (1969) explained how the steady-state rotational speed is determined when the driving torque characteristic $T_{\rm d}(\dot{\psi})$ and the dynamic load torque characteristic L are given as discussed below.

The third equation in Eq. (8.16) shows that the angular acceleration $\ddot{\psi}$ is determined by the difference in the driving torque $T_{\rm d}=A-B\dot{\psi}$ and the dynamic load torque $L=Ker\sin(\psi-\varphi)$. Let us assume A to be a control parameter in Eq. (8.15) and represent it explicitly as $T_{\rm d}(A,\omega)$, where ω is used instead of $\dot{\psi}$. The dynamic load torque is generally a function of the deflection r and the phase $\psi-\varphi$. However, since $r=R_0$ and $\psi-\varphi=-\beta_0$ are determined by ω when the rotor condition follows the resonance curve, we represent the dynamic load torque as $L(\omega)$. Substituting Eqs. (8.17–8.19) into the third equation of Eq. (8.16), we have the following expression:

$$\ddot{\psi}=T_{\rm d}(A,\omega)-L(\omega) \tag{8.20}$$

where

$$\left.\begin{aligned} &T_{\rm d}(A,\omega)=A-B\omega\\ &L(\omega)=KeR\sin(-\beta)=\frac{cKe^2\omega^3}{(1-\omega^2)^2+c^2\omega^2} \end{aligned}\right\} \tag{8.21}$$

Figure 8.8 shows how the equilibrium position is determined when the parameter A in the torque $T_{\rm d}$ is constant. In this figure, three characteristic curves for $A=A_1, A_2, A_3$ $(A_1>A_2>A_3)$ and one curve L are illustrated. For the case $A=A_2$, the rotor is accelerated in the range lower than point c_2 and in the range between points b_2 and a_2 because the driving torque is larger than the dynamic load there. On the contrary, it is decelerated in the range between points c_2 and b_2 and in the range higher than point a_2 because the relationship between the magnitudes of these torques is reversed. When a condition lower than point b_2 is given initially, the rotor approaches point c_2. But when a condition higher than point b_2 is given, the rotor approaches point a_2. From such a consideration, we know that when the

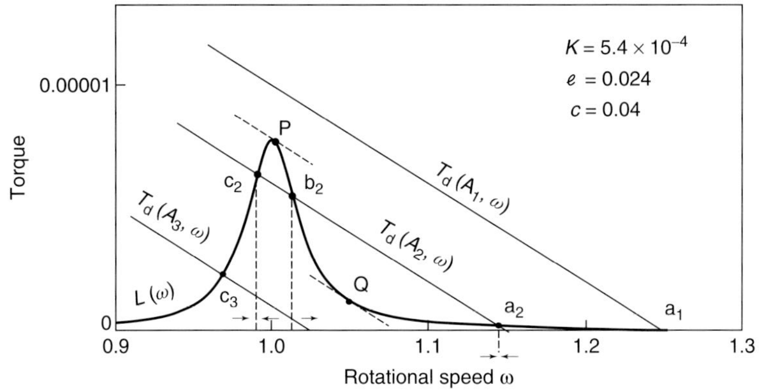

Figure 8.8 Determination of the steady-state oscillation when T_d is constant.

driving source has a characteristic curve $A = A_2$, the rotational speed finally settles down at the speed given by point c_2 or point a_2 after some initial nonstationary vibration. Similarly, a steady-state vibration occurs at point a_1 in the case of a larger driving torque $A = A_1$ and a steady-state vibration occurs at point c_3 in the case of a smaller driving torque $A = A_3$. The steady-state vibration at point b_2 for $A = A_2$ is unstable because the rotational speed leaves it after some disturbances.

Figure 8.9 shows a phenomenon when the driving torque T_d is changed quasistatically. When parameter A in the torque T_d increases, the straight line T_d moves upward and thus the cross point m moves upward on line L. When point m reaches point P_1, the rotational speed changes abruptly, that is, it jumps to point P_2. When the driving torque decreases from a steady-state vibration in the higher speed side, the cross point m moves on curve L to the lower speed side and jumps from point Q_1 to point Q_2. The speed range between points P_1 and Q_1, where no steady-state vibration exists, is called an *unstable range*. However, this unstable range is the same as that defined in Chapter 5 in the sense that no steady-state vibration is attained, but it also differs from it in the sense that the amplitude does not increase.

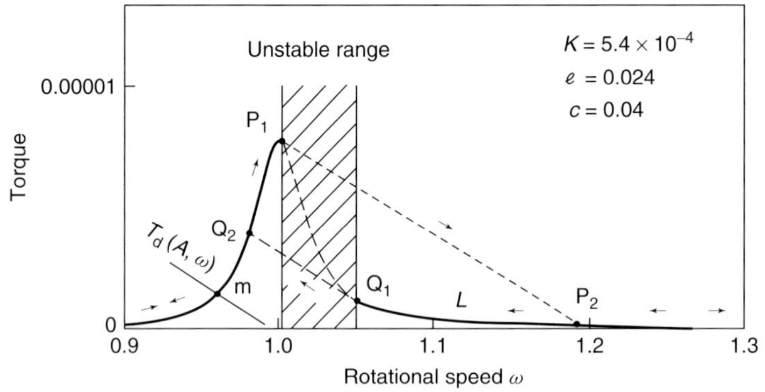

Figure 8.9 Change of the rotational speed when parameter A changed quasistatically.

8.4.3
Stability Analysis

As mentioned above, a steady-state vibration does not appear in a certain speed range in a system with a limited power supply. In this section, we investigate this problem using stability analysis. Although Kononenko (1969) demonstrated a similar stability analysis, we explain in a different way.

Equations of motion (8.16) are used here. Since the steady-state solution is given in the form of Eq. (8.7), we transform the variables r, φ, and ψ into new variables r, β, and ω, respectively, by the following transformation.

$$\dot{\psi} = \omega, \quad \varphi = \psi + \beta \tag{8.22}$$

In this transformation, we consider r, β, and ω as slowly varying functions of time and assume that $\dot{r}, \dot{\beta}$, and $\dot{\omega}$ are of $O(\varepsilon)$ and $\ddot{r}, \ddot{\beta}$, and $\ddot{\omega}$ are of $O(\varepsilon^2)$. Since we consider the phenomenon in the vicinity of the major critical speed, we can assume that the detuning $\sigma = 1 - \omega^2$ is of $O(\varepsilon)$. In addition, we also assume that an unbalance e, a damping coefficient c, and a driving torque $T_{\text{d}} = A - B\omega$ are of $O(\varepsilon)$. Substituting Eq. (8.22) into Eq. (8.16) and considering magnitudes of parameters, we obtain the following equations to an accuracy of $O(\varepsilon)$:

$$\left.\begin{aligned} 2\omega\dot{r} &= -(A - B\omega)r - Ker^2 \sin\beta - cr\omega - e\omega^2 \sin\beta \\ 2r\omega\dot{\beta} &= r\sigma - e\omega^2 \cos\beta \\ \dot{\omega} &= A - B\omega + Ker \sin\beta \end{aligned}\right\} \tag{8.23}$$

A steady-state solution is obtained from Eq. (8.18), which is the result of putting $\dot{\beta} = 0,\ \dot{r} = 0,\ \dot{\omega} = 0,\ r = R_0,\ \beta = \beta_0$, and $\omega = \omega_0$ in these equations.

To investigate the stability of solutions, we consider small deviations ξ, η, and ζ and put

$$r = R_0 + \xi, \quad \beta = \beta_0 + \eta, \quad \omega = \omega_0 + \zeta \tag{8.24}$$

We substitute these into Eq. (8.23) and expand it by the Taylor series. If we consider up to the first-order terms ξ, η, and ζ taking Eq. (8.18) into account, we get

$$\left.\begin{aligned} \dot{\xi} &= a_{11}\xi + a_{12}\eta + a_{13}\zeta \\ \dot{\eta} &= a_{21}\xi + a_{22}\eta + a_{23}\zeta \\ \dot{\zeta} &= a_{31}\xi + a_{32}\eta + a_{33}\zeta \end{aligned}\right\} \tag{8.25}$$

where

$$\left.\begin{aligned} a_{11} &= (-A + B\omega_0 - c\omega_0 - 2KeR_0 \sin\beta_0)/(2\omega_0) \\ a_{21} &= (-KeR_0{}^2 \cos\beta_0 - e\omega_0{}^2 \cos\beta_0)/(2\omega_0) \\ a_{13} &= (BR_0 - cR_0 - 2e\omega_0{}^2 \sin\beta_0)/(2\omega_0) \\ a_{21} &= \sigma_0/(2R_0\omega_0), \quad a_{22} = e\omega_0 \sin\beta_0/(2R_0) \\ a_{23} &= -2(R_0 + e\cos\beta_0)/(2R_0), \quad a_{31} = Ke\sin\beta_0 \\ a_{32} &= KeR_0 \cos\beta_0, \quad a_{33} = -B \end{aligned}\right\} \tag{8.26}$$

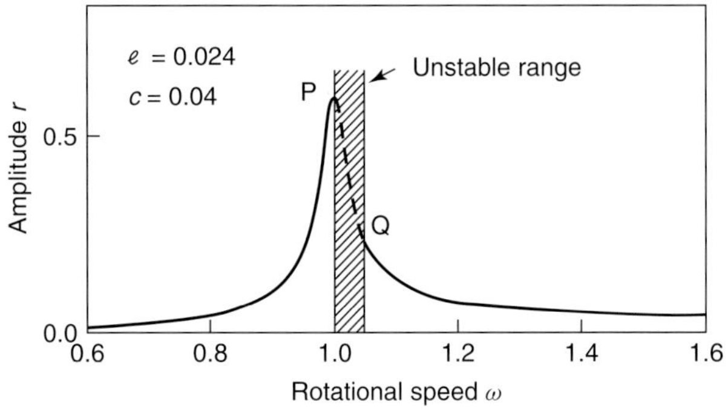

Figure 8.10 Resonance curve and an unstable range.

These coefficients are composed of the terms of $O(\varepsilon)$. Substituting

$$\xi = Ae^{st}, \quad \eta = Be^{st}, \quad \zeta = Ce^{st} \tag{8.27}$$

into Eq. (8.25), we obtain the following characteristic equation:

$$s^3 + A_2 s^2 + A_1 s + A_0 = 0 \tag{8.28}$$

where

$$\left.\begin{aligned} A_2 &= -(a_{11} + a_{22} + a_{33}) \\ A_1 &= a_{11}a_{22} + a_{22}a_{33} + a_{33}a_{11} - a_{12}a_{21} - a_{23}a_{32} - a_{31}a_{13} \\ A_0 &= a_{11}a_{32}a_{23} + a_{22}a_{31}a_{13} + a_{12}a_{21}a_{33} - a_{11}a_{22}a_{33} - a_{12}a_{23}a_{31} - a_{13}a_{21}a_{32} \end{aligned}\right\} \tag{8.29}$$

The solution is stable when

$$A_0 > 0, \quad A_1 > 0, \quad A_2 > 0, \quad A_1 A_2 > A_0 \tag{8.30}$$

The resonance curves of the steady-state solution given by Eq. (8.19) and the result of stability criteria are shown in Figure 8.10. The solution is unstable between points P and Q; therefore, a steady-state solution does not appear there. This range corresponds to that between points P_1 and Q_1 in Figure 8.9.

8.4.4 Nonstationary Vibration

Nonstationary vibration during passage through the major critical speed is studied here by numerical integration. Let us assume that a rotor is whirling steadily at $\omega = \omega_s = 0.8$ in the precritical speed range. If the parameter A is changed into a larger value, the rotor starts to accelerate and a nonstationary vibration appears. The variation in deflection is shown in Figure 8.11a, and the variation in rotational speed is shown in Figure 8.11b. For the values $A = 0.0004$ and $A = 0.002$, which correspond to a comparatively large driving torque $T_d = A - B\dot{\psi}$, the rotor can pass the critical speed $\omega_c = 1.0$. In these cases, the deflection rapidly becomes large in the resonance range and then decreases with vibrating amplitude. The maximum

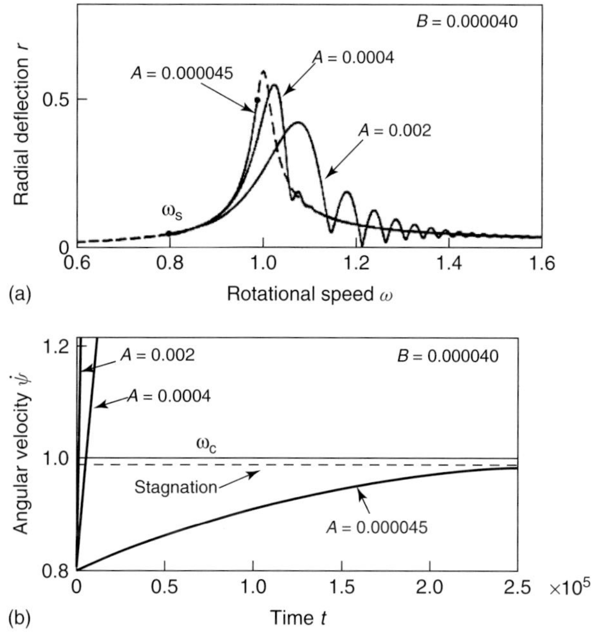

Figure 8.11 Transition through the major critical speed with a limited driving torque. (a) Radial deflection and (b) rotational speed.

amplitude during this transition becomes smaller as the driving torque becomes larger. When $A = 0.000045$, which corresponds to a comparatively small torque, the deflection of the rotor increases at the resonance range and then the rotor stagnates at some rotational speed. In this case, the rotor cannot pass through the major critical speed.

8.5 Analysis by the Asymptotic Method (Nonlinear System, Constant Acceleration)

So far, we obtained time histories $x(t), y(t)$ by integrating equations of motion numerically by the Runge–Kutta method and calculated radial deflection $r(t) = \sqrt{x(t)^2 + y(t)^2}$. This convenient method can be applied only in a case such as the one described in Section 8.4, where only one component is included in the time histories. However, for example, in the case of a subharmonic resonance in a nonlinear system, we cannot know the change in radial deflection of one component by this method because many components are contained and the orbit is not circular. For this purpose, a special method of approach must be adopted. The *asymptotic method* is one such method.

The asymptotic method developed by Bogoliuboff and Mitropol'skii is an approximate method that can be applied to weakly nonlinear systems. Since this method also applicable to systems with slowly varying parameters, it has been used widely in the study of nonstationary vibrations. Refer to the book by Bogoliuboff and Mitropol'skii (1961) for details of this method. The book by Evan-Iwanowski (1976) is recommended for those interested in its application to mechanical systems.

In this section, we explain how to apply the asymptotic method to a study of nonstationary vibrations. A phenomenon during passage through the major critical speed with constant acceleration is used as an example (Ishida, Yasuda, and Murakami, 1997).

8.5.1 Equations of Motion and Their Transformation to a Normal Coordinate Expression

The analytical model is the same 2 DOF inclination model used in Section 2.4. First, we derive equations of motion using some results obtained in the derivation of equations of motion of a 4 DOF rotor system in Section 2.5. The time rates of change of angular momentums in the x-, y-, and z-directions, are given by Eq. (2.40);

$$\left.\begin{aligned} H_x &= -I\ddot{\theta}_{1y} + I_p\frac{d}{dt}(\dot{\Theta}_1\theta_{1x}) \\ H_y &= I\ddot{\theta}_{1x} + I_p\frac{d}{dt}(\dot{\Theta}_1\theta_{1y}) \\ H_z &= I_p\ddot{\Theta}_1 \end{aligned}\right\} \tag{8.31}$$

In the 2 DOF inclination model given by Eq. (6.9), where a disk is mounted at the center of an elastic shaft, the moments M_x and M_y acting from the shaft to the disk in the x- and y-directions are given by

$$M_x = \delta\theta_y + N_{\theta x}, \quad M_y = -\delta\theta_x - N_{\theta y} \tag{8.32}$$

corresponding to Eq. (2.43). Since the time change of Θ_1 is known in this case of constant acceleration, we consider the first and second equations of Eq. (8.31). Therefore, from the relationships of $H_y = M_y$ and $H_x = M_x$, we have

$$\left.\begin{aligned} I\ddot{\theta}_{1x} + I_p\frac{d}{dt}(\dot{\Theta}_1\theta_{1y}) + \delta\theta_x + N_{\theta x} = 0 \\ I\ddot{\theta}_{1y} - I_p\frac{d}{dt}(\dot{\Theta}_1\theta_{1x}) + \delta\theta_y + N_{\theta y} = 0 \end{aligned}\right\} \tag{8.33}$$

Let the angle of the unbalance measured from the x-axis be Ψ. This angle corresponds to φ' in Figure 2.20. From Figure 2.20b and the explanations related to Eq. (2.46), we know that $\Psi = \varphi' = \varphi_1 + 90^\circ + \psi_1' - 180^\circ = \Theta_1 - 90^\circ + \psi_1' - 180^\circ = \Theta_1 - 90^\circ$. Then, we have $\Theta_1 = \Psi + 90^\circ$, $\dot{\Theta}_1 = \dot{\psi}$, and $\ddot{\Theta} = \ddot{\psi}$. Substituting these relationships and

$$\left.\begin{aligned} \theta_{1x} &= \theta_x + \tau\cos\Psi \\ \theta_{1y} &= \theta_y + \tau\sin\Psi \end{aligned}\right\} \tag{8.34}$$

into Eq. (8.33), we obtain

$$\left.\begin{aligned} I\ddot{\theta}_x + I_p\ddot{\Psi}\theta_y + I_p\dot{\Psi}\dot{\theta}_y + \delta\theta_x + N_{\theta x} &= (I - I_p)\tau\dot{\Psi}^2\cos\Psi + (I - I_p)\tau\ddot{\Psi}\sin\Psi \\ I\ddot{\theta}_y - I_p\ddot{\Psi}\theta_x - I_p\dot{\Psi}\dot{\theta}_x + \delta\theta_y + N_{\theta y} &= (I - I_p)\tau\dot{\Psi}^2\sin\Psi - (I - I_p)\tau\ddot{\Psi}\cos\Psi \end{aligned}\right\} \tag{8.35}$$

Let the constant angular acceleration be λ. Then, we have

$$\ddot{\Psi} = \lambda, \quad \dot{\Psi} = \lambda t + \omega_s, \quad \Psi = \frac{1}{2}\lambda t^2 + \omega_s t + \Psi_0 \tag{8.36}$$

Adding damping terms to Eq. (8.35) and using dimensionless quantities given in Eq. (6.8), we arrive at

$$\left.\begin{aligned} \ddot{\theta}_x + i_p\ddot{\Psi}\theta_y + i_p\dot{\Psi}\dot{\theta}_y + c\dot{\theta}_x + \theta_x + N_{\theta x} &= F_0(\dot{\Psi}^2\cos\Psi + \ddot{\Psi}\sin\Psi) \\ \ddot{\theta}_y - i_p\ddot{\Psi}\theta_x - i_p\dot{\Psi}\dot{\theta}_x + c\dot{\theta}_y + \theta_y + N_{\theta y} &= F_0(\dot{\Psi}^2\sin\Psi - \ddot{\Psi}\cos\Psi) \end{aligned}\right\} \tag{8.37}$$

where $F_0 = (1 - i_p)\tau$.

To apply the asymptotic method, we must transform these equations into those in the normal coordinate expression. In Section 6.3, we showed that the transformation is given by Eq. (6.12). However, since the equations of motion transformed by Eq. (5.12) have some integration terms that are unknown beforehand in the case of nonstationary vibrations, such a transformation is not appropriate. For this reason, Agrawal and Evan-Iwanovski (1973) adopted the following transformation:

$$\theta_x = -p_f \int X_1 dt - p_b \int X_2 dt, \quad \theta_y = X_1 + X_2 \tag{8.38}$$

By this expression, the integral forms do not appear in the equations expressed in the normal coordinates. We adopt this transformation in the following analysis.

Let us consider the case where rotational speed changes slowly. It means that the angular acceleration $\ddot{\Psi} = \lambda$ is of $O(\varepsilon)$. For the case of harmonic resonance, we can assume that the unbalance force F_0 is of $O(\varepsilon)$. To show the magnitude of parameters explicitly, we adopt the following expressions using a small parameter ε:

$$\left.\begin{aligned} &c = \varepsilon\bar{c}, \quad \lambda = \varepsilon\bar{\lambda}, \quad F_0 = \varepsilon\bar{F}_0, \quad N_{\theta x} = \varepsilon\bar{N}_{\theta x} \\ &N_{\theta y} = \varepsilon\bar{N}_{\theta y}, \quad \beta^{(0)} = \varepsilon\bar{\beta}^{(0)} \end{aligned}\right\} \tag{8.39}$$

Natural frequencies p_f and p_b change as functions of the rotational speed due to the gyroscopic moment; however, we assume that these frequencies change slowly in this nonstationary process. Here, the concept of *slowly varying time* $\tau = \varepsilon t$ is introduced. Then, a quantity that varies slowly with time is expressed as a quantity dependent on τ. For example, we express natural frequencies by $p_f(\tau)$ and $p_b(\tau)$. (Note that the notation τ is used for slowly varying time and not for dynamic unbalance in this section.)

After transferring small second, fourth, and sixth terms to the right-hand sides and differentiating Eq. (8.37) by time, we get the following to an accuracy of $O(\varepsilon)$:

$$\left.\begin{aligned} \dddot{\theta}_x + i_p\ddot{\Psi}\dot{\theta}_y + i_p\dot{\Psi}\ddot{\theta}_y + \dot{\theta}_x &= \varepsilon\left\{-i_p\bar{\lambda}\dot{\theta}_y - \bar{c}\ddot{\theta}_x - \frac{d}{dt}\bar{N}_{\theta x} - \bar{F}_0(\dot{\Psi}^3\sin\Psi)\right\} \\ \dddot{\theta}_y - i_p\dot{\Psi}\ddot{\theta}_x + \theta_y &= \varepsilon\left\{i_p\bar{\lambda}\dot{\theta}_x - \bar{c}\ddot{\theta}_y - \bar{N}_{\theta y} + \bar{F}_0(\dot{\Psi}^3\sin\Psi)\right\} \end{aligned}\right\} \tag{8.40}$$

We transform the term $i_p\ddot{\Psi}\dot{\theta}_y = i_p\varepsilon\overline{\lambda}\dot{\theta}_y$ in the first expression to the right-hand side and then substitute Eq. (8.38). After some rearrangement to an accuracy of $O(\varepsilon)$, we have

$$\left.\begin{aligned}&(i_p\dot{\Psi} - p_f(\tau))\ddot{X}_1 + (i_p\dot{\Psi} - p_b(\tau))\ddot{X}_2 - p_f(\tau)X_1 - p_b(\tau)X_2\\&= \varepsilon\left\{-2i_p\overline{\lambda}\dot{\theta}_y - \overline{c}\ddot{\theta}_x - \frac{d}{dt}\overline{N}_{\theta x} - \overline{F}_0(\dot{\Psi}^3\sin\Psi)\right\}\\&\ddot{X}_1 + \ddot{X}_2 + i_p\dot{\Psi}(p_f(\tau)X_1 + p_b(\tau)X_2) + X_1 + X_2\\&= \varepsilon\left\{i_p\overline{\lambda}\theta_x - \overline{c}\dot{\theta}_y - \overline{N}_{\theta y} + \overline{F}_0(\dot{\Psi}^2\sin\Psi)\right\}\end{aligned}\right\} \tag{8.41}$$

Considering Eq. (6.11), we have

$$\left.\begin{aligned}&\frac{\ddot{X}_1 + p_f(\tau)^2X_1}{p_f(\tau)} + \frac{\ddot{X}_2 + p_b(\tau)^2X_2}{p_b(\tau)}\\&= \varepsilon\left\{2i_p\overline{\lambda}\dot{\theta}_y + \overline{c}\ddot{\theta}_x + \frac{d}{dt}\overline{N}_{\theta x} + \overline{F}_0\dot{\Psi}^3\sin\Psi\right\}\\&(\ddot{X}_1 + p_f(\tau)^2X_1) + (\ddot{X}_2 + p_b(\tau)^2X_2)\\&= \varepsilon\left\{i_p\overline{\lambda}\theta_x - \overline{c}\dot{\theta}_y - \overline{N}_{\theta y} + \overline{F}_0(\dot{\Psi}^2\sin\Psi)\right\}\end{aligned}\right\} \tag{8.42}$$

Solving these equations for $\ddot{X}_1 + p_f(\tau)^2X_1$ and $\ddot{X}_2 + p_b(\tau)^2X_2$ gives the following equations of motion in the normal coordinate expression:

$$\left.\begin{aligned}&\ddot{X}_1 + p_f(\tau)^2X_1 = \varepsilon M_f(\tau)\left\{i_p\overline{\lambda}(2\dot{\theta}_y + p_f(\tau)\theta_x) + \overline{c}(\ddot{\theta}_x - p_f(\tau)\dot{\theta}_y)\right.\\&\quad\left. + \frac{d}{dt}\overline{N}_{\theta x} - p_f(\tau)\overline{N}_{\theta y} + \overline{F}_o(p_f(\tau) + \dot{\Psi})\dot{\Psi}^2\sin\Psi\right\}\\&\ddot{X}_2 + p_b(\tau)^2X_2 = \varepsilon M_b(\tau)\left\{i_p\overline{\lambda}(2\dot{\theta}_y + p_b(\tau)\theta_x) + \overline{c}(\ddot{\theta}_x - p_b(\tau)\dot{\theta}_y)\right.\\&\quad\left. + \frac{d}{dt}\overline{N}_{\theta x} - p_b(\tau)\overline{N}_{\theta y} + \overline{F}_o(p_b(\tau) + \dot{\Psi})\dot{\Psi}^2\sin\Psi\right\}\end{aligned}\right\} \tag{8.43}$$

where $M_f(\tau) = p_f(\tau)/(1 + p_f(\tau)^2)$ and $M_b(\tau) = p_b(\tau)/(1 + p_b(\tau)^2)$.

8.5.2 Steady-State Solution

Although the steady-state solution in the vicinity of the major critical speed was obtained in Section 6.4.1, here we show how to get it using the asymptotic method. First, the condition of the steady-state solution

$$\lambda = 0, \quad \Psi = \omega t \tag{8.44}$$

is substituted into the first equation of Eq. (8.43). Then, considering that the frequency is close to the forward natural frequency p_f, that is, the detuning

$$p_f^2 - \omega^2 = \varepsilon\sigma \tag{8.45}$$

is small, we transfer this term to the right-hand side and treat it as a perturbation term:

$$\ddot{X}_1 + \omega^2 X_1 = \varepsilon\left[-\sigma X_1 + M_\mathrm{f}\left\{\bar{c}(\ddot{\theta}_x - p_\mathrm{f}\dot{\theta}_y) + \frac{\mathrm{d}}{\mathrm{d}t}\overline{N}_{\theta y} - p_\mathrm{f}\overline{N}_{\theta y} + \overline{F}_0(p_\mathrm{f}+\omega)\omega^2 \sin\omega t\right\}\right] \tag{8.46}$$

When there is no perturbation ($\varepsilon = 0$), the solutions are given by

$$X_1 = a_1 \sin\psi_1, \quad X_2 = 0 \tag{8.47}$$

where

$$\frac{d}{dt}a_1 = 0, \quad \frac{d}{dt}\psi_1 = \omega \tag{8.48}$$

hold. We selected sine instead of cosine as the solution of X_1 so that Eq. (8.38) may express a whirling orbit $\theta_x = a_1\cos\psi_1, \theta_y = a_1 \sin\psi_1$ in physical coordinates. Now, we introduce a new notation φ_1 given by $\psi_1 = \omega t + \varphi_1$ instead of the total phase ψ_1 because the phase difference between the external force and the response is important in the case of steady-state vibration.

When there is perturbation ($\varepsilon \neq 0$), we assume a solution in the following form:

$$\left.\begin{aligned} X_1 &= a_1\sin(\omega t + \varphi_1) + \varepsilon u_1^f(a_1, \varphi_1, \omega t) \\ X_2 &= \varepsilon u_1^b(a_1, \varphi_1, \omega t) \end{aligned}\right\} \tag{8.49}$$

where it is assumed that u_1^f and u_1^b do not contain fundamental components $\sin(\omega t + \varphi_1)$ and $\cos(\omega t + \varphi_1)$. The amplitude a_1 and the phase φ_1 are determined by

$$\frac{d}{dt}a_1 = \varepsilon A_1(a_1, \varphi_1), \quad \frac{d}{dt}\varphi_1 = \varepsilon B_1(a_1, \varphi_1) \tag{8.50}$$

We substitute Eq. (8.49) into Eq. (8.46) and calculate to an accuracy of $O(\varepsilon)$. In this process, we consider that $\mathrm{d}a_1/\mathrm{d}t \approx \varepsilon A_1$ and $\mathrm{d}\varphi_1/\mathrm{d}t \approx \varepsilon B_1$ are of $O(\varepsilon)$ and $\mathrm{d}^2 a_1/\mathrm{d}t^2$ and $\mathrm{d}^2\varphi_1/\mathrm{d}t^2$ are of $O(\varepsilon^2)$. In addition, the solution of $O(\varepsilon^0)$ given by Eq. (8.47), that is, $X_1 = X_{10} = a_1\sin\psi_1, X_2 = X_{20} = 0$, and the solution given by Eq. (8.38), that is, $\theta_x = a_1\cos\psi_1$ and $\theta_y = a_1\sin\psi_1$, are substituted into θ_x and θ_y on the right-hand side. After some calculation and rearrangement, we have

$$\begin{aligned} &\varepsilon\left[2\omega A_1\cos\psi_1 - 2a_1\omega B_1\sin\psi_1 + \partial^2 u_1^\mathrm{f}/\partial t^2 + \omega^2 u_1^\mathrm{f}\right] \\ &= \varepsilon\left[-\sigma a_1 + M_\mathrm{f}\left\{-8\overline{\beta}^{(0)}a_1^3\omega + \overline{F}_0(p_\mathrm{f}+\omega)\omega^2\cos\varphi_1\right\}\right]\sin\varphi_1 \\ &\quad -\varepsilon M_\mathrm{f}\left\{\bar{c}a_1\omega(\omega + p_\mathrm{f}) + \overline{F}_0(p_\mathrm{f}+\omega)\omega^2\sin\varphi_1\right\}\cos\psi_1 \end{aligned} \tag{8.51}$$

From a comparison of the coefficients of $\sin\psi_1$ and $\cos\psi_1$ on both sides, respectively, we obtain expressions for εA_1 and εB_1. Then, from Eq. (8.50), we get

$$\left.\begin{aligned} \frac{d}{dt}a_1 &= -mM_\mathrm{f}\left(c\omega a_1 + F_0\omega^2\sin\varphi_1\right) \\ \frac{d}{dt}\varphi_1 &= \frac{{p_\mathrm{f}}^2 - \omega^2}{2\omega} + M_\mathrm{f}\left(4\beta^{(0)}{a_1}^2 - \frac{F_0 m\omega^2}{a_1}\cos\varphi_1\right) \end{aligned}\right\} \tag{8.52}$$

where $m = (\omega + p_f)/2\omega$ and the parameters with overbars are returned to the original ones through Eq. (8.39).

The steady-state solution is obtained from the following expressions, which are given by putting $da_1/dt = 0$ and $d\varphi_1/dt = 0$:

$$\left.\begin{aligned} &c\omega a_1 + F_0\omega^2 \sin\varphi_1 = 0 \\ &\frac{p_f^2 - \omega^2}{2\omega} + M_f\left(4\beta^{(0)}a_1^2 - \frac{F_0 m\omega^2}{a_1}\cos\varphi_1\right) = 0 \end{aligned}\right\} \tag{8.53}$$

We can ascertain that this expression coincides with Eq. (6.33) if we put $a_1 = P$ and $\varphi_1 = \beta$ and use the relationships $\omega \approx p_f \approx \omega_c = 1/\sqrt{1 - i_p}$, $F_0\omega^2 = F$, and $m \approx 1$, which hold to an accuracy of $O(\varepsilon)$.

Although we get a result where the steady-state solutions are influenced only by the isotropic nonlinear component $N(0)$ to an accuracy of $O(\varepsilon)$, the anisotropic nonlinear components $N(1), \cdots, N(4)$ have limited influence practically, causing little deviation from a circular orbit. We can know their influence if we apply this method with higher accuracy; however, the analysis will become cumbersome.

8.5.3 Nonstationary Vibration

Here, we apply this method to a nonstationary vibration. The directional angle Ψ is given by Eq. (8.36). We assume that the angular velocity varies slowly and express it by $\dot{\Psi} = \omega(\tau)$, where $\tau = \varepsilon t$. The equations of motion are given by Eq. (8.43).

When there is no perturbation and ω is constant, Eq. (8.43) becomes

$$\left.\begin{aligned} \ddot{X}_1 + p_f^2 X_1 = 0 \\ \ddot{X}_2 + p_b^2 X_2 = 0 \end{aligned}\right\} \tag{8.54}$$

Since only mode X_1 is resonant at major critical speed, we set the solution as

$$X_1 = a_1 \sin\psi_1 = a_1 \sin(\psi + \varphi_1), \quad X_2 = 0 \tag{8.55}$$

where

$$\frac{d}{dt}a_1 = 0, \quad \frac{d}{dt}\psi_1 = p_f \tag{8.56}$$

When perturbation exists ($\varepsilon \neq 0$), we look for a solution in the following form:

$$\left.\begin{aligned} X_1 &= a_1 \sin(\Psi + \varphi_1) + \varepsilon u_1^f(\tau, a_1, \Psi, \Psi + \varphi_1) \\ X_2 &= \varepsilon u_1^b(\tau, a_1, \Psi, \Psi + \varphi_1) \end{aligned}\right\} \tag{8.57}$$

where a_1 and φ_1 are determined by

$$\left.\begin{aligned} \frac{d}{dt}a_1 &= \varepsilon A_1(\tau, a_1, \varphi_1) \\ \frac{d}{dt}\varphi_1 &= p_f(\tau) - \omega(\tau) + \varepsilon B_1(\tau, a_1, \varphi_1) \end{aligned}\right\} \tag{8.58}$$

where it is assumed that terms u_1^f and u_1^b do not contain fundamental components $\sin\psi_1$ and $\cos\psi_1$.

Functions on the right-hand sides of Eqs. (8.57) and (8.58) are determined by the same procedure as before. Substituting Eq. (8.57) into Eq. (8.43) and using the relationships

$$\left.\begin{aligned} &\frac{d\tau}{dt} \approx \varepsilon, \quad \frac{da_1}{dt} \approx \varepsilon A_1, \quad \frac{d\psi_1}{dt} \approx p_{\mathrm{f}} + \varepsilon B_1, \quad \frac{d\Psi}{dt} \approx \omega, \\ &\frac{d^2 a_1}{dt^2} \approx \varepsilon \frac{\partial A_1}{\partial \varphi_1}(p_{\mathrm{f}} - \omega), \quad \frac{d^2\Psi}{dt^2} \approx \varepsilon\overline{\lambda}, \quad \frac{d^2\Psi_1}{dt^2} \approx \varepsilon \left(\frac{\partial B_1}{\partial \varphi_1}\right)(p_{\mathrm{f}} - \omega) \end{aligned}\right\} \tag{8.59}$$

we obtain the following expressions to an accuracy of $O(\varepsilon)$:

$$\begin{aligned} &\varepsilon\left[2A_1 p_{\mathrm{f}} + a_1 \frac{\partial B_1}{\partial \varphi_1}(p_{\mathrm{f}} - \omega)\right]\cos\psi_1 - \varepsilon\left[2p_{\mathrm{f}} a_1 B_1 - \frac{\partial A_1}{\partial \varphi_1}(p_{\mathrm{f}} - \omega)\right]\sin\psi_1 \\ &\quad + \varepsilon\left[2\omega p_{\mathrm{f}} \frac{\partial^2 u_1^{\mathrm{f}}}{\partial\varphi_1 \partial\Psi} + p_{\mathrm{f}}^2 \frac{\partial^2 u_1^{\mathrm{f}}}{\partial \varphi_1^2} + \omega^2 \frac{\partial^2 u_1^{\mathrm{f}}}{\partial \Psi^2} + p_{\mathrm{f}}^2 u_1^{\mathrm{f}}\right] \\ &= \varepsilon M_{\mathrm{f}}\left[3 i_{\mathrm{p}} \overline{\lambda} p_{\mathrm{f}} a_1 - 2\overline{c} p_{\mathrm{f}}^2 a_1 - \overline{F}_0 (p_{\mathrm{f}} + \omega)\omega^2 \sin\varphi_1\right]\cos\psi_1 \\ &\quad + \varepsilon M_{\mathrm{f}}\left[-8 p_{\mathrm{f}} \overline{\beta}^{(0)} a_1^3 + \overline{F}_0 (p_{\mathrm{f}} + \omega)\omega^2 \cos\varphi_1\right]\sin\psi_1 \end{aligned} \tag{8.60}$$

From a comparison of the coefficients of $\sin\psi_1$ and $\cos\psi_1$, we have

$$\left.\begin{aligned} &-2p_{\mathrm{f}} a_1(\varepsilon B_1) + \frac{\partial}{\partial \varphi_1}(\varepsilon A_1)(p_{\mathrm{f}} - \omega) = \varepsilon M_{\mathrm{f}}\left\{-8 p_{\mathrm{f}} \overline{\beta}^{(0)} a_1^3 + \overline{F}_0(p_{\mathrm{f}} + \omega)\omega^2 \cos\varphi_1\right\} \\ &2p_{\mathrm{f}}(\varepsilon A_1) + a_1 \frac{\partial}{\partial \varphi_1}(\varepsilon B_1)(p_{\mathrm{f}} - \omega) = \varepsilon M_{\mathrm{f}}\left\{3 i_{\mathrm{p}} \overline{\lambda} p_{\mathrm{f}} a_1 - 2\overline{c} {p_{\mathrm{f}}}^2 a_1 - \overline{F}_0(p_{\mathrm{f}} + \omega)\omega^2 \sin\varphi_1\right\} \end{aligned}\right\} \tag{8.61}$$

To obtain εA_1 and εB_1 from these expressions, we put $\varepsilon A_1 = C_1 + C_2 \cos\varphi_1 + C_3 \sin\varphi_1$ and $\varepsilon B_1 = C_4 + C_5 \cos\varphi_1 + C_6 \sin\varphi_1$ and determine $C_1, \cdots, C_6$. Substituting these results into Eq. (8.58), we get the following equations that govern the amplitude a_1 and the phase angle $\varphi_1$of the harmonic component:

$$\left.\begin{aligned} &\frac{d}{dt}a_1 = M_{\mathrm{f}}(\tau)\left\{\frac{3}{2} i_{\mathrm{p}} \lambda a_1 - c p_{\mathrm{f}}(\tau) a_1 - F_0 \omega(\tau)^2 \sin\varphi_1\right\} \\ &\frac{d}{dt}\varphi_1 = p_{\mathrm{f}}(\tau) - \omega(\tau) + M_{\mathrm{f}}(\tau)\left\{4\beta^{(0)} a_1^2 - \frac{F_0 \omega(\tau)^2}{a_1}\cos\varphi_1\right\} \end{aligned}\right\} \tag{8.62}$$

In these expressions, original parameters without overbars are used. Since only the coefficient $\beta^{(0)}$ is contained in these first approximations, we know that the characteristics of a nonstationary vibration in a system with constant acceleration are also determined predominantly by the isotropic nonlinear component.

Figure 8.12a shows a resonance curve given by Eq. (8.53), an amplitude variation curve given by Eq. (8.62), and a radius variation curve obtained by integrating Eq. (8.37) numerically. Figure 8.12b shows the spectrum of a time history obtained by numerical integration. This is a result obtained in a system with nonlinear components $N(0)$ and $N(2)$. Owing to the anisotropic component $N(2)$, the orbit deviates from a circular orbit, and components occur in addition to harmonics in its spectrum. As a result, the amplitude variation curve $\sqrt{\theta_x^2 + \theta_y^2}$ obtained by

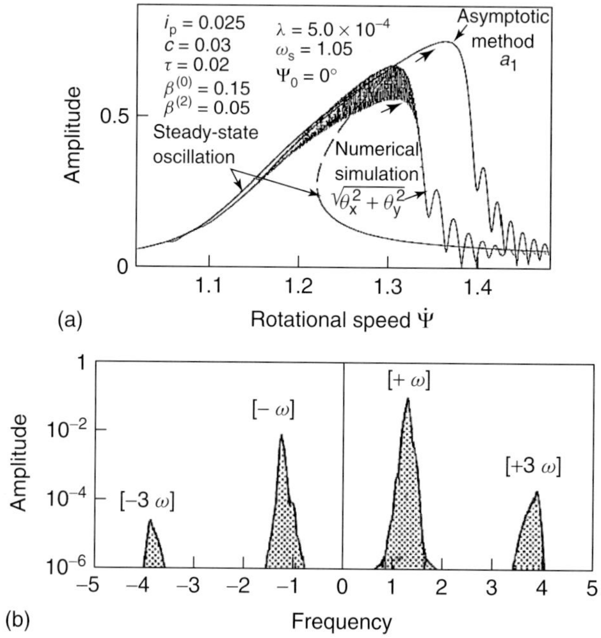

Figure 8.12 Nonstationary vibration during passage through the major critical speed. (a) Amplitude variation and (b) spectrum.

numerical integration shows high-frequency variation, which makes the change unclear. On the contrary, the curve given by the first approximation of the asymptotic method is a smooth curve that shows clearly a change in the radius of the harmonic component. (However, this has a comparatively large quantitative error. The higher approximation of the asymptotic method gives a more accurate curve, close to that of numerical integration.)

For example, in the case of a nonstationary vibration in transition through the subharmonic resonance, where two components of $O(\varepsilon^0)$ coexist, the high-frequency variation in the radius $\sqrt{\theta_x{}^2 + \theta_y{}^2}$ becomes very large; therefore, it is impossible to know the variation of the subharmonic component. In such a case, the asymptotic method is recommended.

The complex-FFT method explained in Chapter 16 is a powerful tool that can be applied not only to numerical data but also to experimental data. This method is also recommended for the analysis of nonstationary vibrations.

References

Agrawal, B.N. and Evan-Iwanovski, R.M. (1973) Resonances in nonstationary, nonlinear, multidegree-of-freedom systems. *AIAA J.*, **11** (7), 907–912.

Bogoliubov, N.N. and Mitropol'skii, Y.A. (1961) *Asymptotic Methods in the Theory of Nonlinear Oscillations*, Gordon & Breach.

Dimentberg, F.M. (1961) *Flexural Vibrations of Rotating Shafts*, Butterworths, London.

Evan-Iwanowski, R.M. (1976) *Resonance Oscillations in Mechanical Systems*, Elsevier, New York.

Ishida, Y., Ikeda, T., and Yamamoto, T. (1987) Transient vibrations of a rotating shaft with nonlinear spring characteristics during acceleration through a major critical speed. *JSME Int. J.*, **30** (261), 458–466.

Ishida, Y., Yasuda, K., and Murakami, S. (1997) Nonstationary vibration of a rotating shaft with nonlinear spring characteristics during acceleration through a major critical speed: a discussion by the asymptotic method and the FFT method. *Trans. ASME, J. Vib. Acoust. Stress Reliab. Des.*, **119** (1), 31–36.

Kononenko, V.O. (1969) *Vibration System with a Limited Power Supply*, Iliffe Books, London. (in Russian).

Lewis, F.M. (1932) Vibrations during acceleration through a critical speed. *Trans. ASME*, **54** (3), 253–261.

Mitropol'skii, Y.A. and (1965) *Problems of the Asymptotic Theory of Nonstationary Vibrations*, Israel Program for Science Translations, Jerusalem.

Nayfeh, A.H. and Mook, D.T. (1979) *Nonlinear Oscillations*, Wiley-Interscience, New York, p. 30.

Neal, H.L. and Nayfeh, A.H. (1990) Response of a single-degree-of-freedom system to a non-stationary principal parametric excitation. *Int. J. Non-Linear Mech.*, **25** (3), 275–284.

Yamada, H. and Tsumura, T. (1951) Vibration analysis of accelerated unbalanced rotor. *Trans. JSME.*, **17** (64), 115–119.

Yanabe, S. and Tamura, A. (1971) Vibration of a shaft passing through a critical speed (1st report, experiments and numerical solutions). *Bull. JSME*, **14** (76), 1050–1058.

Yanabe, S. and Tamura, A. (1972) Vibration of a shaft passing through a critical speed (2nd report, approximate equations). *Bull. JSME*, **15** (89), 1364.

Zeller, W. (1949) Naherungsverfahren zur bestimmung der beim durchfahren der resonanz auftretenden hochstamplitude. *MTZ*, **10** (l), 11–12.

9
Vibrations due to Mechanical Elements

9.1
General Considerations

Rotating machinery has various kinds of mechanical elements such as bearings, pedestals, and joints. The construction itself or manufacturing errors in these elements sometimes cause vibrations. In the process of troubleshooting, we generally investigate amplitudes, frequencies, time histories, phase angles, and so on. The most helpful data are ratios of the vibration frequency and the rotational speed. To find the causes of vibration and eliminate them, we must fully understand the mechanisms by which vibrations occur.

In this chapter, vibrations caused by commonly used mechanical elements in rotor systems are explained.

9.2
Ball Bearings

Bearings are commonly used mechanical elements in many types of rotating machinery. They are classified into *rolling-element bearings* and *plain bearings (fluid-film bearings)*, depending on whether rolling contact or sliding contact occurs. Both bearings have merits and demerits and are better suited to certain purposes. Their characteristics are compared in Table 9.1. In this section, rolling bearings are explained, plain bearings are described in Section 10.2.

Rolling-element bearings are simple and convenient to use. As shown in Figure 9.1, they are generally composed of four parts: the outer ring, the inner ring, the rolling elements (balls, rollers, etc.), and the retainer (separator). They can be subclassified on the basis of certain characteristics: for example, *ball bearings* and *roller bearings* with respect to the type of rolling element, or *radial bearings* and *thrust bearings* with respect to the direction of the load.

9.2.1
Vibration and Noise in Rolling-Element Bearings

Rolling-element bearings produce vibration and noise due to various causes. Although they are the same in principle, bearing engineers customarily discuss

Linear and Nonlinear Rotordynamics: A Modern Treatment with Applications, Second Edition.
Yukio Ishida and Toshio Yamamoto.

Table 9.1 Comparison between rolling-element bearings and plain bearings.

Characteristics	Rolling-element bearing	Fluid-film bearing
Load	Radial load and thrust load	Radial load
Shape	Large outer radius compared to inner radius	Not so large radius compared to inner radius
Damping	Small	Large
Friction coefficient	Static: small (10^{-3}–10^{-2}) Dynamic: small (10^{-3})	Static: large (10^{-2}–10^{-1}) Dynamic: small (10^{-3})
Lubrication	Easy (grease-packed)	Special supplement facility
Noise	Comparatively large	Comparatively small
Life	Limited due to fatigue	No limit
Damage	Seizure in high-speed region	Seizure, wear
After damage	Exchange (simple)	Repair (laborious)

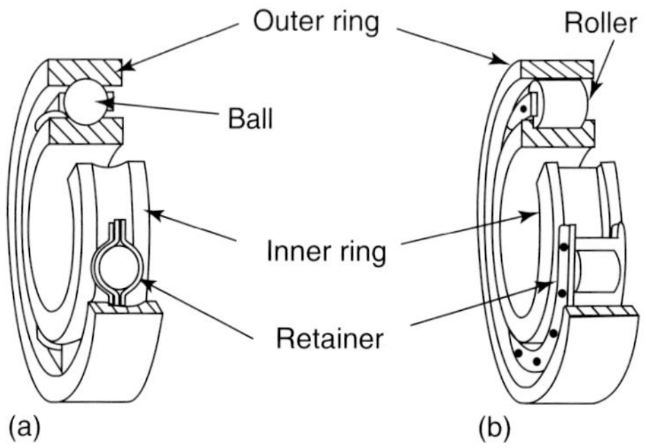

Figure 9.1 Cutaway examples of rolling-element bearings: (a) ball bearing and (b) roller bearing.

vibrations larger than about 1000 Hz as sound and noise. In this section, three representative vibrations and noise of rolling-element bearings are explained (Ohta, 2005).

9.2.1.1 **Vibrations due to the Passage of Rolling Elements**

In a horizontal rotor, a load due to gravity works on the ball as it passes under the rotor. In the case that the number of balls is low, each ball deforms as it supports a comparatively large load. This load distribution changes depending on the position of the balls, and therefore the deformation of the balls changes, as shown in Figure 9.2. As a result, the vertical position of the shaft varies when the shaft rotates. This variation occurs even if there is no imperfection due to manufacturing error in a bearing. Vibration due to the passage of rolling elements was first discovered by Perret (1950), and the corresponding orbit of the shaft center was obtained theoretically by Meldau (1951). Between the upper and lower extremes, rotors have been found to shift in a horizontal direction (Tamura and Taniguchi, 1960).

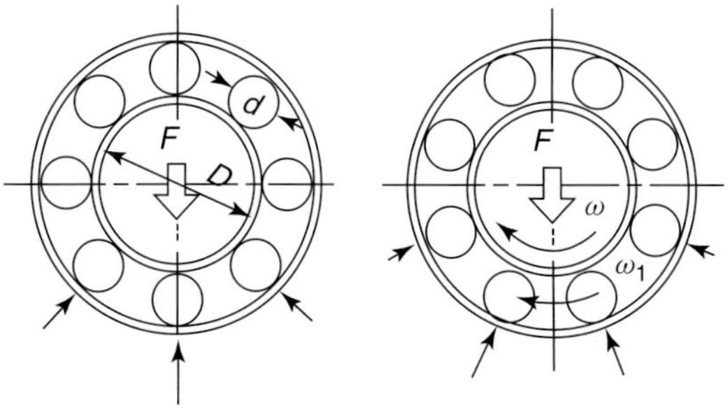

Figure 9.2 Passage of rolling elements.

When the outer ring is fixed and the shaft rotates with ω, the precessional angular velocity ω_1 of the balls is given by

$$\omega_1 = \frac{D}{2(D+d)}\omega (\equiv \alpha_1 \omega) \tag{9.1}$$

where D is the outer diameter of the inner ring and d is the diameter of the balls (refer to "Note: Derivation of Precessional Angular Speed"). If the number of balls is z, the angular velocity of the vibration due to the passage of the balls is $z\omega_1 = z\alpha_1\omega$.

The double-row self-aligning ball bearing (No. 1200) shown in Figure 6.1b has the ratio $\alpha_1 \approx 0.377$ ($\approx 3/8$) and the single-row deep-groove ball bearing (No. 6200) shown in Figures 6.1c and 9.1a has the ratio $\alpha_1 \approx 0.361$ ($\approx 13/36$).

Note: Derivation of Precessional Angular Speed

Expression (9.1) for the precessional angular speed ω_1 can be derived by the following process (refer to Table 9.2):

1) Connect the inner ring IR and a ball B by a bar A, and write the number of rotations for each part of the table when the following operations are completed.
2) *Operation I*: Consider the whole as one body and rotate it clockwise once. Write the number of rotations for each element. The signs + and – denote the rotations in the clockwise and counterclockwise directions, respectively.
3) *Operation II*: Fix the bar A in space and then rotate the outer ring OR counterclockwise once. Write the number of rotations.
4) Calculate the number of rotations from the table when operations I and II are done successively. Derive the value of ω_1 from the relationship (rotation of A) : (rotation of IR) $= \omega_1 : \omega$.

Table 9.2 Derivation of the precessional angular velocity.

Operation	Inner ring (IR)	Ball (B)	Outer ring (OR)	Bar (A)
I	+1	+1	+1	+1
II	$\frac{D+2d}{D}$	$\frac{D+2d}{d}$	−1	0
I + II	$\frac{2(D+d)}{D}$	$-\frac{D+d}{d}$	0	+1

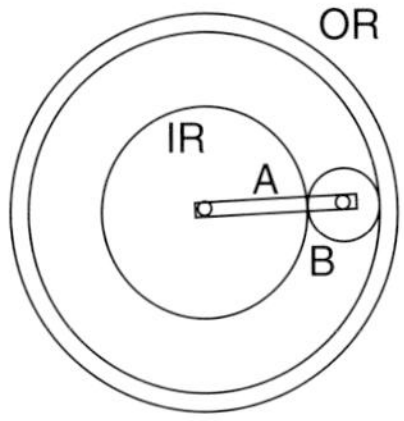

9.2.1.2 **Natural Vibrations of Outer Rings**

Various sounds are generated because of vibration of bearings. The most commonly observable sound is "race noise" which is a smooth and continuous sound like "sha-ahh." We can hear this noise in almost all kinds of rolling bearings and it is caused by the vibration of the outer ring.

In the function of most bearings, the inner ring rotates while the outer ring is fixed. Generally, the inner ring is mounted on the shaft with a shrink fit and the outer ring is inserted into the housing with a loose fit. Therefore, the outer ring can vibrate and this elastic vibration becomes a source of noise (Igarash, 1959). Lohmann (1953) clarified that the frequencies of sound from bearings are generally over 1000 Hz and its spectrum has several peaks that are near to the natural frequencies of the bearing in the radial and axial direction. NSK (1954, 1956) and Tallian and Gustafsson (1965) also observed sounds in an electric motor. They also pointed out that these sounds correspond to natural frequencies of flexible and rigid modes of vibration of the outer rings.

Natural vibrations of outer rings have rigid body modes and flexible body modes. Figure 9.3 illustrates rigid body modes. Figure 9.3a is a mode vibrating in the axial direction, Figure 9.3b is a mode vibrating in the radial direction, and Figure 9.3c is a mode of angular vibration. Figure 9.4 illustrates elastic modes. They are classified into in-plane vibration and out-of-plane vibration. Each of them has many orders of vibration.

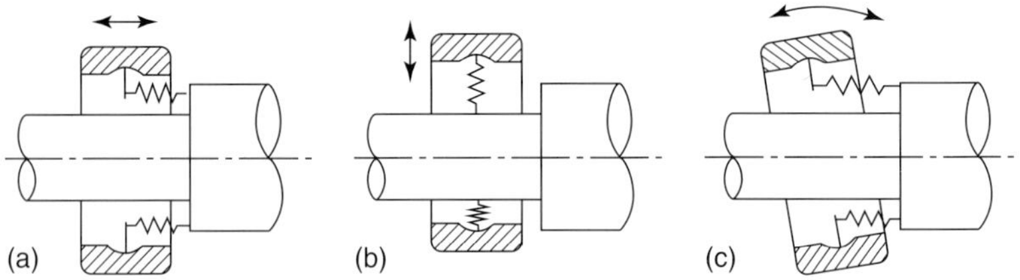

Figure 9.3 Rigid mode shapes of a bearing ring: (a) axial, (b) radial, and (c) angular.

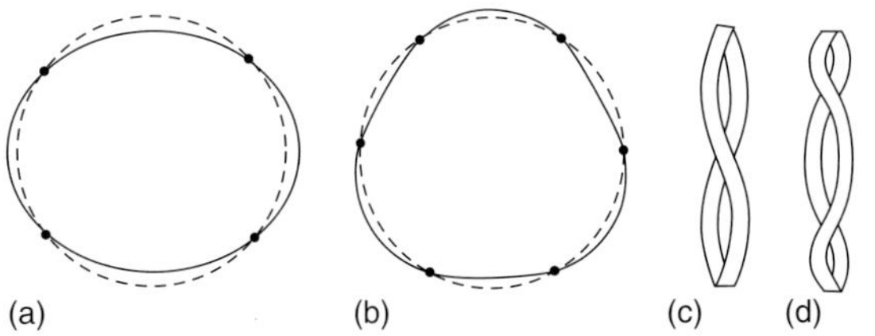

Figure 9.4 Bending mode shapes of an outer ring: (a) radial first, (b) radial second, (c) axial first, and (d) axial second.

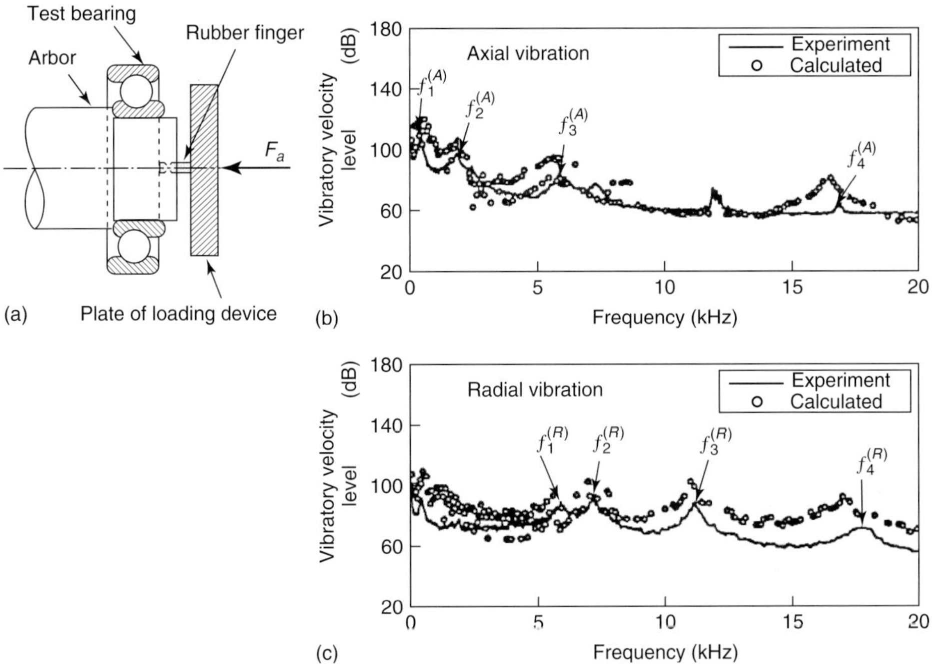

Figure 9.5 Experimental setup and spectra: (a) experimental setup, (b) axial vibration spectra, and (c) radial vibration spectra (Ota and Sakate, 2002, Courtesy of ASME).

Figure 9.5 illustrates the experimental results obtained by Ohta and Satake (2002). The inner ring was mounted in a rotating arbor. An axial load was applied to the outer ring through two rubber fingers attached to the plate of a loading device. The bearing was a single-row deep-groove ball bearing (No. 6206) with inner and outer diameters of 30 and 62 mm and a height of 16 mm. Figure 9.5b,c illustrate radial and axial vibrations obtained at 1800 rpm with the axial load 99.5 N. Figure 9.5b shows out-of-plane natural vibrations of the outer ring. The peak $f_1^{(A)}$ corresponds to the angular natural vibration of the rigid mode. The peak $f_2^{(A)}$ corresponds to the axial natural vibration of the rigid mode. The peaks $f_3^{(A)}$ and $f_4^{(A)}$ correspond to the natural vibrations of the first and second axial bending modes. Figure 9.5c shows in-plane natural vibrations of the outer ring. The peak $f_1^{(R)}$ corresponds to radial natural vibration of the rigid mode. Peaks $f_2^{(R)}$, $f_3^{(R)}$, and $f_4^{(R)}$ correspond to the natural vibrations of the first, second, and third axial bending modes. The frequencies of these peaks do not change even if the rotational speed changes. The plots in these figures are the numerical results obtained by Ohta and Satake (2002).

"Raceway noise" is caused by configuration errors (called *microundulation* or *waviness*) as explained in the following section. Owing to such errors, the contact between the raceway ring and rolling elements behaves like a spring that fluctuates minutely during operation and this variation excites vibrations and noise.

9.2.1.3 **Geometrical Imperfection**

Owing to the limited accuracy achievable in the manufacturing process for bearings, we cannot avoid some waviness of the inner and outer rings. Furthermore, irregularities of the shaft surface and the housing bore cause deformation of the rings and produce similar waviness. Goto and Watanabe (1963) measured the surfaces of outer and inner races of an angular contact ball bearing that was used to support the main shaft of a lathe. Figure 9.6 illustrates the most commonly found shapes. They explained that these shapes can be classified into (1); a circular type, (2) an ellipsoidal type, or simple polygonal shape, and (3) a complex polygonal shape with many corners, for example, 12–15.

Igarashi and Ohta (1990) investigated vibrations of an outer ring of a single-row deep-groove ball bearing (No. 6306) with an axial load of 62.72 N. Figure 9.7 shows a spectrum in a low frequency range (0–200 Hz) obtained at 1800 rpm. The notations f_r and f_c beside the peaks correspond to the frequencies of rotation of the inner ring

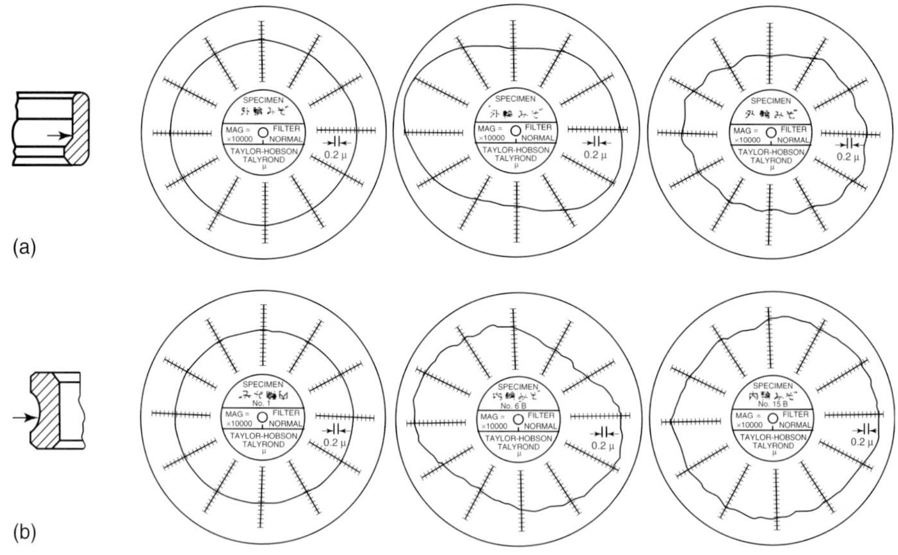

Figure 9.6 Representative shapes of raceways: (a) outer ring and (b) inner ring. (Goto and Watanabe, 1963, Courtesy of JSME.)

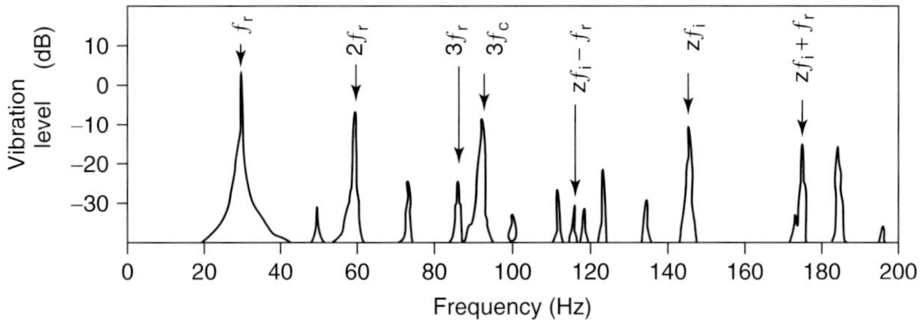

Figure 9.7 Spectra of resultant of the radial and axial vibrations in a low frequency range (axial load 62.72 N, rotational speed 1800 rpm, bearing #6306).

and the retainer, respectively. The value of f_i can be represented by the expression $f_i = f_r - f_c$. The notation z is the number of balls. These frequencies increase in proportion to the rotational speed.

Gustafsson (1962) studied vibrations caused by waviness. He represented the waviness of the surfaces in circumferential direction as the summation of sinusoidal waves. He theoretically obtained the relationship between the number of waviness peaks and the frequency of vibrations.

9.2.1.4 **Other Noises**

The term *cage noise* refers to two kinds of noise generated by cages within a bearing. The first type of cage noise is generated because of the collision of the rotating cage and rolling elements. The other is generated by self-excited vibration of a cage caused by sliding friction.

Flaws such as denting and rust on a raceway surface generate a pulsating noise called a *flaw noise*. Since this flaw noise is generated when the rolling elements hit flaws, the pulses in the noise time history appear repeatedly in the same intervals when the rotational speed is constant.

Foreign particles entered into a bearing also generate noise, which is known as *contamination noise*. The magnitude and the interval of peaks of this noise are not constant.

9.2.2
Resonances of a Rotor Supported by Rolling-Element Bearings

When an elastic rotor is supported by rolling-element bearings, the bearing's structure itself and various kinds of errors cause vibrations of the rotor. Nonlinearity appears due to clearances and contact stiffness in bearings. Resonances due to the former have already been explained in Chapter 6. In the following, linear resonances due to ball bearings are explained.

9.2.2.1 **Resonances due to Shaft Eccentricity**

Rolling elements generally have manufacturing errors of about several micrometers in their size, although it depends on the size and the class of precision of bearings.

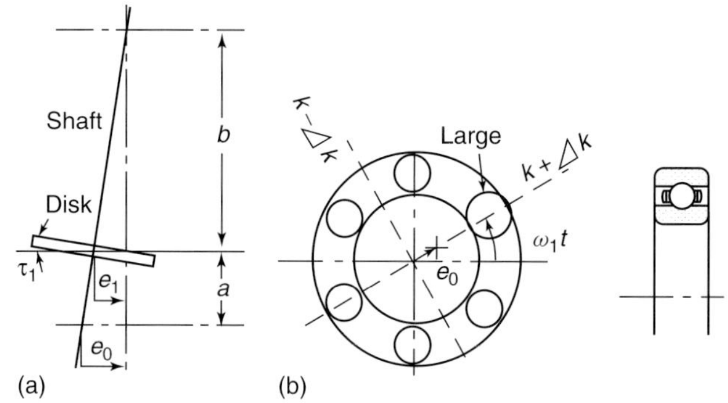

Figure 9.8 Two effects of the irregularity of ball diameters: (a) shift of the shaft center and (b) directional difference in stiffness.

Figure 9.8a illustrates a rotor supported by bearings at both ends. If there is an irregularity in ball diameter in the lower bearing as shown in Figure 9.8b, the shaft center deviates from the bearing centerline that passes through the centers of the outer rings of the upper and lower bearings. Let this magnitude be e_0. This eccentricity e_0 makes an eccentricity e_1 and an inclination τ_1 which work as the static and dynamic unbalances of the rotor. These unbalances produce an excitation force rotating with the angular velocity $\omega_1 = \alpha_1\omega$, and, as a result, a resonance with frequency $\omega_1 = \alpha_1\omega$ occurs (Yamamoto, 1954a).

An experimental result is shown in Figure 9.9. The experimental setup is the same type as that is shown in Figure 9.8a. Double-row self-aligning ball bearings (No. 1200) are used. Figure 9.9a shows a natural frequency diagram and Figure 9.9b shows resonance curves related to ball irregularity. The resonance explained here occurs at the rotational speed ω_a, which corresponds to the cross point A of the straight line $p = \alpha_1\omega$ and p_2-curve in Figure 9.9a. In Figure 9.9b, peak IV in the vicinity of 4200 rpm represents this resonance.

9.2.2.2 **Resonances due to the Directional Difference in Stiffness**

The irregularity of balls makes a directional difference in stiffness of the restoring force, and this difference causes resonances (Yamamoto, 1954b). Suppose that one ball is a little larger than others in Figure 9.8b. Then, the clearance in the direction of this large ball becomes small and sometimes leads to a state of shrink fit, and, as a result, the shaft stiffness increases in this direction. This directional difference in stiffness rotates with angular velocity ω_1. In a procedure similar to that for asymmetrical shaft, we can derive the equations of motion for a 2 DOF (degrees of freedom) inclination model as follows:

$$\left.\begin{aligned} m\ddot{x} + kx - \Delta k(x\cos 2\omega_1 t + y\sin 2\omega_1 t) &= me\omega^2\cos\omega t \\ m\ddot{y} + ky - \Delta k(x\sin 2\omega_1 t - y\cos 2\omega_1 t) &= me\omega^2\sin\omega t \end{aligned}\right\} \tag{9.2}$$

If we neglect small terms of higher orders on the assumption that the quantity $\Delta k/k$ is small, we can obtain a solution for forced vibration in the form

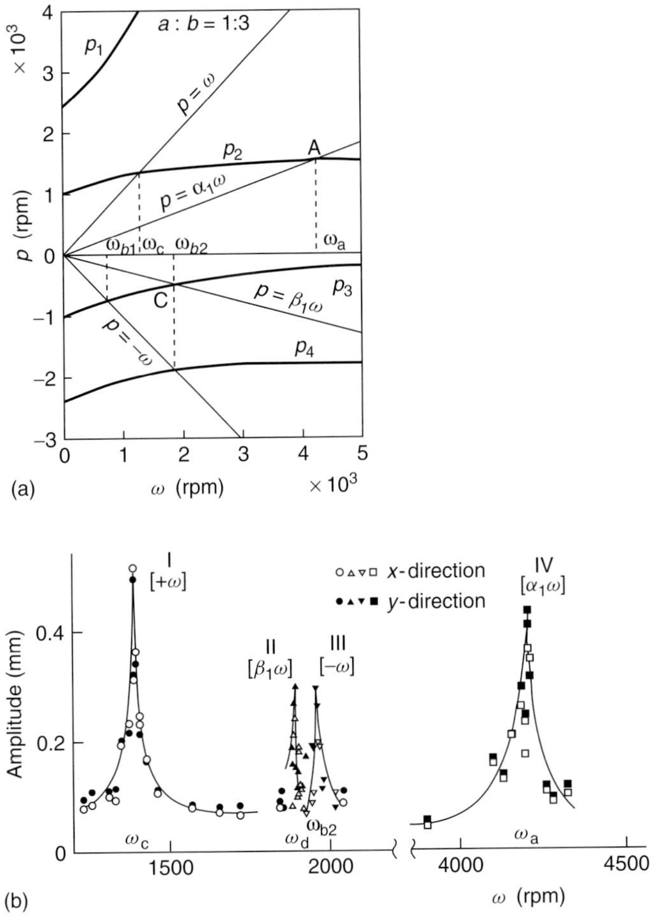

Figure 9.9 Experimental results: (a) natural frequency diagram and (b) resonance curve. Disk: diameter 482.8 mm, thickness 5.2 mm, $a : b = 1 : 3$ (Yamamoto, 1954a, 1954b).

$$\left.\begin{aligned} x &= A\cos\omega t + B\cos\beta_1\omega t \\ y &= A\sin\omega t + B\sin\beta_1\omega t \end{aligned}\right\} \tag{9.3}$$

where $\beta_1 = 2\alpha_1 - 1$. Using the notation $p = \sqrt{k/m}$, the amplitudes are given by

$$A = \frac{e\omega^2}{p^2 - \omega^2}, \quad B = \frac{(\Delta k/m)e\omega^2}{(p^2 - \omega^2)\left\{p^2 - (\beta_1\omega)^2\right\}} \tag{9.4}$$

The second term of Eq. (9.3) is the vibration due to irregularity of balls. This component becomes large in resonance at the rotational speed where $p \approx \beta_1\omega$ holds. In Figure 9.9b, peak II corresponds to this resonance. The ball bearings used in this setup have the ratio $\alpha_1 \approx 0.377$. Since $\beta_1 = 2\alpha_1 - 1 \approx -0.246 < 0$ holds, this vibration is a backward whirling mode. The resonance speeds are given by the cross points of the straight line $p = \beta_1\omega$ and the curves p_3 and p_4 in Figure 9.9a.

9.2.2.3 Vibrations of a Horizontal Rotor due to the Passage of Rolling Elements

When an elastic shaft is horizontally supported by rolling-element bearings, the inner ring deviates periodically because of the passage of rolling elements as mentioned in Section 9.2.1. The frequency of this vibration is $z\omega_1 = z\alpha_1\omega$, where z is the number of balls and ω_1 is the precessional angular speed. Since the movement of the shaft center due to this shift is not circular, two excitation forces with frequencies $+z\alpha_1\omega$ (forward) and $-z\alpha_1\omega$ (backward) act on the system in the same way as that explained in Exercise 2.1. Consequently, resonances with frequencies $+z\alpha_1\omega$ and $-z\alpha_1\omega$ occur at the rotational speed where $p_i = +z\alpha_1\omega$ and $p_i = -z\alpha_1\omega$ hold, respectively (Yamamoto, 1957d).

Figure 9.10 shows an example of resonance curves obtained in experiments. Single-row deep-groove ball bearings (No. 6200) is used, and this type of bearing has the values $z = 6, \alpha_1 \approx 13/36$, and $\pm z\alpha_1\omega \approx \pm 13/6$. Peaks III and IV in the vicinities of 525 and 420 rpm, respectively, are the resonances due to the reasons stated above. This type of resonance also occurs in a vertical rotor system if a lateral force or a moment works because of the misalignment of the upper and lower bearings.

9.2.2.4 Vibrations due to the Coexistence of the Passage of Rolling Elements and a Shaft Initial Bend

If a shaft has an initial bend, the passage of balls causes another kind of resonance (Yamamoto, 1957d). Suppose that a vertical rotor system has a misalignment between the upper and lower bearings and the shaft is initially bent. It is also assumed that bearings that have constraint for inclination, such as a single-row deep-groove ball bearing, are used. If the shaft rotates quasi-statically, the bending

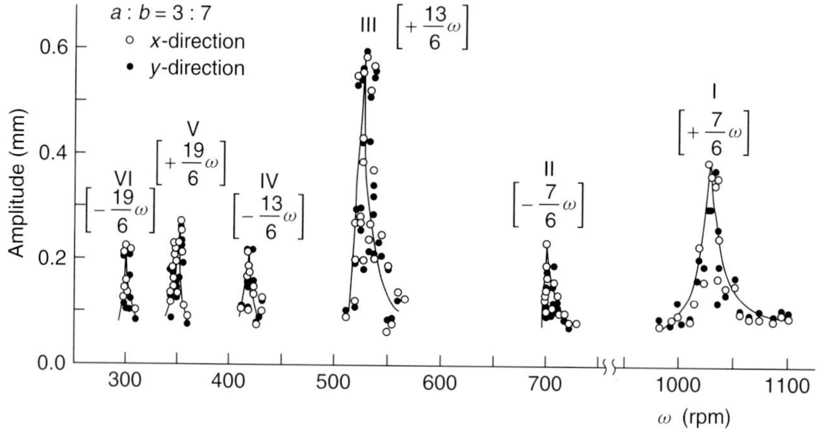

Figure 9.10 Experimental results. Disk: same as that shown in Figure 9.9, shaft: length 508 mm, $a : b = 3 : 7$ (Yamamoto, 1957d).

moment applied to the shaft changes depending on the angular position of the shaft, because the shift of the bearing centerline fixed in space and the bend rotating with the shaft coexist. Then, the radial load applied to the bearing also varies with angular velocity ω. If the variation due to the shaft bend and that due to the passage of rolling elements coexist, the excitation force F is given by

$$\begin{aligned} F &= (a + b\cos\omega t)\cos z\alpha_1\omega t \\ &= F_1 \cos z\alpha_1\omega t + F_2\cos(z\alpha_1 + 1)\omega t + F_3\cos(z\alpha_1 - 1)\omega t \end{aligned} \tag{9.5}$$

The term $b\cos\omega t$ within the parentheses represents the effect of an initial bend of the shaft. This force, working in one direction, is equivalent to two forces of the forward and backward modes. That is, owing to coexistence of the passage of the balls and the initial bend, not only forces with frequencies $\pm z\alpha_1\omega$ but also forces with frequencies $\pm(z\alpha_1 + 1)\omega$ and $\pm(z\alpha_1 - 1)\omega$ work on the rotor.

In a case of bearings with $\alpha_1 \approx 13/36$ and $z = 6$, resonances with frequencies $\pm(z\alpha_1 + 1)\omega = \pm(19/6)\omega$ and $\pm(z\alpha_1 - 1)\omega = \pm(7/6)\omega$ occur. They are shown by peaks I, II, V, and VI in Figure 9.10.

9.3 Bearing Pedestals with Directional Difference in Stiffness

The stiffness of a bearing pedestal often differs depending on the direction. For example, Figure 9.11 shows a bearing pedestal whose stiffness due to bending in the x-direction is smaller than that is due to tension or compression in the y-direction.

Let us consider a 2 DOF deflection model corresponding to the rotor governed by Eq. (2.3). The equations of motion are given by

$$\left.\begin{aligned} m\ddot{x} + (k - \Delta k)x &= me\omega^2\cos\omega t \\ m\ddot{y} + (k + \Delta k)y &= me\omega^2\sin\omega t \end{aligned}\right\} \tag{9.6}$$

where the stiffnesses $k - \Delta k$ and $k + \Delta k$ are determined by both the shaft and the bearing pedestal. Since these two equations do not couple, we can solve them independently. The critical speed in the x- and y-directions are given by

$$\omega_{cx} = \sqrt{\frac{k - \Delta k}{m}}, \quad \omega_{cy} = \sqrt{\frac{k + \Delta k}{m}} \tag{9.7}$$

Resonance curves are given by Figure 9.12. The shape of the orbit changes in the manner "circle → ellipse → line (at $\omega = \omega_{cx}$) → ellipse → circle → ellipse → line (at $\omega = \omega_{cy}$) → ellipse → circle" when the rotational speed increases through these critical speeds. As mentioned in Exercise 2.1, an elliptic orbit implies the simultaneous occurrence of forward and backward whirling motions. For example, since a rotational speed in the range $\omega_{cx} < \omega < \omega_{cy}$ is above the critical speed in the x-direction and is under the critical speed in the y-direction, the solutions in these two directions have a phase difference of 180°. Then, the solution is transformed

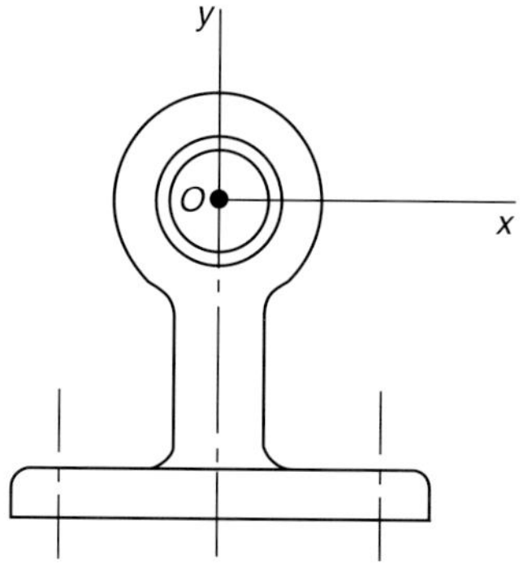

Figure 9.11 Bearing pedestal.

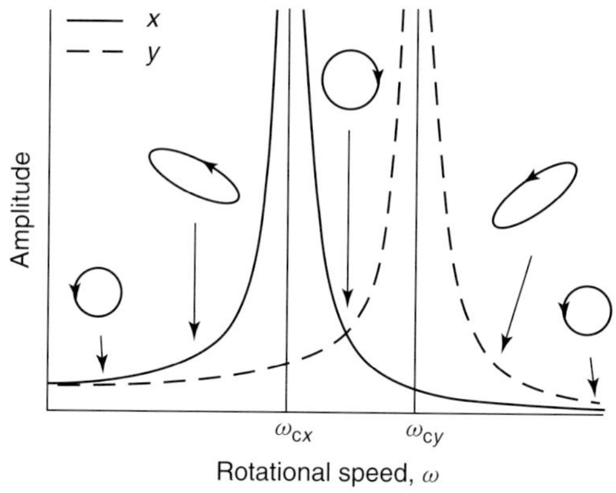

Figure 9.12 Response curve.

into the following form:

$$\left.\begin{aligned} x &= A\cos(\omega t - \pi) = -\frac{1}{2}(A - B)\cos\omega t - \frac{1}{2}(A + B)\cos(-\omega t) \\ y &= B\sin\omega t = -\frac{1}{2}(A - B)\sin\omega t - \frac{1}{2}(A + B)\sin(-\omega t) \end{aligned}\right\} \tag{9.8}$$

This expression shows the existence of a backward component.

The equations of motion of a 4 DOF rotor system where a deflection and an inclination couple with each other are given by

$$\left.\begin{aligned} &m\ddot{x} + (\alpha - \Delta\alpha)x + (\gamma - \Delta\gamma)\theta_x = me\omega^2\cos\omega t \\ &m\ddot{y} + (\alpha + \Delta\alpha)y + (\gamma + \Delta\gamma)\theta_y = me\omega^2\sin\omega t \\ &I\ddot{\theta}_x + I_p\omega\dot{\theta}_y + (\gamma - \Delta\gamma)x + (\delta - \Delta\delta)\theta_x = (I_p - I)\tau\omega^2\cos(\omega t + \beta_\tau) \\ &I\ddot{\theta}_y - I_p\omega\dot{\theta}_x + (\gamma + \Delta\gamma)y + (\delta + \Delta\delta)\theta_y = (I_p - I)\tau\omega^2\sin(\omega t + \beta_\tau) \end{aligned}\right\} \tag{9.9}$$

These equations coincide with Eq. (2.71) when the difference in stiffness becomes zero. By referring to Eq. (9.6) and its solution given by Eq. (9.8), we know that the solution of these equations of motion is composed of forward and backward components that appear in response to each of the unbalance forces with total phase ωt and $(\omega t + \beta_\tau)$. Substituting the assumed solution into Eq. (9.9) and comparing the coefficients of the terms with the same frequency component, we obtain

$$\left.\begin{aligned} x &= \frac{A}{f}\cos\omega t + \frac{B}{f}\cos(\omega t + \beta_\tau) + \frac{\Delta A}{ff'}\cos(-\omega t) + \frac{\Delta B}{ff'}\cos(-\omega t - \beta_\tau) \\ y &= \frac{A}{f}\sin\omega t + \frac{B}{f}\sin(\omega t + \beta_\tau) + \frac{\Delta A}{ff'}\sin(-\omega t) + \frac{\Delta B}{ff'}\sin(-\omega t - \beta_\tau) \end{aligned}\right\} \tag{9.10}$$

where

$$\left.\begin{aligned} f &= (\alpha - m\omega^2)\left\{\delta + (I_\mathrm{p} - I)\omega^2\right\} - \gamma^2 \\ f' &= (\alpha - m\omega^2)\left\{\delta - (I_\mathrm{p} + I)\omega^2\right\} - \gamma^2 \end{aligned}\right\} \tag{9.11}$$

Here, the expressions for θ_x and θ_y are abridged. The expressions for A, B, ΔA, and ΔB are also omitted for simplicity of description. The magnitudes of the quantities ΔA and ΔB are of the same order as $\Delta\alpha, \Delta\gamma$, and $\Delta\delta$. The second expression in Eq. (9.11) has the same form as that is obtained by substituting $p = -\omega$ in Eq. (2.75). Therefore, in Figure 9.9a, the amplitudes of a backward whirling motion represented by the third and fourth terms in Eq. (9.10) become infinite in resonance at the cross points of the straight line $p = -\omega$ and the curves p_3 and p_4 because $ff' = 0$ holds. Thus, a synchronous backward whirling motion occurs because of the directional difference in stiffness of pedestals (Yamamoto, 1954d).

This resonance of a backward whirling mode occurs in the case where a directional difference in stiffness appears because of some cause such as the anisotropic fitting between a bearing and a housing and the coexistence of bearing clearance and misalignment, and so on (Yamamoto, 1954d).

9.4 Universal Joint

Generally speaking, a coupling and a joint are used to connect driving shafts in rotor systems. Figure 9.13 illustrates various misalignments that can appear between two shafts to be coupled. The quantity a is a radial misalignment, b an endplay, and α an angular misalignment. These misalignments generally appear simultaneously, in practice. Different couplings are used depending on the kind of misalignment and the purposes of the rotor. A *rigid coupling* is used to connect two shafts whose centerlines coincide as shown by broken lines in Figure 9.13. Figure 9.14a is a *fixed flange coupling* which is one type of rigid coupling. A *flexible coupling* is used to connect two shafts whose centerlines have slightly deviated; one type of flexible coupling is illustrated in Figure 9.14b. The *Oldham's coupling* shown in Figure 9.14c is used to connect two parallel shafts when the distance between them is small. When the centerlines of two shafts cross with a larger angle, a universal joint is used. There are two types of universal joints. One is a *Hooke's joint* (also called *Cardan joint*), where the angular velocity of the driven shaft changes periodically, and the other is a *universal ball joint*, which can transmit torque with the constant angular speed of a driven shaft.

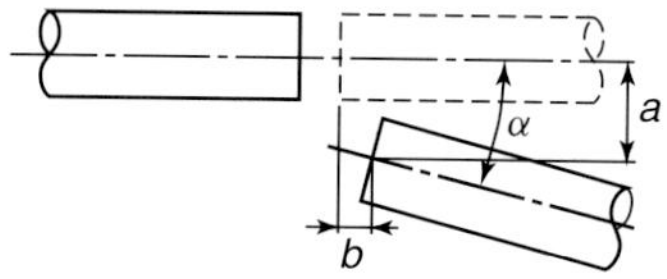

Figure 9.13 Misalignment of shafts.

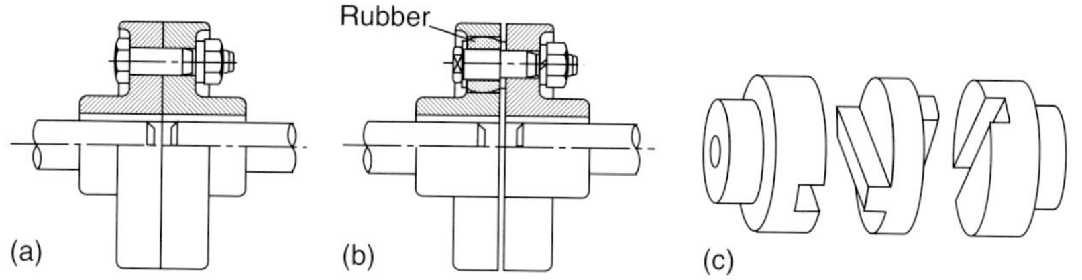

Figure 9.14 Examples of couplings: (a) fixed flange coupling, (b) flexible flange coupling, and (c) Oldham's coupling.

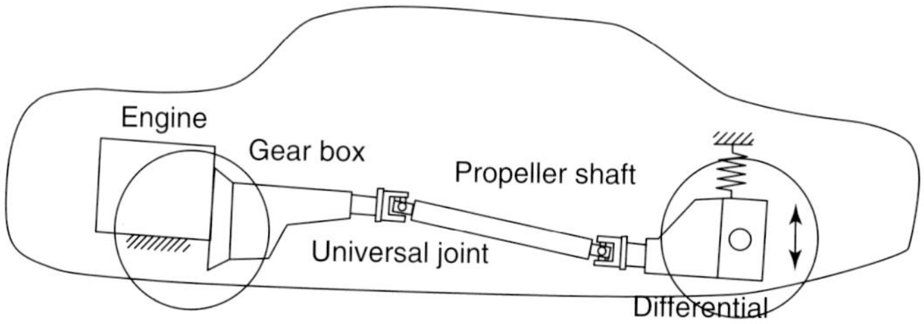

Figure 9.15 Drive train of a car.

Figure 9.15 illustrates a front-engine, rear-wheel drive vehicle. An engine and a transmission are fixed to the frame in the front, the differential and rear axes are suspended from the frame by springs. A propeller shaft (also called a *drive shaft*) transmits the torque from a shaft of a transmission to a shaft of a differential gear. Since the latter moves up and down following a suspension, the two shafts are connected using Hooke's joints.

A Hooke's universal joint is shown in Figure 9.16. In this joint, two yokes are combined by a crosstype pin. Four bearings are used between this pin and yokes. When this joint is used, the rotational speed of the driven shaft varies even if the driver shaft rotates with a constant rotational speed. This fluctuation causes vibrations and noise.

In Figure 9.16a, the driver shaft, rotating with a constant angular velocity ω_1, and the driven shaft are connected by a universal joint with joint angle α. The angular velocity $\omega_2(t)$ of the driven shaft is obtained as follows. Figure 9.16b shows the joint in an instant when the arms of the cross pin are in horizontal and vertical directions. The connected shafts OA and OB are in a horizontal plane. Let the unit vector in the direction of the driver shaft, that in the vertical direction, and that in the direction perpendicular to the other two directions be $\boldsymbol{i}_1, \boldsymbol{j}_1$, and $\boldsymbol{k}$, respectively. Similarly, we define unit vectors $\boldsymbol{i}_2, \boldsymbol{j}_2$, and $\boldsymbol{k}_2$ with respect to the driven shaft. The

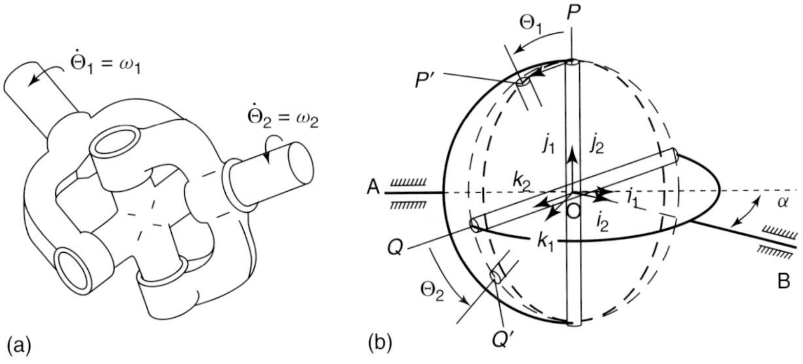

Figure 9.16 Hooke's universal joint: (a) general view and (b) angles.

following relationships hold among these unit vectors:

$$\left.\begin{aligned} \boldsymbol{i}_2 &= \cos\alpha\,\boldsymbol{i}_1 + \sin\alpha\,\boldsymbol{k}_1, \;\; \boldsymbol{j}_2 = \boldsymbol{j}_1, \\ \boldsymbol{k}_2 &= -\sin\alpha\,\boldsymbol{i}_1 + \cos\alpha\,\boldsymbol{k}_1 \end{aligned}\right\} \tag{9.12}$$

Let the driver shaft rotate from OP to OP$'$ by Θ_1 and the driven shaft rotate from OQ to OQ$'$ by Θ_2. Then vectors $\overrightarrow{\mathrm{OP}'}$ and $\overrightarrow{\mathrm{OQ}'}$ are given by

$$\left.\begin{aligned} \boldsymbol{OP}' &= r(\cos\Theta_1\boldsymbol{j}_1 + \sin\Theta_1\boldsymbol{k}_1) \\ \boldsymbol{OQ}' &= r(-\sin\Theta_2\boldsymbol{j}_2 + \cos\Theta_2\boldsymbol{k}_2) \\ &= r(-\sin\alpha\cos\Theta_2\boldsymbol{i}_1 - \sin\Theta_2\boldsymbol{j}_1 + \cos\alpha\cos\Theta_2\boldsymbol{k}_1) \end{aligned}\right\} \tag{9.13}$$

Since pins OP' and OQ' cross at right angle, we set the inner product of these two vectors to zero. Then, we get

$$\tan\Theta_2 = \cos\alpha\tan\Theta_1 \tag{9.14}$$

Differentiating this expression by time, we substitute into it the relationships $d\Theta_1/dt = \omega_1$ and $d\Theta_2/dt = \omega_2$. From the expression obtained in Eq. (9.14) and $\sec^2\Theta_2 = 1 + \tan^2\Theta_2$, we arrive at the following relationship (Burkhalter and Mazziotti, 1956):

$$\omega_2 = \frac{\cos\alpha}{1 - \sin^2\alpha\sin^2\Theta_1}\omega_1 \tag{9.15}$$

When the joint angle α is not zero, the ratio of the angular velocities ω_2/ω_1 varies between the values $\cos\alpha$ and $1/\cos\alpha$, as shown in Figure 9.17.

This angular velocity fluctuation causes not only torsional vibrations but also bending vibrations. It was observed in experiments that resonances with frequencies $N\omega_1$ ($N = 2, 3, \ldots$) occur when the relationships $p = N\omega_1$ hold between the natural frequency p of the driven shaft system and the angular velocity ω_1 of the driver shaft (Fujii, 1956b). These resonance speeds are lower than the major critical speed. The mechanism of occurrences of these vibrations differs depending on the type of resonance. In the following, a resonance with a frequency of three times ω_1 is explained (Rosenberg, 1958).

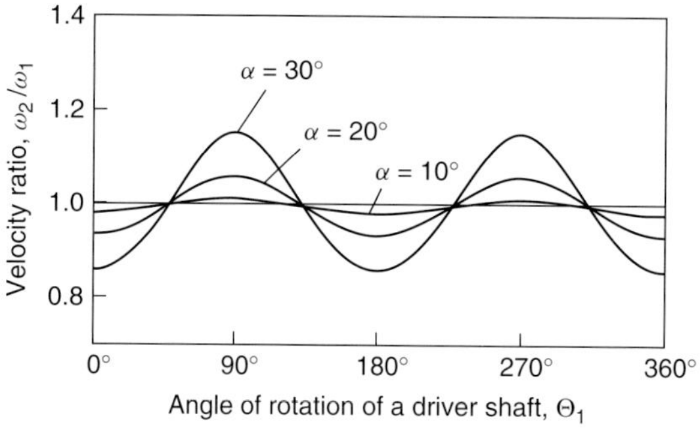

Figure 9.17 Variation of the angular velocity of a driven shaft.

Now we assume that the Jeffcott rotor in Figure 2.4 is driven through a universal joint. The equations of motion are derived by revising Eq. (2.3) as follows:

$$\left.\begin{aligned} m\ddot{x} + kx &= me\omega_2^2 \cos\Theta_2 \\ m\ddot{y} + ky &= me\omega_2^2 \sin\Theta_2 \end{aligned}\right\} \tag{9.16}$$

When the rotational speed $\omega_2(t)$ varies around the average angular speed ω_0 with frequency ν and amplitude $\varepsilon\nu$, that is, $\omega_2(t) = \omega_0 + \varepsilon\nu \sin \nu t$, the rotational angle Θ_2 is given by

$$\Theta_2 = \int \omega_2(t)dt = \omega_0 t - \varepsilon \cos \nu t \tag{9.17}$$

If the quantity ε is small, the excitation term is expressed approximately as follows:

$$\left.\begin{aligned} \cos\Theta_2 &= \cos\omega_0 t \cos(\varepsilon \cos \nu t) + \sin\omega_0 t \sin(\varepsilon \cos\nu t) \\ &\approx \cos\omega_0 t + \sin\omega_0 t(\varepsilon\cos\nu t) \\ &= \cos\omega_0 t + \frac{1}{2}\varepsilon \sin(\omega_0 + \nu)t + \frac{1}{2}\varepsilon\sin(\omega_0 - \nu)t \\ \sin\Theta_2 &= \sin\omega_0 t - \frac{1}{2}\varepsilon\sin(\omega_0+\nu)t - \frac{1}{2}\varepsilon\sin(\omega_0 - \nu)t \end{aligned}\right\} \tag{9.18}$$

This shows that excitations with angular velocities $\omega_0 + \nu$ and $\omega_0 - \nu$ appear in addition to that with ω_0. Therefore, resonances may occur when the natural frequency $p = \sqrt{k/m}$ coincides with one of these frequencies. By Fourier expansion, Eq. (9.15) is expressed by the summations of terms with angles of odd times Θ_1. We know from Figure 9.17 that when the driver shaft rotates with a constant angular velocity ω_0, the angular velocity of the rotor $\omega_2(t)$ has a large component with the angular velocity that changes twice while the shaft rotates once. Therefore, we consider this component and put $\nu = 2\omega_0$. Then, Eq. (9.18) shows that resonance occurs when

$$p = \omega_0 + \gamma = 3\omega_0 \quad \therefore \omega_0 = \frac{1}{3}p \tag{9.19}$$

The frequency of the resonance is $3\omega_0$, which is three times the rotational speed.

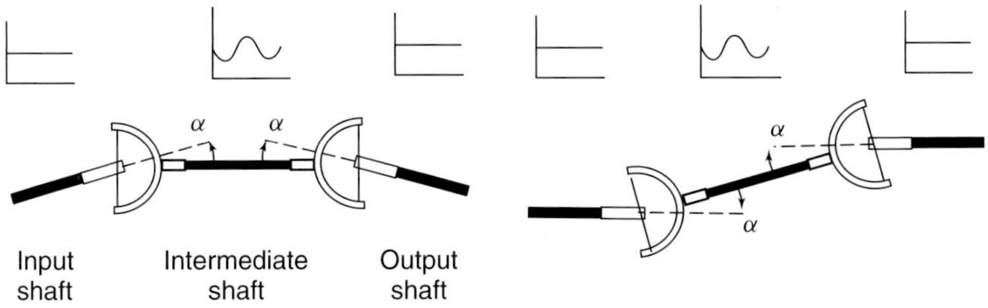

Figure 9.18 Elimination of an angular speed variation.

The jerky rotation shown in Figure 9.17 can be eliminated by using an intermediate shaft and two Hook's joints as shown in Figure 9.18. This assembly is commonly employed in rear-wheel drive vehicles as shown in Figure 9.15. In this configuration, the driving shaft and the driven shaft must be at equal angles with respect to the intermediate shaft. In addition, the two universal joints are 90° out of phase; in other words, two yokes at both ends of the intermediate shaft must be in the same plane. On the contrary, if they make a right angle, the variation is intensified.

A resonance with the frequency of twice the rotational speed occurs at $\omega_0 = p/2$. The mechanism for the occurrence of this vibration is different from the one mentioned above. It is because, when the driver shaft is rotating with a constant angular velocity, a resisting torque that appears in the driven shaft produces a secondary torque, and this torque causes this double-frequency vibration (Ota and Kato, 1984).

9.5 Rubbing

In rotating machinery, small gaps exist, for example, between rotor and housing, rotor and guide, and also in seals. When a rotor makes contact with a stator, violent self-excited oscillations, called *rubbing*, sometimes occur. In fluid machinery, because these gaps are kept small to increase the efficiency, the risk of contact increases.

Among various vibrations due to contact, backward rubbing, which occurs from a low rotational speed, has attracted the attention of engineers at an early stage. In 1926, Newkirk reported that when a shaft made contact with a wooden guard ring, the shaft started to roll around on the inner surface violently. Den Hartog (1956) explained the mechanism of this backward rub as follows. Figure 9.19 shows an overhung rotor with a guard ring to restrict excess deflection. If the rotor touches this guard ring because of disturbances or resonances, a dry friction force $\boldsymbol{F}$ works at the contact point A. Now let us consider two forces $\boldsymbol{F}$ and $-\boldsymbol{F}$ that have the same magnitude but work in opposite directions from point S. Then, we regard these three forces as a moment $\boldsymbol{F} \times \boldsymbol{AS}$ due to ① and ② and a force $\boldsymbol{F}$ due to ③. The

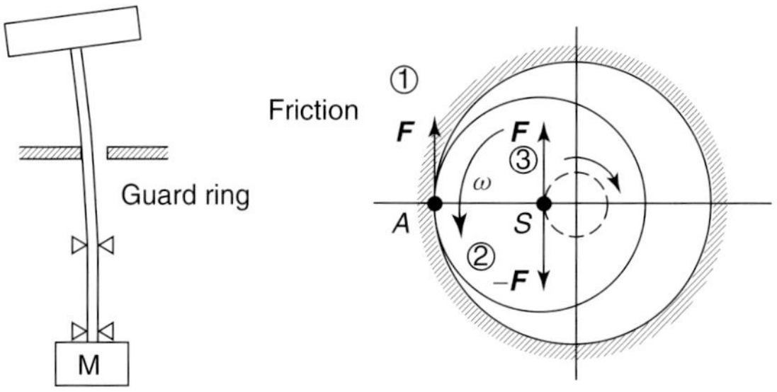

Figure 9.19 Mechanism of a backward rubbing.

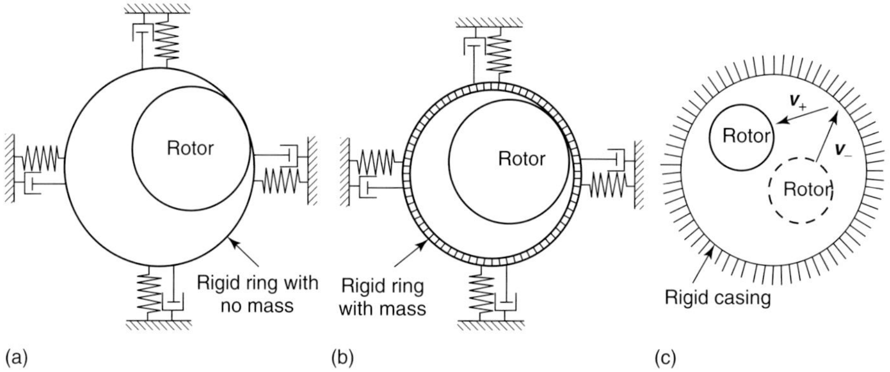

Figure 9.20 Other mathematical models for rubbing: (a) model with a massless ring, (b) model with ring mass, and (c) impacting model.

former works as a load resisting the driving torque and the latter works as a force causing a self-excited vibration of a backward whirling mode (Den Hartog, 1956).

In the analysis of rubbing, the physical models illustrated in Figure 9.20 have been adopted in addition to that of Figure 9.19. In the model shown in Figure 9.20a, the housing is replaced by a massless ring supported by springs and dampers. Goldman and Muszynska (1993) used this model and showed the occurrence of impacting and chaotic motion in addition to backward rubbing by numerical simulation.

In the model shown in Figure 9.20b, the mass of the ring is considered. Ehrich (1969) showed an example of damage of an axial flow compressor due to rubbing and also discussed the stability of the backward rubbing in this model. Black (1968) obtained response curve of a forward synchronous rubbing of an unbalanced rotor theoretically using the model in Figure 9.20b.

In the model shown in Figure 9.20c, the rubbing is considered as an impacting phenomenon. Li and Paidoussis (1994) first adopted this model in their study. In Figure 9.20c, $\boldsymbol{V}_-$ and $\boldsymbol{V}_+$ denote the velocity vectors of the rotor center, right before and after the impact. Li and Paidoussis showed the occurrence of a forward rubbing, a backward rubbing, an impacting, and a chaotic motion numerically in their study.

In the case of forward rubbing whose whirling angular speed is the same as the rotational speed of the shaft, the stress at any point inside the shaft does not change. However, in the case of backward rubbing, the stress inside the shaft changes between alternating states of tension and compression over time. Therefore, the backward rubbing is far more dangerous with respect to fatigue fracture.

In the following, rubbing is explained in terms of the model shown in Figure 9.20a (Ishida *et al.*, 2004).

9.5.1 Equations of Motion

The rotor model and the coordinate system is shown in Figure 9.21. If the equivalent spring constant of the housing is k_e, the radial stiffness force is represented by

$$\boldsymbol{F}_k = \begin{cases} -k_e (r-\delta)\dfrac{\boldsymbol{r}}{|\boldsymbol{r}|} & (\text{for } r \geq \delta) \\ 0 & (\text{for } r < \delta) \end{cases} \tag{9.20}$$

The equivalent spring constant k_e is determined by the contact theory and the material characteristics (Momono, 2000). If the damping coefficient of the housing is c_e, the radial damping force is given by

$$\boldsymbol{F}_c = \begin{cases} -c_e \dot{r}\dfrac{\boldsymbol{r}}{|\boldsymbol{r}|} & (\text{for } r \geq \delta) \\ 0 & (\text{for } r < \delta) \end{cases} \tag{9.21}$$

The equivalent damping coefficient c_e can be determined by measuring the restitution coefficient e in experiments as follows. Let us consider a horizontal surface supported by a spring and a damper with the coefficients k_e and c_e; then, let us also consider a second rigid unsupported surface. Suppose that we drop a rigid ball from the same height above these two surfaces. By making the height expressed by k_e and c_e in the first case and the height expressed by the restitution

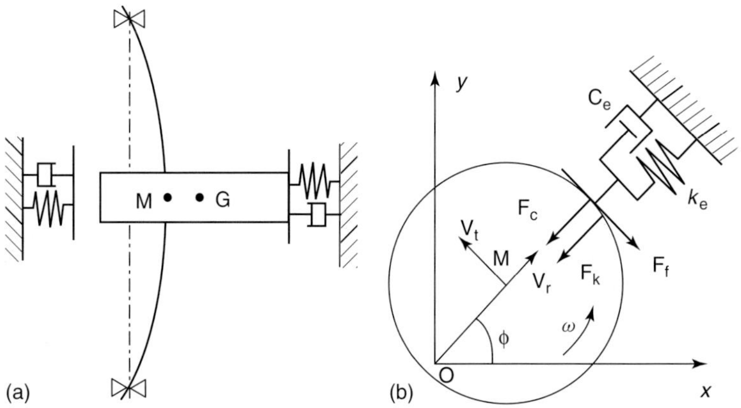

Figure 9.21 Physical models for a rotor-guide system: (a) rotor system and (b) contact model.

coefficient e in the second case equal, we can determine the relationship between c_e, e, and k_e. Since e and k_e can be easily determined independently by other means, we can determine c_e from this relationship.

When the rotor has contact with the ring, the following friction force works in the tangential direction.

$$F_f = |\boldsymbol{F}_f| = \mu\,(|\boldsymbol{F}_k + \boldsymbol{F}_c|) \tag{9.22}$$

Suppose that $\boldsymbol{V}_t$ is the tangential velocity of the rotor center. This friction force works in the opposite direction as V_t when $(V_t + \omega r) > 0$ and in the same direction when $(V_t + \omega r) < 0$.

Let the mass of the rotor be m, the damping coefficient be c, the eccentricity be $\hat{e}$, the spring constant of the shaft be k, the diameter and radius of the shaft be D and R, respectively, and the critical speed be $\omega_n = \sqrt{k/m}$. We adopt the following dimensionless quantities:

$$\begin{aligned} &x' = \frac{x}{D},\quad y' = \frac{y}{D},\quad \delta' = \frac{\delta}{D},\quad r' = \frac{r}{D},\quad \hat{e}' = \frac{\hat{e}}{D},\quad t' = \omega_n t,\\ &c' = \frac{c}{m\omega_n},\quad \omega' = \frac{\omega}{\omega_n},\quad F_k' = \frac{F_k}{m\omega_n D},\quad F_c' = \frac{F_c}{m\omega_n D},\quad F_f' = \frac{F_f}{m\omega_n D} \end{aligned} \tag{9.23}$$

Then, the equations of motion are given by

$$\begin{aligned} &\ddot{x} + c\dot{x} + x + F_{kx} + F_{cx} \mp F_f\frac{y}{r} = \hat{e}\omega^2\cos\omega t\\ &\ddot{y} + c\dot{y} + y + F_{ky} + F_{cy} \pm F_f\frac{x}{r} = \hat{e}\omega^2\sin\omega t \end{aligned} \tag{9.24}$$

where the prime is eliminated for simplicity. The sign before F_f is selected as follows. The upper sign is adopted when $(V_t + \omega r) > 0$ and the lower sign is adopted when $(V_t + \omega r) < 0$.

9.5.2
Numerical Simulation

By integrating Eq. (9.24) numerically, we can obtain a forward synchronous rubbing, a backward rubbing, and an impacting motion depending on the initial conditions. Figure 9.22 shows the amplitude and phase angle of a contact-free synchronous oscillation (symbol ○) and a synchronous forward rubbing with angular frequency ω (symbol •). The shaded range corresponds to the position of the guide. In this case, the clearance $\delta = 0.3$ is larger than the peak of the resonance. The forward rubbing occurs above a certain threshold rotational speed which is in the postcritical range. When the rotational speed decelerates during rubbing, the motion jumps to the contact-free motion. If the clearance is less than the peak of the resonance or the friction is zero ($\mu = 0$), the rubbing occurs from the major critical speed. That is, the response curve of the forward rubbing bifurcates from the synchronous response curve on the left side at the resonance.

Figure 9.23a shows the amplitude and whirling angular speed of a contact-free synchronous oscillation (symbol ○) and the backward rubbing (symbol •). This

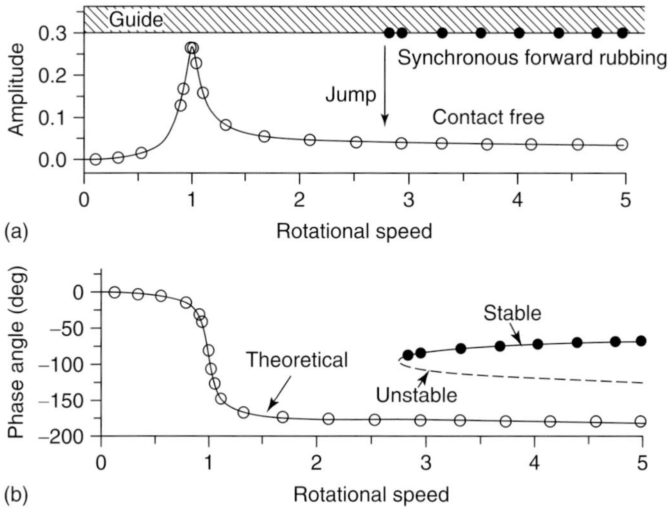

Figure 9.22 Synchronous forward rubbing: (a) amplitude and (b) phase angle. $\hat{e} = 0.035, c = 0.13, k_e = 3.8 \times 10^6, c_e = 4.0 \times 10^2, R = 0.5$.

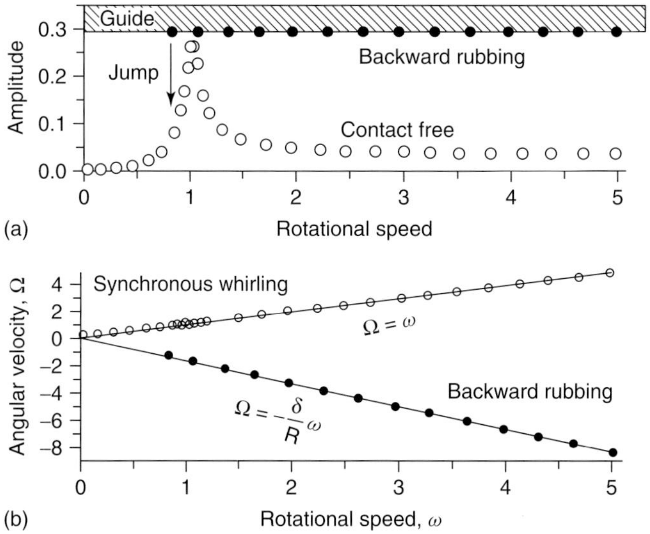

Figure 9.23 Backward rubbing: (a) amplitude and (b) whirling angular speed.

backward rubbing occurs above a certain threshold value which is lower than the critical speed. As shown in Figure 9.23b, the angular velocity of the rubbing is proportional to the rotational speed.

When the friction is small, an impacting motion occurs. Figure 9.24 shows an example of its orbit. Whether rubbing will occur in a forward, backward, or

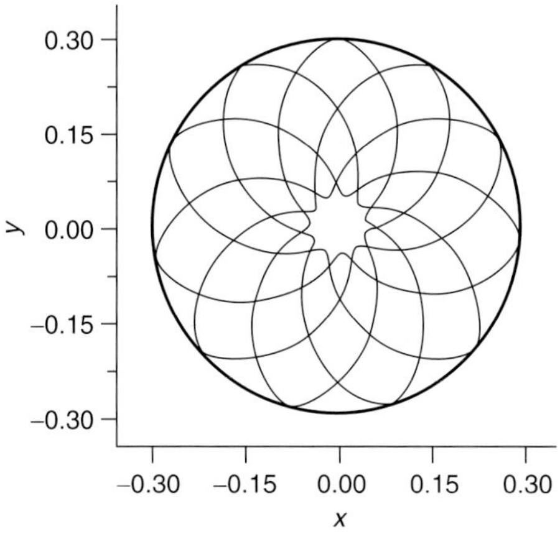

Figure 9.24 Orbit of an Impacting motion ($\mu = 0$, $\omega = 8.0$).

impacting mode depends on the initial conditions of the numerical simulation. Figure 9.25 shows the probability of occurrence for each rubbing motion. At various magnitude of Coulomb friction, numerical simulations are performed for the 48 different initial positions shown Figure 9.25(a) with zero initial velocities. The rotational speed is $\omega = 6.0$. Figure 9.25(b) shows the type of rubbing motion. The notations ∘, •, × and − represent a backward rub, forward rub, impact type rub and no contact motion, respectively. A backward rub is most dangerous with respect from the fatigue of the shaft. From the figure, we see that the backward rub occurs with high probability when the friction is comparatively large. At the friction decreases, the probability of the backward rub decreases and that of the forward rub increases. When the friction is very small, the impact type rub also occurs.

9.5.3 Theoretical Analysis

9.5.3.1 Forward Rubbing

Considering that the whirling speed of the forward rub is ω, we assume the solution of the forward rubbing to be as follows:

$$x = r\cos(\omega t + \beta), \quad y = r\sin(\omega t + \beta) \tag{9.25}$$

Here, we consider the solution in the vicinity of the steady-state forward rubbing and adopt the assumption $\dot{r} = O(\varepsilon)$, $\dot{\beta} = O(\varepsilon)$, and $\ddot{r} = O(\varepsilon^2)$. Then, we obtain the following approximate equations of motion:

$$\left.\begin{aligned} -r\left(\omega^2 + 2\omega\dot{\beta}\right) + r &= \hat{e}\omega^2\cos\beta - k_e\left(r - \delta\right) - c_e\dot{r} \\ -2\dot{r}\omega - cr\omega &= \hat{e}\omega^2\sin\beta + \mu\{k_e(r - \delta) + c_e\dot{r}\} \end{aligned}\right\} \tag{9.26}$$

We can obtain the steady-state solution $r = R_0$ and $\beta = \beta_0$ by putting $\dot{r} = 0$ and $\dot{\beta} = 0$ in this equation. The stability of this solution can be investigated by considering small deviations as $r = R_0 + \xi$ and $\beta = \beta_0 + \eta$, and following the

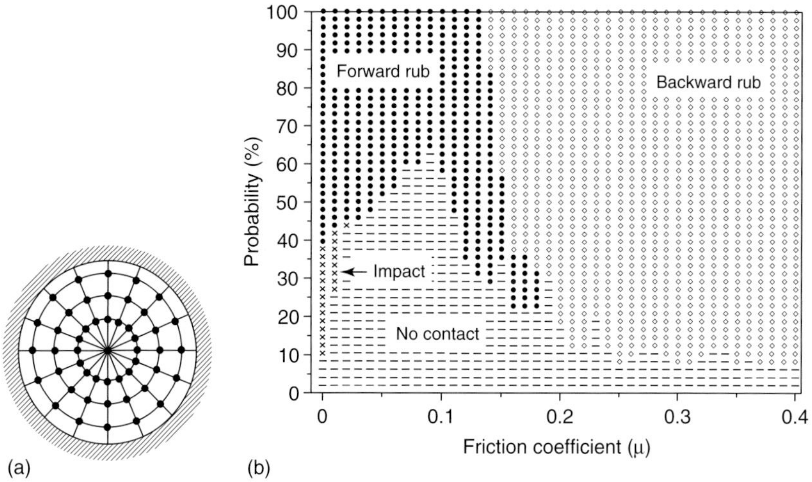

Figure 9.25 Probability of occurrence of each rubbing (a) initial condition and (b) Probability of the occurrence of various rubbing at $\omega \approx 6.0$ (Ishida et al., 2006).

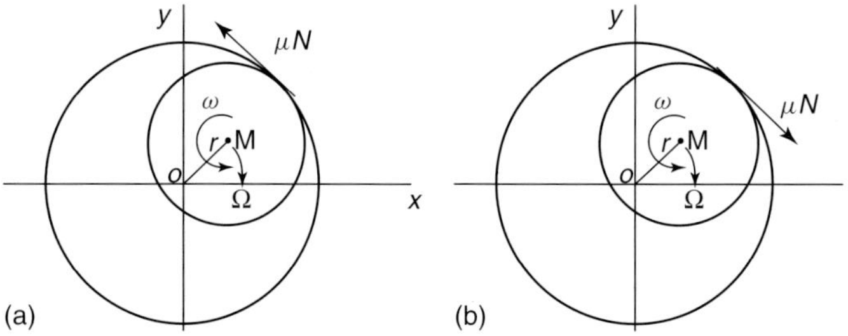

Figure 9.26 Direction of the friction: (a) $0 > \Omega_0 > \Omega$ and (b) $0 > \Omega > \Omega_0$.

same procedure as that has been explained in Chapter 6. The result is shown in Figure 9.22. The solid line represents stable solutions while the broken line represents unstable ones.

9.5.3.2 **Backward Rubbing**

Since backward rubbing occurs without an unbalance, we discuss the case with no unbalance. Substituting $x = r\cos\phi$ and $y = r\sin\phi$ into Eq. (9.24) with $\hat{e} = 0$, we obtain the following equations of motion in a polar form:

$$\left.\begin{aligned} &\ddot{r} - r\dot{\phi}^2 + c\dot{r} + r + k_e(r-\delta) + c_e\dot{r} = 0 \\ &-\left(2\dot{r}\dot{\phi} + r\ddot{\phi}\right) - cr\dot{\phi} \mp \mu\left(k_e(r-\delta) + c_e\dot{r}\right) = 0 \end{aligned}\right\} \qquad (9.27)$$

First, we discuss the angular velocity of the backward rubbing. Let the angular speed of the shaft whirling motion when the shaft rolls without slipping be Ω_0.

This speed is given from Table 9.2 as follows:

$$\Omega_0 = -\frac{R}{\delta}\omega \tag{9.28}$$

where R is the shaft radius and δ the clearance. Since the shaft radius R is generally much larger than the clearance δ, Ω_0 is much faster than the rotational speed ω. From the second expression in Eq. (9.27), we obtain

$$r\ddot{\phi} = -2\dot{r}\dot{\phi} - cr\dot{\phi} \mp \mu\left(k_e(r-\delta) + c_e\dot{r}\right) \tag{9.29}$$

Suppose that the shaft is sliding on the inside wall with an angular velocity Ω and with a constant radius ($\dot{r} = 0$). When $0 > \Omega_0 > \Omega$, the friction works to decelerate the whirling as shown in Figure 9.26a. In this case, Eq. (9.29) becomes

$$r\ddot{\phi} = -cr\dot{\phi} + \mu\left(k_e(r-\delta) + c_e\dot{r}\right) \tag{9.30}$$

since $\dot{\phi} < 0$ and $\ddot{\phi} > 0$ hold. Therefore, the angular velocity $|\Omega|$ decreases and Ω approaches Ω_0. When $0 > \Omega > \Omega_0$, the friction works to accelerate the whirling as shown in Figure 9.26b. In this case, Eq. (9.29) becomes

$$r\ddot{\phi} = -cr\dot{\phi} - \mu\left(k_e(r-\delta) + c_e\dot{r}\right) \tag{9.31}$$

Since the damping c is small, $\ddot{\phi} < 0$ holds. Therefore, the angular velocity $|\Omega|$ increases and Ω approaches Ω_0. In conclusion, we can say that the backward rubbing whirls with an angular velocity $\Omega_0 = -(R/\delta)\omega$. The straight line $\Omega = -(0.5/0.3)\omega$ is illustrated in Figure 9.23b.

Next, we derive the threshold rotational speed from which the backward rubbing occurs. Let us assume that the shaft whirls backward with a constant radius $r = (1+\varepsilon)\delta$ at an angular velocity $\Omega_0 = -(R/\delta)\omega$. Then, the first expression of Eq. (9.27) becomes

$$-(1+\varepsilon)\delta\Omega_0{}^2 + (1+\varepsilon)\delta + k_e(\varepsilon\delta) = 0 \tag{9.32}$$

Since the shaft leaves the guide when $\varepsilon = 0$, we find from this expression that the threshold value is

$$\omega = \delta/R < 1 \tag{9.33}$$

This result shows that the backward rubbing starts from a certain rotational speed lower than the major critical speed.

9.6 Self-Excited Oscillation in a System with a Clearance between Bearing and Housing

In the previous section, the rotor had a clearance between the rotating shaft and the guide. In such a case, the contact occurs between the rotating part and the static part. On the contrary, some rotor systems have a clearance between movable but nonrotating part and static part. For example, a squeeze-film damper bearing has a clearance between movable bearing holder and static housing. A bearing, which is inserted into a hosing loosely to compensate the heat expansion, has a clearance between bearing's outer ring and housing. In the present section, nonlinear resonances and a self-excited oscillation that occur in the latter systems are explained.

9.6.1 Experimental Setup and Experimental Results

Ishida, Imagaki, Ejima and Hayashi (2009) investigated vibrations that occurred in the experimental setup illustrated in Figure 9.27a. There is a clearance between bearing holder and housing. Since double-row self-aligning ball bearings are used at both ends, ball bearings themselves do not produce nonlinear characteristics as mentioned in Figure 6.1. Response curves are illustrated in Figure 9.27b. The system is similar to the system shown in Figure 6.13. However, the kinds of resonance are different from those shown in Figure 6.16, because forward and backward natural frequencies change as a function of the rotational speed in this case. Forward subharmonic resonances of orders 1/2, 1/3, and 1/4, a forward supersubharmonic resonance of order 2/5, and a combination resonance $[2p_2 - p_3]$ occurred. In addition, a self-excited oscillation with the frequency p_2 occurred in a wide speed range above 1150 rpm. This self-excited oscillation was first discovered by Lin and Vance (1994). The self-excited oscillation disappeared in the regions surrounding resonances (this is called an *entrainment phenomena*).

Figure 9.28a shows the natural frequency diagram of this rotor. The curves are obtained theoretically by assuming that the shaft is supported simply without a clearance. The frequencies of self-excited oscillations obtained in experiments are plotted by symbol ○. We can confirm that the self-excited oscillations occurred along the p_2 curve. Although the unbalance is necessary for the occurrence of this

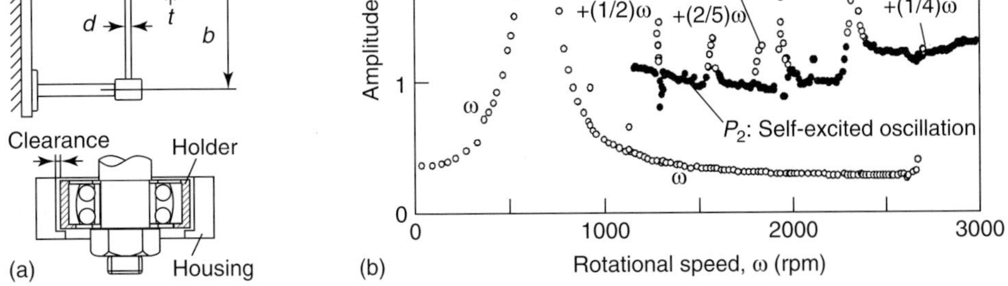

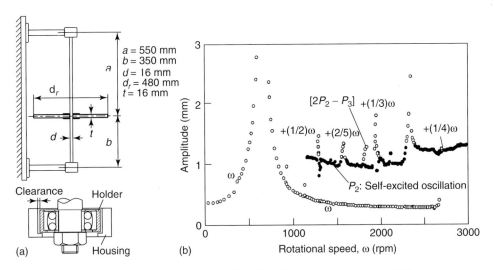

Figure 9.27 Experimental setup and experimental results: (a) experimental setup and (b) response curve.

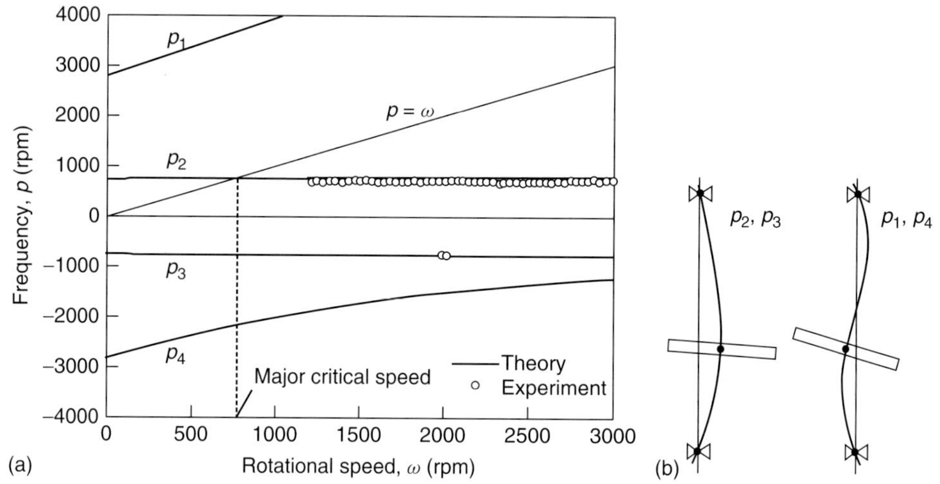

Figure 9.28 Natural frequency diagram and mode shapes: (a) theoretical and experimental results and (b) mode shapes.

self-excited oscillation, we call this a self-excited oscillation because its frequency is equal to a forward natural frequency and has no relationship to the rotational speed.

Figure 9.28b illustrates mode shapes corresponding to p_2, p_3 and p_1, p_4. The radial deflection is predominant in the former and the inclination is predominant in the latter.

9.6.2 Analytical Model and Reduction of Equations of Motion

Figure 9.29 illustrates an analytical model. The clearance between the outer ring of the bearing and the housing is δ. The position of the disk is given by four coordinates $(x_1, y_1, \theta_x, \theta_y)$ and that of the center of the lower bearing is given by (x_2, y_2). The length of the shaft is $l = a + b$. This model has 6 DOF. Since all the resonances and self-excited oscillations observed in Figure 9.27 are related to modes with p_2 and p_3, we focus our discussion on these modes and reduce the DOF (Inagaki and Ishida, 2011). First, we consider the case with no clearance ($\delta = 0$). If there is only an eccentricity as an unbalance and there is no damping, the equations of motion are given from Eq. (2.71) as follows:

$$\left.\begin{aligned}
&m_1\ddot{x}_1 + \alpha x_1 + \gamma\theta_x = m_1 e\omega^2 \cos\omega t\\
&m_1\ddot{y}_1 + \alpha y_1 + \gamma\theta_y = m_1 e\omega^2 \sin\omega t\\
&I\ddot{\theta}_x + I_p\omega\dot{\theta}_y + \gamma x_1 + \delta\theta_x = 0\\
&I\ddot{\theta}_y - I_p\omega\dot{\theta}_x + \gamma y_1 + \delta\theta_y = 0
\end{aligned}\right\} \tag{9.34}$$

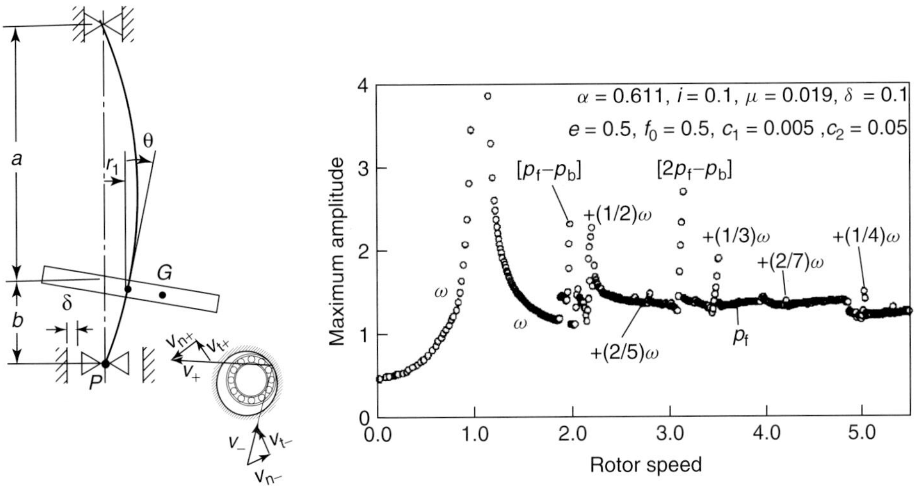

Figure 9.29 Rotor system and numerical results: (a) rotor model and (b) response curve.

From the third and fourth expressions, we see that the following relationships hold for p_2 and p_3 when $\omega = 0$.

$$\theta_x = \frac{\gamma}{\delta} x_1, \qquad \theta_y = \frac{\gamma}{\delta} y_1 \tag{9.35}$$

By assuming that these relationships hold during rotation, we represent the deflection by the following two modes:

$$\phi_x = [1 \;\; 0 - \gamma/\delta \;\; 0]^{\mathrm{T}}, \qquad \phi_y = [0 \;\; 1 \;\; 0 \;\; -\gamma/\delta]^{\mathrm{T}} \tag{9.36}$$

Then, the deflection of the disk is represented by

$$\begin{bmatrix} x_1 \\ y_1 \\ \theta_x \\ \theta_y \end{bmatrix} = \begin{bmatrix} 1 \\ 0 \\ -\gamma/\delta \\ 0 \end{bmatrix} x_1 + \begin{bmatrix} 0 \\ 1 \\ 0 \\ -\gamma/\delta \end{bmatrix} y_1 = \begin{bmatrix} \phi_x & \phi_y \end{bmatrix} \begin{bmatrix} x_1 \\ y_1 \end{bmatrix} \tag{9.37}$$

After representing Eq. (9.34) by matrices, we substitute Eq. (9.37) into Eq. (9.34). By multiplying $\begin{bmatrix} \phi_x & \phi_y \end{bmatrix}^{\mathrm{T}}$ from the left side, we have

$$\begin{bmatrix} m_1 & 0 \\ 0 & m_1 \end{bmatrix} \begin{bmatrix} \ddot{x}_1 \\ \ddot{y}_1 \end{bmatrix} + \begin{bmatrix} 0 & i_p\omega \\ -i_p\omega & 0 \end{bmatrix} \begin{bmatrix} \dot{x}_1 \\ \dot{y}_1 \end{bmatrix} + \begin{bmatrix} c_1 & 0 \\ 0 & c_1 \end{bmatrix} \begin{bmatrix} \dot{x}_1 \\ \dot{y}_1 \end{bmatrix}$$
$$+ \begin{bmatrix} k & 0 \\ 0 & k \end{bmatrix} \begin{bmatrix} x_1 \\ y_1 \end{bmatrix} = \begin{bmatrix} me\omega^2 \cos\omega t \\ me\omega^2 \sin\omega t \end{bmatrix} \tag{9.38}$$

where

$$m_1 = m + \frac{\gamma^2}{\delta^2} I, \quad i_p = \frac{\gamma^2}{\delta^2} I_p, \quad c_1 = c_{11} - \frac{\gamma}{\delta}(c_{12} + c_{21}) + \frac{\gamma^2}{\delta^2} c_{22}, \quad k = \alpha + \frac{\gamma^2}{\delta} \tag{9.39}$$

Equation (9.38) is also represented by

$$\left.\begin{aligned} m_1\ddot{x}_1 + i_p\omega\dot{y}_1 + c_1\dot{x}_1 + kx_1 &= me\omega^2 \cos\omega t \\ m_1\ddot{y}_1 - i_p\omega\dot{x}_1 + c_1\dot{y}_1 + ky_1 &= me\omega^2 \sin\omega t \end{aligned}\right\} \tag{9.40}$$

Next, we consider the case that there is a clearance around the lower bearing. In order to consider the deviation of the neutral position of the lower bearing due to misalignment, we added a lateral force f_0 into this expression. Since the lower end of the shaft deflects, the restoring force working toward mass m_1 becomes $k(x_1 - \alpha x_2)$ and $k(y_1 - \alpha y_2)$ where $\alpha = a/l$. This force is divided into the forces working toward the upper and lower bearings. The latter is given by $-k(x_1 - \alpha x_2)\alpha$ and $-k(y_1 - \alpha y_2)\alpha$. Let the mass of the lower bearing be m_2. Here, we introduce the following dimensionless quantities:

$$t' = \omega_n t, \quad x_1' = \frac{x_1}{d_0}, \quad y_1' = \frac{y_1}{d_0}, \quad x_2' = \frac{x_2}{d_0}, \quad y_2' = \frac{y_2}{d_0}, \quad \mu_2 = \frac{m_2}{m_1}, \quad e' = \frac{e}{d_0},$$
$$i_p' = i_p\frac{\omega'}{m_1}, \quad c_1' = \frac{c_1}{m_1\omega_n}, \quad c_2' = \frac{c_2}{m_1\omega_n}, \quad f_0' = \frac{f_0}{m_1\omega_n^2 d_0}, \quad \delta' = \frac{\delta}{d_0} \tag{9.41}$$

where d_0 is the static deflection of the shaft when it is horizontally supported and $\omega_n = \sqrt{k/m_1}$. Then, the equations of motion are represented by

$$\left.\begin{aligned} &\ddot{x}_1 + i_p\omega\dot{y}_1 + c_1\dot{x}_1 + (x_1 - \alpha x_2) = \mu e\omega^2 \cos\omega t \\ &\ddot{y}_1 - i_p\omega\dot{x}_1 + c_1\dot{y}_1 + (y_1 - \alpha y_2) = \mu e\omega^2 \sin\omega t + f_0 \\ &\mu_2\ddot{x}_2 + c_2\dot{x}_2 + \alpha(\alpha x_2 - x_1) = 0 \\ &\mu_2\ddot{y}_2 + c_2\dot{y}_2 + \alpha(\alpha y_2 - y_1) = 0 \end{aligned}\right\} \tag{9.42}$$

where the primes are omitted for the convenience of expression. In the following, we represent the natural frequencies of a forward and backward modes corresponding to p_2 and p_3 by p_f and p_b, respectively.

9.6.3
Numerical Simulation

The expressions in Eq. (9.42) are integrated numerically (Ishida, Inagaki, Ejina and Hayashi 2009). As shown in Figure 9.20, there are several ways to represent the contact in the clearance. Here, we adopt the model illustrated in Figure 9.20c, which was first adopted by Li and Paidoussis (1994). In Figure 9.29a, the notations V_n and V_t denote the components of the velocity of the bearing center in the normal and tangential directions to the wall. Let the velocities before and after the collision be $V_-(V_{n-}, V_{t-})$ and $V_+(V_{n+}, V_{t+})$, respectively, and the coefficient of restitution be e. We assume that the relationships $V_{n+} = -eV_{n-}$ and $V_{t+} = V_{t-}$ hold among them.

A result of numerical simulation is illustrated in Figure 9.29b. Parameter values are selected so as to explain well the experimental result shown in Figure 9.27b. The result agrees well with that in Figure 9.27b. This result agrees well with the experimental result as a whole; however, there are several differences. In the experimental results shown in Figure 9.27b, two amplitudes exist in the range above 1130 rpm. Depending on the initial condition, the self-excited oscillation occurs together with the harmonic oscillation or only the harmonic oscillation occurs. However, in a numerical simulation, a self-excited oscillation and a harmonic oscillation always occur simultaneously in the corresponding speed range. In addition, although many kinds of nonlinear resonances occurred in the experiments, only the subharmonic resonance of order 1/2 occurred in the numerical simulation. However, the combination resonance of type $[p_f - p_b = \omega]$, which was not observed in the experiments, occurred in the simulation. Since a self-excited oscillation occurred almost in the same speed range as that occurred in the experiments, we see that this theoretical model is appropriate to use in the discussion on the occurrence of a self-excited oscillation.

9.6.4 Self-Excited Oscillations

In order to explain the mechanism underlying the occurrence of self-excited oscillation, we firstly investigate the stability of the synchronous oscillation and then consider the energy balance (Inagaki and Ishida, 2011).

9.6.4.1 Analytical Model and Equations of Motion

We use the model shown in Figure 9.32, which is a simplified model of Figure 9.29. The disk is mounted at the middle position of the shaft. Considering that the lower bearing moves along the inner surface of the housing in Figure 9.29, the bearing is restrained in the smooth circular channel with radius δ in this model. The mass of the bearing is neglected. Let the angular position of the bearing be $\varphi_2 = \omega t + \beta_2$. Then, the equations of motion are expressed as follows:

$$\left.\begin{aligned} &m_1\ddot{x}_1 + c_1\dot{x}_1 + k(x_1 - x_2) = m_1 e\omega^2 \cos\omega t \\ &m_1\ddot{y}_1 + c_1\dot{y}_1 + k(y_1 - y_2) = m_1 e\omega^2 \sin\omega t \\ &c_2\delta\dot{\varphi}_2 + k(x_1 \sin\varphi_2 - y_1\cos\varphi_2) = 0 \end{aligned}\right\} \tag{9.43}$$

where

$$x_2 = \delta\cos\varphi_2 \quad \text{and} \quad y_2 = \delta\sin\varphi_2 \tag{9.44}$$

Although various kinds of friction work in the bearing, we represent them by a viscous damping force with a damping coefficient c_2.

The third equation is obtained as follows. Let the elastic restoring forces of the shaft be $F_x = -k(x_1 - x_2)$ and $F_y = -k(y_1 - y_2)$. Here, we consider the mass of the bearing m_2 temporarily. Then, the equations of motion of the bearing in the tangential direction is given by $m_2\delta\ddot{\varphi}_2 = F_x\cos\varphi_2 - F_y\sin\varphi_2 - c_2\delta\dot{\varphi}_2$. If we put $m_2 = 0$, we can obtain the third expression.

Note: Industrial Case Study of a Self-Excited Oscillation

Figure 9.30 shows a rotor of an automobile gas turbine. The rotor is supported by two bearings. In order to make it easy to assemble and disassemble the system and to compensate the heat expansion, a small clearance often be made between the outer ring of a bearing and the housing. Figure 9.31 shows the Campbell diagram of this rotor. In addition to a harmonic oscillation due to unbalance, a self-excited oscillation with a natural frequency occurred in the supercritical speed range.

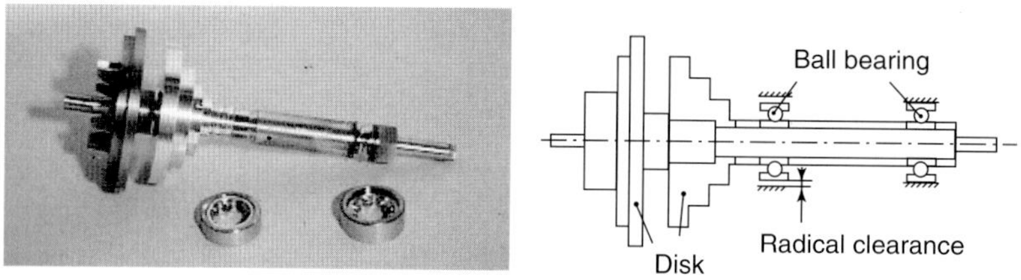

Figure 9.30 Rotor in an automobile gas turbine. (Inagaki and Ishida, 2011)

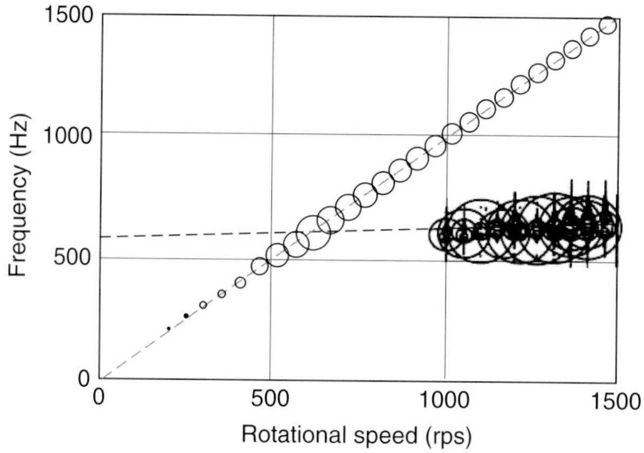

Figure 9.31 Campbell diagram of an automobile gas turbine.

9.6.4.2 **Stability of a Synchronous Whirl**

Now, we assume the synchronous solutions to be

$$x_1 = P\cos(\omega t + \beta_1), \quad y_1 = P\sin(\omega t + \beta_1), \quad \varphi_2 = \omega t + \beta_2 \tag{9.45}$$

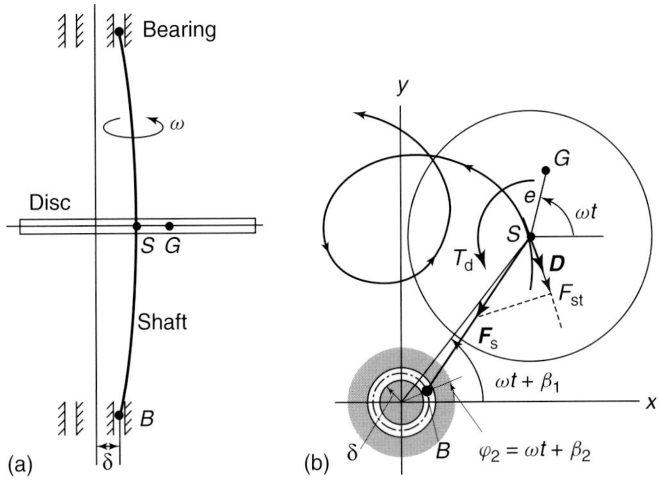

Figure 9.32 Rotor system with a circular movement of bearing: (a) rotor system and (b) forces and work.

Substituting these into Eq. (9.43), we can obtain the following expressions in the accuracy of $O(\varepsilon)$:

$$\left.\begin{aligned} &-m_1(\omega^2 + 2\omega\dot{\beta}_1)P + kP - k\delta\cos(\beta_1 - \beta_2) = m_1 e\omega^2 \cos\beta_1 \\ &2m_1\omega\dot{P} + c_1\omega P - k\delta\sin(\beta_1 - \beta_2) = -m_1 e\omega^2 \sin\beta_1 \\ &c_2\delta(\omega - \dot{\beta}_2) - kP\sin(\beta_1 - \beta_2) = 0 \end{aligned}\right\} \tag{9.46}$$

If we make $c_2 = 0$ in the third equation, we obtain $\beta_1 = \beta_2$. In this case, the first and second equations become the equations of motion for the Jeffcott rotor and, therefore, it is apparent that the self-excited oscillation does not occur. Therefore, we assume that the parameter c_2 can have a value of up to $O(\varepsilon^0)$.

The steady-state solutions can be obtained in the same way as those mentioned in Chapter 6, and the result is shown in Figure 9.33. The results of numerical simulation are also plotted in the same figure. The synchronous steady-state solution is unstable above $\omega \approx 1.5$ in the postcritical speed range. In this range, a self-excited oscillation with frequency p_f occurs together with a synchronous component.

9.6.4.3 Mechanism of a Self-Excited Oscillation

If we consider a driving torque as shown in Figure 9.32b, the equations of motion are given as follows:

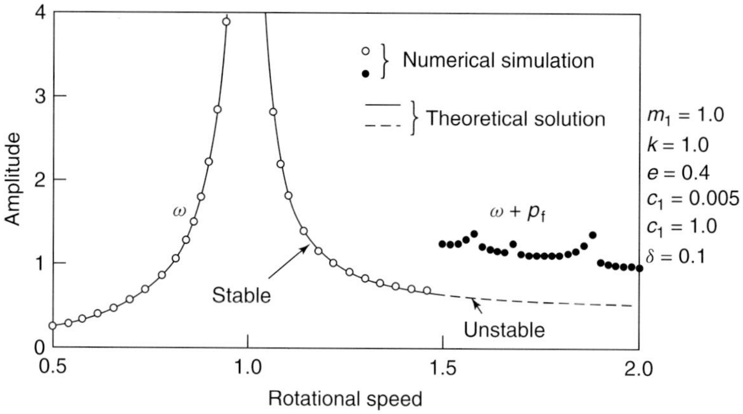

Figure 9.33 Synchronous oscillation.

$$\left.\begin{aligned} m_1\ddot{x}_G &= -k(x_1 - x_2) - c_1\dot{x}_1 \\ m_1\ddot{y}_G &= -k(y_1 - y_2) - c_1\dot{y}_1 \\ I_G\ddot{\psi} &= (y_G - y_1)\{-k(x_1 - x_2) - c_1\dot{x}_1\} - (x_G - x_1)\{-k(y_1 - y_2) - c_1\dot{y}_1\} + T_d \\ c_2\delta\dot{\varphi}_2 &+ k(x_1 \sin\varphi_2 - y_1 \cos\varphi_2) = 0 \end{aligned}\right\} \tag{9.47}$$

where I_G is the polar moment of inertia of the disk. It is assumed that a self-excited oscillation with an "arbitrary" amplitude R occurs together with a harmonic oscillation. Then, the solution is represented by

$$\left.\begin{aligned} x_1 &= P_0\cos(\omega t + \beta_{10}) + R\cos p_f t \\ y_1 &= P_0\sin(\omega t + \beta_{10}) + R\sin p_f t \end{aligned}\right\} \tag{9.48}$$

where P_0 and β_{10} are the steady-state harmonic solutions obtained above. The period of the synchronous oscillation is $T_\omega = 2\pi/\omega$ and that of the self-excited oscillation is $T_p = 2\pi/p_f$. Supposing that the $N_\omega \times T_\omega$ and $N_p \times T_p$ (where N_ω and N_p are integers) are almost equal, we investigate the energy input during this period $T(\approx N_\omega T_\omega \approx N_p T_p)$.

The work W_F done by the elastic restoring force $\boldsymbol{F}_s$ is given by

$$W_F = \int_0^T \boldsymbol{F}_s d\boldsymbol{r} = \int_0^T \left\{-k(x_1 - x_2)\dot{x}_1 - k(y_1 - y_2)\dot{y}_1\right\} dt \tag{9.49}$$

The work W_D done by the viscous damping force $\boldsymbol{D}$ is given by

$$W_D = \int_0^T \boldsymbol{D} d\boldsymbol{r} = \int_0^T (-c_1\dot{x}_1^2 - c_1\dot{y}_1^2)dt \tag{9.50}$$

The work W_T done by the driving torque T_d is given by

$$W_T = \int_0^T T_d d\psi = \int_0^T \left[\begin{array}{l} (y_G - y_1)\{k(x_1 - x_2) + c_1\dot{x}_1\} \\ -(x_G - x_1)\{k(y_1 - y_2) + c_1\dot{y}_1\} \end{array}\right] d\psi \tag{9.51}$$

The total energy charged into the disk during the period T is given by

$$W = W_F + W_D + W_T \tag{9.52}$$

Here, we investigate the total energy input. For that purpose, Eqs. (9.49)–(9.51) are transformed by inserting Eqs. (9.47) and (9.48) as follows:

$$W_F = k\delta P_0\omega \int_0^T \sin(\beta_2(t) - \beta_{10})dt + k\delta Rp_f \int_0^T \sin\left\{(\omega - p_f)t + \beta_2(t)\right\} dt \tag{9.53}$$

$$W_D = -c_1(P_0^2\omega^2 + R^2p_f^2)T \tag{9.54}$$

$$W_T = -me\omega^3 P \int_0^t \sin\beta_{10}(t)dt \tag{9.55}$$

The total energy charged into the disk during the period T is given by

$$\begin{aligned} W = {} & k\delta P_0\omega \int_0^T \sin(\beta_2(t) - \beta_{10}(t))dt - c_1 P^2\omega^2 \\ & - me\omega^3 P \int_0^T \sin\beta_{10}(t)dt + k\delta Rp_f \int_0^T \sin\left\{(\omega - p_f)t + \beta_2(t)\right\} dt - c_1 R^2 p_f^2 \end{aligned} \tag{9.56}$$

where the phase angle $\beta_2(t)$ is determined by the fourth expression in Eq. (9.47). That is,

$$c_2\delta\{\omega + \dot{\beta}_2(t)\} + k[x_1 \sin\{\omega t + \beta_2(t)\} - y_1\cos\{\omega t + \beta_2(t)\}] = 0 \tag{9.57}$$

In Eq. (9.56), the first, second, and third terms are related to the harmonic component and the fourth and fifth are related to the self-excited oscillation. Therefore, the amplitude and phase angle of the harmonic component are determined so as to make the sum of the first three terms zero, and the amplitude of the self-excited oscillation is determined so as to make the sum of the fourth and fifth terms zero.

In order to investigate the existence of a stable self-excited oscillation (stable limit cycle), we calculate the summation of the fourth and fifth terms, that is,

$$W_{\text{self}} = k\delta Rp_f \int_0^T \sin\left\{(\omega - p_f)t + \beta_2(t)\right\} dt - c_1 R^2 p_f^2 \tag{9.58}$$

as a function of amplitude R. The harmonic solution obtained from Eq. (9.46) is used here. By substituting Eq. (9.48) into Eq. (9.57), we can obtain $\beta_2(t)$. This value is then substituted into Eq. (9.58). The amplitude R increases when $W_{\text{self}} > 0$ and decreases when $W_{\text{self}} < 0$.

As an example, the values of W_{self} are illustrated in Figure 9.34 as a function of the amplitude R in the case $\omega = 0.6$ in the undercritical speed range and $\omega = 1.6$ in the supercritical speed range. In the former, since $W_{\text{self}} < 0$ holds for any value of the amplitude R, the energy is always discharged. This means that a self-excited oscillation does not occur. However, in the latter, W_{self} is positive when $R < R_s$ and is negative when $R > R_s$. This means that the amplitude increases when $R < R_s$

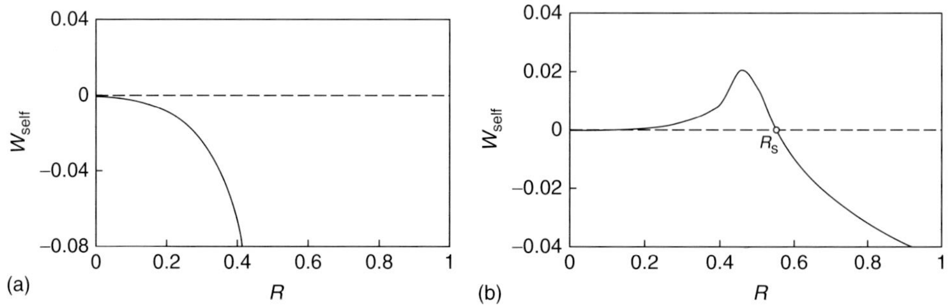

Figure 9.34 Total energy input W_{self} ($m = 1.0, k = 1.0, e = 0.4, \delta = 0.1, c_1 = 0.005, c_2 = 1.0, p_f = 0.93$): (a) $\omega = 0.6 < p_f$ and (b) $\omega = 1.6 > p_f$.

and decreases when $R > R_S$. As a result, the amplitude of a self-excited oscillation converges to the amplitude $R = R_s$ and, as a result, a limit cycle is attained. This value $R_s \approx 0.55$ agrees with the amplitude obtained in Figure 9.33.

References

Black, H.F. (1968) Interaction of a whirling rotor with a vibration stator across a clearance annulus. *J. Mech. Eng. Sci.*, **10** (1), 1–11.

Burkhalter, R. and Mazziotti, P.J. (1956) The low Silhouette drive line. *SAE Trans.*, **64**, 379–393.

Den Hartog, J.P. (1956) *Mechanical Vibration*, 4th edn, McGraw-Hill, New York.

Ehrich, F.F. (1969) The dynamic stability of rotor/stator radial rubs in rotation machinery. *Trans. ASME. J. Eng. Ind.*, **91** (4), 1025–1028.

Fujii, S. (1956b) Whirling of an automotive propeller shaft at lower speed (2nd report). *Trans. JSME*, **22** (119), 489–491. (in Japanese).

Goldman, P. and Muszynska, A. (1993) *Proceedings of ASME Turbo Expo Conference*, ASME, New York, pp. 1–7.

Goto, Y. and Watanabe, T. (1963) Study on precise rolling bearings for machine tools (Observation of the behavior of a main shaft and problems of precise bearings. *J. JSME*, **66** (536), 1171–1182.

Gustafsson, O. (1962), *Study of the Vibration Characteristics of Bearing (Special report on analytical study of the radial, axial and angular vibration of a bearing with flexurally rigid races)*, SKF Reprot, AL62L 005.

Igarash, A. (1959) Noise of a ball bearing (1st report, Case of a single ball bearing). *Trans. JSME*, **25** (158), 1027–1034. (in Japanese).

Igarashi, A. and Ohta, H. (1990) Studies on natural vibrations of ball bearings (1st report, vibration characteristics of ordinary-size ball bearings and natural vibrations on the rigid-body mode of the outer ring. *Trans. JSME, Ser. C*, **56** (528), 2047–2055. (in Japanese).

Ishida, Y., Hossain, M.Z., Inoue, T., and Yasuda, S. (2004) Rubbing due to contact in the rotor-guide system. Proceedings of the EUROMECH 457, pp. 61–64.

Inagaki, M. and Ishida, Y. (2011) Mechanism of occurrence of self-excited oscillations of a rotor with a clearance between bearing holder and housing. ASME, IDETC2011 -48052, CD-Rom.

Ishida, Y., Inagaki, M., Ejima, R., and Hayashi, A. (2009) Nonlinear resonances and self-excited oscillations of a rotor caused by radial clearance and collision. *Nonlinear Dyn.*, **57** (3), 593–605.

Li, G.X. and Paidoussis, M.P. (1994) Impact phenomena of rotor-casing dynamical systems. *Nonlinear Dyn.*, **5** (l), 53–70.

Lin, Y.Q. and Vance, J.M. (1992) Virtual instability in a Rotor-bearing system. Proceedings of 4th International Symposium Transport Phen. and Dynamics of Rotor Machine (ISROMAC-4), pp. 50–58 (Part 1- Measurement), pp. 379–388 (Part 2-Analysis).

Lohmann, G. (1953) Untersuchungen des laufgeräusches von Wälzlagern. *Konstruktion*, **5** (2), 38–46, 85–90.

Meldau, E. (1951) Die bewegung der achse von wälzlagern bei geringen drehzahlen. *Werkstatt Betrieb*, **84** (7), 308–313.

Momono, T. (2000) Rolling bearing stiffness. *NSK J.*, (No. 669), 32–41. (in Japanese).

NSK (1954) On the sound of ball and roller bearings, *NSK Bearing Journal*, No.602, 1–12.

NSK (1956) Conditions of measurement of bearing noise. *NSK Bearing J.*, (No. 606), 1–9.

Ota, H. and Kato, M. (1984) Lateral vibrations of a rotating shaft driven by a universal joint (first report, generation of even multiple vibrations by secondary moment). *Bull. JSME*, **27** (231), 2002–2007.

Ohta, H. and Satake (2002) Vibrations of the all-ceramic ball bearing, *Trans. Asme, J. Tribology*, **124** (3), 448–460.

Ohta, H. (2005) Vibration Analysis of Rolling Bearings, *J. Japan Soc.* Design Eng., **40** (1), 501–507.

Perret, H. (1950) Elastishe spielschwingungen konstant belasteter waelzlager. *Werkstatt Betrieb*, **83** (8), 354–358.

Rosenberg, R.M. (1958) On the dynamical behavior of rotating shafts driven by universal (Hooke) coupling. *Trans. ASME, J. Appl. Mech.*, **25** (l), 47–51.

Tallian, T.E. and Gustafsson, O.G. (1965) Progress in rolling bearing vibration research and control. *ASLE Trans.*, **8** (3), 195–207.

Tamura, A. and Taniguchi, O. (1960) Ball bearing vibrations (1st report, on the radial vibration caused by passing balls. *Trans. JSME*, **26** (161), 19–25. (in Japanese).

Yamamoto, T. (1954a) On the critical speed of a shaft, Chapter III, Pt. I: vibrations of forward precession due to ball bearings. *Mem. Fac. Eng., Nagoya Univ.*, **6** (2), 133–142.

Yamamoto, T. (1954b) On the critical speed of a shaft. Chapter III, Pt. II: Vibrations of backward precession due to ball bearings. *Mem. Fac. Eng., Nagoya Univ.*, **6** (2), 142–152.

Yamamoto, T. (1954c) On the critical speed of a shaft: Chapter IV: forced vibrations of a shaft supported by bearings with radial clearances. *Mem. Fac. Eng., Nagoya Univ.*, **6** (2), 160–170.

Yamamoto, T. (1954d) On the critical speed of a shaft, Chap. II, Synchronous backward precession, Memoirs of Fac. Eng., *Nagoya U.*, Vol. 6, No.2, pp. 119–133.

Yamamoto, T. (1957d) On the vibrations of a rotating shaft, Chapter Vlll: on the critical speeds induced by ball bearings appearing at lower rotating speeds. *Mem. Fac. Eng., Nagoya Univ.*, **9** (1), 86–92.

10
Flow-Induced Vibrations

10.1
General Considerations

Rotating machineries are related to fluid in various ways. For example, a rotor is supported by oil or air in fluid-film bearings, a turbine is rotated by gas or steam power, and a pump induces flow by the rotation of a shaft with blades. In these machines, rotors rotate in fluids. On the contrary, fluid is contained inside a rotor in centrifuges. In aircraft gas turbines, oil and steam condensate are sometimes trapped in a hollow rotor.

Fluid generally has a damping effect on a rotor. However, in some special conditions, fluid in a rotating field causes self-excited vibrations, which tragically lead to unexpected accidents. Besides, the fluid trapped in a hollow rotor works as an unbalance and affects the harmonic response. To prevent such vibrations, we must fully understand the mechanism of their occurrence and adopt an appropriate avoidance measure.

Most vibrations due to fluid are self-excited vibrations. As explained in Chapter 7, the frequency of a self-excited vibration is always equal to one of the natural frequencies of the system. Therefore, unlike the case of a forced vibration whose frequency is proportional to the frequency of excitation, it is generally difficult to find the cause. In addition, unlike the case of resonance, it is difficult to avoid this self-excited vibration by changing the rotational speed of the shaft because it occurs in a wide range of rotational speed. To diminish self-excited vibrations, we must eliminate the mechanism of charging energy into the system. For this purpose, modifications of the dimensions are not enough, and we must often change the construction of the rotor itself.

In this chapter, among various kinds of self-excited vibrations, oil whip due to oil film in journal bearings, unstable vibration due to seals, steam whirl due to steam flow, and vibration of a rotor partially filled with liquid are explained.

10.2
Oil Whip and Oil Whirl

As explained in Section 9.2, bearings are classified into *rolling-element bearings* and *fluid-film bearings*. Fluid-film bearings are widely used in rotating machinery

Linear and Nonlinear Rotordynamics: A Modern Treatment with Applications, Second Edition.
Yukio Ishida and Toshio Yamamoto.

because they have various merits, such as large load capacity, no life limit under good lubrication condition, and large damping. Various types of fluid-film bearings are classified as *oil film bearings* and *gas bearings*, depending on the kind of fluid; *thrust bearings* and *radial bearings*, depending on the direction of the load; and *hydrodynamic bearings* and *hydrostatic bearings*, depending on how the pressure of the lubricant is created. In fluid-film bearings, radial bearings are customarily called *journal bearings*. In this section, a journal bearing with oil film is adopted as a representative fluid-film bearing and vibrations of a rotor supported by this type of bearing are explained.

The self-excited vibrations due to journal bearings are analyzed in the following process. First, the expressions for an elastic force and a damping force produced by oil film are derived, followed by the derivation of the equations of motion of the entire rotor system, and then its stability is investigated. The fundamental framework of this theoretical treatment was given by Hori (1959). In the following, the phenomenon is explained following Hori's theory and its revision by Funakawa and Tatara (1964).

10.2.1 Journal Bearings and Self-Excited Vibrations

Figure 10.1a shows an elastic rotor supported by journal bearings at both ends. The part of the rotor in the bearing is called a *journal* and the cylinder supporting it is called a *bearing*. The ratio of the radial clearance (fluid-film thickness) and the journal diameter when the eccentricity e is zero is generally smaller than 0.001. In a horizontal rotor where gravity works, the rotor moves downward and has contact at point A when the rotor is not rotating. When the rotor rotates, the fluid flows into the wedge-shaped part and high pressure is created. This high pressure supports the shaft and also displaces the rotor from the bottom by a small angle θ. The fluid film works equivalently as a spring and a damper, and the rotor vibrates around this equilibrium position.

Since the damping effect of this journal bearing is pretty large, the resonances in response to the unbalance are well suppressed when this bearing is used. Therefore, journal bearings are generally preferable for vibration prevention but there is a danger of producing violent self-excited vibration above the major critical speed if the design is inappropriate. Figure 10.1b shows the amplitude and frequency of vibration of a rotor supported by journal bearings as functions of a rotational speed. As the rotational speed ω increases, a harmonic resonance due to unbalances occurs in the vicinity of the major critical speed ω_c. When the speed increases further, a self-excited vibration called *oil whirl* occurs above the rotational speed ω_a. The frequency of this vibration is about $\omega/2$. When an oil whirl occurs, the rotor whirls in a forward direction with small amplitude without deformation, like a rigid rotor. Muszynska (1987) reported that the actual frequency is slightly less than half of the rotational speed, although this value varies depending on the type of bearing and static eccentricity conditions resulting from shaft radial load. The threshold speed ω_a sometimes becomes lower than the major critical speed (refer to Figure 10.1c;

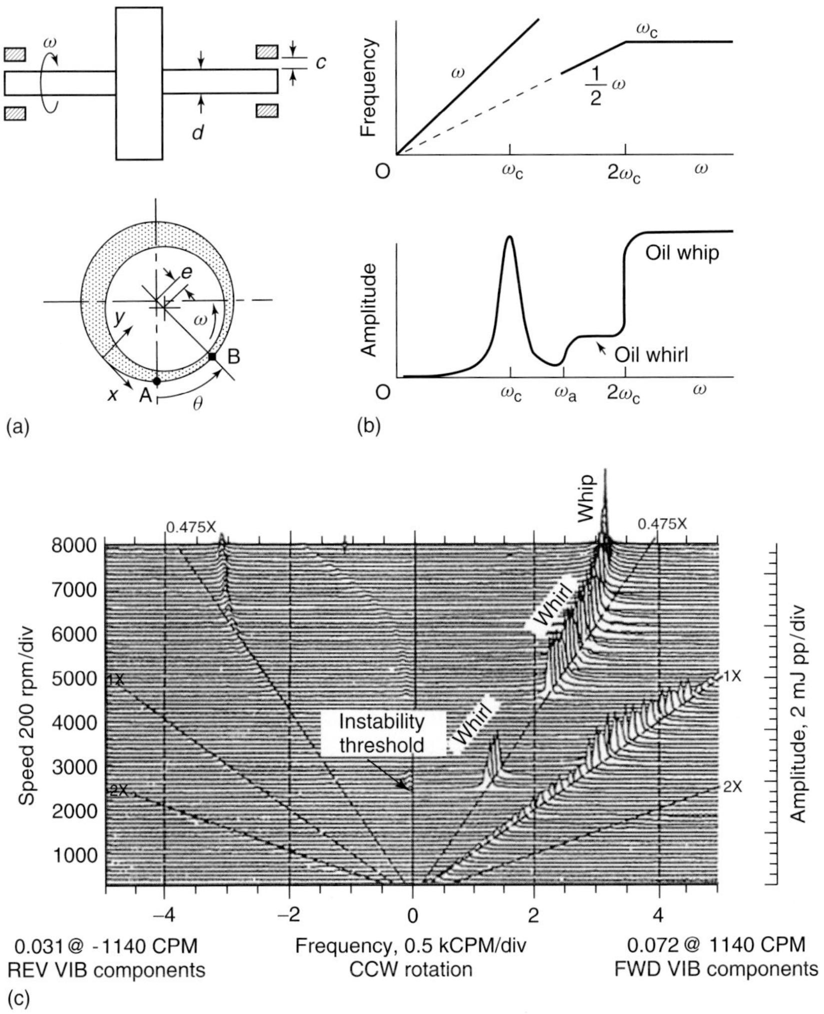

Figure 10.1 Rotor supported by journal bearings. (a) Rotor-bearing system, (b) response curve, and (c) waterfall diagram obtained by an experiment (Courtesy of Dr. Muszynska, 1988).

this three-dimensional plot of spectra at various rotational speed is called a *waterfall diagram*). When the rotational speed increases further, a violent self-excited vibration starts to appear at about twice the major critical speed and persists in a wide speed range above it. This vibration is called *oil whip*. The frequency of an oil whip is almost equal to a natural frequency of the system, and a rotor bends and whirls with large amplitude in a forward directions. In accelerating the rotor, oil whip sometimes does not appear even if the speed exceeds twice the major critical speed. However, once the oil whip occurs, it does not disappear until the speed reaches twice the major critical speed, even if the rotational speed decreases. As a

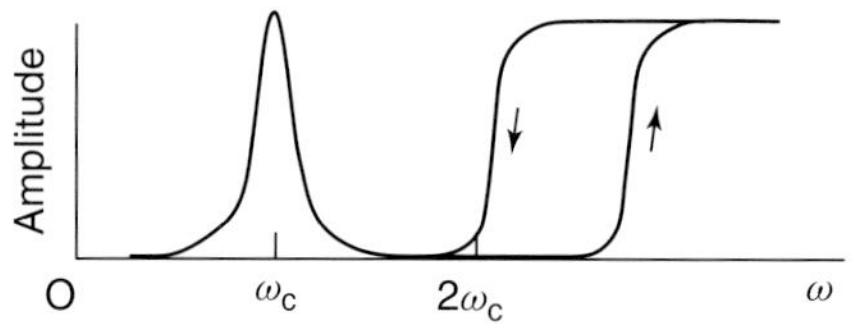

Figure 10.2 Inertia effect.

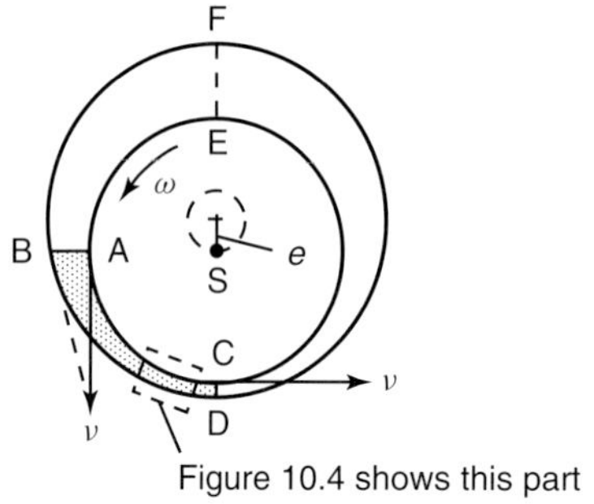

Figure 10.3 Derivation of a whirling velocity.

result, we obtain the amplitude curve shown in Figure 10.2 when we accelerate and then decelerate the rotor. Such a hysteretic phenomenon is called an *inertia effect.*

Oil whip was discovered by Newkirk and Taylor (1925). They gave the following explanation for it. Let us assume that the rotor is whirling with the angular velocity ω and the eccentricity e as shown in Figure 10.3. Let the fluid velocities at the journal surface and the stationary bearing surface be v and zero, respectively. If we assume that the fluid velocity changes linearly in the radial direction considering a thin oil film, the average velocity should be $v/2$. Here, we consider the fluid volume going in and coming out of the shaded part ABDC during a unit time. Let the length, inner radius, and radial clearance of the bearing be l, R, and c, respectively, and the rotor eccentricity be e. Then, the fluid volume passing through sections $\overline{\mathrm{AB}}$ ($=c$) and $\overline{\mathrm{CD}}$ ($=c-e$) per unit time are $cv/2$ and $(c-e)v/2$, respectively. Therefore, the excess of oil equals to $cv/2$. Since oil is incompressible, the rotor must move to increase the volume of the shaded part if the oil does not leak from both sides. The volumes of ABDC and EFBA are $(c-e/2)\times\pi R/2$ and $(c+e/2)\times\pi R/2$, respectively, and their difference is $e\times\pi R/2$. We assume that point S rotates by the angle θ per unit time and the shaded part takes in this increment $ev/2$. Then, from the relationship $\theta:(\pi/2)=(ev/2):(e\pi R/2)$, we have $\theta=(v/R)/2=\omega/2$. This shows that the precessional angular velocity is $\omega/2$; that is, the angular velocity of an oil whirl is half of the rotational speed. When the rotational speed reaches twice the major critical speed during acceleration, it coincides with a forward natural frequency, and an oil whip occurs with large amplitude by this stimulation. However, we should note that this is a very elementary qualitative explanation and is not a complete theory. For example, it cannot explain why oil whip persists above twice the major critical speed and why the inertia effect shown in Figure 10.2 occurs.

Vibrations due to journal bearings are complicated and have various characteristics. Many researchers have tackled this phenomenon, but the details of their results do not agree. For more advanced study, refer to Tondl (1965).

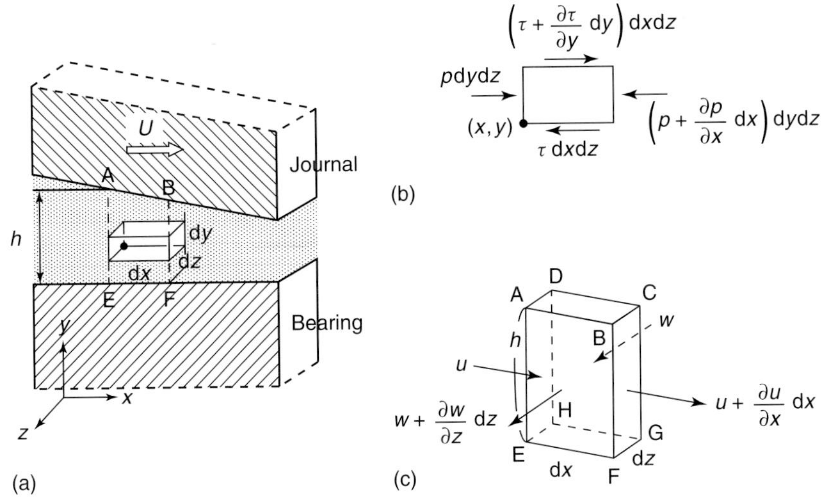

Figure 10.4 Balance of forces in fluid. (a) Fluid element, (b) force, and (c) fluid velocities.

10.2.2 Reynolds Equation

To obtain the force acting on the rotor from the oil film, we must know the pressure distribution in the oil film. For that purpose, the *Reynolds equation* must be derived first. Figure 10.4a is an enlarged view of a part of the oil film in Figure 10.3. It is assumed that the oil is Newtonian fluid, incompressible and with laminar flow. The thickness of oil film $h(x, t)$ is so thin compared with the diameter of the rotor that the curvature can be neglected. The rectangular coordinate system O$-xyz$ is prepared. The surface of the journal rotating with ω moves with the circumferential velocity $U = \omega r$.

A small rectangular parallelepiped with lengths dx, dy, and dz is considered at point (x, y, z) in the oil film as shown in Figure 10.4. We derive the equilibrium equations of pressure p and shearing stress τ. If the inertia of the fluid is small enough to be negligible, the following relationship holds in the x-direction:

$$pdydz - \left(p + \frac{\partial p}{\partial x} dx\right) dydz - \tau dxdz + \left(\tau + \frac{\partial \tau}{\partial y} dy\right) dxdz = 0 \tag{10.1}$$

This reduces to

$$\frac{\partial p}{\partial x} = \frac{\partial \tau}{\partial y} \tag{10.2}$$

Let the fluid velocities in the x-, y-, and z-directions at this point be u, v, and w, respectively, and the coefficient of viscosity be μ. Then, the shearing stress is given by

$$\tau = \mu \frac{\partial u}{\partial y} \tag{10.3}$$

From Eqs. (10.2) and (10.3), we have

$$\frac{\partial p}{\partial x} = \mu \frac{\partial^2 u}{\partial y^2} \tag{10.4}$$

Integrating this with the assumption that the pressure does not change in the y-direction, we have the following expression:

$$u = \frac{1}{2\mu}\left(\frac{\partial p}{\partial x}\right) y^2 + C_1 y + C_2 \tag{10.5}$$

Applying the boundary condition that $u = 0$ at $y = 0$ and $u = U$ at $y = h$ gives

$$u = \frac{1}{2\mu}\left(\frac{\partial p}{\partial x}\right)(y - h)\, y + \frac{U}{h} y \tag{10.6}$$

Similarly, we have the following expression for the velocity w in the z-direction:

$$w = \frac{1}{2\mu}\left(\frac{\partial p}{\partial z}\right)(y - h)\, y \tag{10.7}$$

Next, we consider the continuity of fluid with regard to the element shown in Figure 10.4c. Let the quantities of fluid that pass through the face AEHD per unit time in the x- and z-directions be q_x and q_z, respectively. From Eqs. (10.6) and (10.7), we have

$$\left.\begin{aligned} q_x &= dz \int_0^h u dy = dz\left(-\frac{h^3}{12\mu}\frac{\partial p}{\partial x} + \frac{Uh}{2}\right) \\ q_z &= dx \int_0^h w dy = dx\left(-\frac{h^3}{12\mu}\frac{\partial p}{\partial z}\right) \end{aligned}\right\} \tag{10.8}$$

Since the oil is incompressible, the volume of this element must increase by changing the height h so as to include the increment of the fluid in this element during time dt. Therefore, the following must hold:

$$-\left(\frac{\partial q_x}{\partial x} dxdt + \frac{\partial q_z}{\partial z} dzdt\right) = dxdz\frac{\partial h}{\partial t} dt \tag{10.9}$$

Substituting this into Eq. (10.8), we arrive at the following *Reynolds equation*:

$$\frac{\partial}{\partial x}\left(\frac{h^3}{\mu}\frac{\partial p}{\partial x}\right) + \frac{\partial}{\partial z}\left(\frac{h^3}{\mu}\frac{\partial p}{\partial z}\right) = 6U\frac{\partial h}{\partial x} + 12\frac{\partial h}{\partial t} \tag{10.10}$$

10.2.3
Oil Film Force

Equation (10.10). gives the pressure distribution. If it is known, we can obtain the reaction force acting on the rotor. However, since it is generally impossible to solve Eq. (10.10) analytically, we must use some approximate method or numerical means using computers. Hori (1959) simplified the analysis by assuming that the length of the bearing is infinite (the *long bearing approximation*). By this assumption, the problem is reduced to a two-dimensional one. He obtained the expression for oil film force and discussed the stability of the entire rotor system.

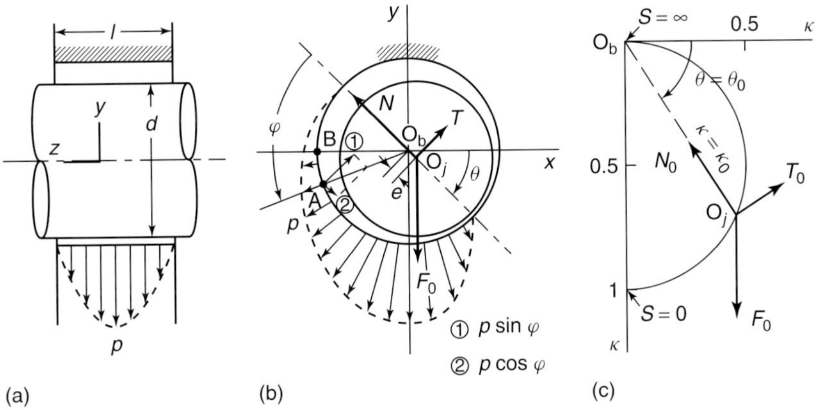

Figure 10.5 Oil film force and journal locus. (a) Pressure distribution, (b) pressure distribution and oil film force, and (c) orbit of an equilibrium position.

Funakawa and Tatara (1964) showed that the theoretical result derived under the assumption that the bearing is infinitely thin can explain the experimental results more accurately from a quantitative viewpoint (the *short bearing approximation*). In practical bearings, the ratio of the bearing length and the diameter (i.e., l/d in Figure 10.5) has a value between 1 and 2. For such a specific value, Eq. (10.10) must be solved numerically using computers (the *finite-length approximation*).

10.2.3.1 Short Bearing Approximation

In the short bearing analysis, we consider that the pressure variation in the z-direction is extensively larger than that in the x-direction (that is, $\partial p/\partial x \ll \partial p/\partial z$) and neglect the first term on the left-hand side of Eq. (10.10). Then, the Reynolds equation is simplified to

$$\frac{h^3}{\mu}\frac{\partial^2 p}{\partial z^2} = 6U\frac{\partial h}{\partial x} + 12\frac{\partial h}{\partial t} \tag{10.11}$$

In Figure 10.5a,b, we adopt the following boundary condition with regard to the pressure distribution in the oil film in the z-direction: $\partial p/\partial z = 0$ for $z = 0$ and $p = 0$ (atmospheric pressure) for $z = \pm l/2$. Putting $x = \varphi R$ and $U = R\omega$ and integrating Eq. (10.11), we get

$$p = \frac{3\mu}{h^3}\left(\omega\frac{\partial h}{\partial \varphi} + 2\frac{\partial h}{\partial t}\right)\left(z^2 - \frac{l^2}{4}\right) \tag{10.12}$$

Let the bearing clearance, the eccentricity, and the eccentricity ratio be c, e, and $\kappa = e/c$, respectively. The oil film thickness is given by

$$h = c\,(1 + \kappa\cos\varphi) \tag{10.13}$$

Substituting this into the expression above gives

$$p = \frac{3\mu}{c^2\,(1 + \kappa\cos\varphi)^3}\left[2\dot{\kappa}\cos\varphi - \kappa\left(\omega + 2\dot{\theta}\right)\sin\varphi\right]\left(z^2 - \frac{l^2}{4}\right) \tag{10.14}$$

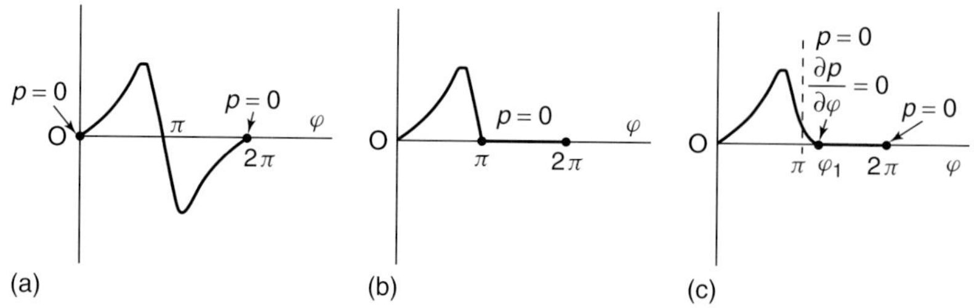

Figure 10.6 Boundary conditions. (a) Sommerfeld condition, (b) Gumbel condition, and (c) Reynolds condition.

where we replace $\dot{\varphi}$ by $\dot{\theta}$ because we know the relationship $\varphi - \angle \mathrm{AO_bB} = \theta$ from Figure 10.5b. We see that the pressure at point A also depends on the rates of change $\dot{\kappa}$ and $\dot{\theta}$ when the rotor is whirling.

We know from this expression that when the journal is rotating in the equilibrium position ($\dot{\kappa} = \dot{\theta} = 0$), the peripheral pressure distribution in the plane at the center $z = 0$ is symmetrical about the point $\varphi = \pi$, and it is positive in the zone from $\varphi = 0$ to π and negative in the zone from $\varphi = \pi$ to 2π in Figure 10.5. This pressure distribution is called the *Sommerfeld condition*. This pressure distribution holds when the pressure is very small. However, in practical journal bearings, the pressure in the zone from $\varphi = \pi$ to 2π is almost zero (i.e., the atmospheric pressure) because evaporation of the lubricant and axial airflow from both ends may occur. Taking this situation into consideration, the pressure in the zone from $\varphi = \pi$ to 2π is set as $p = 0$. This revised condition is called the *Gumbel condition*. This condition is revised further by setting the pressure gradients at the edges of oil film equal to zero. This condition is called the *Reynolds condition*. These conditions are shown in Figure 10.6. The Reynolds condition expresses the practical situation most accurately, but the Gumbel condition is widely used in analysis because of its simplicity.

If the Gumbel condition is adopted, the oil film forces in the directions shown in Figure 10.5b (i.e., the direction opposite the eccentricity and tangential direction) are given by

$$\left.\begin{aligned} N &= -R\int_{-l/2}^{+l/2}\int_0^{\pi} p\cos\varphi d\varphi dz \\ T &= R\int_{-l/2}^{+l/2}\int_0^{\pi} p\sin\varphi d\varphi dz \end{aligned}\right\} \tag{10.15}$$

Substituting Eq. (10.14) into these expressions, we get (Funakawa and Tatara, 1964)

$$\left.\begin{aligned} N &= \frac{1}{2}\mu\left(\frac{r}{c}\right)^2\frac{l^3}{r}\left[\frac{2\kappa^2\left(\omega+2\dot{\theta}\right)}{\left(1-\kappa^2\right)^2}+\frac{\pi\dot{\kappa}\left(1+2\kappa^2\right)}{\left(1-\kappa^2\right)^{5/2}}\right] \\ T &= \frac{1}{2}\mu\left(\frac{r}{c}\right)^2\frac{l^3}{r}\left[\frac{\pi\kappa\left(\omega+2\dot{\theta}\right)}{2\left(1-\kappa^2\right)^{3/2}}+\frac{4\kappa\dot{\kappa}}{\left(1-\kappa^2\right)^2}\right] \end{aligned}\right\} \tag{10.16}$$

where r and l ar e the radius and length of the bearing, respectively.

10.2.3.2 Long Bearing Approximation

In the long bearing approximation analysis, we assume that the pressure does not change in the z-direction (i.e., $\partial p/\partial z \approx 0$) and neglect the second term on the left-hand side in Eq. (10.10). When the Gumbel condition is adopted, the following expression is obtained, corresponding to Eq. (10.16) (Hori, 1959):

$$\left.\begin{aligned} N &= 6\mu\left(\frac{r}{c}\right)^2 rl\left[\frac{2\kappa^2\left(\omega-2\dot{\theta}\right)}{\left(2+\kappa^2\right)\left(1-\kappa^2\right)}+\frac{2\dot{\kappa}}{\left(1-\kappa^2\right)^{2/3}}\left\{\frac{\pi}{2}-\frac{8}{\pi\left(2+\kappa^2\right)}\right\}\right] \\ T &= 6\mu\left(\frac{r}{c}\right)^2 rl\left[\frac{\pi\kappa\left(\omega-2\dot{\theta}\right)}{\left(2+\kappa^2\right)\left(1-\kappa^2\right)^{1/2}}+\frac{4\kappa\dot{\kappa}}{\left(2+\kappa^2\right)\left(1-k^2\right)^{1/2}}\right] \end{aligned}\right\} \tag{10.17}$$

When a rotor rotates, the oil film forces are generated and the journal floats as shown in Figure 10.5b. The equilibrium position $O_j\ (\kappa_0, \theta_0)$ of the journal center is determined by the balancing between the gravity load F_0 working downward and the oil film force (N_0, T_0), which is obtained by setting $\dot{\theta} = 0$ and $\dot{\kappa} = 0$ in Eq. (10.16) or (10.17). The result shows that the equilibrium position is determined by the following dimensionless quantity, called the *Sommerfeld number*:

$$S = \left(\frac{r}{c}\right)^2 \frac{\mu n}{p_m} \tag{10.18}$$

where n (rps) is the rotational speed and $p_m = F_0/(2rl)$ is the average bearing pressure. This means that even if the rotational speed and oil viscosity change, the equilibrium position does not change if the Sommerfeld number is the same. Figure 10.5c shows the locus of this equilibrium position. The radial deflection is shown by the dimensionless eccentricity ratio κ in this figure. When the rotor is not rotating (i.e., $S = 0$), the rotor is at the bottom of the bearing. When the rotational speed increases, that is, when S increases, the rotor center floats up and changes its position along the semicircle line and arrives at the bearing center when the rotational speed becomes extremely high (i.e., $S = \infty$).

10.2.4 Stability Analysis of an Elastic Rotor

To investigate the stability of the equilibrium position, we derive the expressions for the oil film force that acts when the rotor deviates slightly from the equilibrium position and then linearizes them. Since no confusion will occur, we newly define the coordinate system O–xy as shown in Figure 10.7b, which is different from Figures 10.1 and 10.3. Now, let us assume that the equilibrium position moves from $O_j\ (\theta_0, \kappa_0)$ to $O'_j\ (\theta, \kappa)$. We substitute $\kappa = \kappa_0 + \Delta\kappa$ and $\theta = \theta_0 + \Delta\theta$ into Eq. (10.16) or (10.17) and linearize it for $\Delta\kappa, \Delta\theta, \Delta\dot{\kappa}$, and $\Delta\dot{\theta}$. Then, from the transformation $F_x = -N\cos\theta + T\sin\theta$ and $F_y = -N\sin\theta - T\cos\theta$, we obtain the components F_x and F_y of the oil film force in the x- and y-directions, respectively. The components F_x and F_y include signs. For example, in the case of Figure 10.5, $F_x < 0$ and $F_y < 0$ hold. Consequently, the oil film forces are expressed by the

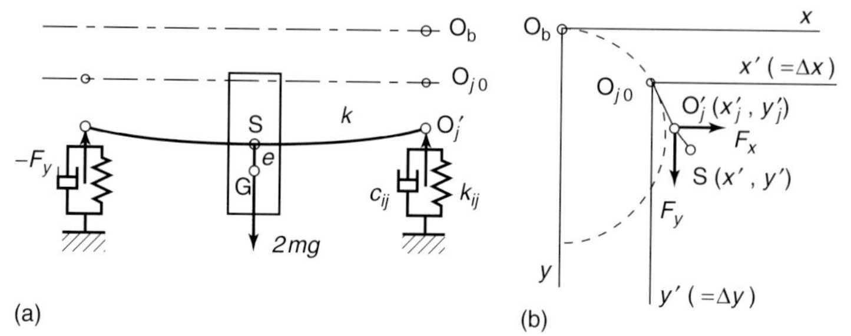

Figure 10.7 Elastic rotor supported by journal bearings. (a) Rotor system and (b) coordinates.

increments Δx, Δy for deflection and $\Delta \dot{x}$, $\Delta \dot{y}$ for velocity as follows:

$$\left\{ \begin{array}{c} -F_x \\ -F_y \end{array} \right\} = \left\{ \begin{array}{c} -F_{x0} \\ -F_{y0} \end{array} \right\} + \left[\begin{array}{cc} k_{xx} & k_{xy} \\ k_{yx} & k_{yy} \end{array} \right] \left\{ \begin{array}{c} \Delta x \\ \Delta y \end{array} \right\} + \left[\begin{array}{cc} c_{xx} & c_{xy} \\ c_{yx} & c_{yy} \end{array} \right] \left\{ \begin{array}{c} \Delta \dot{x} \\ \Delta \dot{y} \end{array} \right\} \tag{10.19}$$

where k_{xx}, k_{xy}, k_{yx}, and k_{yy} are the stiffness coefficients of the oil film and c_{xx}, c_{xy}, c_{yx}, and c_{yy} are its damping coefficients. The coefficients $k_{xx}, k_{yy}, c_{xx}, c_{yy}$, and $k_{xy}, k_{yx}, c_{xy}, c_{yx}$ are also called *diagonal* and *cross-coupled terms*, respectively. These coefficients are determined by the eccentricity ratio κ_0 (i.e., the equilibrium position) alone if the shape and size of the journal bearing are the same. The nondimensionalized values of these coefficients

$$K_{ij} = \frac{ck_{ij}}{F_0}, \quad C_{ij} = \frac{c\omega c_{ij}}{F_0} \; (i, j = x, y) \tag{10.20}$$

are given as functions of κ_0, where F_0 is a load on the bearing. As an example, a result obtained from Eq. (10.16) for a short bearing with the Gumbel condition is illustrated in Figure 10.8.

Now, Let us consider a rotor-bearing system as shown in Figure 10.7a. Since the characteristics of a journal bearing in the vicinity of the equilibrium position are expressed by Eq. (10.19), the bearings are expressed by equivalent springs and equivalent dampers in Figure 10.7a. We prepare the coordinate system O_j–$x'y'$ whose origin coincides with the equilibrium point O_j of the journal. The positions of the journal center O'_j and the rotor center S are represented by $\left(x'_j, y'_j\right)$ and (x', y'), respectively. The x'_j and y'_j are the same as Δx and Δy in Eq. (10.19), respectively. Let the mass and eccentricity of the rotor be $2m$ and e, respectively, the spring constant of the shaft be $2k$, and the rotational speed be ω. The equations of motion are given by

$$\left. \begin{array}{l} m\ddot{x}' + k\left(x' - x'_j\right) = me\omega^2 \cos \omega t \\ m\ddot{y}' + k\left(y' - y'_j\right) = -me\omega^2 \sin \omega t + mg \\ k\left(x' - x'_j\right) = k_{xx}x'_j + k_{xy}y'_j + c_{xx}\dot{x}'_j + c_{xy}\dot{y}'_j - F_{x0} \\ k\left(y' - y'_j\right) = k_{yx}x'_j + k_{yy}y'_j + c_{yx}\dot{x}'_j + c_{yy}\dot{y}'_j - F_{y0} \end{array} \right\} \tag{10.21}$$

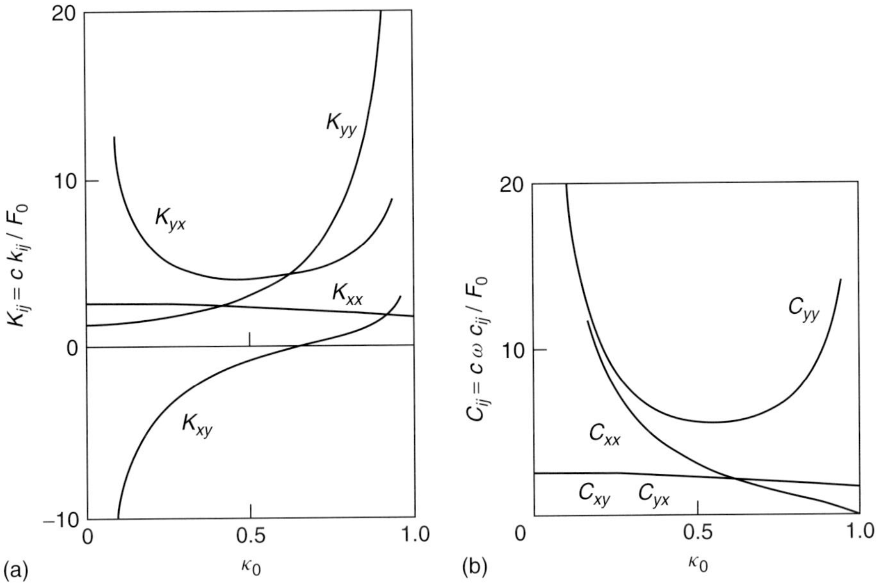

Figure 10.8 Dimensionless coefficients of a short bearing. (a) Stiffness coefficients and (b) damping coefficients. (Funakawa and Tatara, 1964, Courtesy of JSME.)

In the balancing diagram shown in Figure 10.5c, we know that $-F_{x0} = 0$ and $-F_{y0} = mg$ hold. One of the unbalance force terms has a minus sign because the coordinate axes in Figure 10.7 are taken in a manner different from the customary manner.

Here, we consider the stability of the rotor in the case where the unbalance is zero. From the expression obtained by putting $e = 0$ in Eq. (10.21), we get the following characteristic equation:

$$A_6\lambda^6 + A_5\lambda^5 + A_4\lambda^4 + A_3\lambda^3 + A_2\lambda^2 + A_1\lambda + A_0 = 0 \tag{10.22}$$

After the usual analytical process applying the Routh–Hurwitz criterion, we get the following stability criterion (Hori, 1959):

$$\frac{1}{\omega^2}\left(\frac{F_0}{mc}\right) > K_1(\kappa_0)\left[K_2(\kappa_0) + \frac{1}{\omega_c^2}\left(\frac{F_0}{mc}\right)\right] \tag{10.23}$$

where

$$\left.\begin{aligned} &B_6 = C_{xx}C_{yy} - C_{xy}C_{yx}, \quad B_5 = K_{xx}C_{yy} + K_{yy}C_{xx} - K_{xy}C_{yx} - K_{yx}C_{xy} \\ &B_4 = C_{xx} + C_{yy}, \quad B_3 = K_{xx}K_{yy} - K_{xy}K_{yx}, \quad B_2 = K_{xx} + K_{yy} \\ &K_1(\kappa_0) = \frac{B_5^2 + B_4^2 B_3 - B_5 B_4 B_2}{B_6 B_4^2}, \quad K_2(\kappa_0) = \frac{B_4}{B_5} \end{aligned}\right\} \tag{10.24}$$

and ω_c is the critical speed. Figure 10.9 depicts this result on the boundary between the stability and instability for long and short bearings. This figure, called the *stability chart*, shows the boundary graphically on the plane with two axes κ_0 and

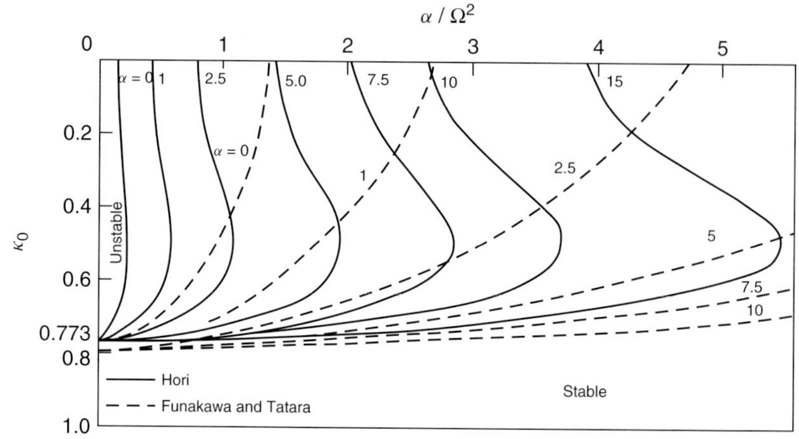

Figure 10.9 Stability chart. (Hori, 1959; Funakawa and Tatara, 1964, Courtesy of JSME.)

α/Ω^2, where $\Omega = \omega\sqrt{m/k}$. Parameter $\alpha = F_0/(ck)$ is related to the inverse of the stiffness. The short bearing approximation gives a much wider stable zone than the long bearing approximation. Experimental data agree well with the former result obtained by the short bearing approximation.

Hori (1959) explained the phenomenon using this chart as follows. Figure 10.10 shows the boundary line of the case of $(1/\omega_c^2)\,(F_0/mc) = 10$ for a long bearing. The dashed lines show the loci of the eccentricity when the rotational speed changes for a light load a_1–a_2, a medium load b_1–b_2, and a heavy load c_1–c_2. In the case of a light load, the following phenomenon occurs in the course of increasing the rotational speed. The equilibrium position of the rotor becomes unstable at point A and oil whirl starts to occur. Then the harmonic resonance occurs and forms a peak at the major critical speed ω_c. After the speed passes through the resonance region, the amplitude decreases to an oil whirl. When the speed arrives at point a_2 where the whirling speed coincides with a forward natural frequency, an oil whip starts to occur. In case of a medium load, an oil whirl starts to occur at point B and an oil whip starts to appear at point b_2. In case of a heavy load, an oil whip starts to appear at point C, which is higher than twice the major critical speed without experiencing the process of oil whirl. In this case, the above mentioned inertia effect occurs because when the rotational speed decreases, the oil whip continues to occur until $2\omega_c$. From these considerations, we know that the three types of response curves shown in Figure 10.10b are obtained for light, medium, and heavy loads.

10.2.5
Oil Whip Prevention

Considering the above-mentioned characteristics, we can use the following means to prevent oil whip:

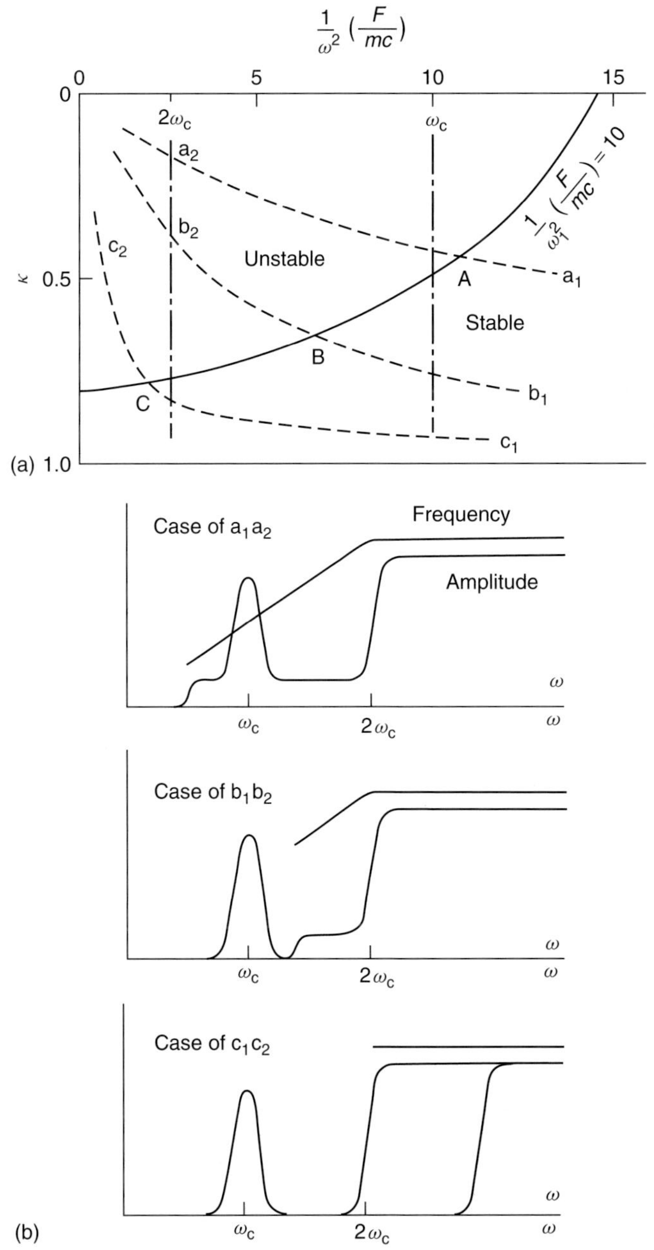

Figure 10.10 Explanation of oil whip and oil whirl using a stability chart. (a) Stability chart and (b) amplitude variation. (From Hori, 1959, Courtesy of ASME.)

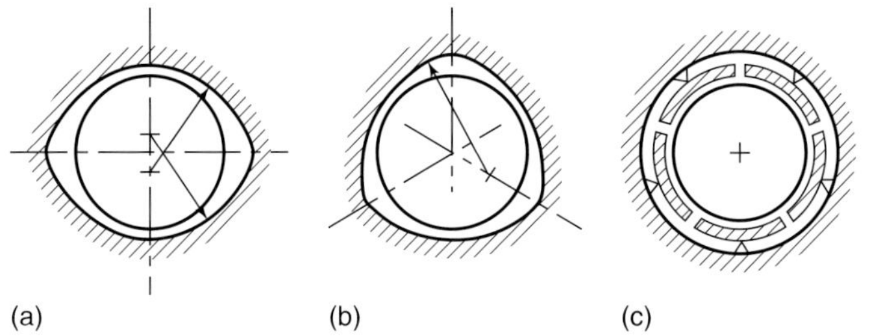

Figure 10.11 Noncircular bearings. (a) Two-lobe bearing, (b) three-lobe bearing, and (c) tilting-pad bearing.

1) **Increase the natural frequencies of the system:** Since oil whip occurs above twice the major critical speed, the oil whip may not occur in the operation range if the rotor diameter is made larger or the rotor length is made shorter.
2) **Expand the stability zone in Figure 10.9:** By adjusting parameters relating to α, we can change this zone.
3) **Increase the load:** In Figure 10.10, the locus c_1-c_2 is preferable to the locus a_1-a_2.
4) **Increase the eccentricity ratio κ_0:** Figure 10.9 teaches us that the rotor is always stable if κ_0 is greater than about 0.8. By changing the length l shorter, the clearance c larger, and the viscosity μ smaller, we can make κ_0 larger.
5) **Use a bearing with a special shape of cross section:** For example, Figure 10.11 shows a two-lobe bearing with a lemon shape, a three-lobe bearing, and a tilting-pad bearing that has movable pads. When these bearings are used, an oil whip appears with less frequency. Especially, use of tilting-pad bearings are recommended because the coupled terms in Eq. (10.19) become zero, and therefore, the tangential force component working as an self-excited force does not appear.

10.3 Seals

In turbomachines, various types of *seals* are used to reduce the leakage of working fluid through the interface between rotors and stators. The fluid in these seals produces forces that make the rotor unstable. In this section, the effects of noncontact seals on rotor characteristics are explained.

10.3.1 Plain Annular Seal

Figure 10.12a illustrates a single-suction, single-stage centrifugal pump that is most widely used as a pump for general purposes. When the shaft rotates, the fluid is sucked from the left side and drained out to the radial direction. The liner ring

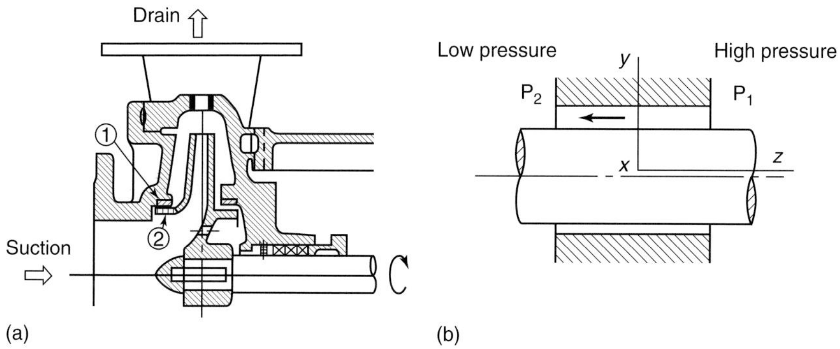

Figure 10.12 Plain annular seal. (a) Single-stage centrifugal pump and (b) physical model of a seal.

①, attached to the casing, and the blades ② form a seal to prevent leakage from the high-pressure side (the drainage side) to the low-pressure side (the suction side). This *plain annular* seal is often used in pumps. A physical model of this seal is shown in Figure 10.12b. The geometrical shape of a seal is similar to that of a journal; however, they are different in the following aspects. To avoid contact between a rotor and a stator, the ratio of the clearance c to the shaft radius r of a seal is made a few times larger than that of a journal. For example, the c/r in the seal is about 0.005, but in the journal it is about 0.001. The flow in a seal is turbulent and that of the latter is laminar. Therefore, we cannot use the Reynolds equation here.

When a rotor vibrates, the fluid in the seal produces a reaction force on the rotor. In case of small vibration around the equilibrium position (x_0, y_0), the fluid force can be linearized on the assumption that deflections Δx and Δy are small. It is expressed as follows:

$$-\left\{\begin{array}{c} F_x \\ F_y \end{array}\right\} = \left[\begin{array}{cc} k_{xx} & k_{xy} \\ k_{yx} & k_{yy} \end{array}\right]\left\{\begin{array}{c} \Delta x \\ \Delta y \end{array}\right\} + \left[\begin{array}{cc} c_{xx} & c_{xy} \\ c_{yx} & c_{yy} \end{array}\right]\left\{\begin{array}{c} \Delta \dot{x} \\ \Delta \dot{y} \end{array}\right\} + \left[\begin{array}{cc} m_{xx} & m_{xy} \\ m_{yx} & m_{yy} \end{array}\right]\left\{\begin{array}{c} \Delta \ddot{x} \\ \Delta \ddot{y} \end{array}\right\} \tag{10.25}$$

where k_{ij}, c_{ij}, and m_{ij} are the stiffness, damping, and added-mass coefficients, respectively. In the case of journal bearings, the third terms (the added-mass terms) are smaller compared to the other terms and are neglected. Owing to such fluid forces, the resonance speed of the rotor system increases and the peak of the resonance becomes low (Black, 1979). Unstable vibrations may appear because of the coupled terms. These coefficients vary depending on the equilibrium position of the rotor (i.e., the magnitude of the eccentricity) and the rotational speed (Nelson, 1985; Marquette, Childs, and San Andres, 1997).

If the eccentricity of the equilibrium position of the rotor is not so large, these coefficients do not differ from those in the case where the rotor is at the center of the stator. In this case, the relationships $k_{xx} = k_{yy}, k_{xy} = -k_{yx}$, and so on hold. Considering these relationships, we use new notations k, k_c, c, c_c, and m and rewrite

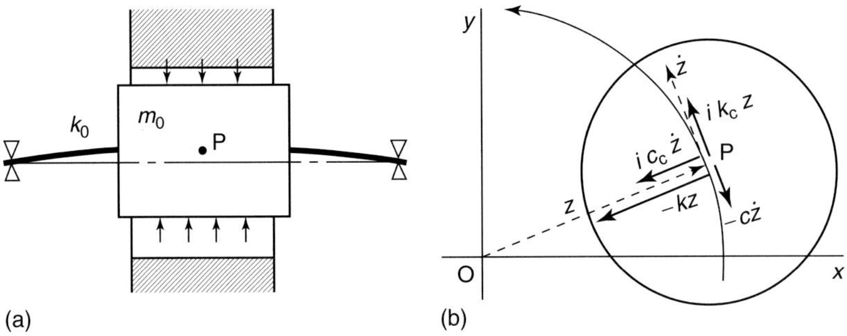

Figure 10.13 Rotor-seal system. (a) Theoretical model and (b) fluid forces.

Eq. (10.25) as follows:

$$-\left\{ \begin{array}{c} F_x \\ F_y \end{array} \right\} = \left[\begin{array}{cc} k & k_c \\ -k_c & k \end{array} \right] \left\{ \begin{array}{c} \Delta x \\ \Delta y \end{array} \right\} + \left[\begin{array}{cc} c & c_c \\ -c_c & c \end{array} \right] \left\{ \begin{array}{c} \Delta \dot{x} \\ \Delta \dot{y} \end{array} \right\} + m \left\{ \begin{array}{c} \Delta \ddot{x} \\ \Delta \ddot{y} \end{array} \right\} \tag{10.26}$$

where the cross-coupled added-mass terms are neglected. We investigate an unstable vibration in a system where the casing and disk of the Jeffcott rotor form a seal as shown in Figure 10.13a (Krämer, 1993). Let the mass of the rotor and the spring constant of the shaft be m_0 and k_0, respectively. The equilibrium position of the rotor is supposed to coincide with the center of the casing. We put $\Delta x = x$ and $\Delta y = y$ for simplicity of expression. From the relationships $m_0\ddot{x} = -k_0 x + F_x$ and $m_0\ddot{y} = -k_0 y + F_y$, we obtain the following equations of motion:

$$\left. \begin{array}{l} M\ddot{x} + c\dot{x} + c_c\dot{y} + Kx + k_c y = 0 \\ M\ddot{y} - c_c\dot{x} + c\dot{y} - k_c x + Ky = 0 \end{array} \right\} \tag{10.27}$$

where $M = m_0 + m$ and $K = k_0 + k$. Using the complex number $z = x + iy$, we can rewrite these equations as follows:

$$M\ddot{z} + c\dot{z} - ic_c\dot{z} + Kz - ik_c z = 0 \tag{10.28}$$

Substituting the assumed solution $z = Ze^{\lambda t}$ into this equation, we get the following characteristic equation:

$$M\lambda^2 + (c - ic_c)\,\lambda + K - ik_c = 0 \tag{10.29}$$

This equation gives two eigenvalues, $\lambda_1 = \alpha_1 + i\omega_1$ and $\lambda_2 = \alpha_2 + i\omega_2$, and we can know the occurrence of an unstable vibration from these values.

The asymmetrical cross-coupled terms with k_c in Eq. (10.27) cause an unstable vibration. Let us assume that the rotor whirls counterclockwise tracking a circular orbit in the xy-plane as shown in Figure 10.13b. Since the multiplication of the quantity $i = e^{i\pi/2}$ to a complex vector makes the vector rotate by 90°, the terms $-c\dot{z}$, $+ic_c\dot{z}$, $-Kz$, and $+ik_c z$ represent the forces shown in Figure 10.13. The force represented by the coupled term $+ik_c z$ points to a forward tangential direction and charges energy into the system, causing an unstable vibration.

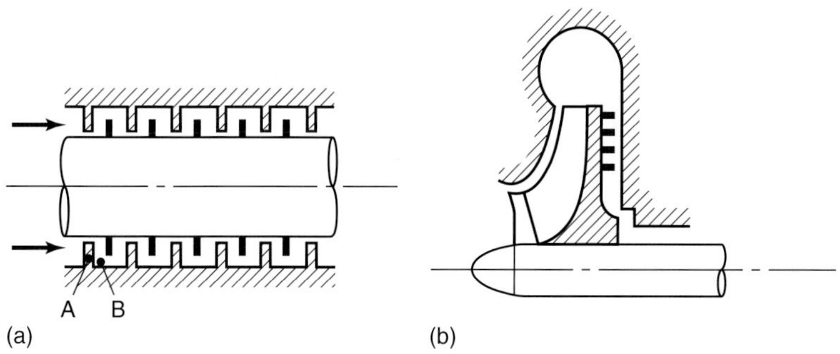

Figure 10.14 Examples of labyrinth seal. (a) Axial flow type and (b) radial flow type.

10.3.2 Labyrinth Seal

Labyrinth seals are often used in compressors, steam turbines, and gas turbines. Figure 10.14 shows labyrinth seals of the axial flow and radial flow types. The basic idea of a labyrinth seal is to control fluid (liquid or gas) flow by combining chalk vanes A and expanding cavities B. Such a complicated construction is thus named "labyrinth." Blades (i.e., chalk vanes) are mounted on the rotor or the stator or on both. In this seal, the high-pressure fluid leaked from inlet vanes makes a vortex in cavities (Iwatsubo and Motooka, 1980). By such a construction, the leakage is less than the case of a simple plain annular seal with the same clearance.

Fluid in a labyrinth generates destabilizing force tangential to the orbit. Alford (1964) first reported an accident due to labyrinth seals. Alford (1965) also explained the mechanism of the occurrence of this unstable vibration by considering energy input due to the phase difference between the rotor deflection and the fluid reaction force. However, vortex and viscosity are not considered in his analysis.

A labyrinth seal generates a circumferential component of a fluid force that induces unstable vibration. As the load increases and the clearance decreases, the magnitude of this destabilizing force becomes large. In this case, the inertia term in Eq. (10.26) can be neglected and the fluid force is expressed as follows:

$$-\left\{\begin{array}{c} F_x \\ F_y \end{array}\right\} = \left[\begin{array}{cc} k & k_c \\ -k_c & k \end{array}\right]\left\{\begin{array}{c} \Delta x \\ \Delta y \end{array}\right\} + \left[\begin{array}{cc} c & c_c \\ -c_c & c \end{array}\right]\left\{\begin{array}{c} \Delta \dot{x} \\ \Delta \dot{y} \end{array}\right\} \tag{10.30}$$

To investigate the dynamics of rotors quantitatively, we must obtain the exact values of the coefficients. Many studies to calculate them have been proposed. Refer to Childs (1993) for details.

10.4 Tip Clearance Excitation

Owing to the aerodynamic force in turbomachinery, a whirling motion called *tip clearance excitation* may occur in addition to the whirling due to labyrinth seals

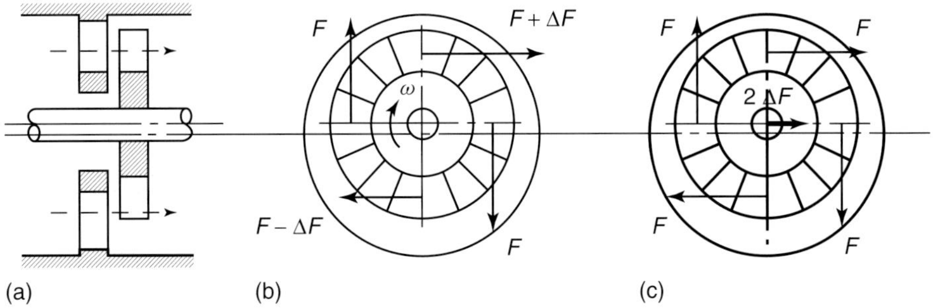

Figure 10.15 Destabilizing force due to steam. (a) Turbine blade, (b) force distribution, and (c) equivalent forces.

mentioned above. This excitation force is due to the circumferential variation of blade-tip clearance caused by eccentricity of a rotor. Since this phenomenon was first found in steam turbines, it is also called *steam whirl*.

On the basis of the fact that the efficiency increases as the clearance between the blades and casing decreases, Thomas (1958) explained this instability in steam turbines as follows. Figure 10.15a shows a model of a turbine rotor with blades. When the rotor has a load, the force of the steam acts on the blades. If the rotor is at the center, the tip clearance is uniform circumferentially; therefore, this force generates no destabilizing force. On the contrary, when the rotor deflects in the deflection, as shown in the figure, the clearance in that direction decreases and that in the opposite direction increases. Consequently, the fluid force in the former increases because of higher efficiency and that in the latter decreases because of lower efficiency. Let the forces acting on a blade in the same side as the rotor deflection be $F + \Delta F$ and that in the opposite side be $F - \Delta F$, as shown in Figure 10.15b, where F is a force that works when the rotor is positioned at the center. The forces acting on a blade perpendicular to the rotor deflection are almost the same as F. Figure 10.15b can be replaced equivalently with Figure 10.15c, which shows the existence of self-excited force $2\Delta F$ charging energy into the system. The torque $4Fr$ works as the driving torque that rotates the turbine, where r is the distance between the force and the center. Since Alford (1965) later explained the occurrence of this self-excited force in a compressor, this force is called the *Thomas–Alford force* (or simply the *Alford force*).

10.5 Hollow Rotor Partially Filled with Liquid

A high-speed rotor with liquid inside has various types of vibration problems. Leblanc (1916) proposed an idea for an automatic dynamic balancer using mercury. This balancer utilizes the fact that the liquid is located on the opposite side of the unbalance in the postcritical speed range. Thearle (1950) revised this idea and designed an automatic dynamic balancer for washing machines. Ehrich (1967)

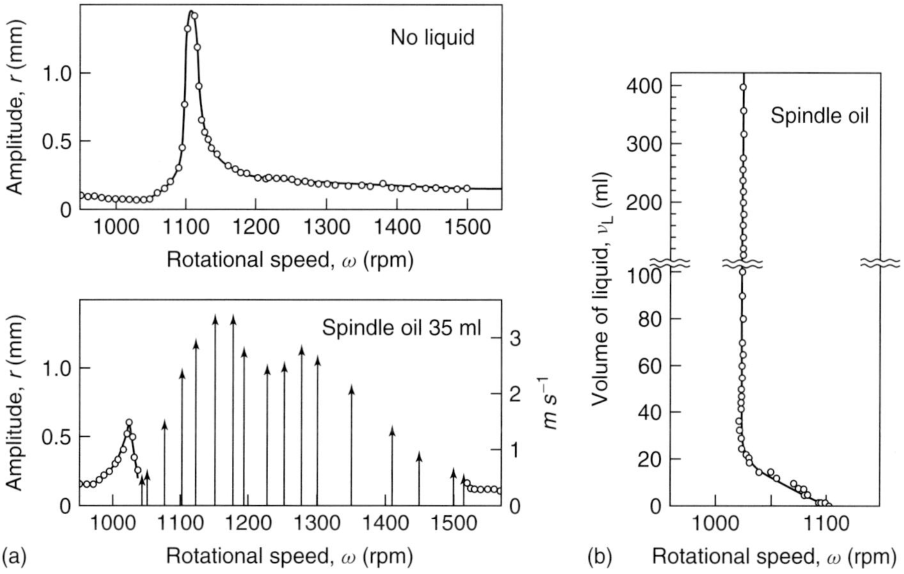

Figure 10.16 Experiments on a rotor with liquid. (a) Response curve and (b) major critical speed (Ota *et al.*, 1986).

reported an asynchronous whiling motion induced by a small amount of oil or condensed water accidentally trapped into a hollow rotor in an aircraft gas turbine. Kollmann (1962) reported an unstable vibration of a centrifuge due to contained liquid.

Figure 10.16a shows response curves of a rotor composed of an elastic shaft and a cylindrical rotor. The upper one is a case of a rotor with no liquid. Only a synchronous vibration (synchronous vibration) appears, and resonance occurs at about 1107 rpm. The lower one is a case of a rotor containing 35 ml of spindle oil. Two remarkable phenomena appeared in this figure. First, asynchronous self-excited vibrations occur in a certain range above the major critical speed. In this unstable range, whirling amplitude $r(t)$ increases exponentially. In this figure, the speed of growth is represented by the length of the arrow corresponding to the parameter m, which is determined by approximating the deflection by $r(t) \approx Ae^{mt}$. Second, the speed and height of resonance curves at the major critical speed changes. Owing to 35 ml being a small amount (only about 0.47% of the disk mass), the major critical speed goes down appreciably (about 80 rpm) and the magnitude of the peak decreases. Figure 10.16b shows this critical speed change versus the amount of the liquid. When the quantity increases, the critical speed changes rapidly in the beginning and then becomes constant.

In this section, the phenomenon about unstable vibration is explained following the study by Wolf (1968), and that about resonance curves is explained following the study by one of the authors (Ota *et al.*, 1986). Refer to these papers for details.

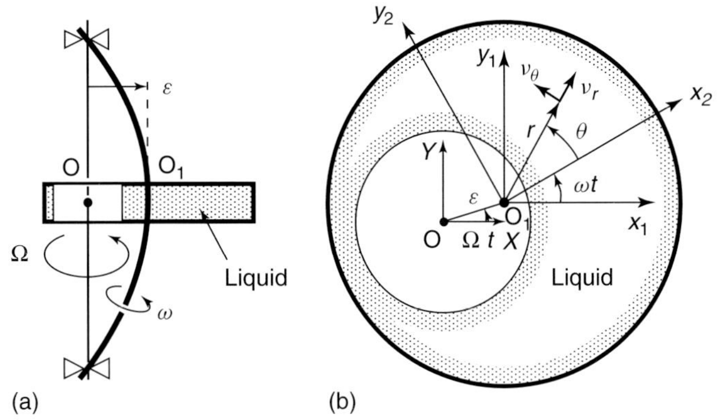

Figure 10.17 Rotor partially filled with liquid. (a) Rotor-liquid system and (b) coordinate systems.

10.5.1
Equations Governing Fluid Motion and Fluid Force

Figure 10.17 shows an analytical model. A hollow rotor partially filled with liquid is mounted at the midspan of an elastic shaft. The fluid is supposed to be inviscid and incompressible, and its distribution is two dimensional. The rotor is rotating at speed ω and is whirling with an (unknown) angular velocity Ω.

As preparation for deriving the equations of motion of this rotor system, we first derive the expression for the fluid force that acts on the disk whirling with radius ε (Wolf, 1968). Let the inner radius and the inner height of the hollow rotor be a and h, respectively. As shown in Figure 10.17b, we prepare the static coordinate system $\mathrm{O}-XY$ whose origin is located on the bearing centerline. From fluid dynamics, we obtain Euler's equation of the liquid inside the rotor as follows:

$$\left.\begin{aligned}\frac{\partial v_X}{\partial t}+v_X\frac{\partial v_X}{\partial X}+v_Y\frac{\partial v_X}{\partial Y}&=-\frac{1}{\rho}\frac{\partial p}{\partial X}\\ \frac{\partial v_Y}{\partial t}+v_X\frac{\partial v_Y}{\partial X}+v_Y\frac{\partial v_Y}{\partial Y}&=-\frac{1}{\rho}\frac{\partial p}{\partial Y}\end{aligned}\right\} \tag{10.31}$$

where v_X and v_Y are the fluid velocities in the X- and Y-directions, respectively, p is the pressure, and ρ is the fluid density. By the transformation

$$\left.\begin{aligned}x_1&=X-\varepsilon\cos\Omega t, &\quad v_X&=v_{X_1}-\varepsilon\Omega\sin\Omega t\\ y_1&=Y-\varepsilon\sin\Omega t, &\quad v_Y&=v_{Y_1}+\varepsilon\Omega\cos\Omega t\end{aligned}\right\} \tag{10.32}$$

we first transform Eq. (10.31) into the expression in the coordinate system $\mathrm{O}_1-x_1y_1$, whose axes are parallel to the X- and Y-axes, respectively, and whose origin O_1 is located at the center of the disk whirling with angular velocity Ω. Next, by the transformations

$$\left.\begin{aligned} x_2 &= x_1 \cos\omega t + y_1 \sin\omega t \\ y_2 &= -x_1 \sin\omega t + y_1 \cos\omega t \\ v_{x_1} &= v_{x_2} \cos\omega t - v_{y_2} \sin\omega t - \omega\left(x_2 \sin\omega t + y_2 \cos\omega t\right) \\ v_{y_1} &= v_{x_2} \sin\omega t + v_{y_2} \cos\omega t + \omega\left(x_2 \cos\omega t - y_2 \sin\omega t\right) \end{aligned}\right\} \tag{10.33}$$

we transform them into the expression in the coordinate system $O_1 - x_2 y_2$, which is fixed to the disk and rotates with the angular velocity ω. Further, by the transformation

$$\left.\begin{aligned} x_2 &= r\cos\theta, \quad v_{x_2} = v_r \cos\theta - v_\theta \sin\theta \\ y_2 &= r\sin\theta, \quad v_{y_2} = v_r \sin\theta + v_\theta \cos\theta \end{aligned}\right\} \tag{10.34}$$

we transform the expression into the polar coordinates (r, θ). Finally, we arrive at the following equations:

$$\left.\begin{aligned} &\frac{\partial v_r}{\partial t} + v_r \frac{\partial v_r}{\partial r} + \frac{v_\theta}{r}\frac{\partial v_r}{\partial \theta} - \frac{{v_\theta}^2}{r} - r\omega^2 - 2v_\theta\omega = -\frac{1}{\rho}\frac{\partial p}{\partial r} + \varepsilon\Omega^2 \cos\left(\sigma t - \theta\right) \\ &\frac{\partial v_\theta}{\partial t} + v_r \frac{\partial v_\theta}{\partial r} + \frac{v_\theta}{r}\frac{\partial v_\theta}{\partial \theta} + \frac{v_r v_\theta}{r} + 2v_r\omega = -\frac{1}{\rho r}\frac{\partial p}{\partial r} + \varepsilon\Omega^2 \sin\left(\sigma t - \theta\right) \end{aligned}\right\} \tag{10.35}$$

where $\sigma = \Omega - \omega$. The continuity equation is given by the following expression in polar coordinates.

$$\frac{\partial\left(r v_r\right)}{\partial r} + \frac{\partial v_\theta}{\partial \theta} = 0 \tag{10.36}$$

We see from Eq. (10.35) that the whirling motion of the rotor induces a forcing term that produces a forced wave in the liquid layer, and as a result, a net force acts on rotor wall.

We obtain the pressure distribution $p(r, \theta, t)$ by solving these equations of motion (10.35) and the continuity Eq. (10.36) by the perturbation method. The boundary conditions are as follows:

$$\left.\begin{aligned} &\text{At the rotor wall } \; [r = a] \quad &&: v_r = 0 \\ &\text{At the free surface } \; \left[r = r_f(\theta, t)\right] \quad &&: p = 0, \; v_r = \frac{\partial r_f}{\partial t} \end{aligned}\right\} \tag{10.37}$$

We assume the perturbation solutions that are represented in powers of whirl amplitude ε as follows:

$$\left.\begin{aligned} v_r &= v_{r0} + \varepsilon v_{r1} + \cdots \\ v_\theta &= v_{\theta 0} + \varepsilon v_{\theta 1} + \cdots \\ p &= p_0 + \varepsilon p_1 + \cdots \end{aligned}\right\} \tag{10.38}$$

We substitute these solutions into Eqs. (10.35) and (10.36) and compare the coefficients of the same order of ε. From the zeroth order, we get

$$v_{r0} = v_{\theta 0} = 0, \quad p_0 = \frac{1}{2}\rho\omega^2\left(r^2 - b^2\right) \tag{10.39}$$

where b is the radius of free surface when the liquid is distributed uniformly (Figure 10.19). If we assume that the free surface $r = r_f$ during the whirling motion is in the form $r = b + \eta(\theta, t)\varepsilon$, we can obtain it in the course of the following analysis. It will become clear that this free surface is a circle with its center at the bearing center, as shown in Figure 10.17.

From the equation obtained from the first-order terms of ε, we get the following expression after some calculations:

$$p_1(r,\theta,t) = -\frac{\rho\Omega^2\omega^2a^2b^2}{\left(a^2-b^2\right)\left(\alpha\sigma^2-2\sigma\omega-\omega^2\right)} \times \left[-\frac{\sigma\,(2\omega-\sigma)}{\omega^2 r}+\frac{\sigma\,(2\omega-\sigma)\,r}{a^2b^2}+\frac{\left(a^2-b^2\right)r}{a^2b^2}\right]\cos\left(\sigma t-\theta\right) \tag{10.40}$$

where $\alpha = [(a/b)^2 + 1]/[(a/b)^2 - 1]$ (Wolf, 1968). Then, from Eqs. (10.39) and (10.40), we get

$$p\,(r,\theta,t) = p_0\,(r,\theta,t) + \varepsilon p_1\,(r,\theta,t) \tag{10.41}$$

Putting $r = a$ gives the pressure distribution $p\,(a,\theta,t)$ along the inside wall.

In the case of a circular distribution shown in Figure 10.17, the x_2- and y_2-components of the net fluid force acting on the rotor are given by

$$\left.\begin{aligned} F_{x_2} &= \int_0^{2\pi} p\,(a,\theta,t)\cos\theta dA = \varepsilon\Omega^2 m_{\mathrm{Lf}}\frac{\left(\sigma^2-2\sigma\omega-\omega^2\right)}{\left(\alpha\sigma^2-2\sigma\omega-\omega^2\right)}\cos\sigma t \\ F_{y_2} &= \int_0^{2\pi} p\,(a,\theta,t)\sin\theta dA = \varepsilon\Omega^2 m_{\mathrm{Lf}}\frac{\left(\sigma^2-2\sigma\omega-\omega^2\right)}{\left(\alpha\sigma^2-2\sigma\omega-\omega^2\right)}\sin\sigma t \end{aligned}\right\} \tag{10.42}$$

where $dA = ahd\theta$ and $m_{\mathrm{Lf}} = \rho\pi a^2 h$ is the mass of the liquid needed to fill the cavity completely. This force points to the same direction as the deflection OO_1 and has the magnitude

$$F_{\mathrm{L}} = \varepsilon\Omega^2 m_{\mathrm{Lf}}\frac{\left(\sigma^2-2\sigma\omega-\omega^2\right)}{\left(\alpha\sigma^2-2\sigma\omega-\omega^2\right)} \tag{10.43}$$

10.5.2
Asynchronous Self-Excited Whirling Motion

We investigate asynchronous self-excited whirling motion, which occurs in a certain range above the major critical speed. It is assumed that the amount of liquid is enough to cover all of the inside wall and forms a complete ring during the whirling motion. In Chapter 5, we explained the occurrence of unstable vibration by calculating natural frequencies and depicting frequency diagrams. In those cases, an unstable vibration occurs in the speed range where some of the natural frequencies become complex numbers.

Wolf (1968) explained the self-excited whirling motion of a rotor partially filled with liquid as an eigenvalue problem as follows. His explanation is similar to that shown in Chapter 5.

The analytical rotor model is shown in Figure 10.17. It is assumed that the system has no damping and no unbalance. Let the mass of the disk be m and the stiffness

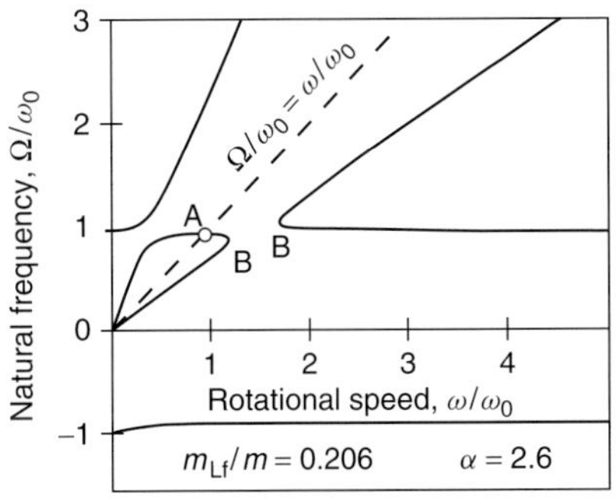

Figure 10.18 Natural frequency diagram. (Wolf, 1968, Courtesy of ASME.)

of the elastic shaft be k. Let us consider a free whirling motion tracing a circular pass of radius ε with the angular velocity Ω. In this case, the elastic restoring force $k\varepsilon$ of the shaft, the centrifugal force $m\varepsilon\omega^2$, and the fluid force F_L are balanced in the rotating coordinate system (Figure 2.5). Therefore, we get

$$k = m\omega^2 + \Omega^2 m_{Lf}\frac{(\sigma^2 - 2\sigma\omega - \omega^2)}{(\alpha\sigma^2 - 2\sigma\omega - \omega^2)} \tag{10.44}$$

This expression gives natural frequencies Ω as a function of the rotational speed ω (to avoid confusion with the pressure, we use the notation Ω instead of p which is used to denote a natural frequency in Chapter 5). A result is shown in Figure 10.18, where the parameters are nondimensionalized using $\omega_0 = \sqrt{k/m}$.

When a value is given for the rotational speed ω, Eq. (10.44) has four roots at most for the whirling velocity Ω. Two of them are complex conjugates in the range BB, and this speed range corresponds to an unstable range. The cross point A between the straight line $\Omega/\omega_0 = \omega/\omega_0$ and a frequency curve gives the major critical speed. The unstable range appears above the major critical speed and its position and width relative to the major critical speed coincide with that in Figure 10.16.

10.5.3
Resonance Curves at the Major Critical Speed (Synchronous Oscillation)

The resonance at the major critical speed is a synchronous vibration that occurs as a response to unbalance excitation. In this whirling motion, liquid does not move relative to the disk and performs as a rigid body. Figure 10.19 shows the geometries of a liquid film. Depending on the amount of liquid, it has an annular or crescent-shaped distribution (Ehrich, 1967).

First, we derive the fluid force and the position of the center of gravity. For an annular distribution, the fluid force is obtained by putting $\Omega = \omega$ (i.e., $\sigma = 0$) in Eq. (10.43) as follows:

$$F_L = \varepsilon\omega^2 m_{Lf} = \varepsilon\omega^2 \rho\pi a^2 h \tag{10.45}$$

This result shows an interesting characteristic of the fluid force. The expression does not have parameter b denoting the radius of free surface of the liquid, and the

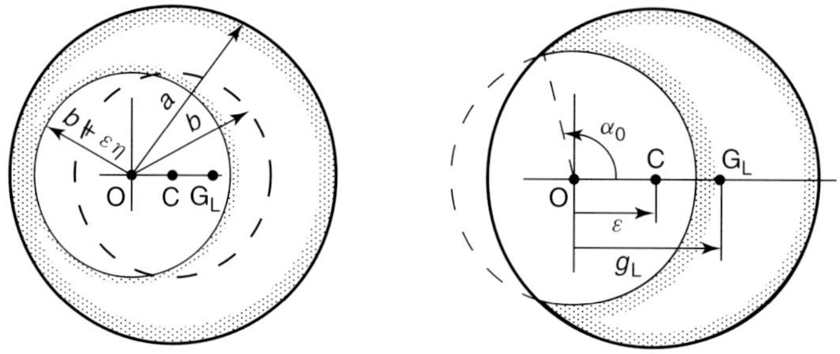

Figure 10.19 Geometry of a fluid film in a synchronous whirl.

quantity m_{Lf} is the mass of the liquid that fills the cavity perfectly. In other words, even if the quantity of the liquid is very small, the effect of liquid is the same as that filling 100% of the cavity as long as it covers the entire circumference. Let us assume that liquid of volume v_L is contained in a rotor. Since this liquid works as a rigid body during a synchronous motion (imagine that the water is frozen at that instant without changing shape), the centrifugal force $\rho v_L g_L \omega^2$ works at the center of gravity G_L, where g_L is the distance from the bearing center to the center of gravity. Putting this centrifugal force equal to Eq. (10.45), we have

$$\rho v_L g_L \omega^2 = \varepsilon \omega^2 \rho \pi a^2 h \tag{10.46}$$

Then, we have

$$g_L = \frac{\pi a^2 h}{v_L} \varepsilon \tag{10.47}$$

For the case of crescent-shaped distribution, we cannot determine the position of the center of gravity analytically. Ehrich (1967) determined this position by assuming the liquid film thickness as a cosine function, and one of the authors (Ota, Ishida, Sato, and Yamada 1986) did it numerically by considering that the free surface of the liquid film is part of a circle whose center is located at the bearing centerline. Here, we cite the result of the latter.

Figure 10.20 shows the position of the center of gravity and the liquid force versus the rotor deflection in the case where liquid with density $d_L = 5.0$ is contained in about 5% of the cavity. Both quantities are shown as dimensionless quantities.

To obtain resonance curves, we first derive the equations of motion of the system. When the rotor does not contain liquid, the equations of motion are given by Eq. (2.10), that is,

$$\left.\begin{aligned} m\ddot{X} + c\dot{X} + kX &= me\omega^2 \cos \omega t \\ m\ddot{Y} + c\dot{Y} + kY &= me\omega^2 \sin \omega t \end{aligned}\right\} \tag{10.48}$$

By the transformation

$$\left.\begin{aligned} X &= \varepsilon \cos (\omega t + \varphi) \\ Y &= \varepsilon \sin (\omega t + \varphi) \end{aligned}\right\} \tag{10.49}$$

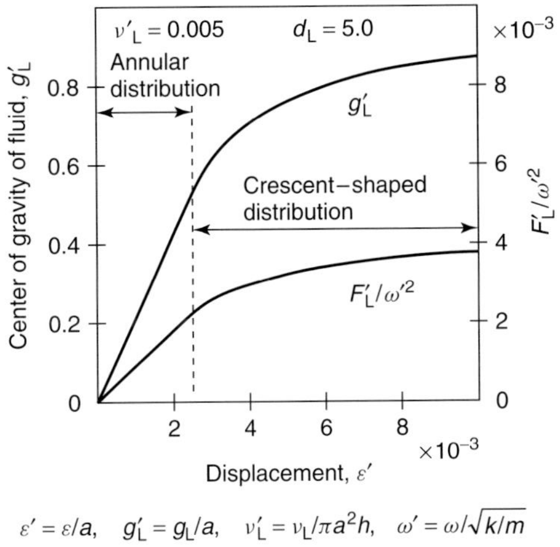

Figure 10.20 Fluid force and the position of the center of gravity.

we change Eq. (10.48) into the coordinates (ε, φ), which are the polar coordinates rotating with ω:

$$\left.\begin{aligned} &m\ddot{\varepsilon} - m\varepsilon(\omega + \dot{\varphi})^2 + c\dot{\varepsilon} + k\varepsilon - me\omega^2\cos(-\varphi) = 0 \\ &m\varepsilon\ddot{\varphi} + 2m\dot{\varepsilon}(\omega + \dot{\varphi}) + c\varepsilon(\omega + \dot{\varphi}) - me\omega^2\sin(-\varphi) = 0 \end{aligned}\right\} \tag{10.50}$$

The resonance curves for the steady-state vibration are obtained by putting $\dot{\varepsilon} = \dot{\varphi} = 0$ in these equations. For the case of a rotor containing liquid, we get the expression for the resonance curve by adding the liquid force $m_L g_L \omega^2$ as follows:

$$\left.\begin{aligned} &-m\varepsilon\omega^2 + k\varepsilon - me\omega^2\cos(-\varphi) - m_L g_L\omega^2 = 0 \\ &c\varepsilon\omega - me\omega^2\sin(-\varphi) = 0 \end{aligned}\right\} \tag{10.51}$$

The balance of forces given by these expressions are illustrated in Figure 10.21a, and resonance curves are shown in Figure 10.21b. The result of the stability criteria is also included. In Figure 10.21b, curve A is the case with no liquid. Curves B and C are cases with small amounts of liquid. The liquid distribution is of annular type when the deflection is small and of crescent-shaped type when it is large. Owing to this change, the resonance curve becomes a hard spring type. In a case with a comparatively large amount of liquid, the liquid distribution is annular in the amplitude range shown in the figure and we have the resonance curve D. Curve D is independent of the liquid quantity if the liquid volume v'_L is greater than 0.0146. The critical speed for this case is given by $\omega = \sqrt{k/(m + m_{Lf})}$, where m_{Lf} is the mass of the liquid filling the cavity completely. This result agrees with the experimental result shown in Figure 10.16b.

Here, we discuss the physical reason why the critical speed and liquid force do not depend on the quantity of liquid. Let us assume that a hollow rotor containing liquid is whirling with angular velocity Ω as shown in Figure 10.17. To take into account the centrifugal force, we observe this motion in the rotating coordinates

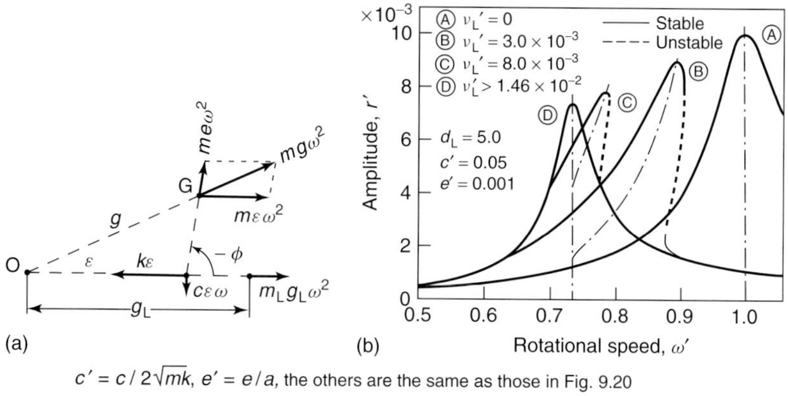

Figure 10.21 Effects of liquid quantity on resonance curves. (a) Forces in balance and (b) resonance curves. (From Ota, Ishida, Sato, and Yamada 1986.)

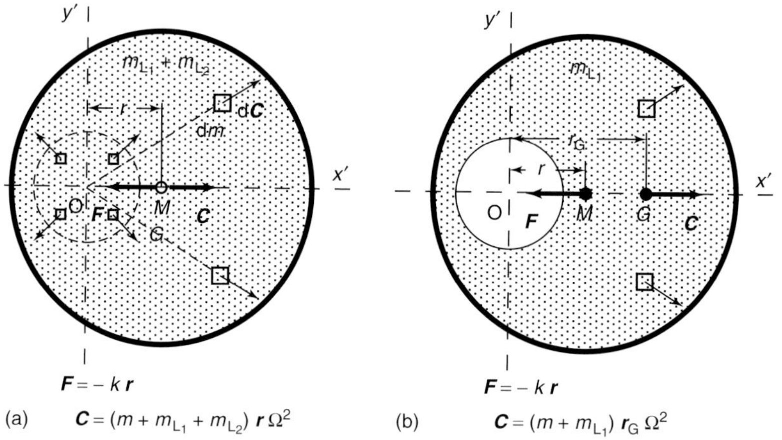

Figure 10.22 Forces working on a rotor with liquid. (a) Completely filled and (b) partially filled.

$O - x'y'$, whose x'-axis coincides with the OO_1 direction. Figure 10.22a is the case where 100% of the cavity is filled with liquid, and Figure 10.22b is the case where the cavity is partially filled with liquid. Let the mass of the rotor be m, the mass of the liquid in Figure 10.22b be m_{L1}, and the mass of the liquid in Figure 10.22a be $m_{Lf} = m_{L1} + m_{L2}$. In Figure 10.22a the elastic restoring force $\boldsymbol{F} = -k\boldsymbol{r}$ and the centrifugal force $\boldsymbol{C} = (m + m_{Lf})\boldsymbol{r}\Omega^2$ work at the center M of the rotor and balance each other. The centrifugal forces working in the liquid is the summation of the small centrifugal force dC working on a small element dm. In Figure 10.22a, the summation of the centrifugal force working on elements in a dashed line circle which corresponds to the cavity in Figure 10.22b clearly becomes zero because the elements are symmetrical about the whirling center O. This means that the magnitude of the centrifugal force working in the liquid in Figure 10.22b is the same as that in Figure 10.22a. We should note that this characteristic comes from the fact that the free surface of the liquid always forms part of a circle whose center is the whirling center O.

References

Alford, J.S. (1964) Protection of labyrinth seals from flexural vibration. *Trans. ASME, J. Eng. Power*, **86** (2), 141–148.

Alford, J.S. (1965) Protecting turbomachinery from self-excited rotor whirl. *Trans. ASME, J. Eng. Power*, **87** (4), 333–344.

Black, H. (1979) Effects of fluid-filled clearance spaces on centrifugal pump and sub-merged motor vibrations. Proceedings of the 8th Turbomachinery Symposium, Texas A&M University, pp. 29–38, Refer to Childs (1993. Fig. 4.3).

Childs, D. (1993) *Turbomachinery Rotordynamics*, John Wiley & Sons, Inc., New York.

Ehrich, F.F. (1967) The influence of trapped fluids on high speed rotor vibration. *Trans. ASME, J. Eng. Ind. Ser. B*, **89** (4), 806–812.

Funakawa, M. and Tatara, A. (1964) Stability criterion of an elastic rotor in journal bearings. *Trans. JSME*, **30** (218), 1238–1244. (in Japanese).

Hori, Y. (1959) A theory of oil whip. *Trans. ASME, J. Appl. Mech.*, **26** (2), 189–198.

Iwatsubo, T. and Motooka, N. (1980) Flow induced forces of labyrinth seal. Rotordynamic Instability Problems in High-Performance Turbomachinery, NASA CP 2250, Proceedings of a Workshop held at Texas A&M University, pp. 205–222.

Kollmann, F.G. (1962) Experimentelle und theoretische Untersuchungen über die kntischen Drehzahlen flüssigkeitsgefüllter Hohlkörper. *Forsch. Geb. Ing. -Wesens*, **28** (4), 147–153. (in German).

Kramer, E. (1993) *Dynamics of Rotors and Foundations*, Springer-Verlag, New York.

Leblanc, M. (1916) Automatic balancer for rotating bodies, U.S. Parent No. 1209730. (Refer to Thearle, 1950, Fig. 5).

Marquette, O.R., Childs, D.W., and San Andres, L. (1997) Eccentricity effects on the rotor-dynamic coefficients of plain annular seals: theory versus experiment. *Trans. ASME, J. Trib.*, **119** (3), 443–448.

Muszynska, A. (1987) Tracking the mystery of oil whirl. *Sound Vib.*, 8–11.

Muszynska, A. (1988) Alford and the destabilizing forces that lead to fluid whirl/whip. *Orbit*, **19** (3), 29–31.

Nelson, C.C. (1985) Rotordynamic coefficients for compressible flow in tapered annular seals. *Trans. ASME, J. Trib.*, **107** (3), 318–325.

Newkirk, B.L. and Taylor, H.D. (1925) Shaft whirling due to oil action in journal bearings. *Gen. Electr. Rev.*, **28** (7), 559–568.

Ota, H., Ishida, Y., Sato, A., and Yamada, T. (1986) Experiments on vibration of a hollow rotor partially filled with liquid. *Bull. JSME*, **29** (256), 3520–3529.

Thearle, E.L. (1950) Automatic dynamic balancing. *Machine Des.*, **22** (9), 119–124.

Thomas, J.J. (1958) *Instabile Eigenschwingungeri von Turbinenlaufern, Angefacht durch die Spaltstromungen, in Stoptbuchsen und Beschauflungen*, AEG-Sonderdruck, pp. 1039–l063.

Tondl, A. (1965) *Some Problems of Rotor Dynamics*, Czechoslovak Academy of Sciences, Prague.

Wolf, J.A. (1968) Whirl dynamics of a rotor partially filled with liquid. *Trans. ASME, J. Appl. Mech.*, **35** (4), 676–682.

11
Vibration Suppression

11.1
Introduction

In order to reduce vibrations of a rotor system, the first consideration in the production of rotors is balancing of the rotor. We can utilize the balancing methods explained in Chapter 4. The second consideration is to determine the dimensions of a rotor not to bring its critical speeds into the speed range of operation. As a general rule of thumb, operation speed should be about 20% way from a resonance. However, we cannot eliminate unbalance completely and it should also be noted that unbalance sometimes changes during the operation. In addition, as we studied in the previous sections, besides forced oscillation due to unbalance, different forms of vibration such as unstable vibration and self-excited oscillation occur in various situations. Therefore, we must apply a different vibration suppression method, which is most effective for each vibration. Vibration suppression methods are classified into the two categories of passive method and active method. From the viewpoint of costing, simplicity, and robustness, passive methods are recommended. In this chapter, various kinds of passive vibration suppression methods are explained.

11.2
Vibration Absorbing Rubber

Rubber mounts are often used to isolate vibration, that is, to reduce vibration which is transmitted to the base. Rubber is installed between the outer ring of the bearing and housing as shown in Figure 11.1. Internal damping of a rubber absorbs the vibration energy. Ito (1962) applied vibration absorbing rubber to an electric motor giving the results as shown in Figure 11.2. It shows vibration spectrums of a motor with and without vibration absorbing rubber at 1000 rpm. Vibration components at higher frequencies are well suppressed.

11.3
Theory of Dynamic Vibration Absorber

Kirk and Gunter (1972) and Ota and Kanbe (1976) applied the *theory of dynamic vibration absorbers* (the *fixed point theory*) to a rotor. Figure 11.3a shows Kirk and

Linear and Nonlinear Rotordynamics: A Modern Treatment with Applications, Second Edition.
Yukio Ishida and Toshio Yamamoto.

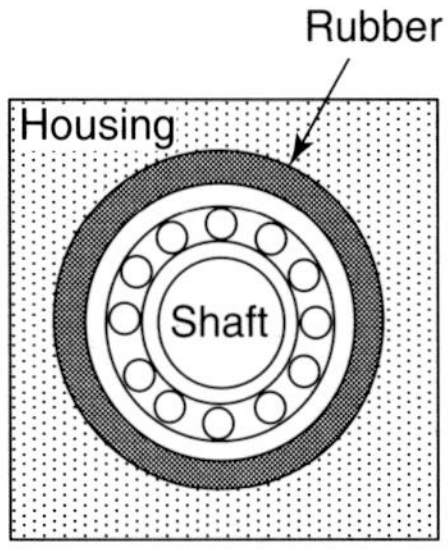

Figure 11.1 Rubber mount.

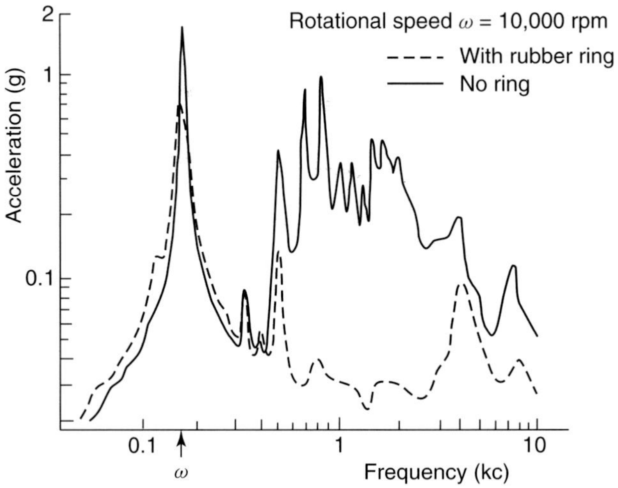

Figure 11.2 Effect of rubber mount. (Ito, 1962, Courtesy of Mechanical Design.)

Gunter's theoretical model. The bearings are supported elastically by springs and dampers, and their parameter values are adjusted based on the theory of dynamic vibration absorbers. Figure 11.3b shows response curves for various values of the damping coefficient c. The curve has one peak for large damping ($c = 50$) and two peaks for small damping ($c = 0.01$). The response curve has a comparatively small amplitude in a wide range for $c = 10$.

11.4 Squeeze-Film Damper Bearing

Figure 11.4 illustrates a *squeeze-film damper bearing* that is used in aircraft gas turbine engines. An oil film in a narrow clearance between the bearing housing and casing gives sufficient damping to the rotor. Morton (1966) obtained the experimental result shown in Figure 11.5 in a Rolls–Royce Avon engine. The original resonance denoted by a dashed line is well suppressed as shown by a solid line. Such an effect is also observed in Figure 6.14 in Section 6.5.

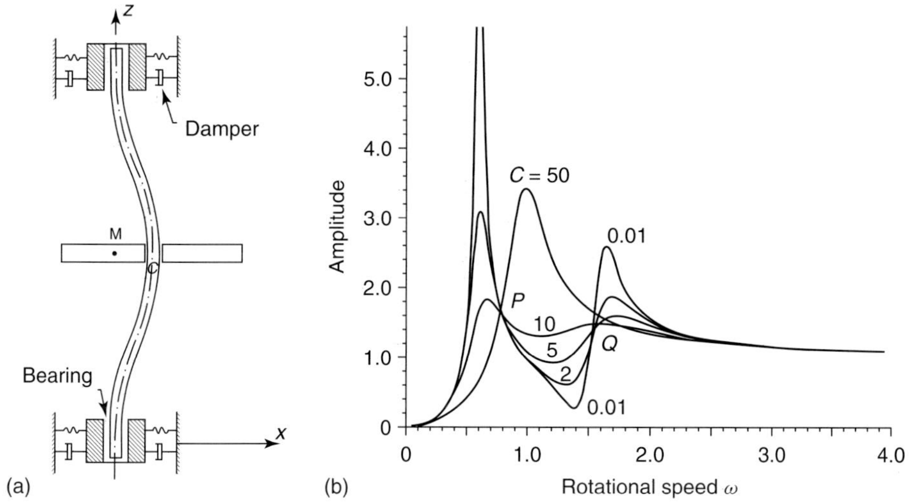

Figure 11.3 Suppression by the fixed point theory. (a) Theoretical model and (b) response curves. (Kirk and Gunter, 1972, Courtesy of ASME.)

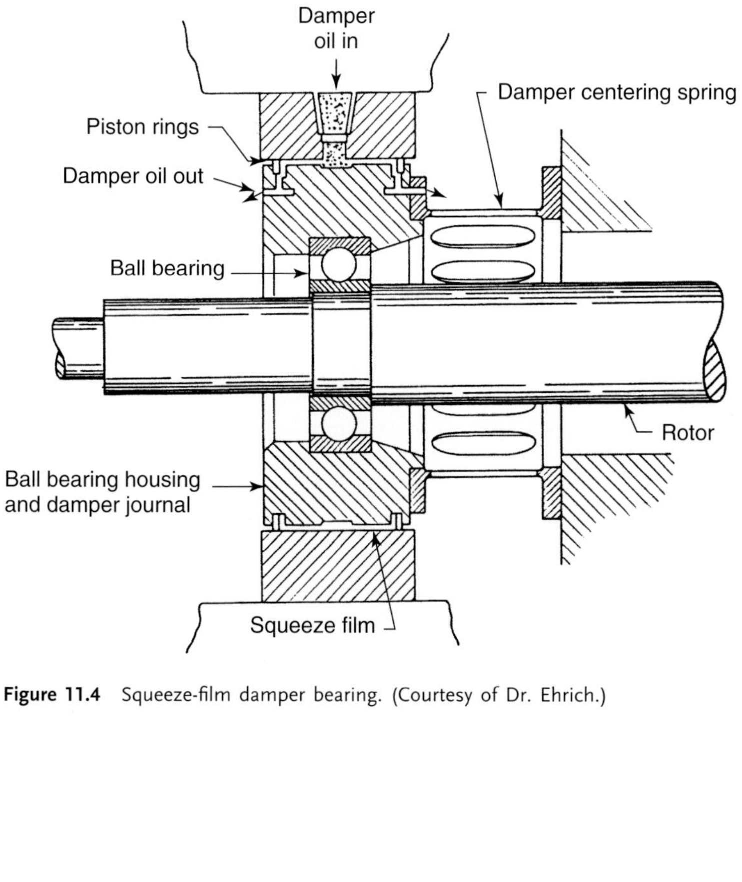

Figure 11.4 Squeeze-film damper bearing. (Courtesy of Dr. Ehrich.)

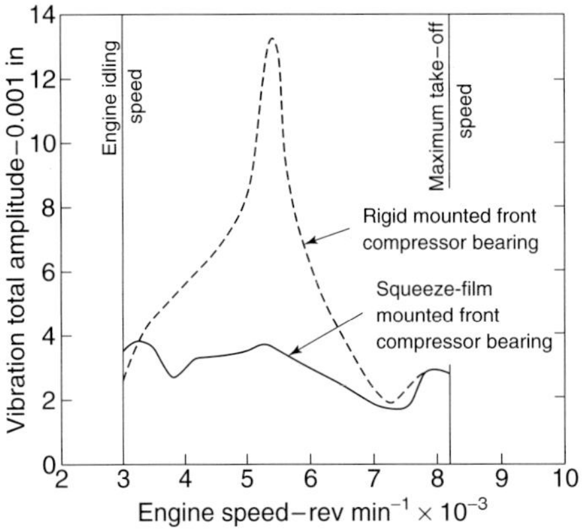

Figure 11.5 Effect of squeeze-film damper bearing. (a) Suppression device and (b) anisotropic support. (Morton, 1965, Courtesy of ASME.)

11.5 Ball Balancer

11.5.1 Fundamental Characteristics and the Problems

In 1932, Thearle (1932) invented a device called an *automatic ball balancer*. This unique device were studied intensively by Inoue, Jinnouchi, and Araki (1978), Innoue, Jinnouchi, Araki, and Nakahara (1979), Inoue, Jinnouchi, and Miyaura (1984), Inoue, Jinnouchi, and Kubo (1983). Automatic ball balancers are already in use in several machines, such as handheld grinders (Lindell, 1996) and hard disk drivers (Van de Wouw, Van den Heuvel, Nijmeijer, and Van Rooij 2005). A simple model of a rotor with a ball balancer is shown in Figure 11.6. Two balls are contained in a hollow disk of a Jeffcott rotor. The response curves and the balls' angular positions of such a 4 DOF system are shown in Figure 11.7. There are several stable (full line) and unstable (broken line) solutions for a given rotational speed. In the precritical speed range, the stable solution (notation ①) appears. In this solution, two balls locate on the same side as the unbalance and therefore the amplitude increases. In the postcritical speed range, two balls locate on the opposite side to the unbalance with an appropriate angle and the stable solution with zero amplitude (notation ⑦) appears above $\omega = 1.121$.

In Figure 11.7, we see that the automatic balancing is attained in the range above $\omega = 1.121$. However, it is difficult to attain such a perfect balancing in practice. Figure 11.8 shows an experimental result that was obtained by an experimental setup similar to that in Figure 11.6. A disk with a diameter of 260 mm and a thickness of 30 mm is mounted at the middle position of an elastic shaft with a length of 700 mm and a diameter of 12 mm. The diameter of balls is 12 mm. The

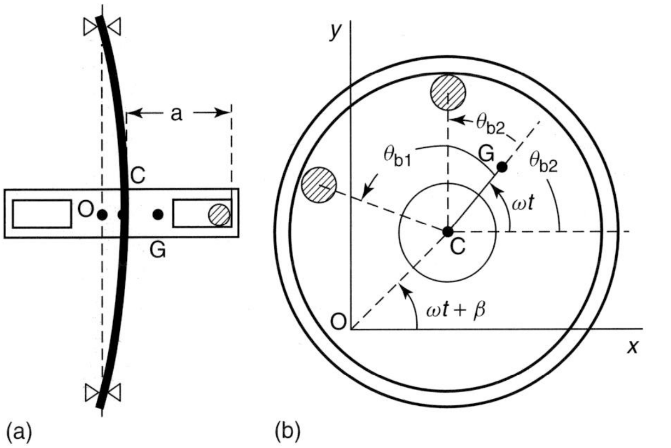

Figure 11.6 Theoretical model of a ball balancer.

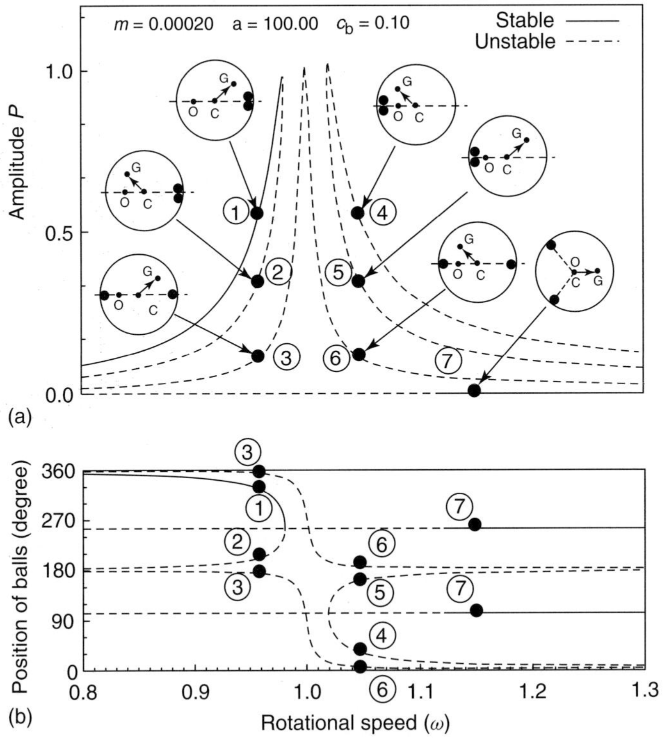

Figure 11.7 Theoretical Response curve, (a) Amplitude, (b) Phase difference (Ishida, Matsuura, and Zhang, 2012).

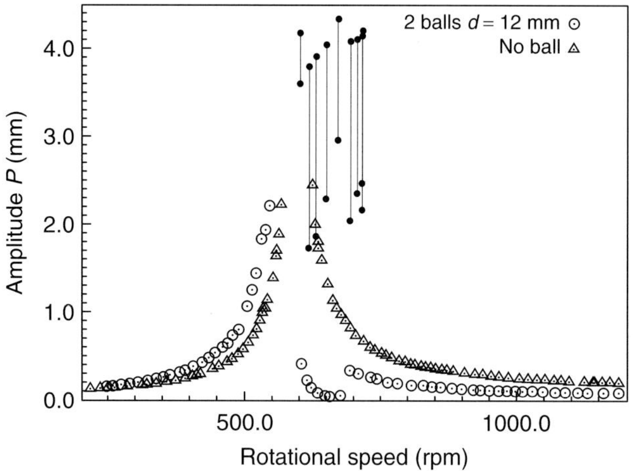

Figure 11.8 Response curve (Experiment : Ishida, Matsuura, and Zhang, 2012).

notation Δ represents the response curve of a system with no ball, and the notation o represents the case with utilization of a ball balancer.

As seen from Figure 11.8, this ball balancer has two problems which reduce its efficiency. The first problem is the occurrence of self-excited oscillations. In the vicinity of the resonance, large whirling motions with varying amplitudes occur. The amplitude variation range is represented by the short lines with dots representing the maximum and minimum amplitudes. In this self-excited oscillation, the two balls roll along the wall inside the rotor. This phenomenon was first reported by Inoue, Jinnouchi, Araki, and Nakahara (1979).

The second problem is the influence of friction. In contrast to the theoretical result shown in Figure 11.7, the amplitude in the postcritical range does not become zero in Figure 11.8. This is because the balls cannot reach their optimal positions due to the inevitable friction. Lindell (1996) reported this phenomenon in a hand grinder. Later, Van de Wouw, Van den Heuvel, Nijmeijer, and Van Rooij (2005) discussed this phenomenon by a model with Coulomb friction, Yang, Ong, Sun, and Guo (2005) explained this phenomenon by considering dry friction, and Chao, Sung, and Leu (2005) discussed by considering the rolling friction of balls.

11.5.2
Countermeasures to the Problems

In order to utilize a ball balancer, overcoming these two problems are important. As the countermeasures to a self-excited oscillation, Innoue, Jinnouchi and kubo (1983) and Wettergren (2002) filled the cavity of a rotor with oil and Jinnouchi *et al.* (1993) used partitions in the cavity.

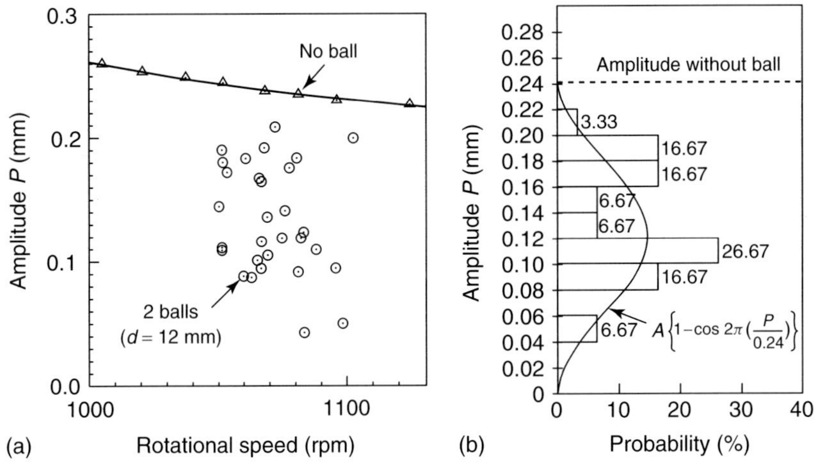

Figure 11.9 Amplitude distribution. (a) Amplitude of 30 trials and (b) probability (Experiment : Ishida, Matsuura, and Zhang, 2012).

As a countermeasure to the problem of friction, Horvath, Flowers, and Fausz (2008) proposed to use an automatic pendulum balancer instead of a ball balancer and showed that sufficient balancing can be attained by it. In a pendulum balancer, the effect of friction is reduced because it works at the hinge that supports the pendulums.

Ishida, Matsuura, and Zhang (2012) proposed a method to use several ball balancers to eliminate the influence of friction in a ball balancer. The repeatability of amplitude was investigated by executing 30 experiments at about 1100 rpm. The distribution of amplitude is shown in Figure 11.9a. The response of an original system with no ball is designated by Δ, and those of a system with a ball balancer are shown by $\circ$. We see that the amplitudes distribute between zero and the original amplitude. Figure 11.9b shows a bar graph representing the probability distribution of these data. This graph was obtained as follows: First, we adopted the amplitude 0.24 mm at 1060 rpm as the original amplitude and divide 0.24 mm into 12 levels. Since there are eight data between 0.10 and 0.12 mm, we determined the probability that the amplitude will be equal to a value within this interval to be $8/30 = 26.7\%$. Figure 11.9b shows that the use of a ball balancer reduced the amplitude by approximately half of the original amplitude in average. Here, we approximate the bar graph in Figure 11.9b by the line that is given by the function $A\left\{1 - \cos 2\pi\left(P/0.24\right)\right\}$.

From this approximation function, we make a bar graph as shown in Figure 11.10a inversely. The value A is determined so as to make the total probability of the bars 100%. Here, we call this distribution the theoretical distribution. If we use several ball balancers, the final effect will be the multiple of the effect of each ball balancer because each set of balls stops one after another eliminating residual unbalance. Considering this, we can obtain the theoretical

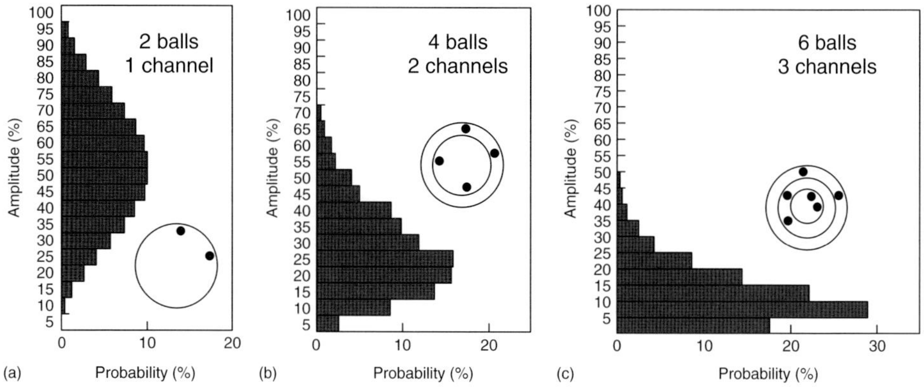

Figure 11.10 (a–c) Comparison of probability among 1, 2, 3 channels. (Ishida, Matsuura, and Zhang, 2012)

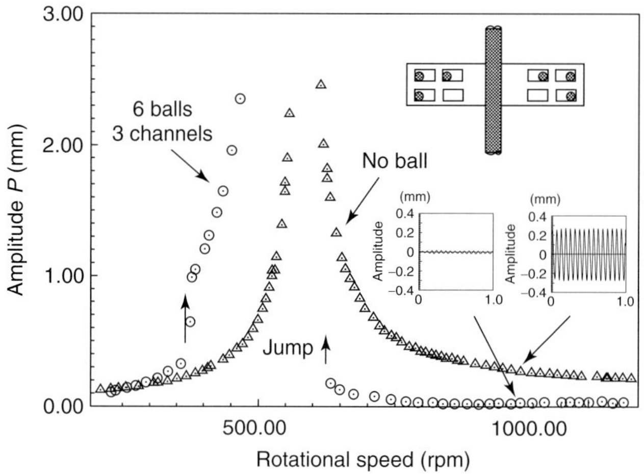

Figure 11.11 Response of a rotor with three-channel ball balancer (Experiment, Ishida, Matsuura, and Zhang, 2012).

distribution of a ball balancer with two channels by calculating the sum of the products of each bar and the corresponding distribution function. The result is shown in Figure 11.10b. Similarly, Figure 11.10c shows probability distribution in the case of a ball balancer with three channels. As the number of channel increases, balancing was seen to have improved.

Figure 11.11 shows a typical resonance curve as observed in an experimental setup utilizing a three-channel ball balancer. The results indicate that balancing in the postcritical speed range was much improved in this case. Ishida, Matsuura, and Zhang (2012) also confirmed that this effect of the multichannel technique is not decreased by partitions that are used to prevent self-excited oscillations.

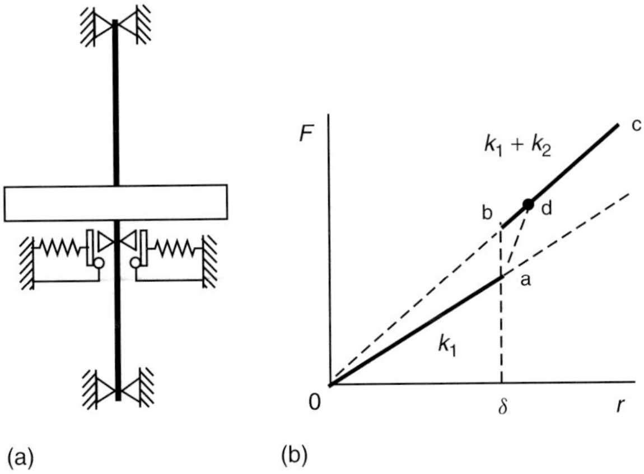

Figure 11.12 Discontinuous spring characteristics. (a) Theoretical model and (b) spring characteristics (Ishida, and Liu (2008)).

11.6 Discontinuous Spring Characteristics[1]

11.6.1 Fundamental Characteristics and the Problems

Ishida and Liu (2008) have proposed a method to suppress rotor vibration using discontinuous spring characteristics. Figure 11.12a shows a theoretical model with a suppression device located under the disk of the Jeffcott rotor. This device is explained in detail later. It is assumed that the discontinuous spring characteristics (lines O-a and b-c) shown in Figure 11.12b are made by mounting such a device. These characteristics are composed of two parts. One is the part with deflection $r < \delta$ and is called *System 1* in the following. The spring constant of this system is k_1. The other is the part with deflection $r > \delta$ and is called *System 2*. The spring constant of this system is $k_1 + k_2$. Each line is proportional to the deflection, but they are discontinuous. When the rotor moves in a circular orbit, the restoring force of the shaft is represented by only one point on this characteristic curve. The response curves denoted by the solid lines in Figure 11.13 can be derived by combining the response curves in System 1 and those in System 2. The dotted lines do not exist practically but are illustrated for reference. Since all the solid lines are stable, every transient solution started from an arbitrary initial condition converges to these solid resonance curves. The small arrows designate the direction of convergence.

In the speed range BE, the deflection decreases, aiming to converge to the imaginary curve GH in System 2, and increases toward BC and DE in System 1. As a result, the amplitude stays around δ. In order to find the whirling motion

1) Patent: PCT/JP2004/013227(International), ZL200480032779X(China).

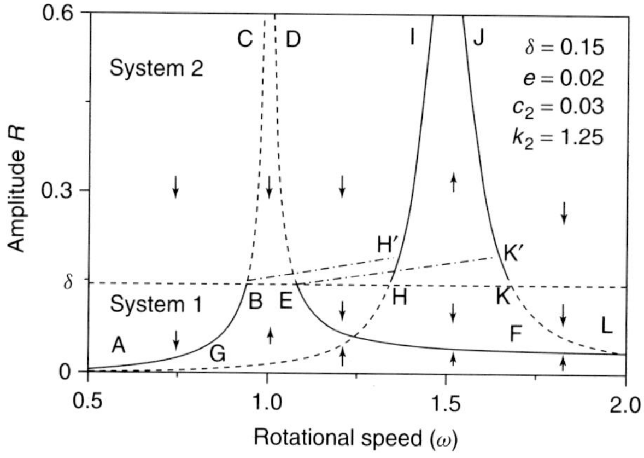

Figure 11.13 Theoretical response curve (Ishida, Matsuura, and Zhang, 2012).

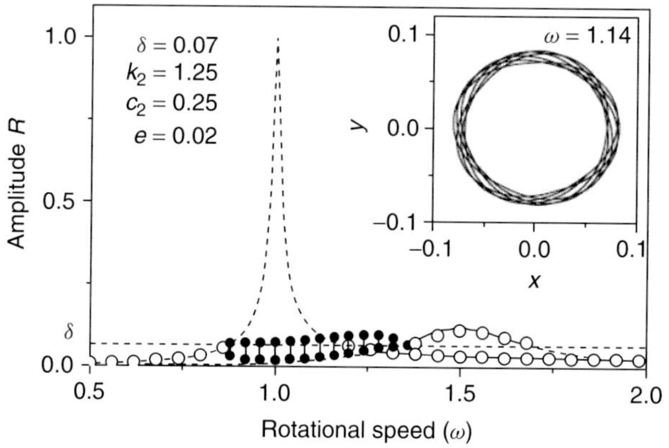

Figure 11.14 Numerical resonance curve (Ishida, Matsuura, and Zhang, 2012).

at the boundary between System 1 and System 2 in the speed range EH, let us consider a similar system that has the piecewise linear spring characteristics given by O-a-d-c in Figure 11.12b. We can imagine the response curve for this spring characteristics from the correspondence to the response curve of a hard spring type shown in Figure 6.7. Namely, the response curves corresponding to the spring characteristics a-d are given by the chain line BH′ (stable) and the chain line EK′ (unstable). By making the spring characteristic O-a-d-c approach O-a and b-c, we see that there exist the stable response curve BH and the unstable response curve EK in the speed range EH. Although these two lines BH and EK have the same amplitude in Figure 11.13, their phase angles are different. In order to confirm this, numerical simulations were carried out and obtained the results shown in Figure 11.14. We can confirm that steady-state oscillations corresponding to the solid lines exist. However, in the vicinity of the boundary of System 1 and System 2, a circular whirling motion whose orbit is shown in Figure 11.14 occurs. This is

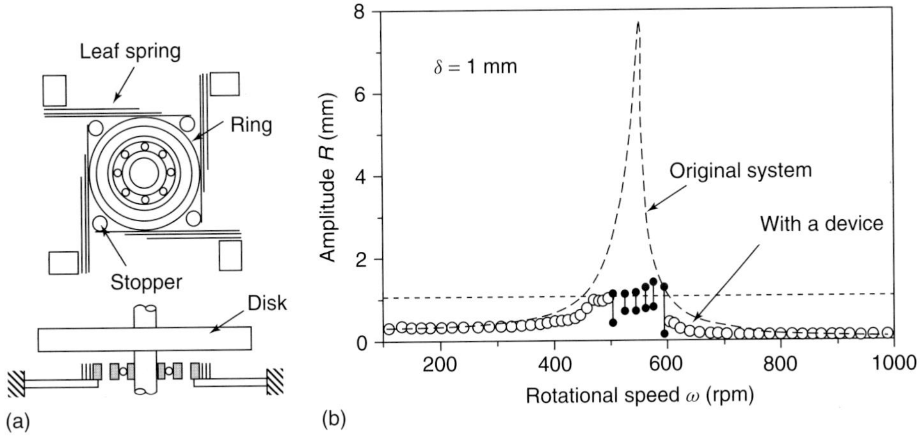

Figure 11.15 Experimental setup and response curves. (a) Suppression device and (b) anisotropic support (Ishida and Liu, 2008).

not a steady-state oscillation but an almost periodic motion whose amplitude varies between the symbols •.

In order to utilize these discontinuous spring characteristics to vibration suppression, we must overcome the following two problems. The first is the existence of the resonance peak in the higher speed range. There is a danger that the rotor jumps to this peak resonance curve by some disturbance. The second is the existence of a stable almost periodic motion between E and H. By this stable oscillation, the vibration is led to the peak of the higher speed side during acceleration.

11.6.2 Countermeasures to the Problems

The first problem can be solved by giving a large damping so as to make the peak value smaller than the magnitude of δ. The second problem can be solved by creating a directional difference in stiffness.

Figure 11.15a shows a device that creates such spring characteristics. A ball bearing is mounted on a shaft, and this bearing is surrounded by a ring with a clearance of $\delta = 1$mm. The ring is supported by leaf springs in the x- and y-directions. These leaf springs work as springs and dampers. Since leaf springs have large damping, it can suppress the peak value of the higher speed side to the level less than δ. Pins are used to stop the deflection of the leaf springs. Owing to these stoppers, leaf springs can bend outward but cannot bend inward. The springs are installed with preload by making initial shrinkage corresponding to δ. When the shaft moves rightward, the bearing contacts with the ring at $r = \delta$ and then starts to push the ring and leaf springs. When this force becomes larger than the preload, the leaf spring starts to bend and ceases to contact with the stopper. Simultaneously, the ring leaves from the leaf spring in the left side.

Figure 11.15b shows experimental results obtained using the setup described above. The broken lines show the magnitude of an original resonance curve. Without the device, the peak amplitude was larger than 8 mm. The small circles represent the resonance curve obtained by using the suppression device. The peak amplitude is suppressed to the value of about 1 mm.

11.6.3
Suppression of Unstable Oscillations of an Asymmetrical Shaft

We can use this method to suppress unstable vibrations of an asymmetrical shaft (Ishida and Liu, 2010). Figure 11.16 shows an asymmetrical rotor with this device, which produces discontinuous spring characteristics. When there is no device, an unstable oscillation occurs in the vicinity of the major critical speed as shown in Figure 11.17. Figure 11.18 shows a natural frequency diagram. The solid lines

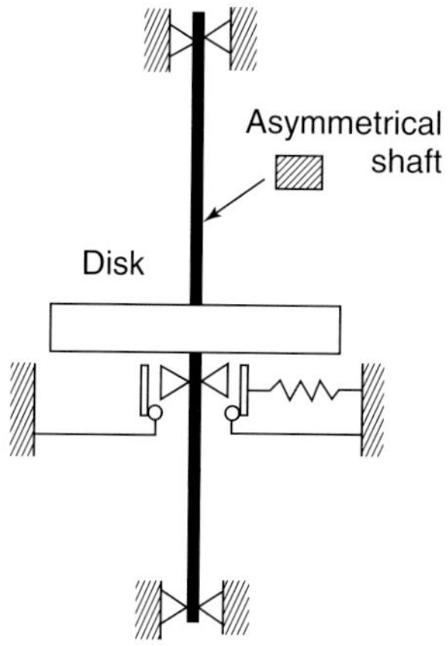

Figure 11.16 Asymmetrical rotor system with a suppression device.

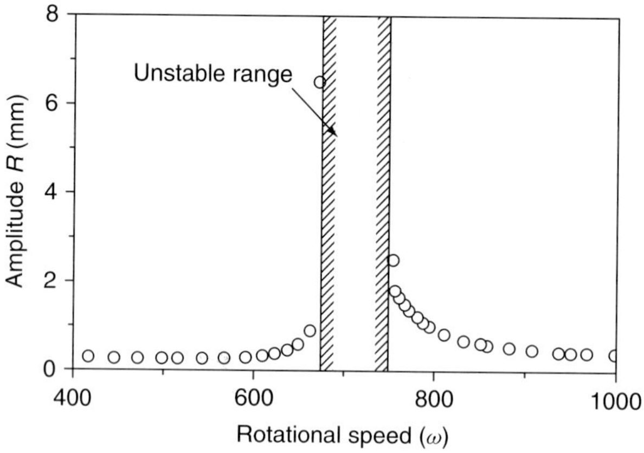

Figure 11.17 Unstable range of an asymmetrical rotor.

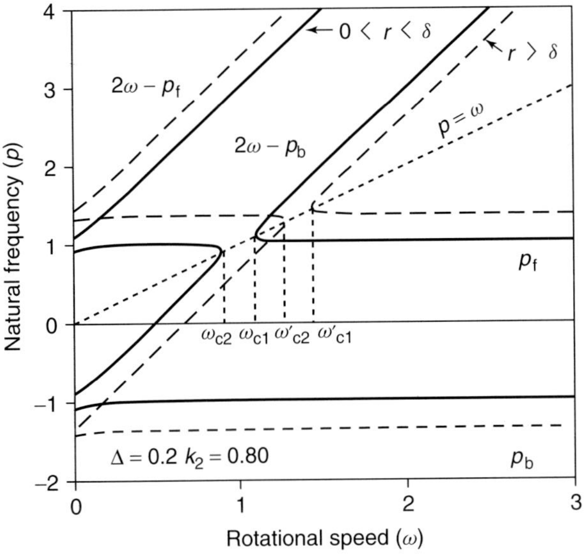

Figure 11.18 Natural frequency diagram.

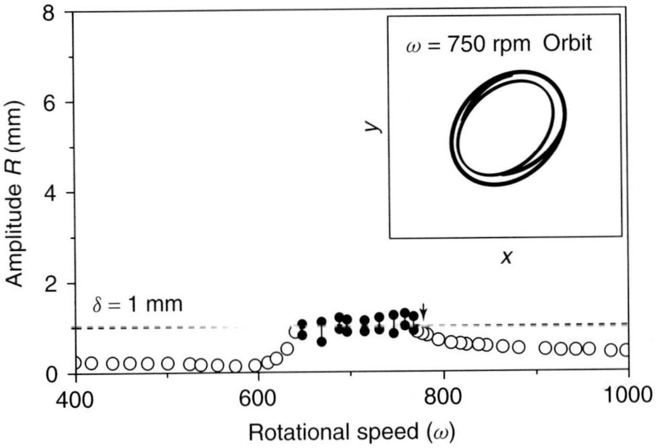

Figure 11.19 Experimental response curve of an asymmetrical rotor with a suppression device. (Ishida and Liu, 2010)

show the natural frequencies of the original system with no device. The unstable range in Figure 11.17 corresponds to the range between ω_{c2} and ω_{c1} where two natural frequencies become complex values. The broken lines show natural frequencies of the system with the device. When the deflection becomes larger than the clearance δ, the unstable range shifts from $\omega_{c2}\sim\omega_{c1}$ to $\omega'_{c2}\sim\omega'_{c1}$. By a similar mechanism as that shown in Figure 11.13, the unstable range can be eliminated. Figure 11.19 shows an experimental result obtained in an experiment where a suppression device with a clearance of 1 mm was used. The unstable vibrations changed to almost periodic motions, and the unstable speed range disappeared).

11.7 Leaf Spring

Tallian and Gustafsson (1965) proposed to use leaf springs instead of rubber in Figure 11.1. They supported a bearing by three leaf springs from three directions. They aimed to dissipate energy by the friction between leaf springs. However, the effect of this kind of damper is not so simple as a rubber because the dry friction is almost constant for velocity.

Ishida and Liu (2004) investigated the effect of leaf springs to suppress vibrations. Their experimental model is shown in Figure 11.20a. Suppose that the damper shown in this figure is mounted on a shaft of an inclination model of 2 DOF in Figure 2.13. Referring to Eq. (6.9), we can obtain the equations of motion as follows:

$$\begin{aligned}\ddot{\theta}_x + i_p\omega\dot{\theta}_y + c\dot{\theta}_x + (1 + k_L)\,\theta_x - D_{Lx} &= e\omega^2 \cos \omega t \\ \ddot{\theta}_y - i_p\omega\dot{\theta}_x + c\dot{\theta}_y + (1 + k_L)\,\theta_y - D_{Ly} &= e\omega^2 \sin \omega t\end{aligned} \qquad (11.1)$$

where k_L represents the increase of spring stiffness and D_{Lx}, D_{Ly} represent damping forces due to leaf springs. Restoring force has a hysteresis characteristic due to the dry friction in the leaf springs as shown in Figure 11.20b. For simplicity, we approximate the dry friction by the Coulomb friction and adopt the expression $D_{Lx} = -h_L\dot{\theta}_x/\left|\dot{\theta}_x\right|$, $D_{Ly} = -h_L\dot{\theta}_y/\left|\dot{\theta}_y\right|$. This damping force increases as the preload between the leaves increases. Viscous damping force increases in proportion to the velocity of the movement; however, Coulomb damping force is independent of the velocity.

Response curves of a rotor with this kind of damping are shown in Figure 11.21a by solid circles. Let the peak amplitude of the resonance curve be R_{max}. Figure 11.21b

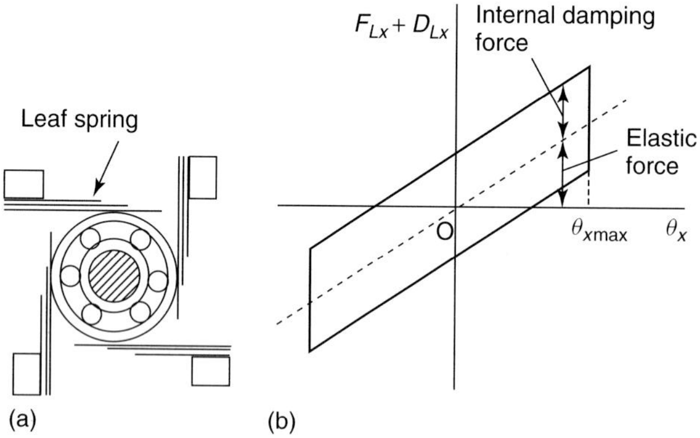

Figure 11.20 Damper utilizing dry friction. (a) Leaf spring and (b) hysteresis.

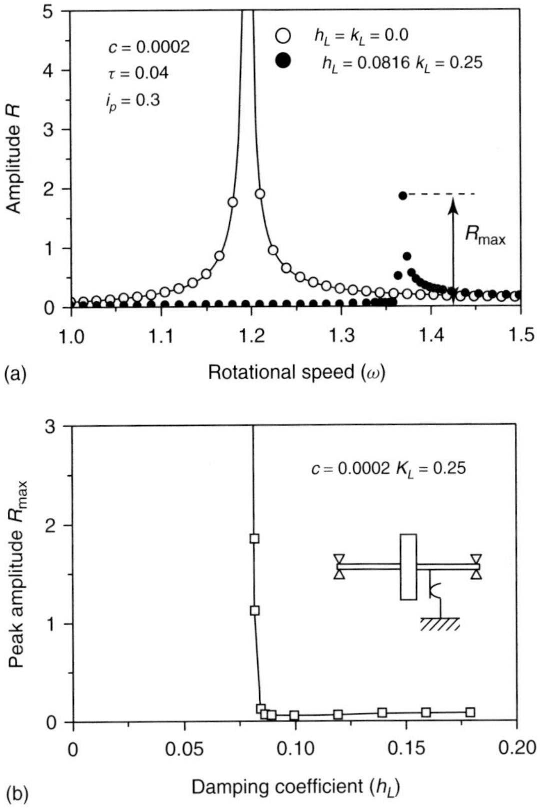

Figure 11.21 Effect of Coulomb friction damping. (a) Response curve and (b) maximum amplitude versus friction damping.

shows the maximum amplitude R_{max} as a function of the magnitude of the Coulomb friction. Coulomb friction cannot reduce the peak value of resonance when the value h_L is less than a critical value ($h_L \approx 0.08$). However, when the friction is larger than this critical value, it diminishes the resonance. Figure 11.21a illustrates a resonance curve that has almost the same value h_L as this critical value.

11.8 Viscous Damper

As mentioned in the "Topic" in Section 3.4, Prause, Meacham, and Voorhees (1967) utilize viscous damping to suppress vibrations of a helicopter power transmission shaft. The sketch of the same type of a viscous damper is shown in Figure 11.22. Figure 11.23 shows the effect of viscous damping on maximum amplitudes in this system. Different from Figure 11.21, the maximum amplitude reduced even for a

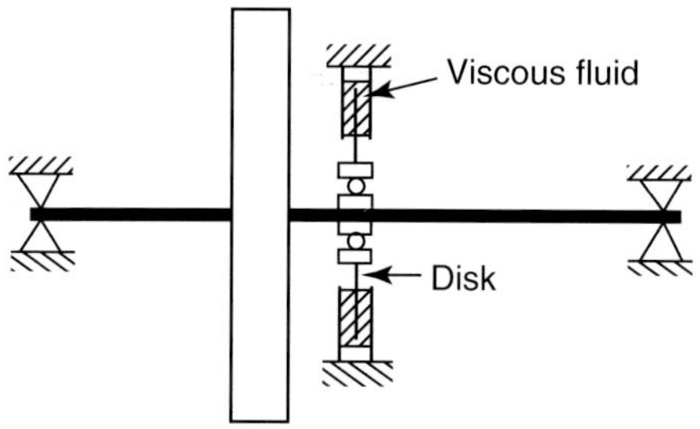

Figure 11.22 Viscous damper.

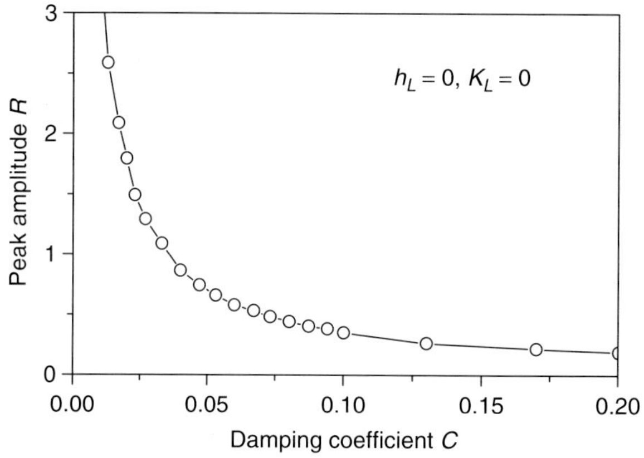

Figure 11.23 Maximum amplitude.

small value of damping coefficients; however, it does not become zero even if its value becomes very large.

11.9 Suppression of Rubbing

As mentioned in Section 9.5, a forward rub, backward rub, and impact type rub occur in a rotor system with, for example, a guide, seal, and casing. It depends on the friction at the contact point and an initial condition what kind of rub occurs as shown in Figure 9.25. On the basis of this result, the rubbing can be eliminated by the following two steps (Ishida, Hossain, and Inoue, 2006; Inoue, Ishida, Fei, and Zahid 2011).

- **Step 1**: Eliminate the possibility of the backward rub by reducing friction. In the case of Figure 9.25, the friction range $\mu \approx 0.04-0.11$ where a contact free motion occurs with high probability is most appropriate. In this range, the forward rub also occur. For example, we can decrease the friction by using a backup bearing in the guide.

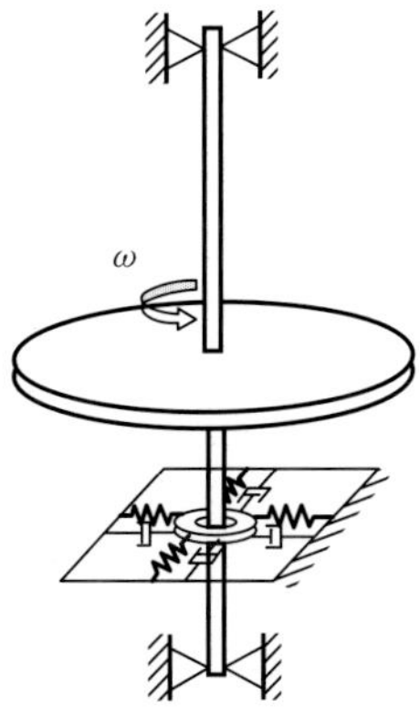

Figure 11.24 Normal contact force.

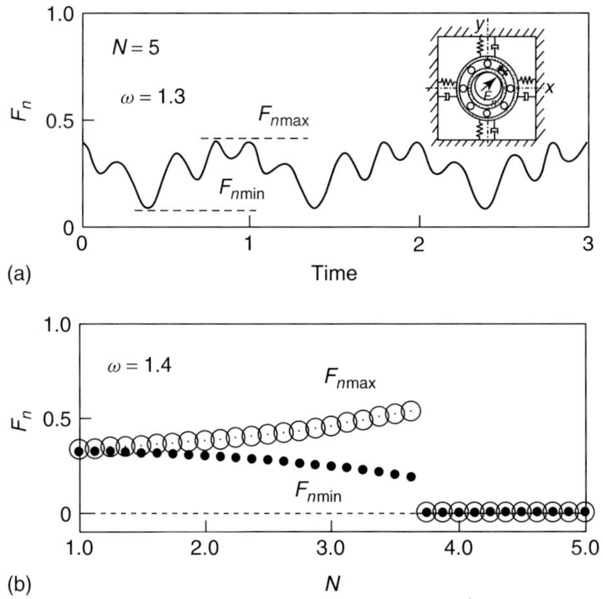

Figure 11.25 Normal component of the contact force. (a) Time history of normal force F_n and (b) maximum and minimum value of F_n (Inoue Ishida, Fei, and Zahid, 2011).

- **Step 2**: Next, we suppress the forward rub by supporting the backup bearing by springs with directional difference in stiffness. Figure 11.24 shows a rotor system with a guide supported elastically. The spring constant k_x in the x-direction and k_y in the y-direction are different. Figure 11.25a shows the magnitude of the normal component of the contact force obtained numerically at $\omega = 1.3$. In this case, the directional difference of stiffness is $N = k_x/k_y = 5$. Owing to this difference, the normal force varies periodically. Figure 11.25b shows the maximum and minimum values of the normal contact force as a function of the ratio N. When the minimum value of the normal contact force becomes zero, the rotor leaves

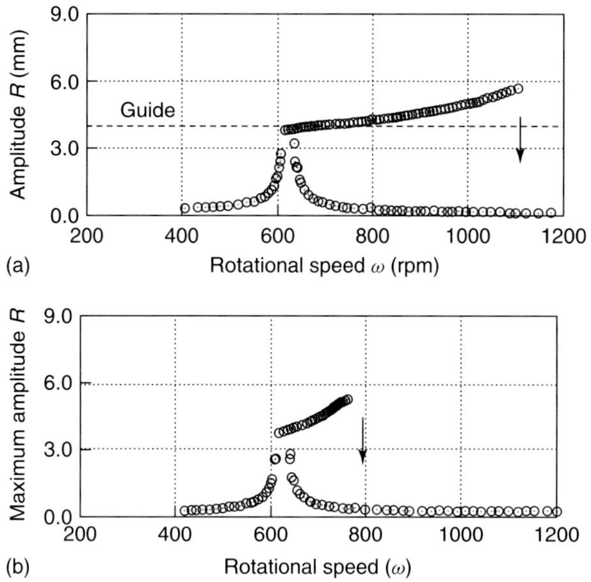

Figure 11.26 Effect of difference in stiffness. (a) Guide with the isotropic stiffness and (b) guide with a directional difference ($N = 5$) (Inoue, Ishida, Fei, and Zahid, 2011).

from the guide. In this figure, when the ratio N is larger than about 3.7, the rotor leaves the guide and converges to the contact-free harmonic motion with small amplitude.

Figure 11.26 shows response curves obtained in experiments. The shaft has contact with the guide when the deflection of the rotor becomes about 4 mm. The peak amplitude is larger than this clearance. When the bearing support is isotropic, the forward rub occurred in a wide range of $\omega = 605 - 1100 rpm$. However, with the anisotropic stiffness of the ratio $N = 5$, the rubbing range decreases to $\omega = 605 - 770 rpm$. If the peak amplitude is smaller than 4 mm, this forward rub disappears.

References

Chao, P.C.-P., Sung, C.-K., and Leu, H.-C. (2005) Effects of rolling friction of the balancing balls on the automatic ball balancer for optical disk drives. *Trans. ASME J. Tribol.*, **127**, 845–856.

Horvath, R., Flowers, G.T., and Fausz, J. (2008) Passive balancing of rotor systems using pendulum balancers. *ASME J. Vib. Acoust.*, **130** (4), 041011-1–041011-11.

Inoue, J., Jinnouchi, Y., and Araki, Y. (1978) Dynamic behavior of automatic ball balancer. Proceedings of the 8th International Conference on Nonlinear Oscillations, pp. 361–366.

Inoue, J., Jinnouchi, Y., Araki, Y., and Nakahara, A. (1979) Automatic ball balancer. *Trans. JSME, Ser. C*, **45** (394), 646–652. (in Japanese).

Inoue, J., Jinnouchi, Y., and Kubo, S. (1983) Dynamic behavior of an automatic ball balancer (application to dynamic unbalance). *Trans. JSME, Ser. C*, **49** (448), 2142–2148. (in Japanese).

Inoue, J., Jinnouchi, Y., and Miyaura, S. (1984) Dynamic behavior of a new type of automatic balancer. Proceedings of the 9th International Conference on Nonlinear Oscillations, Vol. 3, pp. 116–120.

Inoue, T., Ishida, Y., Fei, G., and H.M. Zahid (2011) Suppression of a forward rub in rotating machinery by an asymmetrically supported guide. *Trans. ASME, J. Vib. Acoust.*, **133**, 02115-1–02115-9.

Inoue, T., Ishida, Y., and Niimi, H., (2012) Vibration analysis of the self-excited vibration in the rotor system due to ball balancer. *Trans. ASME, J. Vib. Acoust.*, **134** (2), 021006-1–021006-11.

Ishida, Y., Liu, J. and (2004) Suppression of Self-Excited Vibrations in Rotating Machinery Utilizing Leaf Springs, JSME D&D Conference CD-Rom. (in Japanese).

Ishida, Y. and Liu, J. (2008) Vibration suppression of rotating machinery utilizing discontinuous spring characteristic (stationary and nonstationary vibrations). *Trans. ASME, J. Vib.Acoust.*, **130** (3), 031001-1–031001-7.

Ishida, Y. and Liu, J. (2010) Elimination of unstable ranges of rotors utilizing discontinuous spring characteristics: asymmetrical shaft system, an asymmetrical rotor system and a rotor partially filled with liquid. *Trans. ASME, J. Vib. Acoust.*, **132** (1), 011011-1–011011-7.

Ishida, Y., Matsuura, T., and Zhang, X.L. (2012) Efficiency Improvement of an Automatic Ball Balancer. *Trans. ASME, J. Vib. Acoust.*, **134** (2), 021012-1–021012-10.

Ishida, Y., Hossain, Md.Z., and Inoue, T. (2006) Analysis and suppression of rubbing to contact in rotating machinery. 7th IFToMM Conference on Rotor Dynamics, CD-Rom.

Ito, S. (1962) Sound and vibration of rolling bearings. *Mech. Des.*, **6** (12), 30–36. (in Japanese).

Jinnouchi, Y., Araki, Y., Inoue, J., Ootsuka, Y., and Tan, C. (1993) Automatic balancer (static balancing and transient response of a multi-ball balancer). *Trans. JSME, Ser. C*, **59** (557), 79–84. (in Japanese).

Kirk, R.G. and Gunter, E.J. (1972) The effect of support flexibility and damping on the synchronous response of a single-mass flexible rotor. *Trans. ASME, J. Eng. Ind.*, **94** (1), 221–232.

Lindell, H. (1996) Vibration reduction on hand-held grinders by automatic balancing. *Cent. Eur. J. Public Health*, **4**, 43–45.

Morton, P.G. (1966) On the dynamics of large turbo-generator rotors. *Proc. Inst. Mech. Eng.*, **180** (12, Pt. 1), 295–322.

Ota, H. and Kanbe, Y. (1976) Effects of flexible mounting and damping on synchronous response of a rotor-shaft system. *Trans. ASME, J. Appl. Mech.*, **98** (1), 144–149.

Prause, R.H., Meacham, H.C., and Voorhees, J.E. (1967) The design and evaluation of a supercritical-speed helicopter power-transmission shaft. *Trans. ASME, J. Eng. Ind.*, **89** (4), 719–728.

Tallian, T.E. and Gustafsson, O.G. (1965) Progress in rolling bearing vibration research and control. *ASLE Trans.*, **8** (3), 195–207.

Thearle, E.L. (1932) A new type of dynamic-balancing machine. *Trans. ASME, J. Appl. Mech.*, **54** (12), 131–141.

Van de Wouw, N., Van den Heuvel, M.N., Nijmeijer, H., and Van Rooij, J.A. (2005) Performance of an automatic ball balancer with dry friction. *Int. J. Bifurcat. Chaos*, **15** (1), 65–82.

Wettergren, H.L. (2002) Using guided balls to auto-balance rotors. *Trans. ASME, J. Eng. Gas Turbines Power*, **124** (4), 971–975.

Yang, Q., Ong, E.-H., Sun, J., Guo, G., and Lim, S.-P. (2005) Study on the influence of friction in an automatic ball balancing system. *J. Sound Vib.*, **285**, 73–99.

12
Some Practical Rotor Systems

12.1
General Consideration

So far, vibration phenomena have been explained using physical models and elements of machinery. In this chapter, a steam turbine and a windmill are taken up as examples of rotary machinery and the vibrations occurring in these systems are explained.

12.2
Steam Turbines

12.2.1
Construction of a Steam Turbine

A steam turbine is a mechanical device, which extracts thermal energy from pressurized steam and then converts it into rotary motion. Steam turbines are classified into the *impulse type* and the *reaction type*. Figure 12.1 illustrates their principles. In an impulse type turbine, the pressure energy is converted to the velocity energy in fixed nozzles and the high speed jets from nozzles buffet the bucket-shaped rotor blades. The impulse force caused when steam jets changes direction rotates the shaft. The previously mentioned de Laval's turbine shown in Figure 12.1a, which was invented in 1882, is an example of an impulse turbine. A reaction type turbine utilizes the reaction force produced as the steam accelerates through the blades. In 120 BC, Heron invented "Heron's Aeroliple" as illustrated in Figure 12.1b. This is the earliest example of a reaction turbine. Figure 12.2 shows the blade array in a reaction turbine schematically. A stationary blade (nozzle) in a circumferential array and rotating blades attached to the rotor in a circumferential array are arranged alternately. In this system, stationary blades and rotating blades have a similar shape. The steam pressure energy is converted into velocity energy in the stationary blades, and steam jets spout outwards. The reaction force from the steam expanding in the rotating blades is exerted on the blades, and this force is then converted into rotary motion.

Linear and Nonlinear Rotordynamics: A Modern Treatment with Applications, Second Edition.
Yukio Ishida and Toshio Yamamoto.

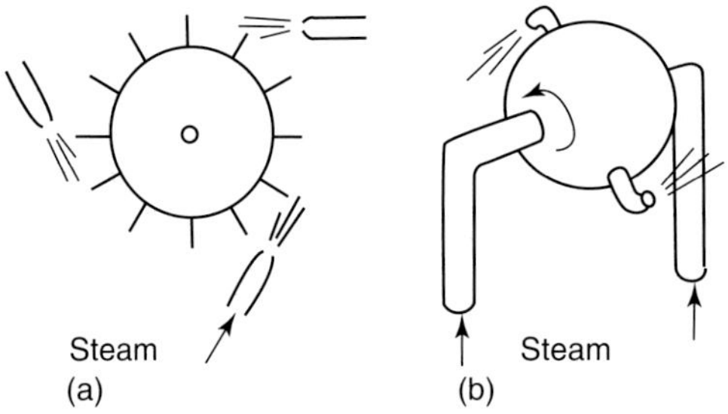

Figure 12.1 Impulse type and reaction type. (a) de Laval turbine and (b) Heron's aerolipile.

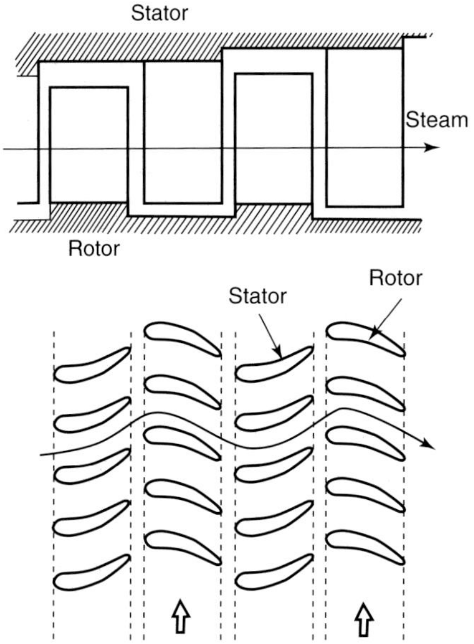

Figure 12.2 Reaction turbine.

The rated rotational speed of steam turbines for power generation is 3000 or 3600 rpm, depending on the frequency of the electric current. In order to best utilize the pressure of the steam, current steam turbine systems combine high-pressure turbines, intermediate-pressure turbines, and low-pressure turbines in the same system. In electric power plants, they are generally connected in a series. However, in the case of turbines for ships, they are sometimes combined parallel to one another because of the constriction of space. Since its pressure drops and its volume expands as the steam flows through each successive stage, the diameter of a turbine increases conically. As an example, Figure 12.3 shows a 75 MW steam turbine (Nakane, Kokubo, Iida, and Takeshita, 1968). This system consists of a high-pressure turbine, a low-pressure turbine, and a generator. Each rotor is supported by two journal bearings. The critical speeds are 1900 rpm (generator), 2000 rpm (high-pressure turbine), and 2700 rpm (low-pressure turbine). The specifications are shown in the figure.

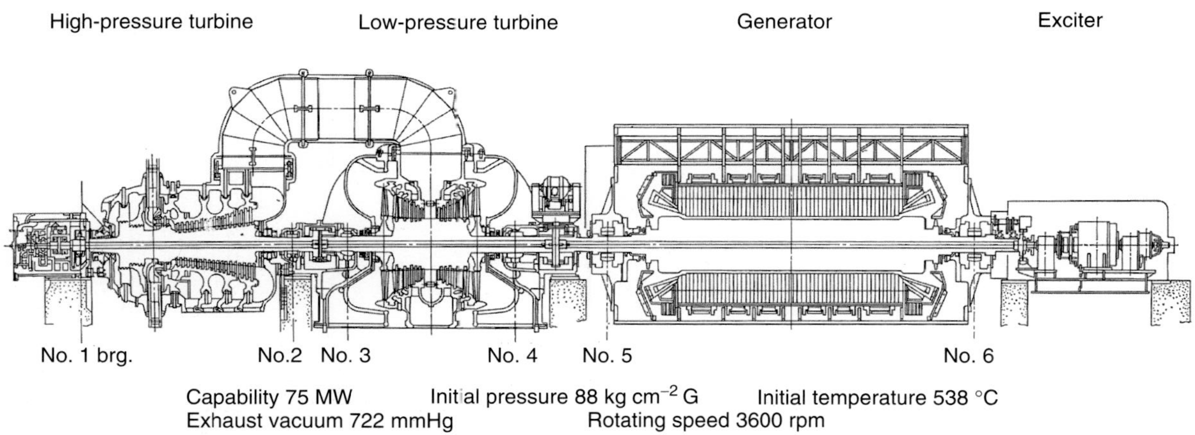

Figure 12.3 Example of a steam turbine. (Source: Nakane, Kokubo, Iida, and Takeshita, 1968, Courtesy of Mitsubishi Heavy Industry.)

12.2.2 Vibration Problems of a Steam Turbine

For the purposes of economy and efficiency, the development of turbines has seen continuously increasing capabilities in terms of power production. And currently, turbines of more than 1000 MW have been developed. In order to meet such economic demands, the scale of rotor, the load to the turbine and number of connected turbines increased. Subsequently, vibration problems have increased. In the case of the turbine in Figure 12.3, many resonances appear below its rated speed of 3600 rpm. In this section, typical vibration problems related to turbines are summarized.

12.2.2.1 Poor Accuracy in the Manufacturing of Couplings

Even if each single rotor is balanced correctly, new vibration problems appear in multirotor systems, which are made by connecting multiple rotors (Shiraki and Kanki, 1974; Gasch and Pfutzner, 1975). Figure 12.4 illustrates couplings and multirotor systems. In Figure 12.4a, inconsistencies in the dimensions of bolts and also the use of keys and keyways create unbalance. In Figure 12.4b, the center of the coupling disk deviates from the shaft centerline. In Figure 12.4c, the coupling disk is mounted obliquely to the shaft centerline. When these couplings are connected rigidly, the shafts are forced to bend and this initial bending produces unbalance, which in turn causes vibration.

Figure 12.5 shows the case that the couplings have been made accurately but the bearings are not aligned precisely. Such a bearing misalignment does not produce unbalance when rotors are coupled rigidly and the centerline of the rotors does not change its position during rotation. This is a similar situation to cases in which the rotor is bent downward by gravity. However, the shear force and bending moments are present inside the shaft. Since the directions of force and movement are fixed in space, they change periodically inside the rotor.

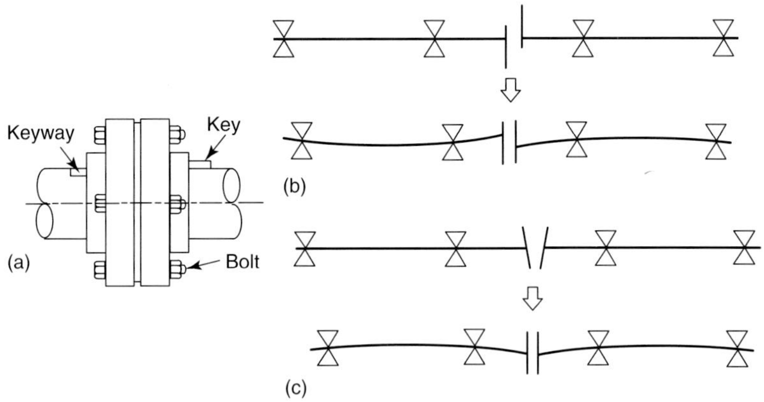

Figure 12.4 Poor accuracy of couplings. (a) Bolt and key, (b) center deviation, and (c) face obliqueness.

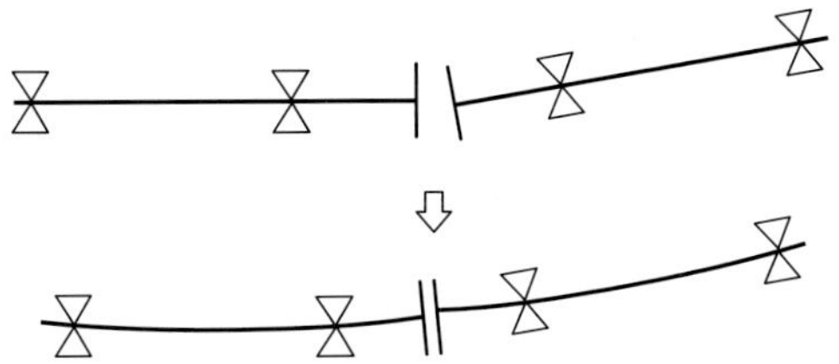

Figure 12.5 Bearing misalignment.

We must keep in mind that the bending due to errors in coupling, as shown in Figure 12.4, is fixed in a rotor and the bending due to bearing misalignment is fixed in space.

Shiraki and Kanki (1974) developed a method to balance such multispan elastic rotors in the field utilizing modal balancing.

12.2.2.2 **Thermal Bow**

When a rotor whirls in the same angular velocity as the rotational speed, the tension side and compression side do not alternate. In such a case, when the rotor makes contact with the stationary part as shown in Figure 12.6a, for example, in a seal, the contacting part on the rotor (called the *hot spot*) expands because of the friction heat and produces unbalance. As the heat accumulates, the amplitude of vibration increases gradually. This unstable vibration was first reported by Newkirk (1926), and the phenomenon is known as the *Newkirk effect*. Similarly, when the rotor is whirling synchronously because of the residual unbalance in a journal bearing as shown in Figure 12.6b, the difference in heat produced in the oil film on the opposing sides creates heat bending. This effect is called the *Morton effect* (Keogh and Morton, 1993, 1994).

12.2.2.3 **Vibrations of Turbine Blades**

In practice, we find that steam pressure fluctuate periodically in turbine systems. Also, the steam flow distribution varies because of the nozzle, stationary bladcs, inlet pipes, and so on. This steam pressure variation has the fundamental frequency equal to the rotational speed and its higher frequency components. These components excite vibration of the blades and disks.

Figure 12.7 shows vibration modes of the longest blade of 30 in in a turbine system of 375 mW in the Chita Electric Power Station in Japan. These following results were obtained by a static shaking test (Kajiyama, 1969). The dotted lines

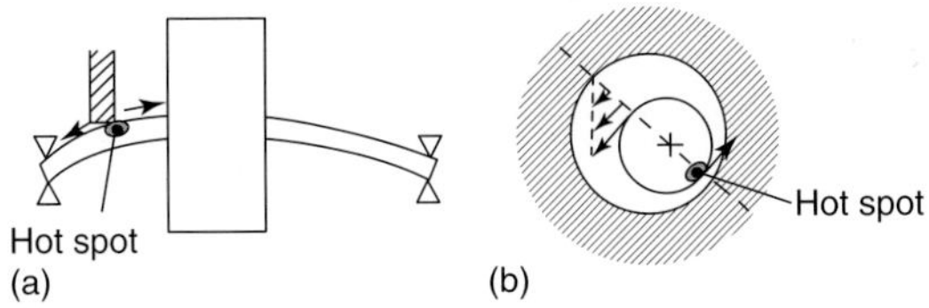

Figure 12.6 Thermal bow. (a) The Newkirk effect and (b) the Morton effect.

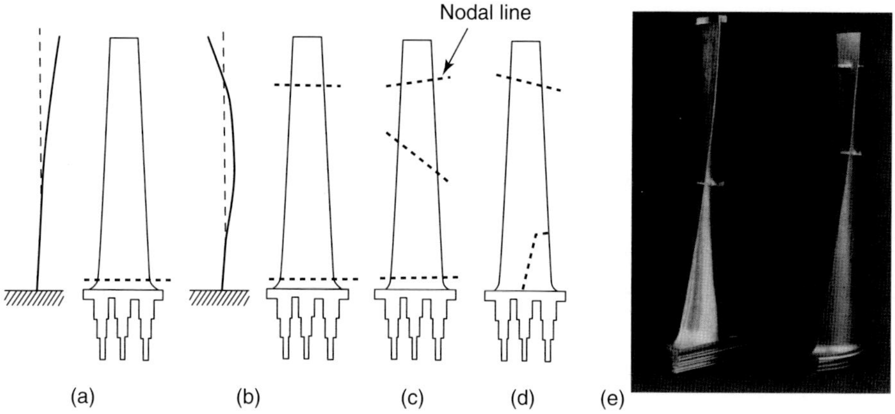

Figure 12.7 (a–d) Modes of blades and (e) Pictures of blades with and without a shroud. (Courtesy of Mitsubishi Heavy Industries, Ltd. Nagasaki Shipyard & Machinery Works.)

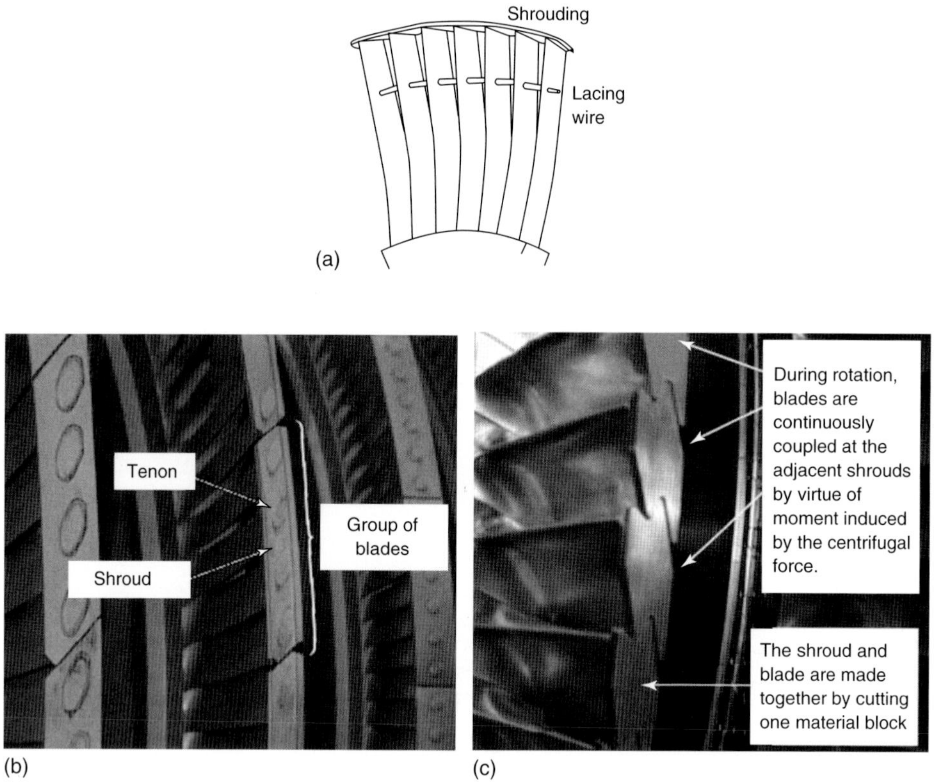

Figure 12.8 Connected blades. (a) Shroud and lacing wire, (b) grouping of blades by shroud, and (c) integral shroud blades. (Courtesy of Mitsubishi Heavy Industries, Ltd. Nagasaki Shipyard & Machinery Works.)

represent nodal lines, where the vibration amplitude is zero. Figure 12.7a is the first bending mode of 60 Hz, Figure 12.7b is the second bending mode of 128 Hz, Figure 12.7c is the bending–torsion coupled mode of 224 Hz, and Figure 12.7d is the bending–torsion coupled mode of 250 Hz. When the nodal line is horizontal, consequently, the blade vibrates in a peripheral direction. When the nodal line is oblique, consequently, the blade vibrates in both a peripheral direction and an axial direction. The natural frequencies of these blades increase as the rotor rotates because of the centrifugal force.

The vibrations of turbine blades are classified into those in the peripheral direction and those in an axial direction. The latter is coupled with the axial vibration of the disk. The long blades for low pressure are arranged in isolation or connected as depicted in Figure 12.8. The tip shroud is used to dampen vibration, and the lacing wire is used to change the vibration modes and their resonance frequency.

In an integral shroud in which all the blades are connected, the whole blades vibrate together as shown in Figure 12.9. Figure 12.9a depicts the first mode in which all blades vibrate in the same phase, and Figure 12.9b illustrates the case that the phase changes once in a peripheral direction. The vibration of disk wheels carrying blades is also of importance. Figure 12.10 illustrates vibration modes that have four or six peripheral nodes, which are obtained by shaking the wheel disk. The central shaded part is thick, and it does not deform during vibration. Such modes are discussed in detail in the paper by Campbell (1924). Generally, disks have sufficient rigidity and do not deform appreciably. However, since the last two stages of turbine system have longer blades, their natural frequency may decrease and enter into the operating range.

Up to this point, the vibration of blades and disks in a turbine system has been discussed independent of one another. However, it should be noted that in practice, they vibrate together in a form of coupled vibration, which is not discussed here.

The resonance speed of these various kinds of modes are found by drawing the Campbell diagram explained in the note "Campbell Diagram" in Section 2.3.2. In the design of a turbine, the rated speed should avoid these resonance speeds.

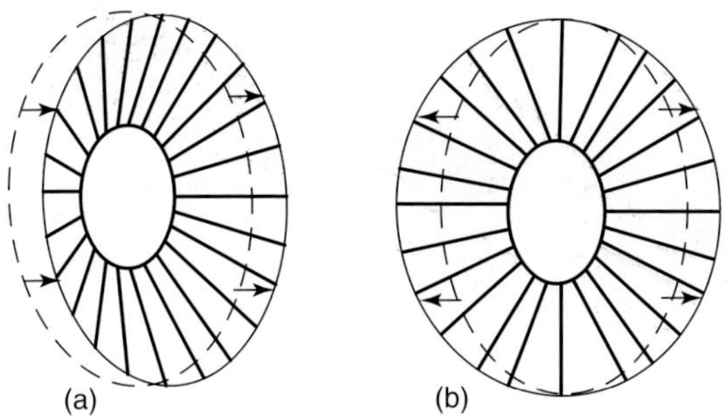

Figure 12.9 Integral shroud blades. (a) First mode and (b) second mode.

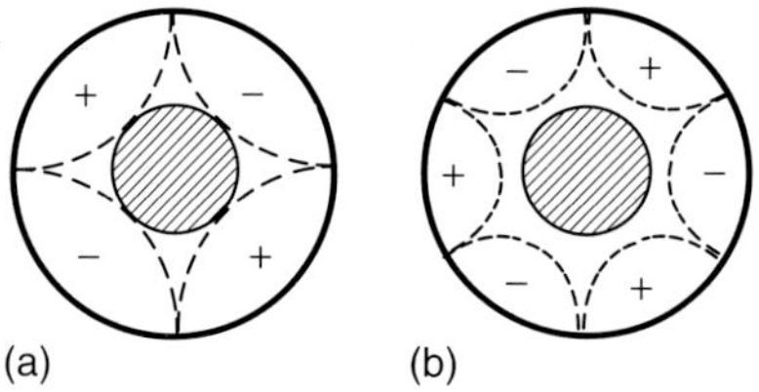

Figure 12.10 Nodes of a wheel. (a) Four nodes and (b) six nodes.

12.2.2.4 **Oil Whip and Oil Whirl**

Refer to Section 10.2.

12.2.2.5 **Labylinth Seal**

Refer to Section 10.3.

12.2.2.6 **Steam Whirl**

Refer to Section 10.4.

12.3 Wind Turbines

12.3.1 Structure of a Wind Turbine

The construction of wind turbines is increasing rapidly in order to reduce carbon dioxide emissions and as a measure to alleviate potential energy shortages in the future. For the purpose of cost reduction, the weight of wind turbines' blades and towers are gradually being reduced. As a result, the turbines' rigidity is decreasing

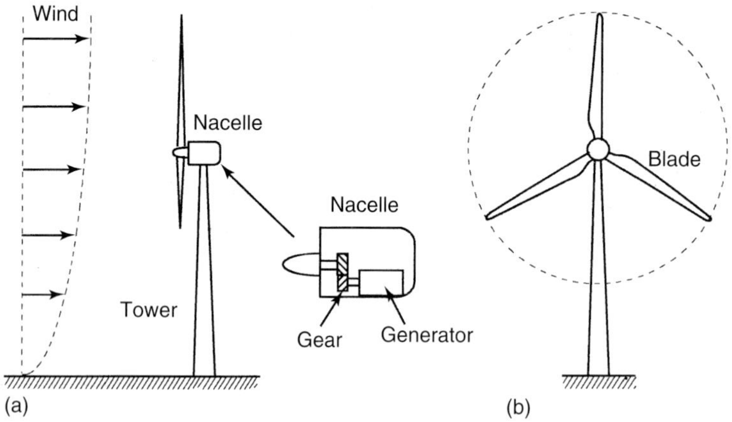

Figure 12.11 Three-bladed wind turbines. (a) Side view and (b) front view.

and consequently increased vibration can be observed. Figure 12.11 illustrates the construction of a three-bladed wind turbine, which is typical among large-scale wind turbines. A housing that contains a rotating shaft, speed-up gear, and generator is mounted on the top of the tower. Blades are mounted on the forefront of the shaft. In the case of large wind turbines, the rotational speed of the main shaft is 10–30 rpm and the rotational speed of the generator shaft is increased by the speed-up gears. For example, in the case of a turbine in which the rotational speed of the generator is 1800 rpm, the speed-up ratio is approximately equal to 100. As shown in Figure 12.11, wind turbines in which the blades face windward are known as *upwind turbines*, conversely, those which have the blades facing in a leeward direction are known as *downwind turbines*.

Next, we discuss power extracted from wind, particularly in terms of the work that is done by wind per unit of time. Let us now consider the stream-tube as it relates to wind-based power. We assume that the stream-tube has a cylindrical shape made by the stream lines. Let the area of the cross section of a stream-tube be A, the air density be ρ, and the air velocity be V. Since the mass of the air that passes the area A per unit of time is represented by the expression ρAV, the power P of the wind is given by the following expression:

$$P = \frac{1}{2}(\rho AV)\,V^2 = \frac{1}{2}\rho AV^3 \tag{12.1}$$

From this expression, we can see that the wind power is proportional to area A and the third power of the wind velocity. Since the wind has velocity after passing through the turbine, the output of the wind turbine is given by the difference between the energy input and the energy output. However, since the wind turbine does not have casing, the practical wind velocity has a radial component and also has various kinds of mechanical losses; its efficiency is approximately equal to 40%. As the velocity of the wind generally increases at greater heights, in recent times, the dimensions of wind turbines have been increased to produce larger amounts of power. At present, wind turbines have been constructed whose blade diameters are over 100 m, the height of the nacelle exceeds 100 m, and energy output exceeds several megawatts. Since the peripheral velocity of a blade tip must be lower than the sound velocity to avoid a shock wave, the rotational speed of the blades is kept quite slow. For example, when a blade with the length $R = 40$ m rotates slowly at 20 rpm, its peripheral speed is $R\omega = 83\ \mathrm{m\ s^{-1}}$, which is quite high.

The speed and direction of the wind generally change at random. The intensity of wind is expressed by the average wind velocity V, which is the average value for 10 min, and the maximum wind velocity, which is the highest speed at the moment. The speed of the wind decreases closer to the ground because of the shearing effect. Therefore, the average wind velocity changes with height as shown in Figure 12.11.

Generally, in the field of meteorology, the wind velocity expresses the average wind velocity at approximately 10 m above ground. However, in the field of wind power generation, the velocity distribution in the vertical direction is also important. Wind velocity also changes throughout a year. Curve ① in Figure 12.12 shows frequency of appearance of the mean wind speed (e.g., at about 10 m above ground) over a year. Its mathematical expression is given in existing literature (Burton, Sharpe, Jenkins and Bossanyi 2001). This curve called the *Rayleigh distribution* is, to a large extent, in agreement with the practical distribution. The vertical axis of this graph represents probability density. Let the probability density function be $p(V)$. Then, for example, the probability that the wind speed is between 5 and 7 $\mathrm{m\,s^{-1}}$ is given by the integration $\int_5^7 p(V)\mathrm{d}V$.

Curve ② represents the energy output of the wind turbine. In this case, the abscissa represents the wind speed at the point of the nacelle. The wind turbine starts to generate power at about 3 $\mathrm{m\,s^{-1}}$ (called the *cut-in* wind speed). As the wind speed increases up to the rated speed of about 12 $\mathrm{m\,s^{-1}}$, the power output increases while the pitch angle explained below is kept constant. This rated speed is determined so as to generate the maximum power in a year period considering the frequency of varying wind speeds over the course of a year. Above the rated speed, the power output is kept constant by controlling the angle of blades. At about 25 $\mathrm{m\,s^{-1}}$ (called the *cut-out* wind speed), the direction of the wing chord length, which is the length from the leading edge to trailing edge of the blade's cross section, is angled to the direction of the wind. In other words, the pitch angle is kept at 90° to reduce the load taken by the turbine. Then, the generator is shut down and the wind turbine is stopped by applying a mechanical brake.

12.3.2
Campbell Diagram of a Wind Turbine with Two Teetered Blades

As an example of a practical wind turbine, the research by Yamane, Matsumiya, Kawamura, Mizutani, Nii, and Gotanda (1990) at Advanced Industrial Science and Technology (AIST) on wind turbines with two teetered blades is introduced in this section. In this type, as shown in Figure 12.13b, in order to reduce the load moment at the root of blades, two blades are connected and can turn freely around the hinge axis, which is perpendicular to the main shaft. Therefore, blades move forward and backward in a similar manner to a seesaw. Yamane, Matsumiya, Kawamura, Mizutani, Nii, and Gotanda obtained the Campbell diagram shown in

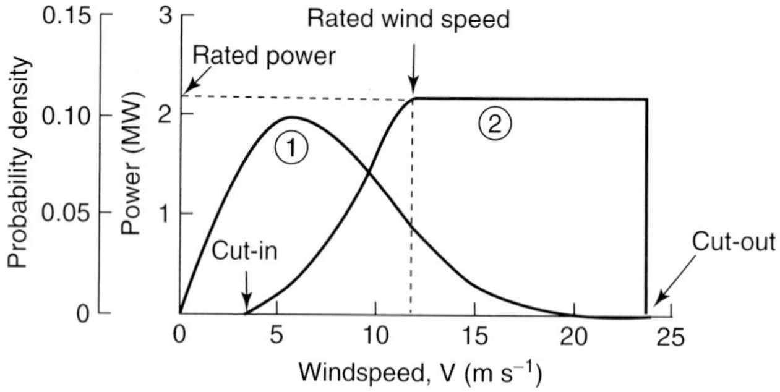

Figure 12.12 Power output and the frequency of appearance of wind speed.

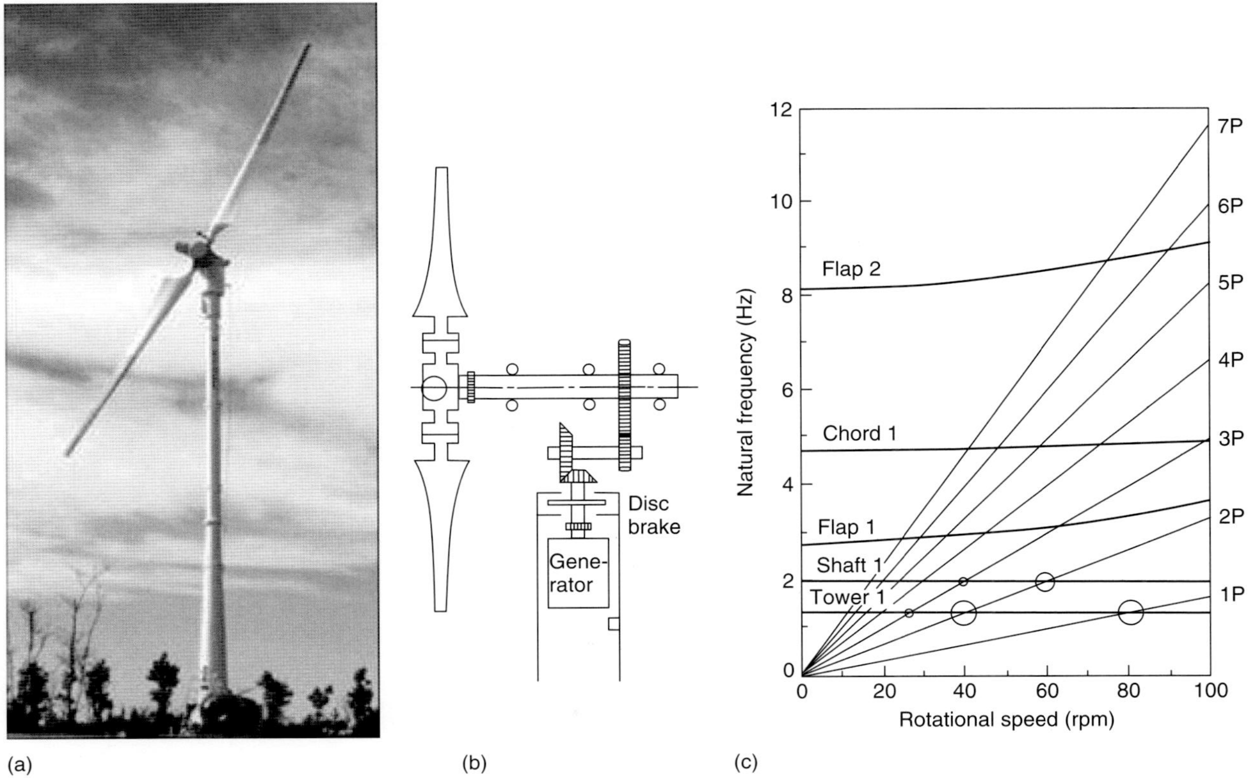

Figure 12.13 Two-bladed wind turbine. (a) Outlook, (b) construction, and (c) Campbell diagram. Rotor diameter is 15 m, hub height 14.4 m, rated speed 81.4 rpm, rated generator speed 1800 rpm, and generator capacity 22 kW. (Courtesy of AIST.)

Figure 12.13b from natural frequency tests and steady-state oscillation tests. The line notated "Tower 1" represents the natural frequency of the first bending mode of the tower. It is often observed that the natural frequencies in the downwind direction and the lateral direction are different. However, such a difference is not clearly observable in this case. The line "Shaft 1" represents the natural frequency of the first torsional mode of the transmission shaft. With regard to the blades, since their cross section is flat, the natural frequencies of flapwise bending are low and those of chordwise (or edgewise) bending are comparatively high. The line "Flap 1" represents the natural frequency of the first mode of blades' flapwise bending, "Chord 1" represents that of the first mode of blades' chordwise (or edgewise) bending, and "Flap 2" is that of the second mode of blade's flapwise bending. The natural frequencies in flapwise bending are easily affected by centrifugal force, and we see that their value increases as the rotational speed increases. Therefore, it is not depicted in Figure 12.13c. The coupled motion of tower–blades also exists but at frequencies that are too high to appear in this range.

12.3.3 Excitation Forces in Wind Turbines

When long blades subject to wind pressure in the vertical plane are rotating, we can observe that various kinds of forces are at work. Figure 12.14 illustrates the external forces involved in the operation of a wind turbine, particularly taking into account the effect of a wind at its various heights. Since wind velocity is always fluctuating, these forces work at random. First, as shown in Figure 12.14a, wind force ① works on the blades. The decomposition of this force is illustrated as in the box. Since the blade has velocity u in the plane perpendicular to the shaft, the wind with relative velocity $\overline{W}$ works to the blade. As a result, drag D works in the same direction as W and lift L works in the direction perpendicular to W. These aerodynamic forces can further be decomposed to the flapwise component ② and chordwise component ③. The other forces mentioned below can also be decomposed into the same kinds of components. When the blades rotate in a wind whose velocity changes according to height because of the shear effect, the magnitude of these forces changes periodically. Therefore, the forces working on the blades and tower result in periodical excitation. In the downwind type shown in Figure 12.14b, since the wind speed becomes small in the tower shadow, an impulsive force ④ works on the blades and tower. This series of pulsation has higher frequency components with frequencies of integral multiples of the rotational speed. These higher frequency components are generally large in the case of a wind turbine. Figure 12.14c illustrates the gravitational force mg working on the blades. Its component ⑤, normal to the blade, is decomposed to a flapwise component and a chordwise component, both of them work to bend the blade. Suppose that the blades rotate in the counterclockwise direction. Then, the gravitational force works in the same direction as the rotation in the left-hand side and works in the opposite direction in the right-hand side. As a result, periodic bending moments can be observed for the blade. The component ⑥ in the longitudinal direction of the blade works as

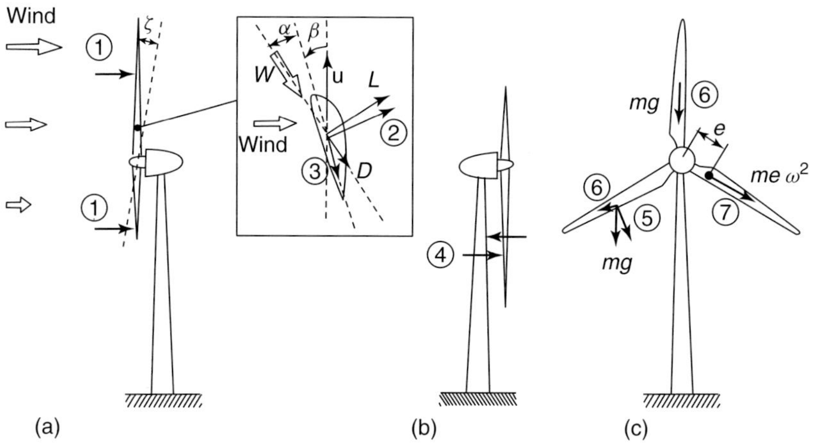

Figure 12.14 Various forces working on turbine blades. (a) Upwind type and (b) downwind type. (c) front view.

tension or compression. When the blade is in the upper position, this force works to reduce the stiffness of the blade, similar to the inverse pendulum. When it is in the lower position, this force works to increase the stiffness. As a result, the blade stiffness varies periodically and the system is classified as a parametrically excited system. The force ⑦ in Figure 12.14c is an unbalanced force. Since the center of gravity of the total blades deviates from the center of rotation, an excitation due to the centrifugal force exists. In a turbine with multiple blades, forces of higher frequencies are at work. For example, the frequency of the force becomes twice and thrice the rotation in a two-bladed and three-bladed wind turbine, respectively. In wind turbine engineering field, the frequency component with n times the rotational speed is customarily expressed by the notation nP. Using this notation, it is expressed that forces P, $2P$, $3P$, $4P$, and so on work on the wind turbine because of the reasons mentioned above. In the Campbell diagram shown in Figure 12.13c, the resonance rotational speeds are obtained from the cross point of the curves representing natural frequencies and the radial straight lines corresponding to P, $2P$, $3P$, $4P$, and so on. We must keep in mind that additionally, the random force is also at work because the direction and intensity of the wind changes at random.

12.3.4
Example: Steady-State Oscillations of a Teetered Two-Bladed Wind Turbine

Yamane, Matsumiya, Kawamura, Mizutani, Nii, and Gotanda (1990) measured steady-state oscillations of a teetered two-bladed wind turbine for various conditions of electricity generation in detail. The obtained spectra are shown in Figure 12.15. The numbers 1, 2, 3, and so on represent the frequencies P, $2P$, $3P$, and so on, respectively. In the following, each of these oscillations is explained.

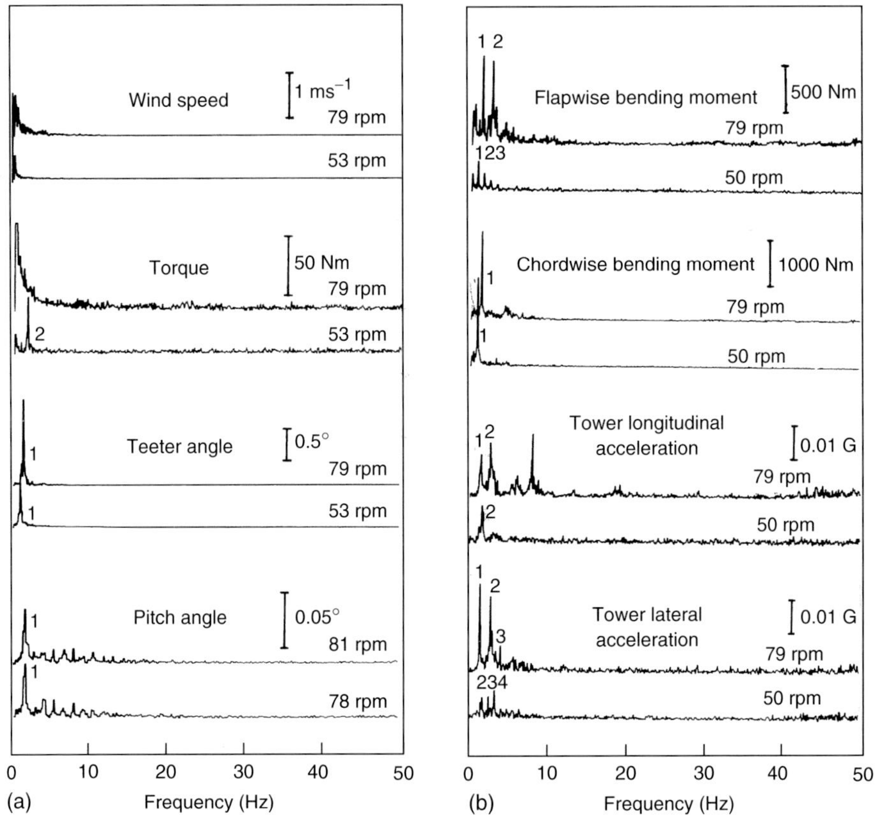

Figure 12.15 (a,b) Spectra of steady-state oscillations. (Courtesy of AIST.)

12.3.4.1 Wind Velocity

The top diagram in Figure 12.15a shows the spectra of wind velocity. There are invariably disturbances in wind. However, we can confirm from these spectra that their frequencies are generally low and do not contain the frequency components that cause vibrations $1P$, $2P$, and so on, which are discussed in the following.

12.3.4.2 Vibration of the Tower

The lower two diagrams in Figure 12.15b illustrate to-and-fro and lateral movement of the tower. Since the damping of the tower is generally very small, its vibration is the most serious problem. Vibrations of $2P$ occur at the low rotational speed (50 rpm $\approx$ 0.83 Hz) and vibrations of $1P$ and $2P$ occur in the vicinity of the rated speed (79 rpm $\approx$ 1.32 Hz) because of the various reasons mentioned above. Since this is a two-bladed wind turbine, the vibration of $2P$ appeared most predominantly. The measured natural frequency of the first bending mode of the tower was 1.28 Hz ($\approx$76.8 rpm). It is near $2P$ in the case of 50 rpm and near P in the case of 79 rpm (refer Figure 12.13b). The symbol ◦ in Figure 12.13b represents amplitude in the same mode relatively (the peak at 8 Hz in the to-and-fro movement is caused by the side wheel mounted to control movement of the shaft in the yaw direction).

12.3.4.3 Flapwise Bending Vibration of the Blade

The top diagram of Figure 12.15b shows flapwise bending vibrations in the same direction as the rotating shaft. Various kinds of vibration such as $1P$, $2P$, $3P$, and so on occurred and their magnitudes depend greatly on the rotational speed.

12.3.4.4 Chordwise Bending Vibration of the Blade

The second diagram of Figure 12.15b shows chordwise bending vibrations. Different from the case of the flapwise bending vibration, only $1P$ occurred predominantly. This may be considered to be due to the effect of periodic bending caused by gravity.

12.3.4.5 Torque Variation of the Low-Speed Shaft

The variation of wind force ① causes the periodical variation of components L and D in the plane of blade rotation. As a result, the torque working on the shaft varies. The second diagram of Figure 12.15a shows the torque variation of the low-speed shaft. The drive train is a torsional vibration system composed of a hub and blades at one end, gears for the speed change in the middle, and a generator at the other end. The natural frequency of this system is approximately 2 Hz. Vibration of $2P$ occurred predominantly at 53 rpm; however, it disappeared in the higher rotational speed range. This is because, as seen from Figure 12.13c, there is a resonance point in the vicinity of 60 rpm (= 1 Hz).

12.3.4.6 Variation of the Teeter Angle

As mentioned before, in this two-bladed wind turbine, blades are mounted on the shaft via a hinge as shown in Figure 12.13b and they can incline as shown in Figure 12.14a. This angle ζ is called a *teeter angle*. In the third diagram in Figure 12.15b, the vibration $1P$ occurred in the teeter angle. This indicates that the plane of blade rotation inclines to the shaft.

12.3.4.7 Variation of the Pitch Angle

In Figure 12.14a, the angle β between the direction of blades' rotation and the blade chord line is called a *pitch angle*. The angle α between the resultant velocity W and the blade chord line is called an *attack angle*. The lowest diagram of Figure 12.15b shows a variation of this angle. The vibration P occurred because the gravitational force working on the unbalanced weight changes the pitch angle slightly in this wind turbine.

12.3.4.8 Gear

The shaft speed is increased using a two-stage gear box in the nacelle. The gear ratio in the first stage is 80 : 17 and that in the second stage is 61 : 13. Therefore, when the rotor is operated at the rated rotational speed, the gear mesh frequency or tooth mesh frequency L, that is, equal to the number of teeth on the gear multiplied

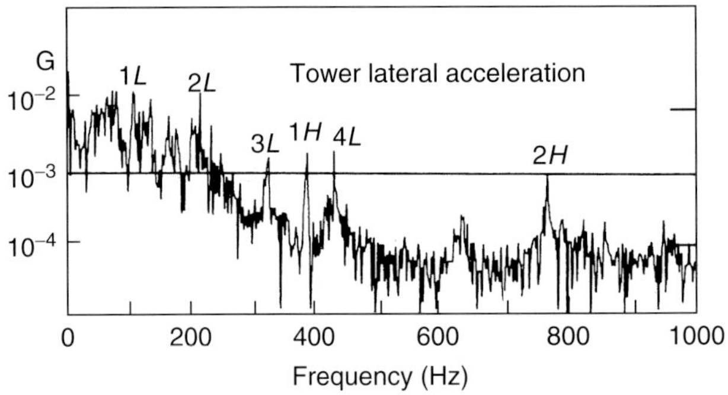

Figure 12.16 Frequency spectra of a vibration of a tower. (Courtesy of AIST.)

by the rpm of the gear, in the first stage is 109 Hz and the gear mesh frequency H in the second stage is 390 Hz. Figure 12.16 illustrates spectra of acceleration of the tower operated with no load. Vibrations with the gear mesh frequency and its higher components occurred.

12.3.5 Balancing of a Rotor

There are following three kind of unbalances in wind turbines: (i) mass unbalance, (ii) aerodynamic unbalance, and (iii) unbalance in drive train. Lösl and Becker (2009) identified their causes as follows. The causes of category (i) includes uneven rotor blade masses, uneven mass distribution in the rotor blade, flange and pitch errors in the hub, hub unbalances, eccentricities of the entire rotor, bent shaft, water penetration, and icing. The causes of category (ii) include blade angle errors, uneven rotor blade profile forms, rotor blade damage and effects of repairs on rotor blades, pitch/cone error, indirect incident flow, and external location-related excitations (gusts, lee turbulence from obstructions). The causes of category (iii) include mass and moment unbalance in the generator, coupling, or brake disk. Another study of interest in this area is the work of Gasch and Twele (2002) who studied the resonance of a tower using 1 DOF (degree of freedom) model where the masses of a nacelle and a blade are concentrated at the top position of the tower.

In some wind turbines, the balancing is not taken into account in the production process. In such a case, generally, blades of similar weight are selected and used in the construction of the turbines. However, the field balancing is strongly recommended in order to reduce vibration. Lösl and Becker (2009) in Prüftechnik Corporation recommended that rotors be balanced to the level of permissible residual imbalance, ideally to the "G16" standard (Table 4.1). Figure 12.17 illustrates the field balancing carried out by Hillmann in "cp.max Rotortechnik Gmbh & Co. KG" Corporation (Hillmann, Bauer, and Rische, 2004). The specifications of his three-bladed wind turbine are nominal power 1500 kW, rated speed 20.4 rpm, rotor diameter 77 m, and hub height 85 m. Hillman's work is an example of successful

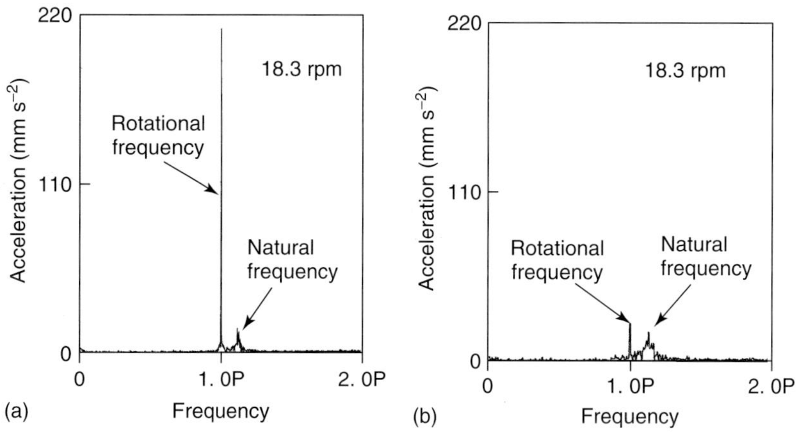

Figure 12.17 An example of field balancing. (a) Before balancing and (b) after balancing. (Source: Courtesy of Mr. Hillmann.)

field balancing suggesting that the process has significant merit in terms of the operation of wind turbines.

However, we must keep in mind that wind turbine differ from ordinary machines in that they must often operate in various unfavorable climate conditions. Especially, in very cold climates, icing may occur and subsequently produce huge unbalance.

12.3.6 Vibration Analysis of a Blade Rotating in a Vertical Plane

In reality, the vibration of both the tower and blades of a wind turbine affects one another, and their vibration modes are very complicated. For the sake of simplicity, Ishida, Inoue, and Nakamura (2009) carried out linear analysis of one rigid rotor, while Inoue, Ishida and Kiyohara (2012) have undertaken nonlinear analysis of one flexible blade. In this section, the vibration of a flexible rotor is explained in terms of linear analysis. By using this method of analysis, we can explain the first mode of flapwise vibration of a single blade in the wind turbine shown in Figure 12.14 and that its natural frequency corresponds to the curve "Flap 1" in Figure 12.13c.

The physical model and coordinate systems are illustrated in Figure 12.18. One blade is attached to a rigid rotating shaft and rotates in the vertical plane with the rotational speed ω. The static coordinate system O-xyz whose x, y, and z axes are taken to be in the horizontal, vertical, and shaft directions, respectively. A rotating coordinate system O-ξz whose ξ axis is taken to be in the direction of the blade is also considered.

12.3.6.1 Derivation of Equations of Motion

It is assumed that the first mode of this rotating blade can be approximated by the first mode of a cantilever. The shape of the first mode of this beam with length l is given by the following modal function in which λ is the eigen value

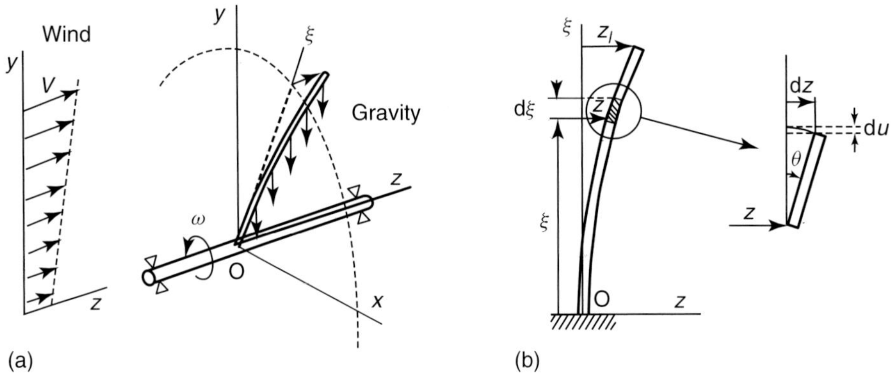

Figure 12.18 Physical model of a turbine blade. (a) Model of a blade and (b) rotating coordinates.

corresponding to this first mode (refer to note "Model Function of a cantilever" in section 12.3.6.4):

$$h(\xi) = (\sin\lambda + \sinh\lambda)\left(\cos\lambda\frac{\xi}{l} - \cosh\lambda\frac{\xi}{l}\right) - (\cos\lambda + \cosh\lambda)\left(\sin\lambda\frac{\xi}{l} - \sinh\lambda\frac{\xi}{l}\right) \tag{12.2}$$

Let the deflection of the tip of the blade in the z-direction be $z_l(t)$. Then, the deflection $z(\xi, t)$ at the distance ξ from the origin O is given as follows by using Eq. (12.2):

$$z(\xi, t) = \frac{h(\xi)}{h(l)} z_l(t) \tag{12.3}$$

When a small element of length dξ inclines, its tip moves by dz in the z-direction and by du in the negative ξ-direction. When the inclination angle θ is small, this length du is approximately expressed as follows:

$$du(\xi) = d\xi(1 - \cos\theta) \approx d\xi\left[1 - \left(1 - \frac{1}{2}\theta^2\right)\right] = \frac{1}{2}\left(\frac{dz(\xi, t)}{d\xi}\right)^2 d\xi \tag{12.4}$$

By integrating this from 0 to ξ, the deflection toward the origin of the point ξ on the blade is expressed as follows:

$$u(\xi, t) = \int_0^{\xi} \frac{1}{2}\left(\frac{\partial z(\xi, t)}{\partial \xi}\right)^2 d\xi = \alpha(\xi) z_l(t)^2 \tag{12.5}$$

where

$$\alpha(\xi) = \frac{1}{2h(l)^2}\int_0^{\xi}\left(\frac{dh(\xi)}{d\xi}\right)^2 d\xi \tag{12.6}$$

Let us suppose that the blade coincides with the ξ axis at $t = 0$. Then, the coordinates of the point ξ on the blade are expressed as follows:

$$\left.\begin{aligned} x_\xi &= (\xi - u(\xi))\cos\omega t = \left(\xi - \alpha(\xi) z_l(t)^2\right)\cos\omega t \\ y_\xi &= (\xi - u(\xi))\sin\omega t = \left(\xi - \alpha(\xi) z_l(t)^2\right)\sin\omega t \\ z_\xi &= \frac{h(\xi)}{h(l)} z_l(t) \end{aligned}\right\} \tag{12.7}$$

Next, we derive the equations of motion using Lagrange's equation. First, we derive expressions for the kinetic energy. Let ρ be the density of the blade and A be the area of its cross section. Since z_l/l and $\dot{z}_l/l$ are small quantities, we do not consider their higher orders. Then, we obtain

$$\begin{aligned} T &= \int_0^l \frac{1}{2}\rho A\left(\dot{x}_\xi^2 + \dot{y}_\xi^2 + \dot{z}_\xi^2\right) d\xi \\ &= \frac{1}{2}\rho A\omega^2 l^3\left[\frac{1}{3} - \left(\frac{2\alpha_1}{l^2}\right)\left(\frac{z_l(t)}{l}\right)^2 + \left(\frac{\alpha_2}{\omega^2 l}\right)\left(\frac{\dot{z}_l(t)}{l}\right)^2\right] \end{aligned} \tag{12.8}$$

where

$$\alpha_1 = \int_0^l \xi\alpha(\xi)d\xi, \quad \alpha_2 = \frac{1}{h(l)^2}\int_0^l h(\xi)^2 d\xi \tag{12.9}$$

Next, we derive the expression of the potential energy which is produced by the bending stiffness and gravity. Let E be Young's modulus, I be the second moment of area, and g be the gravitational constant. Then, the potential energy is expressed as follows:

$$\begin{aligned} U &= \int_0^l \frac{1}{2} EI\left(\frac{\partial^2 z(\xi,t)}{\partial\xi^2}\right)^2 d\xi + \int_0^l \rho A g y_\xi d\xi \\ &= \frac{1}{2} EI\beta_1 z_l(t)^2 + \rho A g\left(\frac{l^2}{2} - \beta_2 z_l(t)^2\right)\sin\omega t \end{aligned} \tag{12.10}$$

where

$$\beta_1 = \frac{1}{h(l)^2}\int_0^l \left(\frac{d^2 h(\xi)}{d\xi^2}\right)^2 d\xi, \quad \beta_2 = \int_0^l \alpha(\xi)d\xi \tag{12.11}$$

The dissipation energy is created by damping forces. We represent it by using the coordinate z of the blade's tip as follows:

$$D = \frac{1}{2} c_l \dot{z}_l^2 \tag{12.12}$$

A periodic wind force works on the blade that rotates in the wind whose intensity varies depending on its height. It is assumed that the virtual work due to the wind force is given as follows:

$$\delta W = Q_l \delta z_l = \left(Q_0 + \Delta Q\sin\omega t\right)\delta z_l \tag{12.13}$$

We obtain the following equation of motion by using Lagrange's equation:

$$\rho A\alpha_2 \ddot{z}_l + c_l \dot{z}_l + \left(2\rho A\omega^2\alpha_1 + EI\beta_1 - 2\rho A g\beta_2\sin\omega t\right) z_l = Q_0 + \Delta Q\sin\omega t \tag{12.14}$$

This expression has the following characteristics: (i) the coefficient is the function of time and its constant term increases as the angular velocity ω increases; (ii) because of the presence of gravity g, the parametric excitation term appears; and (iii) the excitation force is composed of a constant force and a periodic force.

For the sake of convenience, we introduce the following dimensionless quantities:

$$\left.\begin{array}{l} \bar{t} = p_0 t, \quad \bar{z}_l = \dfrac{z_l}{h}, \quad \overline{\alpha}_1 = \dfrac{\alpha_1}{\alpha_2}, \quad \overline{c}_l = \dfrac{c_l}{\sqrt{EI\beta_1 \rho A \alpha_2}}, \\ \overline{\omega} = \dfrac{\omega}{p_0}, \quad \overline{g} = \dfrac{2\rho A g \beta_2}{EI\beta_1}, \quad \overline{Q_c} = \dfrac{Q_c}{EI\beta_1 h}, \quad \overline{\Delta Q} = \dfrac{\Delta Q}{EI\beta_1 h} \end{array}\right\} \tag{12.15}$$

where h is the thickness of the blade and $p_0 = \sqrt{EI\beta_1/(\rho A \alpha_2)}$ is the natural frequency of the blade in nonrotating conditions. Using the notation of Eq. (12.15), the equations of motion becomes

$$\ddot{z}_l(t) + c_l \dot{z}_l + \left(2\omega^2\alpha_1 + 1 - g\sin\omega t\right) z_l = Q_0 + \Delta Q \sin\omega t \tag{12.16}$$

where over bar " $^-$ " is omitted for the sake of simplicity.

12.3.6.2 Natural Frequencies

If we neglect the terms in the right-hand side and damping term, we obtain the expression

$$\ddot{z}_l(t) + \left(\delta(\omega) - g\sin\omega t\right) z_l(t) = 0 \tag{12.17}$$

where $\delta(\omega) = 2\omega^2\alpha_1 + 1$, which is a function of time. Here, under the assumption that the term related to the gravity is small compared to the stiffness $\delta(\omega)$ of the blade, we can determine that the gravitational constant g is small compared to 1. If this variation term is neglected, the equation of motion becomes $\ddot{z}_l(t) + \delta z_l(t) = 0$. Its natural frequency is given by the expression

$$p = \sqrt{\delta} = \sqrt{1 + 2\omega^2\alpha_1} \tag{12.18}$$

The results of numerical simulation can be found in Figure 12.19. As the rotational speed increases, the natural frequency also increases. In the case of Figure 12.19, we adopted parameter values based on the assumption that a steel plate of thickness 2.5 mm and length $l = 1$ m is rotating.

12.3.6.3 Forced Oscillation

In Eq. (12.16), we see that there is both constant force and periodic excitation force with the frequency ω. Therefore, we can assume that the vibration is also composed of a constant component and a periodic component with frequency ω. If these components are inserted into the parametric term $g\sin\omega t z_l(t)$, the constant term and the terms with frequencies of ω and 2ω emerge. Therefore, the solution can be expressed as follows with the same accuracy as the small quantity g:

$$z_l(t) = z_{l0} + z_{lc1}\cos\omega t + z_{ls1}\sin\omega t + z_{lc2}\cos 2\omega t + z_{ls2}\sin 2\omega t \tag{12.19}$$

The components ω and 2ω become large because of the resonance at the cross points between the curve representing the natural frequency and the straight lines

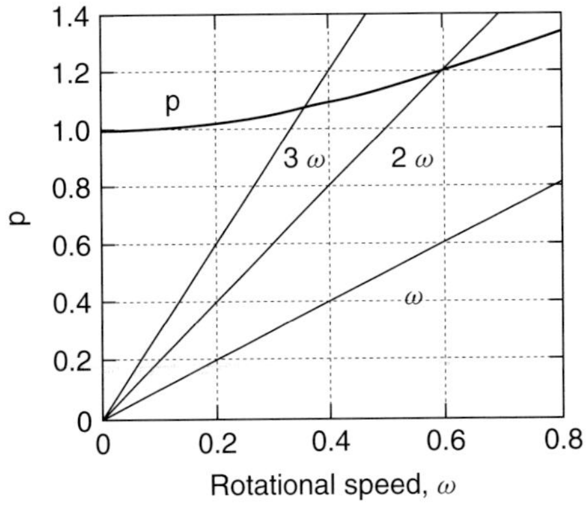

Figure 12.19 Natural frequency diagram.

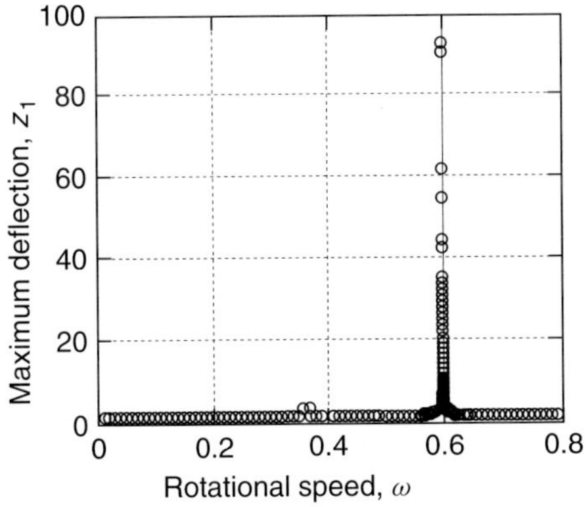

Figure 12.20 Response curve.

$p = \omega$ and $p = 2\omega$ in Figure 12.19. However, since there is no cross point between the straight line $p = \omega$ and the curve of the natural frequency, only the super harmonic resonance 2ω occurs. Indeed, in Figure 12.20, which shows a result of numerical simulations, the resonance 2ω occurs in the vicinity of $p = 2\omega$, which corresponds to the cross point of the curve p and the straight line $p = 2\omega$ in Figure 12.19. Refer to Inoue, Ishida, and Kiyohara (2012) with regard to the method for obtaining this response curve.

12.3.6.4 **Parametrically Excited Oscillation**

Equation (12.16) bears some resemblance to the Matheiu equation $m\ddot{x} + (k + \Delta k \sin \omega t)\, x = 0$. It is well known that unstable oscillations with frequency $\omega/2$ occur in the vicinity of the frequency in which the relationship $\omega/2 = p$ holds. However, in contrast to the Matheiu equation, there is no cross point for $\omega/2 = p$ in this system. Therefore, such unstable oscillations do not appear. The unstable range for the higher-frequency components $2\omega, 3\omega$, and so on exists but is very narrow.

Note: Modal Function of a Cantilever

Let the deflection of the beam be $z(x, t)$. Then, the equations of motion is expressed by

$$\rho A \frac{\partial^2 z}{\partial t^2} + \frac{\partial^2}{\partial x^2}\left(EI\frac{\partial^2 z}{\partial x^2}\right) = 0 \tag{1}$$

We assume the solution to be $z(x, t) = U(x)e^{ipt}$, where p is the natural frequency. Substituting this expression into Eq. (1), we obtain the equation for $U(x)$.

$$\frac{\partial^4 U}{\partial x^4} - \frac{\omega^2}{c^2}U = 0 \qquad \left(c^2 = \frac{EI}{\rho A}\right) \tag{2}$$

Next, we assume that $U(x) = Ae^{\lambda x}$. Substituting this into Eq. (2), we obtain four roots $\lambda = \pm i\nu_0, \pm\nu_0$ (where $\nu_0 = \sqrt{|p|/c}$). The general solution is expressed by the linear combination of the solutions corresponding to these four roots. After rearrangement, we obtain the following expression:

$$U(x) = C_1 \cos \nu_0 x + C_2 \sin \nu_0 x + C_3 \cosh\nu_0 x + C_4 \sinh\nu_0 x \tag{3}$$

The boundary conditions of this cantilever are $z = 0,\ \partial z/\partial x = 0$ at the fixed end $(x = 0)$, and $\partial^2 z/\partial x^2 = 0,\ \partial^4 z/\partial x^4 = 0$ at the free end $(x = l)$. Applying these conditions to Eq. (3), we can obtain the following relationship:

$$\left.\begin{array}{c} C_1 + C_3 = 0 \\ C_2 + C_4 = 0 \\ -C_1 \cos \nu_0 l - C_2 \sin \nu_0 l + C_3 \cosh\nu_0 l + C_4 \sinh\nu_0 l = 0 \\ C_1 \cos \nu_0 l - C_2 \sin \nu_0 l + C_3 \cosh\nu_0 l + C_4 \sinh\nu_0 l = 0 \end{array}\right\} \tag{4}$$

From the determinant of the coefficient matrix of these expressions, we obtain the following frequency equation:

$$1 + \cos \lambda \cosh\lambda = 0 \tag{5}$$

where $\lambda = \nu_0 l$. This equation can be solved graphically as shown in Figure 12.21. The cross points in Figure 12.21 give the roots of Eq. (5), and they are given as follows:

$$\lambda_1 = 1.8751, \quad \lambda_2 = 4.6941, \quad \lambda_3 = 7.8548, \ldots \tag{6}$$

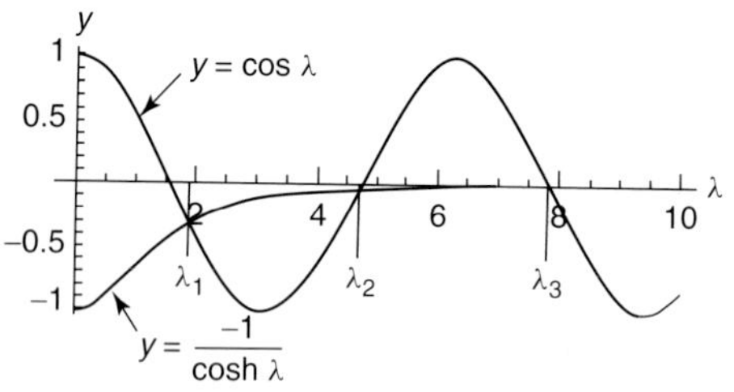

Figure 12.21 Graphical Method for eigen values.

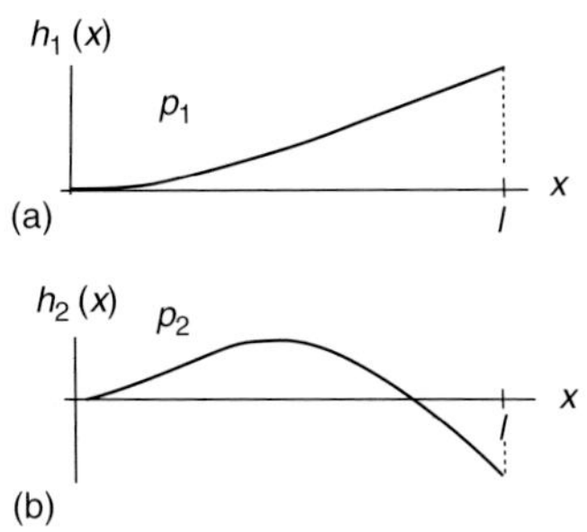

Figure 12.22 Shapes of the (a) first and (b) second modes.

From these values, we can obtain ν_{01}, ν_{02}, ν_{03}, and so on and then the natural frequencies p_1, p_2, p_3, and so on. Corresponding to each $\nu_{0i} (i = 1 - 4)$, we can obtain the ratio of amplitudes C_1:C_2:C_3:C_4 from Eq. (4). As a result, we can determine the modal functions as follows:

$$h_i(x) = \left(\sin\lambda_i + \sinh\lambda_i\right)\left(\cos\lambda_i\frac{x}{l} - \cosh\lambda_i\frac{x}{l}\right) - \left(\cos\lambda_i + \cosh\lambda_i\right)\left(\sin\lambda_i\frac{x}{l} - \sinh\lambda_i\frac{x}{l}\right)$$

The shapes of the first two modes are shown in Figure 12.22.

References

Burton, T., Sharpe, D., Jenkins, N., and Bossanyi, E. (2001) *Wind Energy Handbook*, John Wiley & Sons, Ltd.

Campbell, W. (1924) The Protection of steam-turbine disk wheels from axial vibration. *Trans. ASME*, **46**, 31–160.

Gasch, R. and Pfutzner, H. (1975) *Rotordynamik: Eine Einfuhrung*, Springer-Verlag, Berlin.

Gasch, R. and Twele, J. (2002) *Wind Power Plank, Chapter 12 (Dynamic problems of wind turbines)*, Solarpraxis AG.

Hillmann, M., Bauer, A., and Rische, T. (2004) Vibration analysis of rotor blades and balancing of rotors on wind turbines. Proceedings of DEWEK 2004, Poster 1.

Inoue, T., Ishida, Y., and Kiyohara, T. (2012) Nonlinear vibration analysis of the wind turbine blades (occurrence of the superharmonic resonance in the out of plane vibration of the elastic blades), *Trans. ASME, J. Vib. and Acoust.*, **134** (3), 031009-1–031009-13.

Ishida, Y., Inagaki, M., Ejima, R., and Hayashi, A. (2009b) Nonlinear resonances and self-excited oscillations of a rotor caused by radial clearance and collision. *Nonlinear Dyn.*, **57** (3), 593–605.

Ishida, Y., Inoue, T., and Nakamura, K. (2009a) Vibrations of a wind turbine blade (theoretical analysis and experiment using a single rigid blade mode). *JSME, J. Environ. Eng.*, **4** (2), 443–454.

Kajiyama, Y. (1969) *Dynamics for a Million People*, Agune Publisher, pp. 229–244. (in Japanese).

Keogh, P.S. and Morton, P.G. (1993) Journal bearing differential heating evaluation with influence on rotor dynamic behavior. *Proc. R. Soc. Lond., Ser. A*, **441**, 527–548.

Keogh, P.S. and Morton, P.G. (1994) The dynamic nature of rotor thermal bending due to unsteady lubricant shearing within a bearing. *Proc. R. Soc. London, Ser. A*, **445**, 273–290.

Lösl, J. and Becker, E. (2009) Reducing Vibration by Balancing Rotor Blades, Erneuerbare Energien, August-issue.

Nakane, H., Kokubo, H., Iida, S., and Takeshita, K. (1968) *1/2-Order Subharmonic Resonance of Rotating Shaft Supported by Sliding Journal Bearings*, Technical Review, Mitsubishi Heavy Industries, Ltd, January 1968.

Newkirk, B.L. (1926) Shaft rubbing. *Mech. Eng.*, **48** (8), 830–832.

Shiraki, K. and Kanki, H. (1974) *IUTUM Symposium, Lyngby, Denmark*, Springer-Verlag, pp. 494–521.

Yamane, T., Matsumiya, H., Kawamura, S., Mizutani, H., Nii, Y., and Gotanda, T. (1990) Vibration characteristics of an experimental wind turbine with a 15m teetered rotor. *Trans. JSME*, **56** (530), 7–12. (in Japanese).

13
Cracked Rotors

13.1
General Considerations

Fatigue cracks are one of the primary causes of damage of rotors in accidents. When a material is subjected to cyclic loading, a crack occurs because of the effects of fatigue. When a rotor is supported horizontally, the tension and compression work in the upper and lower sides of the rotor. Since these stresses change periodically as the rotor rotates, there is a high probability that a crack may appear. Figure 13.1 shows the surface of a cracked section of a ship's steam turbine propeller shaft with a diameter of 414 mm (JSPS1970). There are three stages: initiation, propagation, and fracture. Initially, a crack starts from a surface scratch or a sharp corner where stress concentrates. The crack then continues to grow. Finally, fracture occurs when the material cannot withstand the applied stress. In the crack propagation stage, *beachmarks* or *clamshell patterns* are formed by stress cycles that vary in magnitude. One can determine whether a material has fractured due to a crack if this pattern is found *after* failure. Such beachmarks can be observed in Figure 13.1a. The lower part with rough surface shows the fractured part, which comprises approximately 45% of the cross section. Figure 13.1b is a sketched representation of this cracked section.

In order to prevent a catastrophic accident, we must find cracks *before* fractures occur. For example, aircraft gas turbines are commonly inspected at regular intervals on the basis of the number of flight hours. In such a case, machines are disassembled and examined carefully to identify the occurrence of surface cracks. Since cracks are not visible in most cases, the following static nondestructive methods are used:

1) *Colored dye penetrative testing*
   First, a dye liquid is applied to the surface of the rotor and the liquid is allowed to soak into any cracks for a set period. Second, the excess dye is removed. Third, the rotor is rotated in order to detect cracks by checking the rotor for the remaining dye, which, if cracks are present, bleeds out from those cracks.
2) *Florescent dye testing*
   The process is the same. However, ultraviolet light (black light) is used to illuminate the florescent dye in the crack.

Linear and Nonlinear Rotordynamics: A Modern Treatment with Applications, Second Edition.
Yukio Ishida and Toshio Yamamoto.

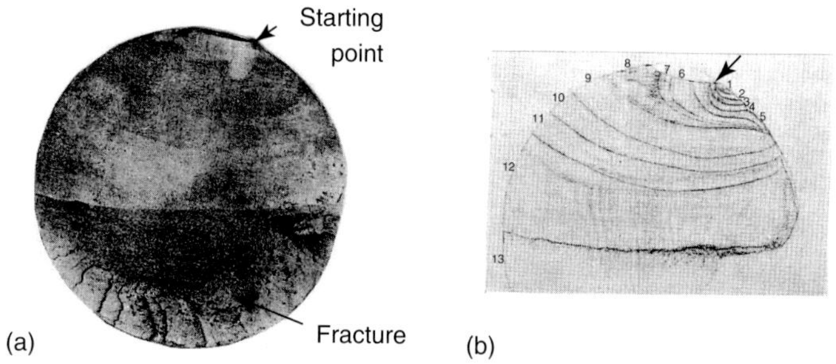

Figure 13.1 Cracked section. (a) Photograph and (b) sketch. (Source: Courtesy of JSPS 129 Committee (JSPS1970).)

3) *Ultrasonic testing*
   A transponder generates an acoustic signal that passes into and through the material. If a crack is present, it will reflect some of the sound waves back to the detector. This method has the additional advantage of being able to find internal cracks.
4) *Magnetic particle testing*
   This technique can be used to find cracks in ferromagnetic materials such as cast iron and steel. A magnetic field is created using a permanent magnet or an electromagnet in the part of the rotor to be tested. Magnetic particles are then applied to the surface. Since the magnetic field leaked from a crack attracts magnetic particles, the presence of cracks becomes easily detectable as the particles concentrate around any existing cracks in a pattern several times larger than the cracks themselves.
5) *Eddy current testing*
   This test uses electromagnetic induction. A circular coil carrying current is moved along the electrically conductive test specimen. The alternating current generates a changing magnetic field, which generates eddy currents. By monitoring any changes of the magnitude of these eddy currents, we can detect any crack that is present.

These methods are useful, but in the case of rapidly developing cracks, fractures can occur in the time between the above types of periodic inspections. Also, disassembling machinery is generally time consuming and costly. In order to detect a crack at the early stage of its propagation, on-line vibration monitoring systems are used in large rotor systems. In order to develop such monitoring systems, vibration characteristics of cracked rotors have been investigated by many researchers. The study of vibrations of cracked rotors developed from the mid-1970s. Gasch and Pfutzner (1975) and Henry and Okah-Avae (1976) investigated vibration phenomena of a cracked shaft using a cracked Jeffcott rotor model. A thorough review of the studies in this area up until 1989 has been presented by Bauer (1990). For an excellent explanation of the dynamics of cracked rotors, please refer to Bachschmid, Pennacchi, and Tanzi (2010). Also, the chapter on cracked rotors in Dimarogonas and Paipetis (1983) may be of some use.

When a horizontal rotor has a transverse crack, the crack area opens or closes because of the self-weight bending as it rotates. This is called *breathing*, and this makes the characteristics of a cracked rotor nonlinear. Gasch (1976a) and Henry and Okah-Avae (1976) considered this breathing to be a nonlinear mechanism by using different flexibilities for open and closed conditions and solved the equations of motion using analog computers. Since then, many researchers have investigated in detail the vibration characteristics of cracked rotors qualitatively and quantitatively using various kinds of physical models. In this chapter, the essential characteristics of a cracked rotor are explained from the viewpoint of nonlinear vibration. So far, we have mainly considered vertical rotor systems. However, in this chapter, we discuss horizontal rotor systems because most of the industrial rotors are horizontal and more kinds of resonances appear in them than in vertical rotor systems.

13.2
Modeling and Equations of Motion

13.2.1
Piecewise Linear Model (PWL Model)

Figure 13.2a shows a Jeffcott rotor with a crack. Cracks in practical machinery have various shapes; however, it is common to have a crack that is orthogonal to the shaft center line. Such a crack is called a *transverse crack*, and it is often represented by a model with a semicircular crack as shown in the figure. Bending stiffness of a cracked rotor differs depending on the direction of deflection. The stiffness level is low when the crack opens and high when it closes. Under these circumstances, if we consider the coordinate system O-$x'y'$ rotating with the shaft, the spring characteristic of the cracked rotor in the y'-direction is represented by the piecewise linear characteristics. Gasch and Henry and Okah-Avae derived equations of motion with this model and solved them using analog computers. However, as explained so far, a rotor system has resonances of forward and backward whirling modes that generally show different characteristics even if their frequencies are the same. In addition, the effect of the internal resonance must be considered in the Jeffcott rotor in a nonlinear system. In this section, in order to avoid the effect of internal resonance, we use a 2 DOF inclination model with a gyroscopic moment as shown in Figure 13.2b. It's piecewise linear spring characteristic in the $\theta_y{}'$-direction is shown in the figure. The spring characteristic in the $\theta_x{}'$-direction is linear. The spring characteristics are represented by

$$
\begin{aligned}
&-M_x{}' = \delta_1 \theta_x{}' \\
&-M_y{}' = (\delta_2 - \Delta\delta_2)\,\theta_y{}' \quad (\theta_y{}' > 0) \\
&-M_y{}' = (\delta_2 + \Delta\delta_2)\,\theta_y{}' \quad (\theta_y{}' < 0)
\end{aligned}
\tag{13.1}
$$

where $M_x{}'$ and $M_x{}'$ are the components of the restoring moment in the $x'z$- and $y'z$-planes.

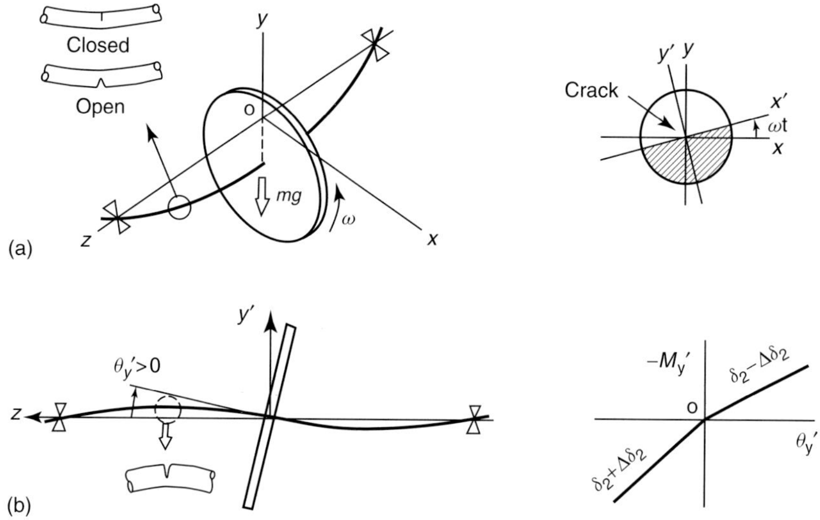

Figure 13.2 Cracked rotor models. (a) 2 DOF deflection model and (b) 2 DOF inclination model.

Now, we revise these equations of motion for a cracked rotor. By transforming this into an expression in static coordinates and connecting them with the equations of motion in Section 2.4, we can obtain the equations of motion of a cracked rotor. Now, in addition to the nondimensional quantities of Eq. (6.8), we adopt nondimensional quantities $\Delta_1^* = \Delta\bar{\delta}/\bar{\delta}$ and $\Delta_2^* = \Delta k_2/2\bar{\delta}$, which are defined by the average of spring constants $\bar{\delta} = (\delta_1 + \delta_2)/2$ and the difference between them $\Delta\bar{\delta} = (k_1 - k_2)/2$. Then, the equations of motion are represented as follows:

$$\left.\begin{aligned} &\ddot{\theta}_x + i_p\omega\dot{\theta}_y + c\dot{\theta}_x + ((1 \mp \Delta_2)\,\theta_x \\ &\quad + (\Delta_1 \pm \Delta_2)\left(\theta_x \cos 2\omega t + \theta_y \sin 2\omega t\right) = M\cos(\omega t + a) \\ &\ddot{\theta}_y - i_p\omega\dot{\theta}_x + c\dot{\theta}_y + (1 \mp \Delta_2)\,\theta_y \\ &\quad + (\Delta_1 \pm \Delta_2)\left(\theta_x \sin 2\omega t - \theta_y \cos 2\omega t\right) = M\sin(\omega t + a) + M_0 \end{aligned}\right\} \tag{13.2}$$

where the asterisk representing nondimensional parameters is eliminated. The notation M_0 is a constant moment corresponding to gravity. Regarding the symbol "$\pm$" in these equations, we use all the upper signs for $\theta_y' > 0$ and all the lower signs for $\theta_y' < 0$. In the following discussion, we call this a *piecewise linear* (*PWL*) *model*.

These equations of motion possess the following dynamic characteristics: (i) time-dependent coefficients similar to an asymmetrical rotor (parametrically excited system), (ii) rotating nonlinearity of the piecewise linear type (nonlinear system), and (iii) excitation due to the unbalance (forced oscillation system). Since a cracked rotor has these three characteristics simultaneously, many unique phenomena appear.

13.2.2 Power Series Model (PS Model)

The piecewise linear expression of Eq. (13.1) is inconvenient for theoretical treatments because we must use different expressions for the restoring force depending on the deflection. Therefore, we approximate such spring characteristics by a power series (PS) as follows:

$$\left.\begin{aligned} -M_{\theta x}' &= \delta_x'\theta_x' \\ -M_{\theta y}' &= \delta_y'\theta_y' + \varepsilon_2\theta_y'^2 + \beta_3\theta_y'^3 + \varepsilon_4\theta_y'^4 + \cdots \end{aligned}\right\} \tag{13.3}$$

Although a more accurate result can be obtained if higher power terms are used in the θ_y'-direction, for simplicity, we have limited the expression to the second power terms. After transforming Eq. (13.3) by the expression obtained by replacing x', y' with θ_x', θ_y', in Eq. (5.2) and substituting them into Eq. (13.2), we get the following expressions with nondimensional quantities:

$$\left.\begin{aligned} \ddot{\theta}_x + i_p\omega\dot{\theta}_y + c\dot{\theta}_x + \Delta\left(\theta_x\cos 2\omega t + \theta_y \sin 2\omega t\right) + N_{\theta x} &= M\cos(\omega t + a) \\ \ddot{\theta}_y - i_p\omega\dot{\theta}_x + c\dot{\theta}_y + \Delta\left(\theta_x\sin 2\omega t - \theta_y \cos 2\omega t\right) + N_{\theta y} &= M\sin(\omega t + a) + M_0 \end{aligned}\right\} \tag{13.4}$$

where $\Delta = \left(\delta_x' - \delta_y'\right)/\left(\delta_x' + \delta_y'\right)$ represents the directional difference in stiffness. The nonlinear terms $N_{\theta x}$ and $N_{\theta y}$ are given by

$$\left.\begin{aligned} N_{\theta x} &= \frac{1}{4}\varepsilon_2\left\{(-3S_1 + S_3)\,\theta_x^2 + 2\,(C_1 - C_3)\,\theta_x\theta_y - (S_1 + S_3)\,\theta_y^2\right\} \\ N_{\theta y} &= \frac{1}{4}\varepsilon_2\left\{(C_1 - C_3)\,\theta_y^2 - 2\,(C_1 + C_3)\,\theta_x\theta_y + (3C_1 + C_3)\,\theta_y^2\right\} \end{aligned}\right\} \tag{13.5}$$

where the notations $C_n = \cos n\omega t$ and $S_n = \sin n\omega t$ are used for convenience of representation. In the following, we call this a *power series (PS) model*. In Figure 13.3, a piecewise linear spring characteristic is compared with its approximation by a power series up to the second power. These values are used in the following numerical calculations.

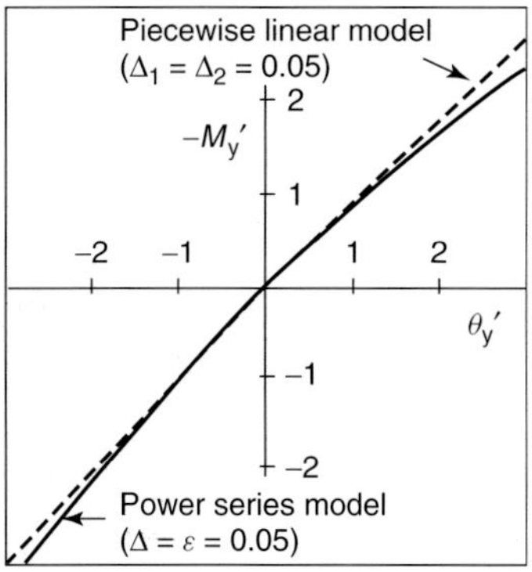

Figure 13.3 Comparison of spring characteristics.

13.3
Numerical Simulation (PWL Model)

13.3.1
Horizontal Rotor

Figure 13.4 shows the resonance curves obtained by calculating Eq. (13.2) numerically by the Runge–Kutta method. The parameter in the gyroscopic term is $i_p = 0.1$, and the corresponding natural frequency diagram is Figure 6.24a. Figure 13.4a shows the case where an unbalance is located on the same side as the crack, and Figure 13.4b shows the case where this unbalance is located on the opposite side of the crack. The circles denote amplitude values obtained numerically, and the lines are obtained by connecting them smoothly. A symmetrical shaft with no crack has resonance only at the major critical speed; however, this cracked rotor has the following resonances: (i) unstable oscillation at the major critical speed, depending on the direction of the unbalance, (ii) harmonic

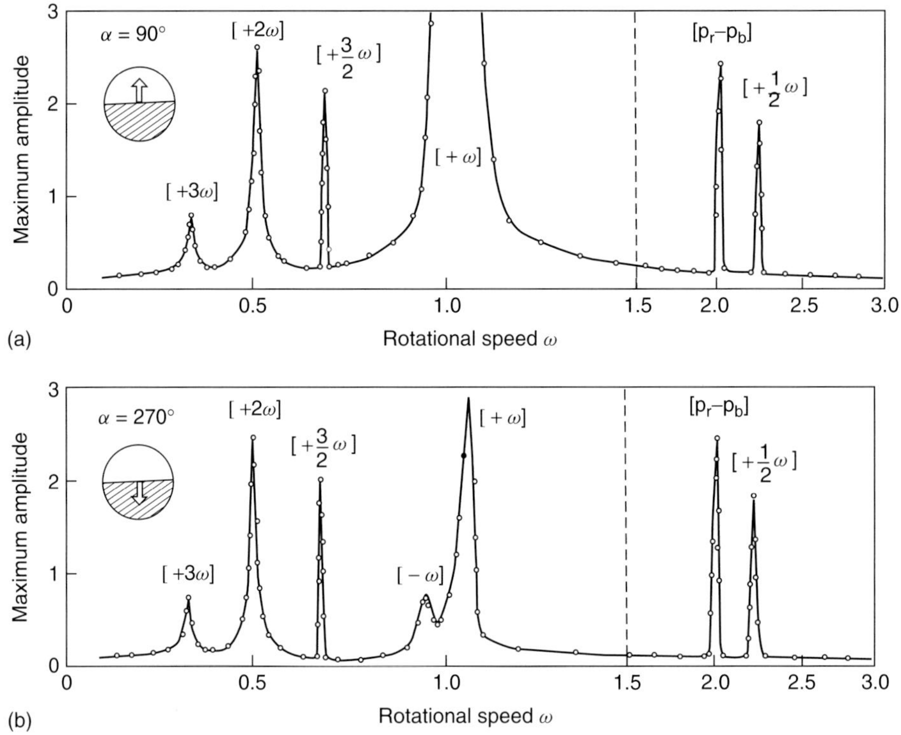

Figure 13.4 Response curve of a PWL model of a cracked rotor ($i_p = 0.1$, $\tau = 0.1$). (a) Case of the unbalance located on the same side as the crack and (b) case of the unbalance located on the opposite side of the crack.

resonance of a backward whirling mode $[-\omega]$, (iii) superharmonic resonances of a forward whirling mode $[+\omega]$ and $[+\omega]$, (iv) subharmonic resonance of a forward whirling mode $[+(1/2)\omega]$, (v) super-subharmonic resonance of a forward whirling mode $[+(3/2)\omega]$, and (vi) combination resonance $[p_f - p_b]$. Although these resonances change depending on the direction and magnitude of the unbalance, it is possible to determine the occurrence of a crack by monitoring these resonances.

13.3.2 Vertical Rotor

The response of a vertical cracked rotor is obtained by putting $M_0 = 0$ in Eq. (13.2). In this case, only the resonance at the major critical speed appears. Similar to the case of a horizontal rotor, an unstable range appears when the unbalance is located on the same side as the crack and it disappears when the unbalance is located on the opposite side of the crack.

13.4 Theoretical Analysis (PS Model)

As shown in Figure 13.2, various kinds of resonances appear in cracked rotors. In this section, these resonances are explained by theoretical analyses. Experimental results are shown selectively as necessary.

13.4.1 Forward Harmonic Resonance $[+\omega]$ (Horizontal Rotor)

In many turbine systems, vibrations are monitored at the bearing pedestals. The occurrence of a crack can often be found by investigating the increase of harmonic vibrations (Ziebarth and Baumgartner, 1981; Inagaki and Hirabayashi, 1990; Sanderson, 1992). In this section, response curves of a harmonic resonance are explained (Ishida, Yamamoto, and Hirokawa, 1994).

Since the rotor is horizontal, we assume the harmonic solution to an accuracy of $O(\varepsilon)$ as follows:

$$\left.\begin{aligned} \theta_x &= P\cos(\omega t+\beta) + A_x + \varepsilon\left\{a\cos(\omega t+\beta) + b\sin(\omega t+\beta)\right\} \\ \theta_y &= P\sin(\omega t+\beta) + A_y + \varepsilon\left\{a'\sin(\omega t+\beta) + b'\cos(\omega t+\beta)\right\} \end{aligned}\right\} \tag{13.6}$$

This expression is obtained by adding constant terms to Eq. (6.17). Substituting this solution into Eq. (13.4) and using the harmonic balance method to an accuracy of $O\left(\varepsilon^0\right)$, we obtain $A_x = 0$ and $A_y = M_0$. Similarly, to an accuracy of $O(\varepsilon)$, we

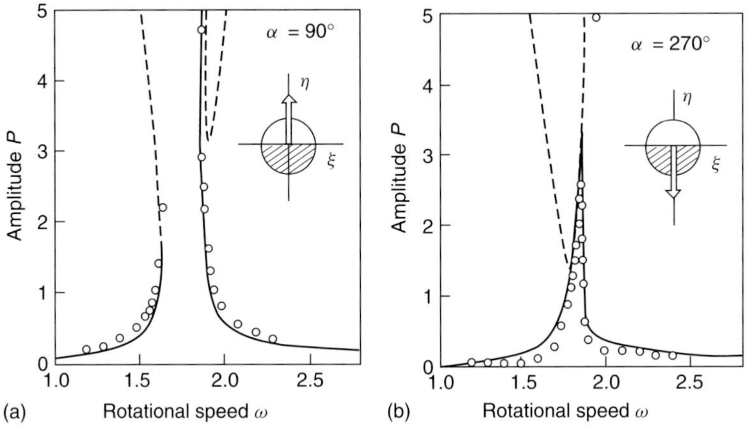

Figure 13.5 (a) Theoretical results (PS model) and (b) numerical results (PWL model) in the case of large unbalance ($\tau = 0.1$).

have

$$\left.\begin{aligned} A_f P\dot{\beta} &= GP + \Delta P\cos 2\beta + \frac{1}{4}\varepsilon_2\left(3P^2 + 2M_0^2\right)\sin\beta \\ &\quad -\frac{1}{4}\varepsilon_2 P^2\sin 3\beta - M\cos(\alpha - \beta) \\ A_f \dot{P} &= -c\omega P + \Delta P\sin 2\beta + \frac{1}{4}\varepsilon_2\left(P^2 + 2M_0^2\right)\cos\beta \\ &\quad +\frac{1}{4}\varepsilon_2 P^2\cos 3\beta + M\sin(\alpha - \beta) \end{aligned}\right\} \tag{13.7}$$

where $A_f = (2 - i_p)\,\omega > 0$ and $G(\omega) = 1 + i_p\omega^2 - \omega^2$. In a procedure similar to that explained in Section 6.4.1, we can obtain the harmonic solutions and investigate their stability.

The resonance curves are shown in Figures 13.5 and 13.6. Figure 13.5 is a case with a comparatively large unbalance and corresponds to Figure 13.4. In this case, the shape of the resonance curves changes remarkably, depending on the direction of the unbalance. When an unbalance is located on the same side as the crack, the resonance curves spread in the large-amplitude region and, as a result, an unstable zone appears. On the other hand, when an unbalance is located on the opposite side of the crack, the unstable region disappears and only steady-state vibrations appear. Figure 13.6 is a case with a comparatively small unbalance. An unstable zone appears regardless of the direction of unbalance. This is because the effect of parametric terms on the equations of motion becomes predominant in cases where the unbalance is small. If the case with no unbalance ($M = 0$) is calculated, resonance curves with an unstable zone are also obtained (Ishida, Yamamoto, and Hirokawa, 1994).

In these figures, the results of numerical simulation using a PWL model are illustrated by small open circles. From the comparison, we can ascertain that the second power approximation gives a fairly workable result.

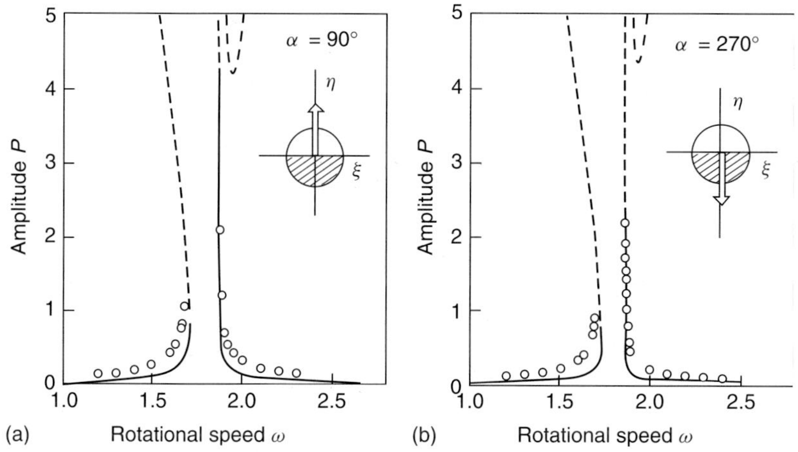

Figure 13.6 (a) Theoretical results (PS model) and (b) numerical results (PWL model) in the case of small unbalance ($\tau = 0.01$).

13.4.2
Forward Harmonic Resonance [$+\omega$] (Vertical Rotor)

If the rotor is supported vertically, the static deflection due to gravity does not appear. The responses at the major critical speed are similar to the case of Figure 13.5 for large unbalance (Ishida, Ikeda, Yamamoto, and Masuda, 1988). The shape depends on whether the unbalance is located on the same side or the opposite side of the crack. An unstable range appears when the unbalance is located on the same side as the crack, and it disappears when the unbalance is located on the opposite side of the crack.

13.4.3
Forward Superharmonic Resonance [$+2\omega$] (Horizontal Rotor)

Another important symptom of the occurrence of a crack is the superharmonic resonance of the second order, which is also called *twice per revolution resonance* or *double-frequency resonance.* Occurrences of this kind of resonance in industrial machinery were reported, for example, by Mayes and Davis (1980) in a 100 MW generator rotor, Inagaki and Hirabayashi (1990) in a wind tunnel fan, and Sanderson (1992) in a 935 MW four-pole turbine generator at Darlington Nuclear Generator. Figure 13.7 is a natural frequency diagram corresponding to Figure 13.4. This resonance appears at the rotational speed at which the natural frequency p_f curve and the straight line $p = 2\omega$ cross.

Since a forward superharmonic component [$+2\omega$] occurs in addition to the harmonic component in this resonance range, we assume the approximate solution

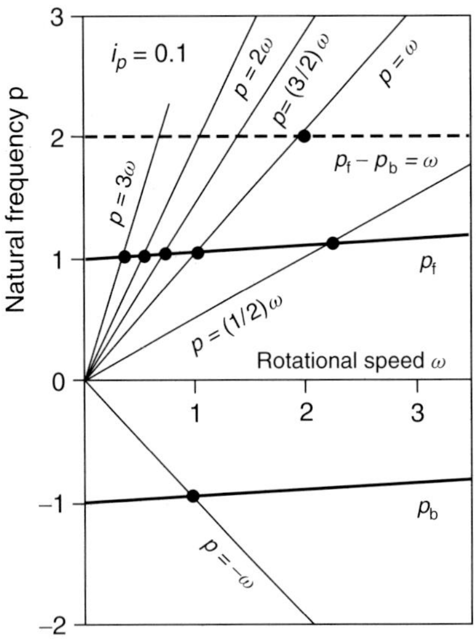

Figure 13.7 Resonance speeds.

to an accuracy of $O(\varepsilon)$ as follows (Ishida, Yamamoto, and Murakami, 1992):

$$\left.\begin{aligned}\theta_x &= R\cos(2\omega t+\delta)+P\cos(\omega t+\beta)+A_x+\varepsilon\left\{a\cos(2\omega t+\delta)+b\sin(2\omega t+\delta)\right\}\\ \theta_y &= R\sin(2\omega t+\delta)+P\sin(\omega t+\beta)+A_y+\varepsilon\left\{a'\sin(2\omega t+\delta)+b'\cos(2\omega t+\delta)\right\}\end{aligned}\right\} \tag{13.8}$$

In a similar procedure, we can obtain steady-state solutions $R = R_0$, $\delta = \delta_0$, $A_x = A_{x0}$, and $A_y = A_{y0}$ and can determine their stability. For example, the amplitude is given as follows:

$$R_0^2 = \frac{\left\{\dfrac{(\Delta-\varepsilon_2 P_s)}{(1+\varepsilon_2 P)}\right\}^2 M_0^2}{(2c\omega)^2+\left[G_f+\varepsilon_2 P_s-\left(\dfrac{(\Delta-\varepsilon_2 P_s)^2}{(1+\varepsilon_2 P_s)}\right)\right]^2} \tag{13.9}$$

where $P_s = P\sin\alpha$ and $G_f = 1 + i_p\omega(2\omega) - (2\omega)^2$.

Resonance curves are shown in Figure 13.8 where a theoretical result of a PS model and a numerical result of a PL model are illustrated. Since a cracked shaft has stiffness asymmetry similar to asymmetrical rotors explained in Chapter 5, it is expected that twice per revolution resonance may occur when a rotor is supported horizontally. However, the characteristics of this resonance in a cracked rotor are different from those in an asymmetrical rotor explained in Section 5.4. Unlike the case of a linear asymmetrical rotor, the amplitude changes remarkably depending on the angular position of the unbalance as shown in Figure 13.8. (This phenomenon depends on parameter values. In Figure 13.4, for $i_p = 0.1$, this difference is not clear.) The amplitude is maximum when the unbalance is located at the center of the crack and reaches its minimum when it is located at the center of the uncracked area. Therefore, we must keep in mind that the resonance is not always observed in a diagnosis system.

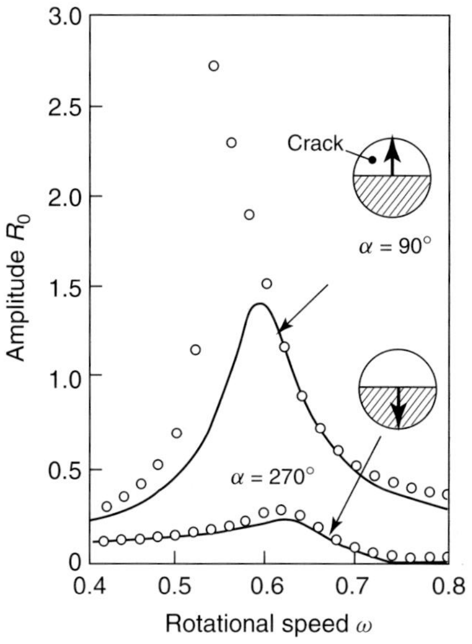

Figure 13.8 Superharmonic resonance $[+2\omega]$ ($i_p = 0.7$).

The backward superharmonic resonance of the second order $[-2\omega]$ does not occur due to a crack as shown in Figure 13.2.

13.4.4 Other Kinds of Resonance

As shown in Figure 13.4, other kinds of resonance appear due to cracks. However, these resonances are, generally speaking, not so large. Therefore, only the results obtained by Ishida and Hiromawa (1995) by theoretical analyses with a horizontal PS model and numerical analyses with a horizontal PWL model (and some of them with PS model) are illustrated in the following discussion. The occurrence of these resonances has also been confirmed experimentally.

13.4.4.1 Backward Harmonic Resonance $[-\omega]$

In Figures 13.4 and 13.7, when the relationship $p_b = -\omega$ holds, a backward harmonic resonance with frequency $-\omega$ occurs. Its resonance curve as obtained by Ishida and Hirokawa (1995) is shown in Figure 13.9. The shape of this resonance curve does not change appreciably depending on the angular position of the unbalance.

13.4.4.2 Forward Superharmonic Resonance $[+3\omega]$

In Figures 13.4 and 13.7, when the relationship $p_f = +3\omega$ holds, a forward superharmonic resonance with frequency $+3\omega$ occurs. Gasch (1976a, 1976b), Henry and Okah-Avae (1976), and Bachschmid, Pennachi, and Tanzi (2010). reported the occurrence of this kind of resonance by numerical simulation. The

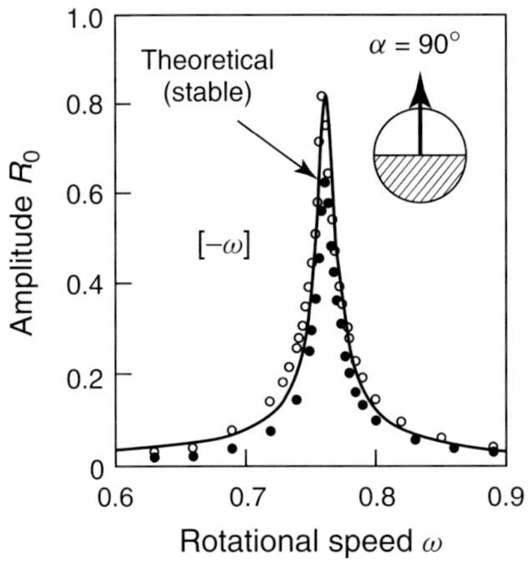

Figure 13.9 Harmonic resonance $[-\omega]$ ($i_p = 0.7$).

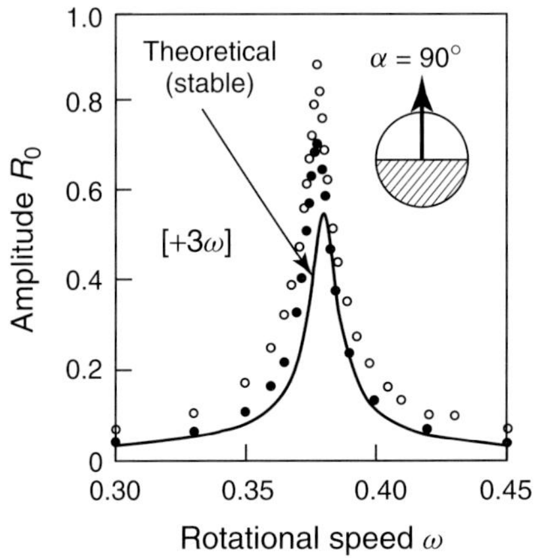

Figure 13.10 Superharmonic resonance $[+3\omega]$ ($i_p = 0.7$).

result obtained by Ishida and Hirokawa (1995) is shown in Figure 13.10. Differing from the forward superharmonic resonance $[+2\omega]$, the shape of this resonance curve does not change appreciably according to the direction of the unbalance.

13.4.4.3 Forward Subharmonic Resonance $[+(1/2)\omega]$

In Figures 13.4 and 13.7, when the relationship $p_f = +(1/2)\omega$ holds, a forward subharmonic resonance with frequency $+(1/2)\omega$ occurs. Gasch (1976b) showed the occurrence of this kind of oscillation numerically. Dimarogonas and Papadopoulos (1988) observed the frequency spectrum of a cracked 300 MW turbine and found that a $(1/2)\omega$ component was contained in addition to various superharmonic components. The result obtained by Ishida and Hirokawa (1995) is shown in Figure 13.11. In this approximate analysis, the second and fourth power terms are considered. Unlike the harmonic and superharmonic resonances, the resonance curve bifurcates from the trivial solution with zero amplitude. Although they are related to the parametric resonance, its amplitude does not increase infinitely but converges to a finite amplitude because of nonlinearity.

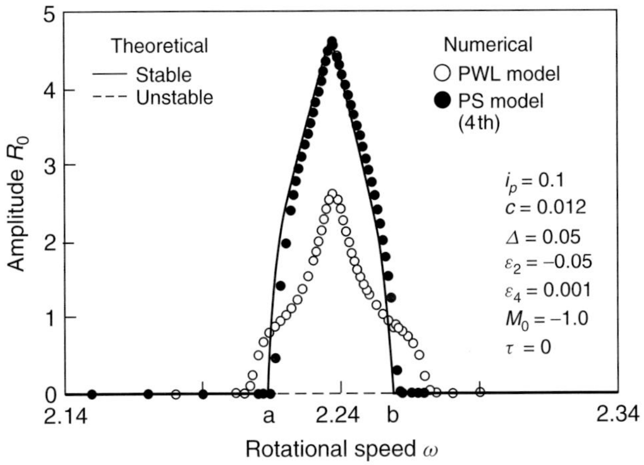

Figure 13.11 Subharmonic resonance $[+(1/2)\omega]$.

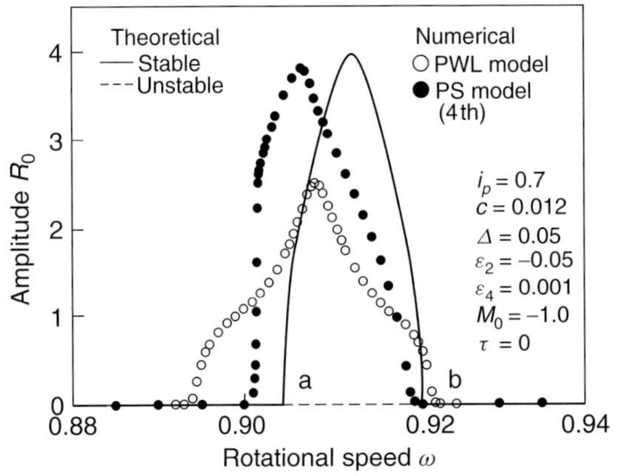

Figure 13.12 Super-Subharmonic resonance $[+(3/2)\omega]$.

13.4.4.4 **Forward Super-Subharmonic Resonance $[+(3/2)\omega]$**

In Figures 13.4 and 13.7, when the relationship $p_f = +(3/2)\omega$ holds, a forward super-subharmonic resonance with frequency $+(3/2)\omega$ occurs. Gasch (1976b) illustrated the occurrence of this kind of oscillation numerically, and Tamura, Iwata, and Sato (1988) observed this resonance in an experimental setup. The result obtained by Ishida and Hirokawa (1995) is shown in Figure 13.12. In this approximate analysis, the second and fourth power terms are considered. The shape of this resonance curve shown in Figure 13.12 is similar to that of subharmonic resonance curve illustrated in Figure 13.11. It is noted that, among various kinds of super-subharmonic resonance of types $[\pm(m/n)\omega]$, only this kind of resonance occurs.

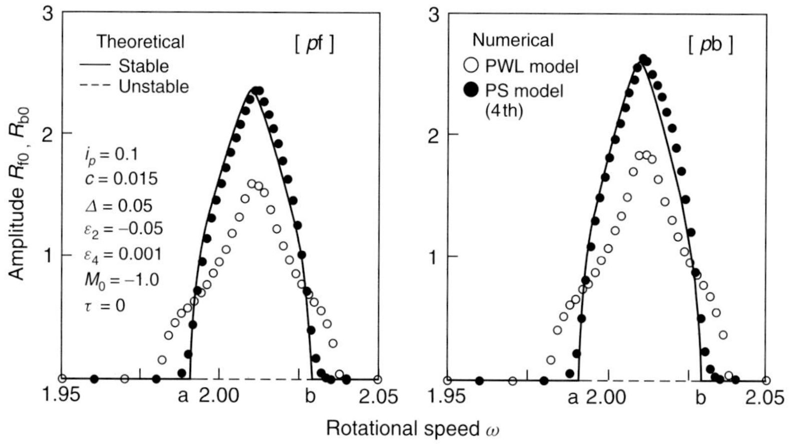

Figure 13.13 Combination resonance $[p_f - p_b]$.

13.4.4.5 Combination Resonance

In Figures 13.4 and 13.7, when the relationship $p_{\mathrm{f}} - p_{\mathrm{b}} = \omega$ holds among the natural frequencies p_{f} and p_{b} and the rotational speed ω, a combination resonance of the type $[p_{\mathrm{f}} - p_{\mathrm{b}}]$ occurs. This type of resonance does not occur in the Jeffcott rotor model since its resonance speed coincides with that of a subharmonic resonance $[+(1/2)\omega]$. The responses of each component whose frequencies are near p_{f} and p_{b} are shown in Figure 13.13. The characteristics of this resonance are almost the same as those of the subharmonic resonance of order 1/2 (Ishida and Hirokawa, 1995).

Note: Condition Monitoring System

Detection of changes in vibration above normal values can indicate change in the condition of rotors. Therefore, monitoring vibration levels is critical to ensure safety. Many rotor systems such as power generation steam turbines use a *diagnostic system*, which is also called *a machinery health monitoring system*. These systems have, for example, the following functions:

- Measuring and recording vibrations automatically.
- Processing and analyzing data.
- Displaying vibration conditions in various forms in real time.
- Monitoring vibration conditions in comparison to a predetermined reference.
- Alerting of abnormal conditions.
- Diagnosing the cause of abnormal conditions.
- Calculating optimal balancing weights.

A sketch shown in Figure 13.14 is an example of a monitor display of a diagnostic system for a steam turbine. Such data is renewed, for example, every 15 s. In this system, a high-pressure (HP) turbine, a low-pressure (LP)

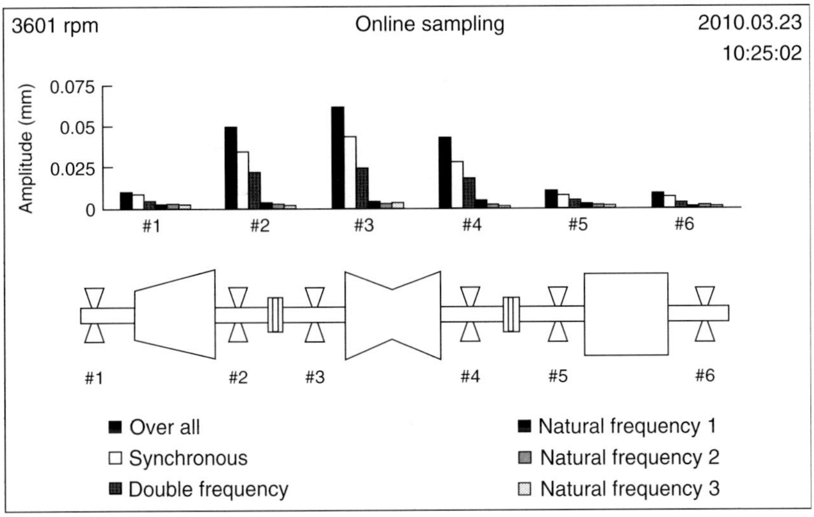

Figure 13.14 Display example showing vibration levels.

turbine, and a generator are connected in series. The vibrations in the x- and y-directions are measured at the bearing positions (#1–6) by contact acceleration sensors or noncontact deflection sensors. The extracted components of the overall vibration, the synchronous component, the double-frequency component, and components with frequencies equal to natural frequencies are displayed. Frequency spectrum, waterfall diagrams, and orbits can also be shown, if necessary. The analysis of the vibration components is conducted in real time. When changes in vibration above normal values are detected, alarms are activated and the machine is taken out of service for investigation. As mentioned above, synchronous, double-frequency, and overall component values will increase when a crack appears. Natural frequency components will increase when self-excited oscillations such as an oil whip and rubbing occur.

13.5 Case History in Industrial Machinery

Since the 1950s, a large number of cracks have been found in industrial machinery. In some cases, it is possible that the reported tragic machinery accidents could be attributed to cracked turbine shafts or blades. However, the problems associated with analyzing destroyed machinery make it difficult to attribute accidents to such causes with certainty. There are some reports on the survey of the accidents of rotors due to cracks. Yoshida (1976) reported four cases of cracks being found in steam turbine rotors from 1970 to 1971 in Japan. The cracks can be attributed to the cyclic thermal stress that appeared due to the frequent shutdown and restarting of the machinery. Muszynska (1982) reported that at least 28 cracked failures happened within the period of the 1970s. Laws (1986) at Bently Nevada Co. reported four cracked rotor incidents in the UK during the period 1971–1981. These incidents include a 40% transverse crack of a 500 MW turbine generator, a circumferential 30 in. long crack of an MP rotor of a 60 MW turbine generator, a 45% transverse crack of the main generator rotor of a 660 MW turbine generator, and a crack extending around complete circumference of the MP rotor of 350 MW turbine at the main HP coupling end. In the following discussion, major cases are listed and briefly explained (Ishida, 2008).

- *1953 Tanner Creek Station: (USA)*
  In January 1953, the 1800 rpm low-pressure steam turbine at the Tanners Creek power station of the Indiana and Michigan Electric Company, USA, suddenly began to vibrate. When the turbine was opened for inspection, it was found that a segment of approximately 160° had broken out of the first-stage wheel because of a crack. The crack originated in the region of the hole in the notch opening and progressed toward the shaft (Rankine and Seguin, 1956).
- *1954 Arizona: (USA)*
  In March 1954, a 3600 rpm generator rotor being manufactured for the Arizona Public Service Company, USA, burst suddenly while being balanced in the factory. Several cracks were found before bursting in this case (Schabtach, Fogleman, Rankin and Winne 1956).
- *1954 Cromby: (USA)*
  In September 1954, a generator rotor in Cromby Station of the Philadelphia Electric Company, USA, burst while running at 3780 rpm. There was no evidence of a fatigue or progressive-type fracture in this case (Schabtach, Fogleman, Rankin and Winne 1956).
- *1954 Ridgeland: (USA)*
  In December 1954, the LP turbine spindle in Ridgeland Station of the Commonwealth Edison Company, USA, burst during a routine overspeed trip test. The cause of this accident was determined to be flakes or thermal cracks that had developed in the shaft during heat treatment (Emmert, 1956).
- *1956–1957: Castle Donnington (UK)*
  In 1956–1957, the shafts of three large turbine rotors developed fatigue cracks under very low nominal gravitational bending stresses in Castle Donnington Power Station, UK. When the shafts were taken out of service, the following cracks were found. In one case (60 MW machine), the fatigue cracks propagated over approximately 75% of the shaft section. In the other case (60 MW machine), cracking started around the circumference and propagated no deeper than 1/16th of an inch. In the third case (100 MW machine), the crack had propagated over approximately 75% of the section (Grieves, Parish, and Hall, (1961); Coyle and Watson, 1963).

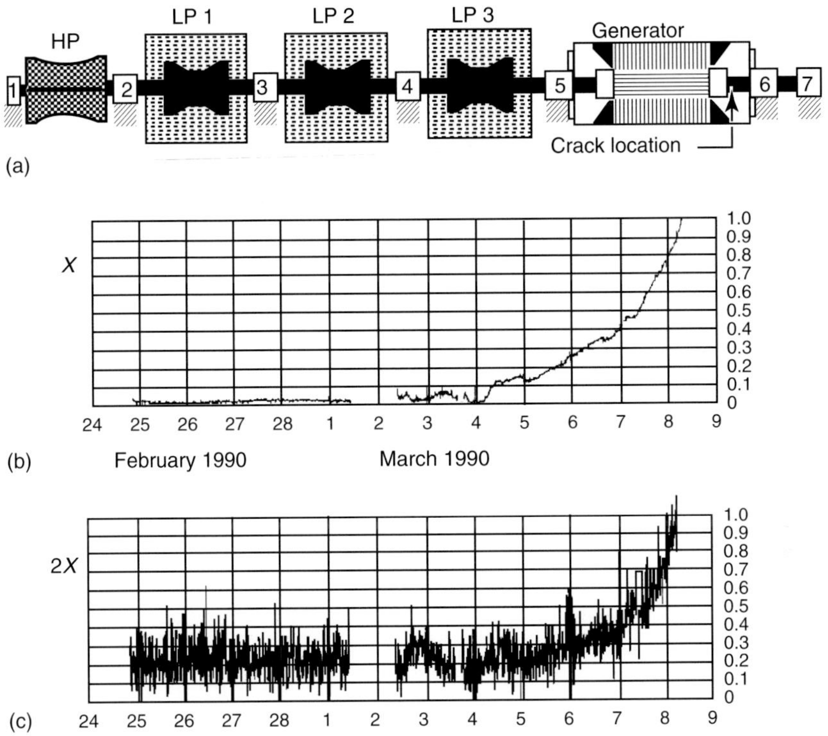

Figure 13.15 Monitored time history. (a) Rotor system, (b) harmonic component X, and (c) double-frequency component 2X (Sanderson 1992, Courtesy of IMechE).

- *1972–1974: Ferrybridge Power Station C (UK)*
  Crack was discovered in three LP shafts in the 500 MW turbines at Ferrybridge "C" Power Station during 1972–1974. This crack was detected early by monitoring the system. The crack in the center groove of each of the three shafts initiated at the corrosion pits. A beachmark analysis suggested that the crack expanded due to high cycle fatigue from a depth of 10 to 90 mm over 720 h of operation (Jack and Paterson, 1977).
- *1983: Lavrion Power Plant (Greece)*
  A crack of approximately 120° angle of the cross section was found in the 300 MW turbine in Lavrion Power Plant of the Public Power Corporation of Greece (Papadopoulos and Dimarogonas, (1987); Dimarogonas and Papadopoulos, 1988).
- *Wind Tunnel Fan (Japan)*
  A rotor crack of approximately 120° angle of the cross section of a shaft (a tube about 80 m long) of a wind tunnel fan made for car testing was reported in 1990. Synchronous and double-frequency vibrations increased due to the crack in this case (Inagaki and Hirabayashi, 1990).

- *Cumberland Power Plant (USA)*
  Cracks occurred in the 1300 MW turbine in the Cumberland Power Plant, USA. Early indications appeared and began to increase three days before the shutdown. A double-frequency excitation was noticed on coastdown (Ziebarth and Baumgartner, 1981).
- *Wuergassen Nuclear Power Plant (Germany)*
  In the Wuergassen Nuclear Power Plant in Germany, the horizontal and vertical vibrations of the LP turbine started to increase from six days before the shutdown. In this case, a double-frequency excitation was also noticed on coastdown (Ziebarth and Baumgartner, 1981).
- *1986 Crystal River Nuclear Power Plant (USA)*
  A pump shaft in the Crystal River Nuclear Power Plant broke due to a fatigue crack in a groove in 1986 and again due to a crack in another part in 1988 (The Tampa Tribune, Friday, 20 January 1989).
- *1989 Fukushima Nuclear Power Plant (Japan)*
  In 1989, initially, the monitoring system of the plant alerted the staff on New Year's Day to increased vibration. Six days later, a ring of the recurrent pump broke due to cracks made by the increased vibration (Quark, 1990).
- *1990 Darlington Nuclear Power Plant*
  As shown in Figure 13.15a, a crack comprising 25% of the cross section was found in a generator rotor of the 935 MW turbine generator of Darlington Nuclear Power Plant. Figure 13.15b,c shows time histories of fundamental component X and double-frequency component 2X recorded in the monitoring system. The vibration started to increase about five days before the shutdown, and the amplitudes of the harmonic resonance and the double-frequency resonance increased due to the occurrence of a crack (Sanderson, 1992).

References

Bachschmid, N., Pennacchi, P., and Tanzi, E. (2010) *Cracked Rotors*, Springer.

Bauer, J. (1990) On the dynamics of cracked rotors: a literature survey. *Appl. Mech. Rev.*, **43** (1), 13–17.

Coyle, M.B. and Watson, S.J. (1963) Fatigue strength of turbine shaft with shrunk-on disks. *Proc. Inst. Mech. Eng.*, **178** (6, Pt. 1), 147–160.

Dimarogonas, A.D. and Paipetis, S.A. (1983) *Analytical Methods in Rotor Dynamics*, Applied Science Publishers, Barking, Essex.

Dimarogonas, A.D. and Papadopoulos, A.D. (1988) Crack detection in turbine rotors. Proceedings of the Second International Symposium on Transport Phenomena, Dynamics, and Design of Rotating Machinery, Vol. 2, pp. 286–298.

Emmert, H.D. (1956) Investigation of large steam-turbine spindle failure. *Trans. ASME*, **78** (10), 1547–1565.

Gasch, R. (1976a) *Proceedings of the International Conference on Vibrations in Rotating Machinery*, Institute of Mechanical Engineers, New York, pp. 123–128.

Gasch, R. (1976b) Dynamisches verhalten des Lavalläufers mit angerissenem wellenquerschnitt. *VDI Ber.*, **269**, 183–188.

Gasch, R. and Pfutzner, H. (1975) *Rotordynamik: Eine Einfuhrung*, Springer-Verlag, Berlin.

Grieves, J.O., Parish, A.H., and Hall, J.S. (1961) Some features of the design and operation of the castle donington power station. *Proc. Inst. Mech. Eng.*, **175** (16), 807–852.

Henry, T.A. and Okah-Avae, B.E. (1976) *Proceedings of the International Conference on Vibrations in Rotation Machinery*, Institute of Mechanical Engineers, New York, pp. 15–17.

Inagaki, T. and Hirabayashi, M. (1990) Transverse vibrations of cracked rotors (examples of crack detection and vibration analysis). *Trans. JSME*, **56** (523), 582–588. (in Japanese).

Ishida, Y. and Yamamoto, T. (1992) Vibrations of a Rotating Shaft containing a Transverse Crack (Vibrations at the Secondary Critical Speed) 4th Int. Sympo.. *Transport Pheno. and Dynamics of Rot. Machinery*, pp. 341–350.

Ishida, Y. and Hirokawa, K. (1995) Internal resonance of a cracked rotor (major critical speed and critical speeds in precritical speed range). *Trans. JSME*, **61** (586), 49–97. (in Japanese).

Ishida, Y., Ikeda, T., Yamamoto, T., and Masuda, N. (1988) Vibrations of a rotating shaft containing a transverse crack. *Bull. JSME*, **31** (1), 22–29.

Ishida, Y., Yamamoto, T., and Murakami, S. (1992) Nonstationary oscillation of a rotating shaft with nonlinear spring characteristics during acceleration through a critical speed (a critical speed of a 1/3-Order subharmonic oscillation). *JSME Int.J. Ser. III*, **35** (3), 360–368.

Ishida, Y., Yamamoto, T., and Hirokawa, I. (1994) Vibrations of a rotating shaft containing a transverse crack (a major critical speed of a horizontal shaft). Proceedings of the IFToMM 4th International Conference on Rotor Dynamics, pp. 47–52.

Ishida, Y. (2008) Cracked rotors: industrial machine case histories and nonlinear effects shown by simple Jeffcott rotor. *Mech. Syst. Signal Process.*, **22** (4), 805–817.

Jack, A.R. and Paterson (1977) Cracked in 500 MW L.P. rotor shafts. *Proc. IMech E*, 75–83.

Laws, C.W. (1986) A brief history of cracked rotor saves – turbine generators in the U.K. rated 60MW to 660MW (1971–1981). Bently Rotor Dynamics Corporation Seminar on Shaft Crack Detection.

Mayes, L.W. and Davis, W.G.R (1980) A method of calculating the Vibrational Behaviour of coupled rotating shafts containing a transverse crack. *Proc. Int. Conf. Vib. rotating Machinery, IMech E*, 17–27.

Muszynska, A. (1982) Shaft crack detection. The 7th Machinery Dynamics Seminor, Canada.

Papadopoulos, C.A. and Dimarogonas, W. (1987) Stability of cracked rotors in the coupled vibration mode. *ASME Rotating Machinery Dyn.*, **1**, 1.

Quark (1990) Why the nuclear power plant stopped? Quark, 1990 March, pp. 94–103.

Rankine, A.W. and Seguin, B.R. (1956) Report of the investigation of the turbine wheel fracture at tanners creek. *Trans. ASME*, **78** (10), 1527–1546.

Sanderson, A.F.P. (1992) The vibration behavior of a large steam turbine generator during crack propagation through the generator rotor. Proceedings of IMechE, pp. 263–273.

Schabtach, C., Fogleman, E.L., Rankin, A.W., and Winne, D.H. (1956) Report of the investigation of two generator rotor fractures. *Trans. ASME*, **78** (10), 1567–1584.

Tamura, A., Iwata, Y., and Sato, H. (1988) Unstable vibration of a rotor with a transverse crack. Vibration in Rotating Machinery, IMechE 1988 C322/88, pp. 647–653.

Yoshida, M. (1976) Steam turbine rotor accidents and its countermeasure. *Turbomachinery*, **4** (11), 728. (in Japanese).

Ziebarth, H. and Baumgartner, R.J. (1981) Early detection of cross-sectional rotor cracks by turbine shaft vibration monitoring techniques. ASME Paper 81-JPGC-Pwr-26.

14
Finite Element Method

14.1
General Considerations

Practical rotating machines are generally very complicated. For example, many flexible disks and blades are mounted on a flexible shaft with a variable cross section. Bearing pedestals and foundations also deform during vibration. For analytical convenience, these rotors are often modeled as simple rotors, such as a concentrated mass and massless elastic shaft system or a uniform rotor system. However, in the design of a practical rotor, it often becomes necessary to know quantitatively natural frequencies, vibration modes, and response to unbalance excitations. The *finite element method* is one method used for this purpose.

The finite element method is a discretization method of a continuous structure and has been used successfully in the design and analysis of practical rotors with a complicated and irregular shape. In this chapter, application of the finite element method to rotor systems is explained. The explanation basically follows the paper by Nelson and McVaugh (1976).

14.2
Fundamental Procedure of the Finite Element Method

In the finite element method, a complicated rotor with finite degrees of freedom is discretized following the principles of dynamics, and then the multi-degree-of-freedom system obtained is analyzed. Since the dimensions of the approximated system are very large, matrix representation is used. For example, Eq. (2.48) is represented in matrix form as follows:

$$[M]\{\ddot{U}\}+[G]\{\dot{U}\}+[C]\{\dot{U}\}+[K]\{U\}=\{F\} \tag{14.1}$$

where

$$\{U\}=\begin{Bmatrix} x \\ y \\ \theta_x \\ \theta_y \end{Bmatrix},\quad [M]=\begin{bmatrix} m & 0 & 0 & 0 \\ 0 & m & 0 & 0 \\ 0 & 0 & I & 0 \\ 0 & 0 & 0 & I \end{bmatrix},\quad [G]=\begin{bmatrix} 0 & 0 & 0 & 0 \\ 0 & 0 & 0 & 0 \\ 0 & 0 & 0 & I_p\omega \\ 0 & 0 & -I_p\omega & 0 \end{bmatrix},$$

Linear and Nonlinear Rotordynamics: A Modern Treatment with Applications, Second Edition.
Yukio Ishida and Toshio Yamamoto.

$$[c] = \begin{bmatrix} c_{11} & 0 & c_{12} & 0 \\ 0 & c_{11} & 0 & c_{12} \\ c_{21} & 0 & c_{22} & 0 \\ 0 & c_{21} & 0 & c_{22} \end{bmatrix}, \quad [K] = \begin{bmatrix} \alpha & 0 & \gamma & 0 \\ 0 & \alpha & 0 & \gamma \\ \gamma & 0 & \delta & 0 \\ 0 & \gamma & 0 & \delta \end{bmatrix}, \text{and}$$

$$\{F\} = \begin{Bmatrix} me\omega^2 \cos \omega t \\ me\omega^2 \sin \omega t \\ \left(I - I_{\mathrm{p}}\right) \tau \omega^2 \cos(\omega t + \beta_\tau) \\ \left(I - I_{\mathrm{p}}\right) \tau \omega^2 \sin(\omega t + \beta_\tau) \end{Bmatrix} \tag{14.2}$$

The notations $\{U\}$ and $[F]$ are called *displacement vector* and *force vector*, respectively, and $[M]$, $[G]$, $[C]$, and $[K]$ are called *mass matrix* or *inertia matrix*, *gyroscopic matrix*, *damping matrix*, and *stiffness matrix*, respectively.

Before we provide a detailed analysis, the procedure of the finite element method is summarized.

1) Divide the rotor into simple, uniform components, called *finite elements*, by fictitious cuts.
2) Define nodes in each element and represent the displacement of an arbitrary point in the element by the displacements of these nodes. Nodes are generally taken at the joints. A system with infinite degrees of freedom is reduced to a system with finite degrees of freedom by such an operation.
3) Derive expressions for the kinetic energy, potential energy, and virtual work of an element.
4) Derive expressions for the element stiffness matrix $[k]$, element mass matrix $[m]$, element gyroscopic matrix $[g]$, and element force vector $\{f\}$.
5) Derive the stiffness matrix [K], mass matrix $[M]$, gyroscopic matrix $[G]$, and force vector $\{F\}$ of the complete system.
6) Derive the equations of motion of the complete system by Lagrange's equation or Hamilton's principle. The equations obtained represent the relationship between the nodal displacement and the nodal forces and have the same form as the equations for a multi-degree-of-freedom system.
7) Solve these equations for the given boundary conditions and obtain natural frequencies and vibration modes (eigenvalue problem).
8) Obtain the responses to unbalance excitation.

14.3 Discretization of a Rotor System

14.3.1 Rotor Model and Coordinate Systems

Figure 14.1 is a rotor system composed of a rigid disk and a flexible shaft with distributed mass and distributed stiffness. We prepare the static coordinate

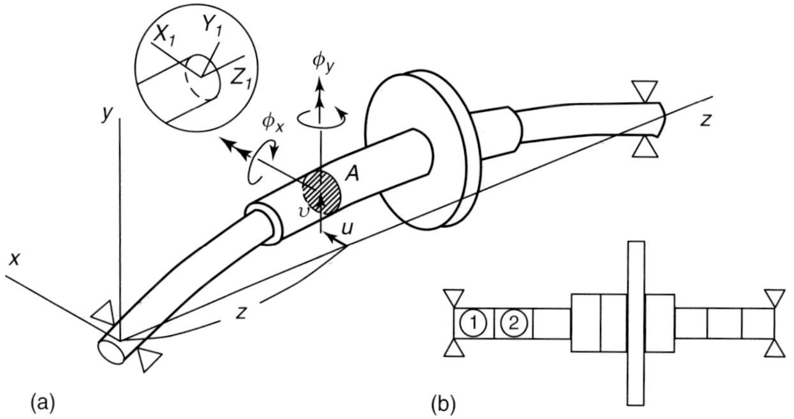

Figure 14.1 Rotor model and finite elements. (a) Coordinate system and (b) division into elements.

system O-xyz whose z-axis coincides with the bearing centerline. Let us consider a representative plane A perpendicular to the shaft centerline at position z. The translations of this plane in the x- and y-directions are denoted by u and v, respectively, and its rotation angles are denoted by ϕ_x and ϕ_y, respectively. These angles ϕ_x and ϕ_y are small angles and have the relationships $\phi_x = -\theta_y$ and $\phi_y = \theta_x$ with the inclination angles θ_x and θ_y, which were used in earlier chapters. Further, we consider the rectangular coordinate system O-$X_1Y_1Z_1$, which is fixed on plane A (Figure 2.16).

This rotor system is divided into the finite number of elements shown in Figure 14.1b. The appropriate number of elements are determined depending on the order of the vibration modes expected to be known and the shape of the rotor. It is assumed that the shape of the cross section, diameter, and material constants are uniform in each element.

14.3.2
Equations of Motion of an Element

In this section, equations of motion for elements are developed by Lagrange's equation.

14.3.2.1 Rigid Disk

Let the mass, polar moment of inertia, and diametral moment of inertia of the disk be m_d, I_p, and I, respectively. For simplicity, it is assumed that the center of gravity coincides with the center of an elastic shaft. When the disk displaces by u and v and inclines by ϕ_x and ϕ_y, the kinetic energy T_d is obtained by putting $e = \tau = 0$,

$\Theta_1 + \pi/2 - \beta = \omega t, \quad x = u, \quad y = v, \quad \theta_y = -\phi_x$, and $\theta_x = \phi_y$ into Eqs. (2.61) and (2.67) as follows:

$$
\begin{aligned}
T_{\mathrm{d}} &= \frac{1}{2} m_{\mathrm{d}} \left(\dot{u}^2 + \dot{v}^2\right) + \frac{1}{2} I_{\mathrm{p}} \left\{\omega^2 + \omega \left(\dot{\phi}_x \phi_y - \dot{\phi}_x \dot{\varphi}_y\right)\right\} + \frac{1}{2} I \left(\dot{\phi}_x^2 + \dot{\phi}_y^2\right) \\
&= \frac{1}{2} \begin{Bmatrix} \dot{u} \\ \dot{v} \end{Bmatrix}^T \begin{bmatrix} m_{\mathrm{d}} & 0 \\ 0 & m_{\mathrm{d}} \end{bmatrix} \begin{Bmatrix} \dot{u} \\ \dot{v} \end{Bmatrix} + \frac{1}{2} \begin{Bmatrix} \dot{\phi}_x \\ \dot{\phi}_y \end{Bmatrix}^T \begin{bmatrix} I & 0 \\ 0 & I \end{bmatrix} \begin{Bmatrix} \dot{\phi}_x \\ \dot{\phi}_y \end{Bmatrix} \\
&+ \frac{1}{2} I_{\mathrm{p}} \left\{\omega^2 + \omega \left(\dot{\phi}_x \phi_y - \phi_x \dot{\phi}_y\right)\right\}
\end{aligned}
\tag{14.3}
$$

Substituting this expression into Lagrange's equation (Eq. 5.59), we can obtain the equations of motion. For uniformity of symbols we put $u = q_1^{\mathrm{d}}, v = q_2^{\mathrm{d}}, \phi_x = q_3^{\mathrm{d}}$,, and $\phi_y = q_4^{\mathrm{d}}$. Then, the equations of motion of the rigid disk are given as follows:

$$
[m^{\mathrm{d}}]\left\{\ddot{q}^{\mathrm{d}}\right\} + [g^{\mathrm{d}}]\left\{\dot{q}^{\mathrm{d}}\right\} = \left\{f^{\mathrm{d}}\right\}
\tag{14.4}
$$

where $\left\{q^{\mathrm{d}}\right\} = \lfloor q_1^{\mathrm{d}}\, q_2^{\mathrm{d}}\, q_3^{\mathrm{d}}\, q_4^{\mathrm{d}} \rfloor^T = \lfloor u \;\, v \;\, \phi_x \;\, \phi_y \rfloor^T$ is a displacement vector and $\left\{f^d\right\}$ represents various kinds of forces working on a disk, such as unbalance forces given by Eq. (14.2) and forces from the adjacent elements. The superscript d in these notations shows that the quantities are related to a disk. The element mass matrix $[m^d]$ and the element gyroscopic matrix $[g^d]$ are given as follows:

$$
\left[m^{\mathrm{d}}\right] = \begin{bmatrix} m_{\mathrm{d}} & 0 & 0 & 0 \\ 0 & m_{\mathrm{d}} & 0 & 0 \\ 0 & 0 & I & 0 \\ 0 & 0 & 0 & I \end{bmatrix}, \quad \left[g^{\mathrm{d}}\right] = \omega \begin{bmatrix} 0 & 0 & 0 & 0 \\ 0 & 0 & 0 & 0 \\ 0 & 0 & 0 & I_{\mathrm{p}} \\ 0 & 0 & -I_{\mathrm{p}} & 0 \end{bmatrix}
\tag{14.5}
$$

14.3.2.2 Finite Rotor Element

A typical element obtained by dividing the shaft appropriately is shown in Figure 14.2. The length of this element is l. We consider the coordinates O$-xys$ whose origin is taken at the left end of this element. The above-mentioned coordinate system O$-xyz$ considered for the complete system in Figure 14.1 is called *global reference system*, while the coordinate system O$-xys$ fixed to the element is called a *local coordinate system*. Let us consider a sliced disk with thickness ds at the position s. The time-dependent displacements u, v, ϕ_x, ϕ_y also change as functions of position s, and its rotations ϕ_x, ϕ_y and translations u, v have the following relationships:

$$
\phi_x = -\frac{\partial v}{\partial s}, \quad \phi_y = \frac{\partial u}{\partial s}
\tag{14.6}
$$

The displacements u, v, ϕ_x, ϕ_y of the nodal points at the left and right ends are denoted by $q_1^{\mathrm{s}}, q_2^{\mathrm{s}}, q_3^{\mathrm{s}}, q_4^{\mathrm{s}}$ and $q_5^{\mathrm{s}}, q_6^{\mathrm{s}}, q_7^{\mathrm{s}}, q_8^{\mathrm{s}}$, respectively, where the superscript s denotes a shaft. The displacement of an arbitrary point is determined from these node displacements by the interpolation. By such an operation, the problem of a continuous system is reduced to the problem of a multi-degree-of-freedom system. Since the rotations are given by Eq. (14.6), the quantities u and v are given by linear combinations of the node displacements as follows:

$$\left.\begin{aligned} u(s,t) &= \psi_{x1}q_1^s + \psi_{x2}q_2^s + \cdots + \psi_{x8}q_8^s \\ v(s,t) &= \psi_{y1}q_1^s + \psi_{y2}q_2^s + \cdots + \psi_{y8}q_8^s \end{aligned}\right\} \tag{14.7}$$

These are expressed by matrices as follows:

$$\left\{ \begin{array}{c} u(s,t) \\ v(s,t) \end{array} \right\} = [\Psi(s)]\{q^s\} \tag{14.8}$$

where

$$\left.\begin{aligned} \{q^s\} &= \lfloor q_1^s \quad q_2^s \cdots q_8^s \rfloor^T \\ [\Psi(s)] &= \left[\begin{array}{c} \lfloor \Psi_x \rfloor \\ \lfloor \Psi_y \rfloor \end{array} \right] = \left[\begin{array}{c} \psi_{x1}\psi_{x2} \cdots \psi_{x8} \\ \psi_{y1}\psi_{y2} \cdots \psi_{y8} \end{array} \right] \end{aligned}\right\} \tag{14.9}$$

The symbol $\{\bullet\}$ represents a column vector and $\lfloor \bullet \rfloor$ represents a row vector. The functions $\psi_{xi}, \psi_{yi}\quad (i = 1, \cdots, 8)$ used for interpolation are called *shape functions.* Here, we use the static deflections that occur when a unit displacement is given to one of the coordinates at the left or right end (Petyt, 1990). For example, we know from the theory of strength of materials that when the values $q_1^s = 1, q_2^s = \cdots = q_8^s = 0$ are given, the static deflection is represented by $u(s) = 1 - 3(s/l)^2 + 2(s/l)^3$. Therefore, the matrix of translation shape function is given by

$$\left.\begin{aligned} &[\Psi(s)] = \left[\begin{array}{cccccccc} \psi_1 & 0 & 0 & \psi_2 & \psi_3 & 0 & 0 & \psi_4 \\ 0 & \psi_1 & -\psi_2 & 0 & 0 & \psi_3 & -\psi_4 & 0 \end{array} \right] \\ &\psi_1 = 1 - 3\left(\frac{s}{l}\right)^2 + 2\left(\frac{s}{l}\right)^3, \psi_2 = s\left[1 - 2\left(\frac{s}{l}\right) + \left(\frac{s}{l}\right)^2\right] \\ &\psi_3 = 3\left(\frac{s}{l}\right)^2 - 2\left(\frac{s}{l}\right)^3, \quad \psi_4 = l\left[-\left(\frac{s}{l}\right)^2 + \left(\frac{s}{l}\right)^3\right] \end{aligned}\right\} \tag{14.10}$$

From Eqs. (14.6), (14.8), and (14.10), the rotations are given by

$$\left\{ \begin{array}{c} \phi_x(s,t) \\ \phi_y(s,t) \end{array} \right\} = [\Phi(s)]\{q_s\} \tag{14.11}$$

where

$$[\Phi(s)] = \left[\begin{array}{c} \lfloor \Phi_x \rfloor \\ \lfloor \Phi_y \rfloor \end{array} \right] = \left[\begin{array}{cccccccc} 0 & -\psi_1' & \psi_2' & 0 & 0 & -\psi_3' & \psi_4' & 0 \\ \psi_1' & 0 & 0 & \psi_2' & \psi_3' & 0 & 0 & \psi_4' \end{array} \right] \tag{14.12}$$

is a matrix of rotation shape functions and ' denotes the differentiation by s.

Next, we derive the energy expressions. First, we consider the potential energy. From knowing the theory of strength of materials, we can derive the expressions of the potential energy stored in a beam with length l deflecting in the xz-plane. It is given by

$$U = \int_0^l dU, \quad dU = \frac{1}{2}EI_a\left(\frac{\partial^2 u}{\partial s^2}\right)^2 ds \tag{14.13}$$

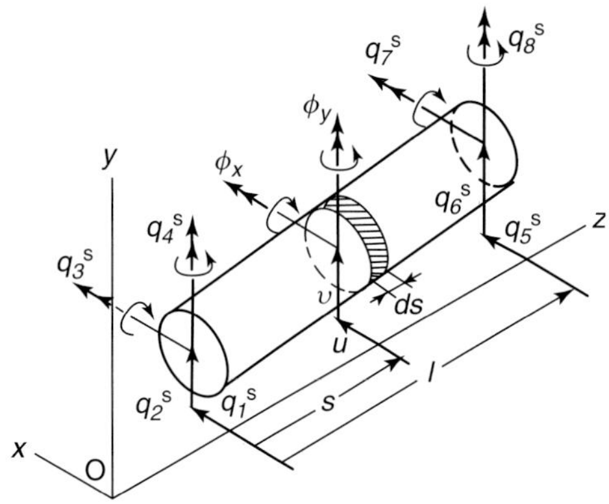

Figure 14.2 Typical rotor element and coordinates.

where I_a is the cross-sectional area moment of inertia of the beam about the y-axis (Petyt, 1990). Referring to this, we can derive the expression for the potential energy dU_s stored in a sliced disk with length ds shown in Figure 14.2 as follows

$$dU_s = \frac{1}{2} \begin{Bmatrix} u'' \\ v'' \end{Bmatrix}^T \begin{bmatrix} EI_a & 0 \\ 0 & EI_a \end{bmatrix} \begin{Bmatrix} u'' \\ v'' \end{Bmatrix} ds \tag{14.14}$$

Equation (14.14) is rewritten as follows using Eq. (14.8):

$$dU_s = \frac{1}{2} EI_a \{q^s\}^T [\Psi'']^T [\Psi''] \{q^s\} \, ds \tag{14.15}$$

The potential energy of the element is obtained by integrating this from 0 to l:

$$U_s = \frac{1}{2} \{q^s\}^T [k^s] \{q^s\} \tag{14.16}$$

where $[k^s]$ is the element stiffness matrix given by

$$[k^s] = \int_0^l EI_a [\Psi'']^T [\Psi''] \, ds \tag{14.17}$$

Next, we consider the kinetic energy. From Eq. (A.9) of Appendix A and considering that the sliced disk shown in Figure 14.2 is very thin, we can obtain the expressions for the polar moment of inertia dI_p and the diametral moment of inertia dI as follows:

$$dI_p = \frac{1}{2}\mu r^2 ds, \quad dI = \mu \left(\frac{r^2}{4} + \frac{ds^2}{12} \right) ds \approx \frac{1}{2} dI_p \tag{14.18}$$

where μ is mass of the shaft per unit length and r is the shaft radius. Let the polar moment of inertia per unit length be $\bar{I}_p (= \mu r^2/2)$. Then the polar moment of inertia and the diametral moment of inertia of the sliced disk are given by $\bar{I}_p ds$ and $\bar{I}_p ds/2$, respectively. Consequently, from Eq. (14.3), the kinetic energy of the sliced disk is given as follows:

$$dT_{\mathrm{s}}=\frac{1}{2}\left\{\begin{array}{c}\dot{u}\\ \dot{v}\end{array}\right\}^{T}\left[\begin{array}{cc}\mu & 0\\ 0 & \mu\end{array}\right]\left\{\begin{array}{c}\dot{u}\\ \dot{v}\end{array}\right\}ds+\frac{1}{2}\omega^{2}\bar{I}_{\mathrm{p}}ds$$
$$+\frac{1}{4}\left\{\begin{array}{c}\dot{\phi}_{x}\\ \dot{\phi}_{y}\end{array}\right\}^{T}\left[\begin{array}{cc}\bar{I}_{\mathrm{p}} & 0\\ 0 & \bar{I}_{\mathrm{p}}\end{array}\right]\left\{\begin{array}{c}\dot{\phi}_{x}\\ \dot{\phi}_{y}\end{array}\right\}ds+\frac{1}{2}\bar{I}_{\mathrm{p}}ds\omega\left(\dot{\phi}_{x}\phi_{y}-\phi_{x}\dot{\phi}_{y}\right) \tag{14.19}$$

Using Eqs. (14.8) and (14.11), this expression is transformed into the following form:

$$dT_{\mathrm{s}}=\frac{1}{2}\mu\left\{\dot{q}^{\mathrm{s}}\right\}^{T}[\Psi]^{T}[\Psi]\left\{\dot{q}^{\mathrm{s}}\right\}ds+\frac{1}{2}\omega^{2}\bar{I}_{\mathrm{p}}ds+\frac{1}{4}\bar{I}_{\mathrm{p}}\left\{\dot{q}^{\mathrm{s}}\right\}^{T}[\Phi]^{T}[\Phi]\left\{\dot{q}^{\mathrm{s}}\right\}ds$$
$$+\frac{1}{2}\bar{I}_{\mathrm{p}}ds\omega\left\{\dot{q}^{\mathrm{s}}\right\}^{T}\lfloor\Phi_{x}\rfloor^{T}\lfloor\Phi_{y}\rfloor\left\{q^{\mathrm{s}}\right\}-\frac{1}{2}\bar{I}_{\mathrm{p}}ds\omega\left\{\dot{q}^{\mathrm{s}}\right\}^{T}\lfloor\Phi_{y}\rfloor^{T}\lfloor\Phi_{x}\rfloor\left\{q^{\mathrm{s}}\right\} \tag{14.20}$$

In this derivation, we used the relationship

$$\begin{aligned}\dot{\phi}_{x}\phi_{y}-\phi_{x}\dot{\phi}_{y}&=\lfloor\Phi_{x}\rfloor\left\{\dot{q}^{\mathrm{s}}\right\}\lfloor\Phi_{y}\rfloor\left\{q^{\mathrm{s}}\right\}-\lfloor\Phi_{x}\rfloor\left\{q^{\mathrm{s}}\right\}\lfloor\Phi_{y}\rfloor\left\{\dot{q}^{\mathrm{s}}\right\}\\&=\left\{\dot{q}^{\mathrm{s}}\right\}^{T}\lfloor\Phi_{x}\rfloor^{T}\lfloor\Phi_{y}\rfloor\left\{q^{\mathrm{s}}\right\}-\lfloor\Phi_{y}\rfloor\left\{\dot{q}^{\mathrm{s}}\right\}\lfloor\Phi_{x}\rfloor\left\{q^{\mathrm{s}}\right\}\\&=\left\{\dot{q}^{\mathrm{s}}\right\}^{T}\lfloor\Phi_{x}\rfloor^{T}\lfloor\Phi_{y}\rfloor\left\{q^{\mathrm{s}}\right\}-\left\{\dot{q}^{\mathrm{s}}\right\}^{T}\lfloor\Phi_{y}\rfloor^{T}\lfloor\Phi_{x}\rfloor\left\{q^{\mathrm{s}}\right\}\end{aligned} \tag{14.21}$$

The kinetic energy of the element is obtained by integrating Eq. (14.19) from 0 to l. Then, we have

$$T_{\mathrm{s}}=\frac{1}{2}\left\{\dot{q}^{\mathrm{s}}\right\}^{T}[m^{\mathrm{s}}]\left\{\dot{q}^{2}\right\}+\frac{1}{2}\omega^{2}\bar{I}_{\mathrm{p}}l+\frac{1}{2}\omega\left\{\dot{q}^{\mathrm{s}}\right\}^{T}[n^{\mathrm{s}}]\left\{q^{\mathrm{s}}\right\} \tag{14.22}$$

where

$$\left.\begin{array}{l}[m^{\mathrm{s}}]=\left[m_{t}^{\mathrm{s}}\right]+\left[m_{r}^{\mathrm{s}}\right]\\ \left[m_{t}^{\mathrm{s}}\right]=\displaystyle\int_{0}^{l}\mu[\Psi]^{T}[\Psi]ds,\quad\left[m_{r}^{\mathrm{s}}\right]=\frac{1}{2}\int_{0}^{l}\bar{I}_{\mathrm{p}}[\Phi]^{T}[\Phi]ds\\ [n^{\mathrm{s}}]=\left[n_{a}^{\mathrm{s}}\right]-\left[n_{b}^{\mathrm{s}}\right]\\ \left[n_{a}^{\mathrm{s}}\right]=\displaystyle\int_{0}^{l}\bar{I}_{\mathrm{p}}\lfloor\Phi_{x}\rfloor^{T}\lfloor\Phi_{y}\rfloor ds\quad\left[n_{b}^{\mathrm{s}}\right]=\int_{0}^{l}\bar{I}_{\mathrm{p}}\lfloor\Phi_{y}\rfloor^{T}\lfloor\Phi_{x}\rfloor ds\end{array}\right\} \tag{14.23}$$

Next, we consider the virtual work. Let the distributed nonconservative forces in the x- and y-directions be $f_x(s,t)$ and $f_y(s,t)$, respectively, and the distributed nonconservative moments around the x- and y-axes be $m_x(s,t)$ and $m_y(s,t)$, respectively. Further, let the forces and moments exerted by the adjacent elements on the left and right ends be $f_1^*,\cdots,f_4^*$ and $f_5^*,\cdots,f_8^*$, respectively. The virtual work δW done by these forces for the virtual displacements $\delta u(s,t), \delta v(s,t), \delta\phi_x(s,t), \delta\phi_y(s,t)$ at s and $\delta q_i^{\mathrm{s}}(i=1-8)$ at the ends of the elements are given as follows:

$$\begin{aligned}\delta W=&\int_{0}^{l}f_{x}(s,t)\delta u(s,t)ds+\int_{0}^{l}f_{y}(s,t)\delta v(s,t)ds\\&+\int_{0}^{l}m_{x}(s,t)\delta\phi_{x}(s,t)ds+\int_{0}^{l}m_{y}(s,t)\delta\phi_{y}(s,t)ds\\&+\sum_{i=1}^{8}f_{i}^{*}(t)\delta q_{i}^{\mathrm{s}}(t)\end{aligned} \tag{14.24}$$

Substituting the relationships

$$\left.\begin{aligned} \delta u(s,t) &= \lfloor \Psi_x(s) \rfloor \{\delta q^s\} \\ \delta v(s,t) &= \lfloor \Psi_y(s) \rfloor \{\delta q^s\} \\ \delta \phi_x(s,t) &= \lfloor \Phi_x(s) \rfloor \{\delta q^s\} \\ \delta \phi_y(s,t) &= \lfloor \Phi_y(s) \rfloor \{\delta q^s\} \end{aligned}\right\} \tag{14.25}$$

obtained from Eqs. (14.18) and (14.11) into Eq. (14.24), we have

$$\delta W(s,t) = \lfloor f^s(t) \rfloor \{\delta q^s\} \tag{14.26}$$

where $\lfloor f^s(t) \rfloor$ is a row vector composed of the nodal forces $f_i (i = 1, \ldots, n)$ which are given by

$$f_i(t) = \int_0^l f_x \psi_{xi} ds + \int_0^l f_y \psi_{yi} ds + \int_0^l m_x \phi_{xi} ds + \int_0^l m_y \phi_{yi} ds + f_i^* \tag{14.27}$$

We substitute Eqs. (14.16), (14.22), and (14.27) into Lagrange's equation given by Eq. (5.59).

Using the relationships

$$\left.\begin{aligned} &\left\{\frac{\partial U_s}{\partial q}\right\} = [k^s]\{q^s\} \\ &\left\{\frac{d}{dt}\left(\frac{\partial T_s}{\partial \dot{q}}\right)\right\} = [m^s]\{\ddot{q}^s\} + \frac{1}{2}\omega [n^s]\{\dot{q}^s\} \\ &\left\{-\frac{\partial T_s}{\partial q}\right\} = -\frac{1}{2}\omega \{\dot{q}^s\}^T [n^s] = -\frac{1}{2}\omega [n^s]^T \{\dot{q}^s\} \end{aligned}\right\} \tag{14.28}$$

we arrive at the following element equations of motion:

$$[m^s]\{\ddot{q}^s\} + [g^s]\{\dot{q}^s\} + [k^s]\{q^s\} = \{f^s\} \tag{14.29}$$

where

$$[g^s] = \frac{1}{2}\omega \left([n^s] - [n^s]^T\right) \tag{14.30}$$

Matrices in Eq. (14.29) are obtained by Eqs. (14.17) and (14.23). For example, from Eqs. (14.12) and (14.23), we have

$$[m_r^s] = \frac{1}{2}\int_0^l \bar{I}_p [\Phi]^T [\Phi] \, ds = \frac{1}{2}\int_0^l \bar{I}_p \begin{bmatrix} 0 & \psi_1' \\ * & * \\ \vdots & \vdots \\ * & * \end{bmatrix} \begin{bmatrix} 0 & * \cdots * \\ \psi_1' & * \cdots * \end{bmatrix} ds \tag{14.31}$$

The element of the first row and first column is

$$\frac{1}{2}\int_0^l \bar{I}_p \psi_1'^2 ds = \frac{1}{2}\int_0^l \bar{I}_p \left(6\frac{s}{l^2} - 6\frac{s}{l^3}\right)^2 ds = \frac{36\bar{I}_p}{60l} = \frac{36\mu r^2}{120l} \tag{14.32}$$

From such calculations, we have

$$[m_t^s] = \frac{\mu l}{420}\begin{bmatrix} 156 & & & & & & & \\ 0 & 156 & & & & \text{Sym.} & & \\ 0 & -22l & 4l^2 & & & & & \\ 22l & 0 & 0 & 4l^2 & & & & \\ 54 & 0 & 0 & 13l & 156 & & & \\ 0 & 54 & -13l & 0 & 0 & 156 & & \\ 0 & 13l & -3l^2 & 0 & 0 & 22l & 4l^2 & \\ -13l & 0 & 0 & -3l^2 & -22l & 0 & 0 & 4l^2 \end{bmatrix} \tag{14.33}$$

$$[m_r^s] = \frac{\mu r^2}{120l}\begin{bmatrix} 36 & & & & & & & \\ 0 & 36 & & & & \text{Sym.} & & \\ 0 & -3l & 4l^2 & & & & & \\ 3l & 0 & 0 & 4l^2 & & & & \\ -36 & 0 & 0 & -3l & 36 & & & \\ 0 & -36 & 3l & 0 & 0 & 36 & & \\ 0 & -3l & -l^2 & 0 & 0 & 3l & 4l^2 & \\ 3l & 0 & 0 & -l^2 & -3l^2 & 0 & 0 & 4l^2 \end{bmatrix} \tag{14.34}$$

where "Sym." means that the matrix is symmetric. The element mass matrix is obtained by summing up these matrices:

$$[m^s] = [m_t^s] + [m_r^s] \tag{14.35}$$

Similarly, the element gyroscopic matrix is given by

$$[g^s] = \frac{\omega \bar{I}_p}{30l}\begin{bmatrix} 0 & & & & & & & \\ -36 & 0 & & & & \text{Skew}-\text{sym.} & & \\ 3l & 0 & 0 & & & & & \\ 0 & 3l & -4l^2 & 0 & & & & \\ 0 & -36 & 3l & 0 & 0 & & & \\ 36 & 0 & 0 & 3l & -36 & 0 & & \\ 3l & 0 & 0 & -l^2 & -3l & 0 & 0 & \\ 0 & 3l & l^2 & 0 & 0 & -3l & -4l^2 & 0 \end{bmatrix} \tag{14.36}$$

where "Skew-sym." means that the matrix is skew-symmetric. Moreover, the element stiffness matrix is given by

$$[k^s] = \frac{EI_a}{l^3}\begin{bmatrix} 12 & & & & & & & \\ 0 & 12 & & & & \text{Sym.} & & \\ 0 & -6l & 4l^2 & & & & & \\ 6l & 0 & 0 & 4l^2 & & & & \\ -12 & 0 & 0 & -6l & 12 & & & \\ 0 & -12 & 6l & 0 & 0 & 12 & & \\ 0 & -6l & 2l^2 & 0 & 0 & 6l & 4l^2 & \\ 6l & 0 & 0 & 2l^2 & -6l & 0 & 0 & 4l^2 \end{bmatrix} \tag{14.37}$$

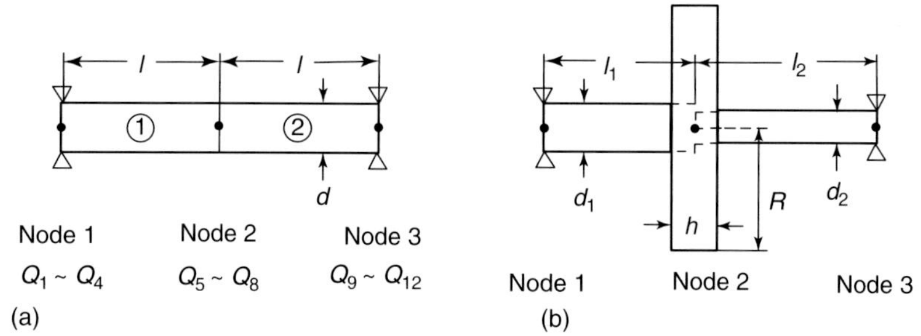

Figure 14.3 Two analytical rotor models. (a) Model I (uniform rotor) and (b) model II (disk-shaft system).

In the following, we denote the elements of the ith row and jth column of matrices $[m^s]$, $[g^s]$, and $[k^s]$ by m_{ij}^s, g_{ij}^s, and k_{ij}^s, respectively.

14.3.3 Equations of Motion for a Complete System

In this section, we derive the equations of motion for the complete system by assembling the results for elements. For simplicity, we explain using examples shown in Figure 14.3. Figure 14.3a shows a uniform continuous elastic rotor divided into two elements with length l. Since the exact solution of this model is known theoretically, we can ascertain the accuracy of the solution obtained by the finite element method. Figure 14.3b shows a rotor composed of a rigid disk and a flexible shaft with variable cross section. In the following discussion, we call the former "model I" and the latter "model II."

14.3.3.1 Model I: (Uniform Elastic Rotor)

To derive the equations of motion for the complete system, we must prepare the global reference system. Here the global reference system O$-xyz$ shown in Figure 14.1 and the local coordinate system O$-xys$ shown in Figure 14.2 are adopted. Since the axes of these two coordinate systems are parallel, the relationship between them is very simple because the direction cosines have values 0 or 1. The displacements in the corresponding axes in these two systems have the same magnitude and are in the same direction. A new displacement vector $\{Q\} = \lfloor Q_1, Q_2, Q_3, \ldots, Q_{12} \rfloor^T$, which has all the node displacements as its components is defined, where $Q_1, \ldots, Q_4$ correspond to the displacements $q_1^s, \ldots, q_4^s$ of the left end of element 1, and $Q_5, \ldots, Q_8$ correspond to both the displacement $q_5^s, \ldots, q_8^s$ of the right end of element 1 and the displacement $q_1^s, \ldots, q_4^s$ of the left end of element 2. For the kth element, the element displacement vector $\{q^s\}_k = \lfloor q_1^s, \ldots, q_8^s \rfloor^T$ and the displacement vector $\{Q\}$ of the complete system are related by the 8×12 matrix $[A]_k$ as follows:

$$\{q^s\}_k = [A]_k \{Q\} \quad (k = 1, 2) \tag{14.38}$$

or

$$\begin{Bmatrix} q_1^s \\ q_2^s \\ \cdots \\ q_8^s \end{Bmatrix}_k = \begin{bmatrix} 0 & 0 & \cdots & 0 & 1 & 0 & \cdots & 0 & 0 & \cdots & 0 \\ 0 & 0 & \cdots & 0 & 0 & 1 & \cdots & 0 & 0 & \cdots & 0 \\ \cdots & \cdots & \cdots & \cdots & \cdots & \cdots & \cdots & \cdots & \cdots & \cdots & \cdots \\ 0 & 0 & \cdots & 0 & 0 & 0 & \cdots & 1 & 0 & \cdots & 0 \end{bmatrix} \begin{Bmatrix} Q_1 \\ Q_2 \\ \cdots \\ \cdots \\ Q_{12} \end{Bmatrix}$$

$$\uparrow \qquad\qquad \uparrow$$

$$4(k-1)+1\text{st} \qquad 4(k-1)+8\text{th}$$

(14.39)

where the subscript k indicates the element number. In each row of $[A]_k$, only one of the elements is 1 and the rest are 0.

Next we derive energy expressions for the complete system. The total energy of the system is given by summing up the corresponding energies of elements. From Eq. (14.22), the kinetic energy T for the complete system is given by

$$\begin{aligned} T &= \frac{1}{2}\sum_{k=1}^{2}\{\dot{q}^s\}_k^T[m^s]_k\{\dot{q}^s\}_k + \omega^2\bar{I}_p l + \frac{1}{2}\omega\sum_{k=1}^{2}\{\dot{q}^2\}_k^T[n^s]_k\{q^s\}_k \\ &= \frac{1}{2}\{\dot{Q}\}^T[M]\{\dot{Q}\} + \omega^2\bar{I}_p l + \frac{1}{2}\omega\{\dot{Q}\}^T[N]\{Q\} \end{aligned} \tag{14.40}$$

where the subscript k denotes that these quantities pertain to the kth element and where

$$[M] = \sum_{k=1}^{2}[A]_k^T[m^s]_k[A]_k, \quad [N] = \sum_{k=1}^{2}[A]_k^T[n^s]_k[A]_k \tag{14.41}$$

The matrix $[M]$ is the mass matrix of the complete system. From Eqs. (14.23) and (14.30), the gyroscopic matrix $[G]$ of the complete system is obtained by

$$[G] = \frac{1}{2}\omega\left([N]-[N]^T\right) \tag{14.42}$$

From Eq. (14.16), the potential energy is expressed as

$$U = \frac{1}{2}\{Q\}^T[K]\{Q\} \tag{14.43}$$

where the stiffness matrix $[K]$ of the complete system is given by

$$[K] = \sum_{k=1}^{2}[A]_k^T[k^s]_k[A]_k \tag{14.44}$$

Substituting Eq. (14.38) into Eq. (14.26), we get the following expression for the virtual work:

$$\delta W = \sum_{k=1}^{2}\{f^s\}_k^T[\delta q^s]_k = \{F\}^T\{\delta Q\} \tag{14.45}$$

where

$$\{\mathrm{F}\}^T = \sum_{k=1}^{2} \{f^s\}_k^T [A]_k \tag{14.46}$$

The force vector {F} has 12 elements. Among them, the elements relating to the center node 2 are given by the summation of the nodal forces at the right end of element 1 and those at the left end of element 2, that is,

$$\{\mathrm{F}\} = \left\lfloor f_1^{(1)} \cdots f_4^{(1)} f_5^{(1)} + f_1^{(2)} \cdots f_8^{(1)} + f_4^{(2)} f_5^{(2)} \cdots f_8^{(2)} \right\rfloor^T \tag{14.47}$$

where the numbers in the superscript denote element numbers. The forces f_i^* produced by the adjacent elements are canceled at the center node, and the forces f_i^* at the left end (node 1) and the right end (node 3) correspond to the reaction forces from the bearings. Substituting Eqs. (14.40), (l4.43), and (14.47) into Lagrange's equation, we obtain the equations of motion for the complete system.

$$[M]\{\ddot{Q}\} + [G]\{\dot{Q}\} + [K]\{Q\} = \{F\} \tag{14.48}$$

In the following discussion, we calculate the concrete forms of the matrices $[M]$, $[G]$, and $[K]$. First, the mass matrix is derived from Eq. (14.41). In that expression $[m^s]_k$ is an 8×8 matrix and $[A]_k$ is an 8×12 matrix. Let $[I]$ be an 8×8 unit matrix and [0] be a null matrix. Then $[A]_1 = [[\mathrm{I}]\ [\mathrm{O}]]$ and $[A]_2 = [[\mathrm{O}]\ [\mathrm{I}]]$, and we get

$$\begin{aligned}
[M] &= \begin{bmatrix} [I] \\ [0] \end{bmatrix} [m^s]_1 [[I] \quad [0]] \quad + \quad \begin{bmatrix} [0] \\ [I] \end{bmatrix} [m^s]_2 [[I] \quad [0]] \\
&= \begin{bmatrix} [m^s]_1 & [0] \\ [0] & [0] \end{bmatrix} + \begin{bmatrix} [0] & [0] \\ [0] & [m^s]_2 \end{bmatrix} = \begin{bmatrix} [m^s]_1 & [0] \\ [0] & [m^s]_2 \end{bmatrix}
\end{aligned} \tag{14.49}$$

Equation (14.49) shows that the process given by Eq. (14.41) to calculate the mass matrix for the complete system is equivalent to the following two steps: first, extend the element mass matrix to a 12×12 matrix by adding zero elements, and then, add these two extended matrices. In this example, elements are duplicated in the zone concerning rows 5–8 and columns 5–8. The result is

$$[M] = \begin{bmatrix}
m_{11}^{(1)} & \cdots & m_{14}^{(1)} & \cdots & \cdots & m_{18}^{(1)} & 0 & \cdots & 0 \\
\cdots & \cdots & \cdots & \cdots & \cdots & \cdots & \cdots & \cdots & \cdots \\
\cdots & \cdots & \cdots & \cdots & \cdots & \cdots & 0 & \cdots & 0 \\
\cdots & \cdots & \cdots & m_{55}^{(1)} + m_{11}^{(2)} & \cdots & m_{58}^{(1)} + m_{14}^{(2)} & m_{15}^{(2)} & \cdots & m_{18}^{(2)} \\
\cdots & \cdots & \cdots & \cdots & \cdots & \cdots & \cdots & \cdots & \cdots \\
m_{81}^{(1)} & \cdots & m_{84}^{(1)} & m_{85}^{(1)} + m_{41}^{(2)} & \cdots & m_{88}^{(1)} + m_{44}^{(2)} & \cdots & \cdots & \cdots \\
0 & \cdots & 0 & \cdots & \cdots & \cdots & \cdots & \cdots & \cdots \\
\cdots & \cdots & \cdots & \cdots & \cdots & \cdots & \cdots & \cdots & \cdots \\
0 & \cdots & 0 & m_{81}^{(2)} & \cdots & m_{84}^{(2)} & m_{85}^{(2)} & \cdots & m_{88}^{(2)}
\end{bmatrix} \tag{14.50}$$

Since the dimensions of elements 1 and 2 are the same, the relationship $m_{ij}^{(1)} = m_{ij}^{(2)}$ holds. From Eq. (14.35), we can obtain these quantities by adding the terms in Eq. (14.33) and those in Eq. (14.34). The result has the following form, where null elements are denoted explicitly.

$$[M] = \left[\begin{array}{cccc|cccc|cccc}
m_{11} & & & & & & & & & \text{Sym} & & \\
0 & m_{11} & & & & & & & & & & \\
0 & -m_{41} & m_{33} & & & & & & & & & \\
m_{41} & 0 & 0 & m_{33} & & & & & & & & \\
\hline
m_{51} & 0 & 0 & -m_{81} & m_{55} & & & & & & & \\
0 & m_{51} & m_{81} & 0 & 0 & m_{55} & & & & & & \\
0 & -m_{81} & m_{84} & 0 & 0 & 0 & m_{77} & & & & & \\
m_{81} & 0 & 0 & m_{84} & 0 & 0 & 0 & m_{77} & & & & \\
\hline
0 & 0 & 0 & 0 & m_{51} & 0 & 0 & -m_{81} & m_{11} & & & \\
0 & 0 & 0 & 0 & 0 & m_{51} & m_{81} & 0 & 0 & m_{11} & & \\
0 & 0 & 0 & 0 & 0 & -m_{81} & m_{84} & 0 & 0 & m_{41} & m_{33} & \\
0 & 0 & 0 & 0 & m_{81} & 0 & 0 & m_{84} & -m_{41} & 0 & 0 & m_{33}
\end{array}\right] \tag{14.51}$$

where

$$\left.\begin{array}{llll}
m_{11} = m_{11}^{(1)}, & m_{41} = m_{41}^{(1)}, & m_{51} = m_{51}^{(1)}, & m_{81} = m_{81}^{(1)} \\
m_{33} = m_{33}^{(1)}, & m_{84} = m_{84}^{(1)}, & m_{55} = 2m_{11}^{(1)}, & m_{77} = 2m_{33}^{(1)}
\end{array}\right\} \tag{14.52}$$

Similarly, we get the gyroscopic matrix $[G]$ as follows:

$$[G] = \begin{bmatrix}
0 & & & & & & & & & \text{Skew sym.} & & \\
g_{21} & 0 & & & & & & & & & & \\
g_{31} & 0 & 0 & & & & & & & & & \\
0 & g_{31} & g_{43} & 0 & & & & & & & & \\
0 & g_{21} & g_{31} & 0 & 0 & & & & & & & \\
-g_{21} & 0 & 0 & g_{31} & g_{65} & 0 & & & & & & \\
g_{31} & 0 & 0 & -g_{83} & 0 & 0 & 0 & & & & & \\
0 & g_{31} & g_{83} & 0 & 0 & 0 & g_{87} & 0 & & & & \\
0 & 0 & 0 & 0 & 0 & g_{21} & g_{31} & 0 & 0 & & & \\
0 & 0 & 0 & 0 & -g_{21} & 0 & 0 & g_{31} & g_{21} & 0 & & \\
0 & 0 & 0 & 0 & g_{31} & 0 & 0 & -g_{83} & -g_{31} & 0 & 0 & \\
0 & 0 & 0 & 0 & 0 & g_{31} & g_{83} & 0 & 0 & -g_{31} & g_{43} & 0
\end{bmatrix} \tag{14.53}$$

Denoting the elements of Eq. (14.36) for rotor element 1 by $g_{ij}^{(1)}$, the elements of this matrix are given by

$$\left.\begin{array}{llll}
g_{21} = g_{21}^{(1)}, & g_{31} = g_{31}^{(1)}, & g_{43} = g_{43}^{(1)}, & g_{83} = g_{83}^{(1)} \\
g_{65} = 2g_{21}^{(1)}, & g_{87} = 2g_{43}^{(1)} & &
\end{array}\right\} \tag{14.54}$$

Moreover, the stiffness matrix is given as follows:

$$[K] = \begin{bmatrix}
k_{11} & & & & & & & & & & & \\
0 & k_{11} & & & & & & & & \text{Sym.} & & \\
0 & -k_{41} & k_{33} & & & & & & & & & \\
k_{41} & 0 & 0 & k_{33} & & & & & & & & \\
-k_{11} & 0 & 0 & -k_{41} & k_{55} & & & & & & & \\
0 & -k_{11} & k_{41} & 0 & 0 & k_{55} & & & & & & \\
0 & -k_{41} & k_{84} & 0 & 0 & 0 & k_{77} & & & & & \\
k_{41} & 0 & 0 & k_{84} & 0 & 0 & 0 & k_{77} & & & & \\
0 & 0 & 0 & 0 & -k_{11} & 0 & 0 & -k_{41} & k_{11} & & & \\
0 & 0 & 0 & 0 & 0 & -k_{11} & k_{41} & 0 & 0 & k_{11} & & \\
0 & 0 & 0 & 0 & 0 & -k_{41} & k_{84} & 0 & 0 & k_{41} & k_{33} & \\
0 & 0 & 0 & 0 & k_{41} & 0 & 0 & k_{84} & -k_{41} & 0 & 0 & k_{33}
\end{bmatrix} \tag{14.55}$$

where

$$\left.\begin{array}{l} k_{11} = k_{11}^{(1)}, \quad k_{41} = k_{41}^{(1)}, \quad k_{84} = k_{84}^{(1)}, \quad k_{33} = k_{33}^{(1)} \\ k_{55} = 2k_{11}^{(1)}, \quad k_{77} = 2k_{33}^{(1)} \end{array}\right\} \tag{14.56}$$

14.3.3.2 **Model II: Disk–Shaft System**

For the shaft of model II in Figure 14.3b, we can utilize the known results for the shaft element of model I. Therefore, we need to derive the energy expressions only about the rigid disk.

From Eq. (14.3), the kinetic energy T_d for a rigid disk is expressed as follows:

$$\begin{aligned} T_d &= \frac{1}{2}\begin{Bmatrix} \dot{q}_1^d \\ \dot{q}_2^d \end{Bmatrix}^T \begin{bmatrix} m_d & 0 \\ 0 & m_d \end{bmatrix} \begin{Bmatrix} \dot{q}_1^d \\ \dot{q}_2^d \end{Bmatrix} + \frac{1}{2}\begin{Bmatrix} \dot{q}_3^d \\ \dot{q}_4^d \end{Bmatrix}^T \begin{bmatrix} I & 0 \\ 0 & I \end{bmatrix} \begin{Bmatrix} \dot{q}_3^d \\ \dot{q}_4^d \end{Bmatrix} \\ &\quad + \frac{1}{2} I_p \omega^2 + \frac{1}{2} I_p \omega \left(\dot{q}_3^d q_4^d - q_3^d \dot{q}_4^d \right) \\ &= \frac{1}{2} \left\{ \dot{q}^d \right\}^T \left[m^d \right] \left\{ \dot{q}^d \right\} + \frac{1}{2} I_p \omega^2 + \frac{1}{2} \omega \left\{ \dot{q}^d \right\}^T \left[n^d \right]^T \left[n^d \right] \left\{ q^d \right\} \end{aligned} \tag{14.57}$$

where

$$\left.\begin{array}{l} \left\{ q^d \right\} = \left[q_1^d \quad q_2^d \quad q_3^d \quad q_4^d \right]^T \\ \left[m^d \right] = \begin{bmatrix} m_d & 0 & 0 & 0 \\ 0 & m_d & 0 & 0 \\ 0 & 0 & I & 0 \\ 0 & 0 & 0 & I \end{bmatrix}, \quad \left[n^d \right] = \begin{bmatrix} 0 & 0 & 0 & 0 \\ 0 & 0 & 0 & 0 \\ 0 & 0 & 0 & I_p \\ 0 & 0 & -I_p & 0 \end{bmatrix} \end{array}\right\} \tag{14.58}$$

To add this energy expression to Eq. (14.40), we extend this expression to that using the displacement vector for the complete system as follows, noting the relationships

$Q_5 = q_1^{\mathrm{d}}, Q_6 = q_2^{\mathrm{d}}, Q_7 = q_3^{\mathrm{d}}$, and $Q_8 = q_4^{\mathrm{d}}$:

$$T_{\mathrm{d}} = \frac{1}{2}\{\dot{Q}\}^T \begin{bmatrix} [0] & [0] & [0] \\ [0] & [m^{\mathrm{d}}] & [0] \\ [0] & [0] & [0] \end{bmatrix} \{\dot{Q}\} + \frac{1}{2} I_{\mathrm{p}}\omega^2 + \frac{1}{2}\omega\{\dot{Q}\}^T \begin{bmatrix} [0] & [0] & [0] \\ [0] & [n^{\mathrm{d}}] & [0] \\ [0] & [0] & [0] \end{bmatrix} \{Q\} \tag{14.59}$$

where the matrices in the first and third terms include Eq. (14.58) in rows 5–8 and columns 5–8. Comparing this result with Eq. (14.40), we know that matrices $[M]$ and $[G]$ for model II are obtained by adding the elements in Eq. (14.58) to the elements of rows 5–8 and columns 5–8 in matrices $[M]$ and $[G]$, after $[M]$ and $[G]$ are revised considering the difference in dimensions of elements ① and ②.

The expression of the virtual work is obtained if Eq. (14.4) (i.e., the contribution of $\{f^{\mathrm{d}}\}$) is added to Eq. (14.45), because shaft element ①, rigid disk, and shaft element ① are joined at node 2. Since the reaction forces cancel each other out, only the unbalance forces due to the rigid disk are added to components $F_5, \ldots, F_8$ of the force vector.

After these revisions, matrix $[M]$ of the equations of motion (14.48) for model II is given as follows:

$$[M] = \begin{bmatrix}
m_{11} & & & & & & & & & & & \\
0 & m_{11} & & & & & & & \text{Sym.} & & & \\
0 & -m_{41} & m_{33} & & & & & & & & & \\
m_{41} & 0 & 0 & m_{33} & & & & & & & & \\
m_{51} & 0 & 0 & -m_{81} & m_{55} & & & & & & & \\
0 & m_{51} & m_{81} & 0 & 0 & m_{55} & & & & & & \\
0 & -m_{81} & m_{84} & 0 & 0 & -m_{85} & m_{77} & & & & & \\
m_{81} & 0 & 0 & m_{84} & m_{85} & 0 & 0 & m_{77} & & & & \\
0 & 0 & 0 & 0 & m_{95} & 0 & 0 & -m_{125} & m_{99} & & & \\
0 & 0 & 0 & 0 & 0 & m_{95} & m_{125} & 0 & 0 & m_{99} & & \\
0 & 0 & 0 & 0 & 0 & -m_{125} & m_{128} & 0 & 0 & -m_{129} & m_{1111} & \\
0 & 0 & 0 & 0 & m_{125} & 0 & 0 & m_{128} & m_{129} & 0 & 0 & m_{1111}
\end{bmatrix} \tag{14.60}$$

where, if the mass and the diametral moment of inertia of the disk are denoted by m_{d} and I, respectively, the elements are given by

$$\left.\begin{array}{l}
m_{11} = m_{11}^{(1)}, \quad m_{41} = m_{41}^{(1)}, \quad m_{51} = m_{51}^{(1)}, \quad m_{81} = m_{81}^{(1)} \\
m_{33} = m_{33}^{(1)}, \quad m_{84} = m_{84}^{(1)}, \quad m_{95} = m_{51}^{(2)}, \quad m_{125} = m_{81}^{(2)} \\
m_{128} = m_{84}^{(2)}, \quad m_{129} = m_{85}^{(2)}, \quad m_{99} = m_{11}^{(2)}, \quad m_{1111} = m_{33}^{(1)} \\
m_{55} = m_{11}^{(1)} + m_{11}^{(2)} + m_d, \quad m_{77} = m_{33}^{(1)} + m_{33}^{(2)} + I \\
m_{85} = -m_{41}^{(1)} + m_{41}^{(2)}
\end{array}\right\} \tag{14.61}$$

and matrix $[G]$ is given by

$$[G] = \left[\begin{array}{cccccccccccc}
0 & & & & & & & & & & & \\
g_{21} & 0 & & & & & & & & & & \\
g_{31} & 0 & 0 & & & & & & \text{Skew} & \text{sym.} & & \\
0 & g_{31} & g_{43} & 0 & & & & & & & & \\
0 & g_{21} & g_{31} & 0 & 0 & & & & & & & \\
-g_{21} & 0 & 0 & g_{31} & g_{65} & 0 & & & & & & \\
g_{31} & 0 & 0 & -g_{83} & g_{75} & 0 & 0 & & & & & \\
0 & g_{31} & g_{83} & 0 & 0 & g_{75} & g_{87} & 0 & & & & \\
0 & 0 & 0 & 0 & 0 & g_{96} & g_{115} & 0 & 0 & & & \\
0 & 0 & 0 & 0 & -g_{96} & 0 & 0 & g_{115} & g_{96} & 0 & & \\
0 & 0 & 0 & 0 & g_{115} & 0 & 0 & -g_{127} & -g_{115} & 0 & 0 & \\
0 & 0 & 0 & 0 & 0 & g_{115} & g_{127} & 0 & 0 & -g_{125} & g_{1211} & 0
\end{array}\right] \tag{14.62}$$

where if the polar moment of inertia of the rigid disk is denoted by I_{p}, the elements are given by

$$\left.\begin{array}{l}
g_{21} = g_{21}^{(1)}, \quad g_{31} = g_{31}^{(1)}, \quad g_{43} = g_{43}^{(1)}, \quad g_{83} = g_{83}^{(1)} \\
g_{96} = g_{52}^{(2)}, \quad g_{115} = g_{31}^{(2)}, \quad g_{127} = g_{83}^{(2)}, \quad g_{1211} = g_{43}^{(2)} \\
g_{65} = g_{21}^{(1)} + g_{21}^{(2)} \\
g_{75} = -g_{31}^{(1)} + g_{31}^{(2)} \\
g_{87} = g_{43}^{(1)} + g_{43}^{(2)} - I_p
\end{array}\right\} \tag{14.63}$$

and finally, matrix $[K]$ is given by

$$[K] = \left[\begin{array}{cccccccccccc}
k_{11} & & & & & & & & & & & \\
0 & k_{11} & & & & & & & & & & \\
0 & -k_{41} & k_{33} & & & & & & & \text{Sym.} & & \\
k_{41} & 0 & 0 & k_{33} & & & & & & & & \\
-k_{11} & 0 & 0 & -k_{41} & k_{55} & & & & & & & \\
0 & -k_{11} & k_{41} & 0 & 0 & k_{55} & & & & & & \\
0 & -k_{41} & k_{84} & 0 & 0 & -k_{85} & k_{77} & & & & & \\
k_{41} & 0 & 0 & k_{84} & k_{85} & 0 & 0 & k_{77} & & & & \\
0 & 0 & 0 & 0 & -k_{99} & 0 & 0 & -k_{125} & k_{99} & & & \\
0 & 0 & 0 & 0 & 0 & -k_{99} & k_{125} & 0 & 0 & k_{99} & & \\
0 & 0 & 0 & 0 & 0 & -k_{125} & k_{128} & 0 & 0 & k_{125} & k_{1111} & \\
0 & 0 & 0 & 0 & k_{125} & 0 & 0 & k_{128} & -k_{125} & 0 & 0 & k_{1111}
\end{array}\right] \tag{14.64}$$

where

$$\left.\begin{aligned} &k_{11}=k_{11}^{(1)},\quad k_{41}=k_{41}^{(1)},\quad k_{84}=k_{84}^{(1)},\quad k_{33}=k_{33}^{(1)},\\ &k_{99}=k_{11}^{(2)},\quad k_{1111}=k_{33}^{(2)},\quad k_{125}=k_{41}^{(2)},\quad k_{128}=k_{84}^{(2)}\\ &k_{55}=k_{11}^{(1)}+k_{11}^{(2)},\quad k_{77}=k_{33}^{(1)}+k_{33}^{(2)},\quad k_{85}=-k_{41}^{(1)}+k_{41}^{(2)} \end{aligned}\right\} \tag{14.65}$$

14.3.3.3 **Variation of Equations of Motion**

Nelson and McVaugh (1976) derived and analyzed first-order differential equations from Eq. (14.48), as shown in Section 14.6. However, here we adopt a different procedure, which simplifies the analysis using physical characteristics of rotors (Sato, 1992).

As we learned in Chapter 2, the whirling motion of a rotor occurs as a result of the combination of motions in the xz-plane and in the yz-plane through the gyroscopic moment. While paying attention to this, we reform the equations of motion as discussed below.

The variables used in this chapter and the variables x, y, θ_x, and θ_y in Chapter 2 have the relationship $x = u,\quad y = v,\quad \theta_x = \varphi_y$, and $\theta_y = -\varphi_x$. Considering these correspondences, we group the variables as follows:

$$\left.\begin{aligned} &\{\overline{Q}_x\}=\left[u_1\quad \phi_{y1}\quad u_2\quad \phi_{y2}\quad u_3\quad \varphi_{y3}\right]^T,\\ &\{\overline{Q}_y\}=[v_1\quad -\phi_{x1}\quad v_2\quad -\phi_{x2}\quad v_3\quad -\varphi_{x3}]^T,\\ &\{\overline{F}_x\}=[F_{x1}\quad F_{x4}\quad F_{x5}\quad F_{x8}\quad F_{x9}\quad F_{x12}]^T,\\ &\{\overline{F}_y\}=\left[F_{y2}\quad -F_{y3}\quad F_{y6}\quad -F_{y7}\quad F_{y10}\quad -F_{x11}\right]^T \end{aligned}\right\} \tag{14.66}$$

Substituting matrices $[M], [G]$, and $[K]$ into the equations of motion (14.48) and rearranging rows and columns, we have the following expression:

$$\begin{aligned} &\begin{bmatrix} [\overline{M}] & [0] \\ [0] & [\overline{M}] \end{bmatrix}\begin{Bmatrix} \ddot{\overline{Q}}_x \\ \ddot{\overline{Q}}_y \end{Bmatrix}+\begin{bmatrix} [0] & [\overline{G}] \\ [-\overline{G}] & [0] \end{bmatrix}\begin{Bmatrix} \dot{\overline{Q}}_x \\ \dot{\overline{Q}}_y \end{Bmatrix}\\ &+\begin{bmatrix} [\overline{K}] & [0] \\ [0] & [\overline{K}] \end{bmatrix}\begin{Bmatrix} \overline{Q}_x \\ \overline{Q}_y \end{Bmatrix}=\begin{Bmatrix} \overline{F}_x \\ \overline{F}_y \end{Bmatrix} \end{aligned} \tag{14.67}$$

For model I, we have

$$\left.\begin{aligned}
[\overline{M}] &= \begin{bmatrix} m_{11} & & & & & \\ m_{41} & m_{33} & & & \text{Sym.} & \\ m_{51} & -m_{81} & m_{55} & & & \\ m_{81} & m_{84} & 0 & m_{77} & & \\ 0 & 0 & m_{51} & -m_{81} & m_{11} & \\ 0 & 0 & m_{81} & m_{84} & -m_{41} & m_{33} \end{bmatrix} \\
[\overline{G}] &= \begin{bmatrix} -g_{21} & & & & & \\ g_{31} & -g_{43} & & & \text{Sym.} & \\ g_{21} & -g_{31} & -g_{65} & & & \\ g_{31} & -g_{83} & 0 & -g_{87} & & \\ 0 & 0 & g_{21} & -g_{31} & -g_{21} & \\ 0 & 0 & g_{31} & -g_{83} & -g_{31} & -g_{43} \end{bmatrix} \\
[\overline{K}] &= \begin{bmatrix} k_{11} & & & & & \\ k_{41} & k_{33} & & & \text{Sym.} & \\ -k_{11} & -k_{41} & k_{55} & & & \\ k_{41} & k_{84} & 0 & k_{77} & & \\ 0 & 0 & -k_{11} & -k_{41} & k_{11} & \\ 0 & 0 & k_{41} & k_{84} & -k_{41} & k_{33} \end{bmatrix}
\end{aligned}\right\} \tag{14.68}$$

For model II we have

$$\left.\begin{aligned}
[\overline{M}] &= \begin{bmatrix} m_{11} & & & & & \\ m_{41} & m_{33} & & & \text{Sym.} & \\ m_{51} & -m_{81} & m_{55} & & & \\ m_{81} & m_{84} & m_{85} & m_{77} & & \\ 0 & 0 & m_{95} & -m_{125} & m_{99} & \\ 0 & 0 & m_{125} & m_{128} & m_{129} & m_{1111} \end{bmatrix} \\
[\overline{G}] &= \begin{bmatrix} -g_{21} & & & & & \\ g_{31} & -g_{43} & & & \text{Sym.} & \\ g_{21} & -g_{31} & -g_{65} & & & \\ g_{31} & -g_{83} & g_{75} & -g_{87} & & \\ 0 & 0 & g_{96} & -g_{115} & -g_{96} & \\ 0 & 0 & g_{115} & -g_{127} & -g_{115} & g_{1211} \end{bmatrix} \\
[\overline{K}] &= \begin{bmatrix} k_{11} & & & & & \\ k_{41} & k_{33} & & & \text{Sym.} & \\ -k_{11} & -k_{41} & k_{55} & & & \\ k_{41} & k_{84} & k_{85} & k_{77} & & \\ 0 & 0 & -k_{99} & -k_{125} & k_{99} & \\ 0 & 0 & k_{125} & k_{128} & -k_{125} & k_{1111} \end{bmatrix}
\end{aligned}\right\} \tag{14.69}$$

In the following sections of this chapter, we use Eq. (14.67).

14.4 Free Vibrations: Eigenvalue Problem

In this section we investigate natural frequencies and mode shapes. It is assumed that models I and II are both supported simply at nodes 1 and 3. This boundary condition is expressed by setting deflection and moment to zero.

The equations of motion governing free vibrations are obtained by setting the external excitations to zero in Eq. (14.67) as follows:

$$[\overline{M}]\{\ddot{\overline{Q}}_x\} + [\overline{G}]\{\dot{\overline{Q}}_y\} + [\overline{K}]\{\overline{Q}_x\} = \{\overline{F}_x\} \tag{14.70a}$$

$$[\overline{M}]\{\ddot{\overline{Q}}_y\} - [\overline{G}]\{\dot{\overline{Q}}_x\} + [\overline{K}]\{\overline{Q}_y\} = \{\overline{F}_y\} \tag{14.70b}$$

In these force vectors, only the dynamic reaction forces caused by constraint remain at nodes 1 and 3.

Let us assume that the rotor of model I or II is whirling with a frequency p. The deflections and inclinations at each node are shown in Table 14.1. Then the displacement vectors of Eq. (14.66) are expressed as follows:

$$\{\overline{Q}_x\} = \begin{Bmatrix} u_1 \\ \phi_{y1} \\ u_2 \\ \phi_{y2} \\ u_3 \\ \phi_{y3} \end{Bmatrix} = \begin{Bmatrix} 0 \\ \theta_1 \\ r_2 \\ \theta_2 \\ 0 \\ \theta_3 \end{Bmatrix} \cos pt, \quad \{\overline{Q}_y\} = \begin{Bmatrix} v_1 \\ -\phi_{x1} \\ v_2 \\ -\phi_{x2} \\ v_3 \\ -\phi_{x3} \end{Bmatrix} = \begin{Bmatrix} 0 \\ \theta_1 \\ r_2 \\ \theta_2 \\ 0 \\ \theta_3 \end{Bmatrix} \sin pt \tag{14.71}$$

We substitute these assumed solutions into Eq. (14.70a) or (14.70b). Let the reaction forces from the bearings in the radial directions at node 1 be F_{x1} and F_{y2} and those at node 3 be F_{x9} and F_{y10}. Then, from Eq. (14.70a), we have

$$\left([\overline{K}] + p[\overline{G}] - p^2[\overline{M}]\right) \begin{Bmatrix} 0 \\ \theta_1 \\ r_2 \\ \theta_2 \\ 0 \\ \theta_3 \end{Bmatrix} \cos pt = \begin{Bmatrix} F_{x1} \\ 0 \\ 0 \\ 0 \\ F_{x9} \\ 0 \end{Bmatrix} \tag{14.72}$$

From Eq. (14.70b) we have the same expression (except for cos pt and the right-hand side). Let the matrices obtained by eliminating rows 1 and 5 and columns 1 and 5 from $[\overline{K}]$, $[\overline{G}]$, and $[\overline{M}]$ be $[\overline{K}']$, $[\overline{G}']$, and $[\overline{M}']$, respectively, and let us introduce the displacement vector $\{\overline{Q}'\} = \lfloor \theta_1 \ r_2 \ \theta_2 \ \theta_3 \rfloor^T$. From rows 2, 3, 4, and 6 of Eq. (14.72), we have

$$\left([\overline{K}'] + p[\overline{G}'] - p^2[\overline{M}']\right)\{\overline{Q}'\} = \{0\} \tag{14.73}$$

Table 14.1 Displacements at nodes in free vibration.

Node 1	Node 2	Node 3
$u = 0$	$u_2 = r_2 \cos pt$	$u_3 = 0$
$v_1 = 0$	$v_1 = r_2 \sin pt$	$v_3 = 0$
$\theta_{x1} = \theta_1 \cos pt = \phi_{y1}$	$\theta_{x2} = \theta_2 \cos pt = \phi_{y2}$	$\theta_{x3} = \theta_3 \cos pt = \phi_{y3}$
$\theta_{y1} = \theta_1 \sin pt = -\phi_{x1}$	$\theta_{y2} = \theta_2 \sin pt = -\phi_{x2}$	$\theta_{y3} = \theta_3 \sin pt = -\phi_{x3}$

To have a nontrivial solution, the determinant of the coefficients must be zero, that is,

$$\det\left(\left[\overline{K}'\right] + p\left[\overline{G}'\right] - p^2\left[\overline{M}'\right]\right) = 0 \tag{14.74}$$

This expression is the frequency equation and gives natural frequencies p. Substituting the values obtained in Eq. (14.72) gives the mode shapes. Rows 1 and 5 of Eq. (14.71) give reaction forces.

A result for the natural frequencies of model I is shown in Figure 14.4. The same type of scale as Figure 3.8 is used for the coordinate axis. In this case, eight natural frequencies are obtained. Four positive values correspond to forward whirling motions, and four negative values correspond to backward whirling motions. Because model I is a uniform rotor, we can solve it analytically as shown in Chapter 3. This theoretical result is also illustrated in this figure for comparison.

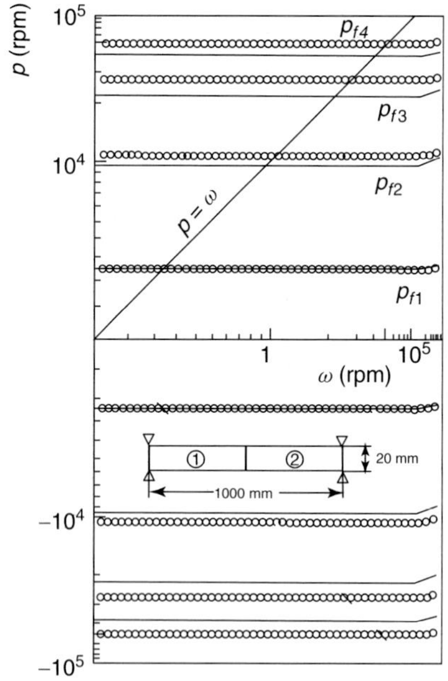

Figure 14.4 Natural frequencies (model I).

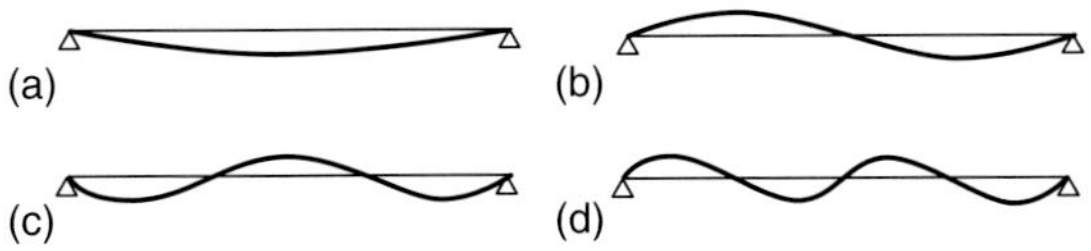

Figure 14.5 Mode shapes (model I). (a) First mode (p_{f1}), (b) second mode (p_{f2}), (c) third mode (p_{f3}), and (d) fourth mode (p_{f4}).

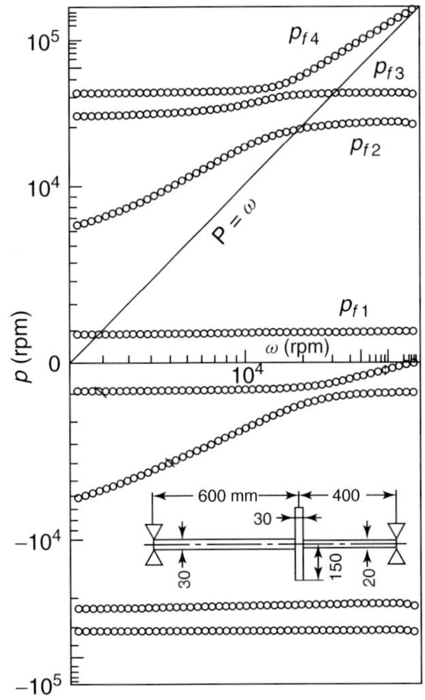

Figure 14.6 Natural frequencies (model II).

Although this finite element model has only two elements, it gives a fairly good result. In particular, we could obtain the first mode with enough accuracy. If we use a larger number of elements in practical application, we can obtain satisfactory results for higher modes. Figure 14.5 shows the result for mode shapes.

A result for model II with a rigid disk is shown in Figure 14.6. Owing to the large gyroscopic moment, natural frequencies change remarkably.

14.5 Forced Vibrations

If a shaft and/or a disk has an unbalance, forced vibrations occur. Distributed unbalance forces in the shaft can be replaced by nodal forces using Eqs. (14.27) and (14.47), and those in the rigid disk are given by Eq. (14.2).

As an example, we study model II with unbalance forces given by Eq. (14.2). If these unbalance forces acting on node 2 are considered together with nodal forces at nodes 1 and 3, the external force vector of Eq. (14.47) is given by

$$\{F\} = [F_{x1} \quad F_{y2} \quad 0 \quad 0 \quad F_{x5} \quad F_{y6} \quad F_{y7} \quad F_{x8} \quad F_{x9} \quad F_{y10} \quad 0 \quad 0]^T \tag{14.75}$$

where

$$\left.\begin{aligned} F_{x5} &= m_{\mathrm{d}} e\omega^2 \cos \omega t, \quad F_{y7} = -\left(I - I_{\mathrm{p}}\right) \tau\omega^2 \sin\left(\omega t + \beta_\tau\right) \\ F_{x6} &= m_{\mathrm{d}} e\omega^2 \sin \omega t, \quad F_{y8} = -\left(I - I_{\mathrm{p}}\right) \tau\omega^2 \cos\left(\omega t + \beta_\tau\right) \end{aligned}\right\} \tag{14.76}$$

If the rows including reaction forces in the bearings are eliminated in Eq. (14.72), we have

$$\left.\begin{aligned} \left[\overline{M}'\right]\left\{\ddot{\overline{Q}}'_x\right\} + \left[\overline{G}'\right]\left\{\dot{\overline{Q}}'_y\right\} + \left[\overline{K}'\right]\left\{\overline{Q}'_x\right\} = \left\{\overline{F}'_c\right\}\cos\omega t - \left\{\overline{F}'_s\right\}\sin\omega t \\ \left[\overline{M}'\right]\left\{\ddot{\overline{Q}}'_y\right\} - \left[\overline{G}'\right]\left\{\dot{\overline{Q}}'_x\right\} + \left[\overline{K}'\right]\left\{\overline{Q}'_y\right\} = \left\{\overline{F}'_c\right\}\sin\omega t - \left\{\overline{F}'_s\right\}\cos\omega t \end{aligned}\right\} \tag{14.77}$$

where $\left[\overline{M}'\right], \left[\overline{G}'\right] \left[\overline{K}'\right]$ are the matrices in Eq. (14.72). Corresponding to these matrices, we define the following displacement vectors and force vectors:

$$\left.\begin{aligned} &\left\{\overline{Q}'_x\right\} = \begin{Bmatrix} \phi_{y1} \\ u_2 \\ \phi_{y2} \\ \phi_{y3} \end{Bmatrix}, \left\{\overline{Q}'_y\right\} = \begin{Bmatrix} -\phi_{x1} \\ v_2 \\ -\phi_{x2} \\ -\phi_{x3} \end{Bmatrix}, \\ &\left\{\overline{F}'_c\right\} = \begin{Bmatrix} 0 \\ m_{\mathrm{d}} e\omega^2 \\ \left(I - I_{\mathrm{p}}\right)\tau\omega^2\cos\beta_\tau \\ 0 \end{Bmatrix}, \left\{\overline{F}'_s\right\} = \begin{Bmatrix} 0 \\ 0 \\ \left(I - I_{\mathrm{p}}\right)\tau\omega^2\sin\beta_\tau \\ 0 \end{Bmatrix} \end{aligned}\right\} \tag{14.78}$$

In response to unbalance forces with the frequency ω, we assume a harmonic response as follows:

$$\left.\begin{aligned} \left\{\overline{Q}'_x\right\} = \left\{\overline{Q}'_c\right\}_1 \cos\omega t - \left\{\overline{Q}'_s\right\}_1 \sin\omega t \\ \left\{\overline{Q}'_y\right\} = \left\{\overline{Q}'_c\right\}_2 \sin\omega t + \left\{\overline{Q}'_s\right\}_2 \cos\omega t \end{aligned}\right\} \tag{14.79}$$

Substituting these equations into Eq. (14.77) and comparing the coefficients of $\cos\omega t$ and $\sin\omega t$, we can obtain the following result:

$$\left.\begin{aligned} \left\{\overline{Q}'_c\right\}_1 = \left\{\overline{Q}'_c\right\}_2 = \left[\left[\overline{K}'\right] + \omega\left[\overline{G}'\right] - \omega^2\left[\overline{M}'\right]\right]^{-1}\left\{\overline{F}'_c\right\} \\ \left\{\overline{Q}'_s\right\}_1 = \left\{\overline{Q}'_s\right\}_2 = \left[\left[\overline{K}'\right] + \omega\left[\overline{G}'\right] - \omega^2\left[\overline{M}'\right]\right]^{-1}\left\{\overline{F}'_s\right\} \end{aligned}\right\} \tag{14.80}$$

For example, the amplitude $R_2 = \sqrt{u_2^2 + v_2^2}$ at node 2 can be obtained from the second row of these expressions. Since the inverse matrices in these expressions are obtained by replacing p with ω in the frequency Eq. (14.74), we see that the resonance occurs at the cross points of the straight line$p = \omega$ and natural frequency curves p_{f} in the frequency diagram. Figure 14.7 shows a resonance curve for model

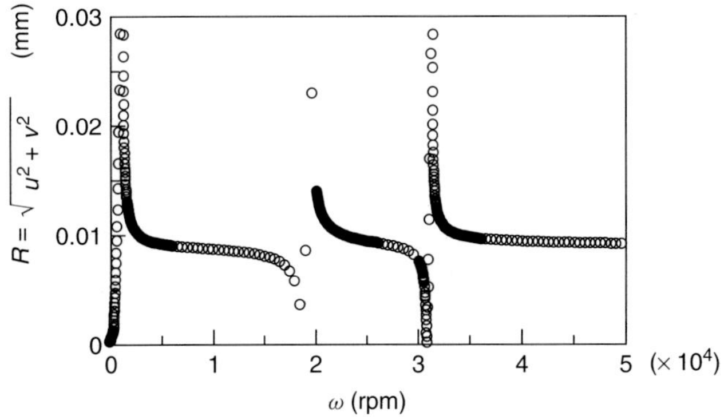

Figure 14.7 Response curves.

II. Since this is a case with no damping, the resonance peaks are very sharp. Note that unlike Figure 14.6, this figure is illustrated by linear scales.

14.6 Alternative Procedure

For reference, we summarize the outline of the procedure used by Nelson and McVaugh (1976). They change the equations of motion (14.48) to the following first-order equations of motion:

$$\begin{bmatrix} [0] & [M] \\ [M] & [G] \end{bmatrix} \left\{ \dot{h} \right\} + \begin{bmatrix} [-M] & [0] \\ [0] & [K] \end{bmatrix} \left\{ h \right\} = \{H\} \tag{14.81}$$

where

$$\left\{ h \right\} = \left\{ \begin{array}{c} \{\dot{Q}\} \\ \{Q\} \end{array} \right\}, \quad \{H\} = \left\{ \begin{array}{c} \{0\} \\ \{F\} \end{array} \right\} \tag{14.82}$$

In order to obtain a solution for a free vibration, we substitute the assumed solution $\left\{ h \right\} = \left\{ h_0 \right\} e^{at}$ into the homogeneous equation obtained from Eq. (14.81) and reduce it to the following eigenvalue problem:

$$\begin{bmatrix} [0] & [I] \\ -[K]^{-1}[M] & -[K]^{-1}[G] \end{bmatrix} \left\{ h_0 \right\} = \frac{1}{\alpha} \left\{ h_0 \right\} \tag{14.83}$$

Then we can obtain the natural frequencies and vibration modes.

As for the forced solution, we assume that the unbalance force

$$\{F\} = \{F_c\} \cos \omega t + \{F_s\} \sin \omega t \tag{14.84}$$

produces the steady-state solution

$$\left\{ Q \right\} = \left\{ Q_c \right\} \cos \omega t + \left\{ Q_s \right\} \sin \omega t \tag{14.85}$$

By substituting this into Eq. (14.48) and comparing the coefficients of $\cos \omega t$ and $\sin \omega t$ on both sides, we can determine Q_c and Q_s.

References

Nelson, H.D. and McVaugh, J.M. (1976) The dynamics of rotor bearing systems, using finite elements. *Trans. ASME, J. Eng. Ind.*, **98** (2), 593–600.

Petyt, M. (1990) *Introduction to Finite Element Vibration Analysis*, Cambridge University Press, Cambridge.

Sato, S. (1992) The elements of numerical analysis of a rotating shaft (finite element method, part 3). *J. JSDE*, **27** (6), 227–230. (in Japanese).

15
Transfer Matrix Method

15.1
General Considerations

Another representative numerical method for analyzing rotor vibration is the *transfer matrix method*. The fundamental idea of this method is given in a paper by Myklestad (1944) on the calculation of natural frequencies of a wing, and in a paper by Prohl (1945) on the calculation of critical speeds of a multi-rotor system. This analytical procedure was refined and completed later and is widely used in the study of dynamic characteristics of structures. The advantage of this method is that the necessary computer memory capacity is much less than that for other methods, such as the finite element method. For a general study of this method, the book by Pestal and Leckie (1963) is recommended.

In the design of rotating machines, we must know natural frequencies, mode shapes, critical speeds, unbalance responses, and so on. This transfer matrix method provides a comparatively simple means for this purpose. This method is suitable for a structure having divided elements in a linear arrangement and is therefore suited for rotors. In this chapter, application of this method to rotor systems is explained (Kikuchi, 1969; Hahn, 1989).

15.2
Fundamental Procedure of the Transfer Matrix Method

To demonstrate this method, the general procedure is explained using an elastic bar with axial vibration.

15.2.1
Analysis of Free Vibration

The steps for the analysis of free vibrations are as follows:

1) **Division of the system**: Figure 15.1a shows a thin rod vibrating axially. Let us assume that this rod is divided into n elements. The displacement x_i and the associated force N_i represent a "state" of the system and are therefore called

Linear and Nonlinear Rotordynamics: A Modern Treatment with Applications, Second Edition.
Yukio Ishida and Toshio Yamamoto.

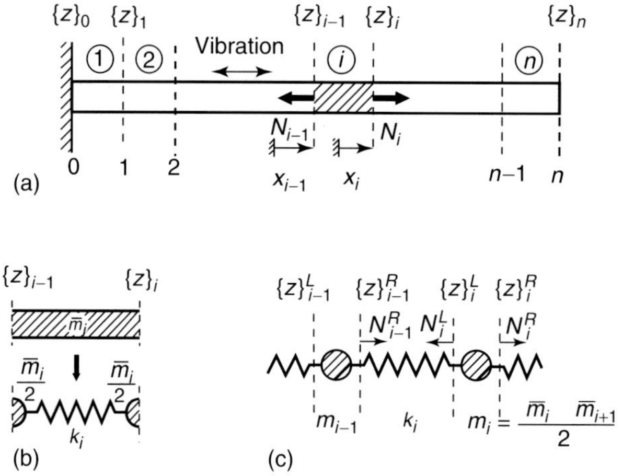

Figure 15.1 Discretization of the system and state vectors. (a) Elastic bar with axial vibration, (b) ith element, and (c) definition of new state variables.

state variables. We define the following column vector called a *state vector*, which has these state variables as its components:

$$\{z\}_i = \begin{Bmatrix} x \\ N \end{Bmatrix}_i \tag{15.1}$$

Let the mass of the ith element be $\overline{m}_i$. We divide this mass into two equal parts and position them at both ends as concentrated masses, and connect them by the spring with a spring constant k_i as shown in Figure 15.1b. Then we define the state vectors $\{z\}_i^{\mathrm{R}}$ and $\{z\}_i^{\mathrm{L}}$ set very close to the right and left sides of mass m_i, as shown in Figure 15.1c. Clearly, the following relationship holds:

$$x_i^{\mathrm{R}} = x_i^{\mathrm{L}} = x_i \tag{15.2}$$

2) **Determination of the transfer matrix of an element**: Since the forces N_i^{L} and $N_{i-1}{}^{\mathrm{R}}$ pulling the spring at both ends are balanced, the relationship

$$N_i^{\mathrm{L}} = N_{i-1}^{\mathrm{R}} = k_i\left(x_i^{\mathrm{L}} - x_{i-1}^{\mathrm{R}}\right) \tag{15.3}$$

holds. Therefore, we get

$$\left.\begin{aligned} x_i^{\mathrm{L}} &= x_{i-1}^{\mathrm{R}} + \frac{1}{k_i} N_{i-1}^{\mathrm{R}} \\ N_i^{\mathrm{L}} &= N_{i-1}^{\mathrm{R}} \end{aligned}\right\} \tag{15.4}$$

This relationship is expressed in matrix form as

$$\begin{Bmatrix} x \\ N \end{Bmatrix}_i^{\mathrm{L}} = \begin{bmatrix} 1 & 1/k \\ 0 & 1 \end{bmatrix}_i \begin{Bmatrix} x \\ N \end{Bmatrix}_{i-1}^{\mathrm{R}} \tag{15.5}$$

and is also expressed by notations as follows:

$$\{z\}_i^{\mathrm{L}} = [F]_i \{z\}_{i-1}^{\mathrm{R}} \tag{15.6}$$

The matrix $[F]_i$ is called a *field transfer matrix*. For convenience, we write the subscript i outside the brackets and parentheses. Next, we consider the mass m_i. Its equation of motion is given by

$$m_i\ddot{x}_i^{\mathrm{L}} = N_i^{\mathrm{R}} - N_i^{\mathrm{L}} \tag{15.7}$$

If the mass vibrates freely in the form $x_i = A\sin pt$, we get the following expression after substitution:

$$N_i^{\mathrm{R}} = -m_i p^2 x_i^{\mathrm{L}} + N_i^{\mathrm{L}} \tag{15.8}$$

From this and Eq. (15.2), we have

$$\left\{ \begin{array}{c} x \\ N \end{array} \right\}_i^{\mathrm{R}} = \left[\begin{array}{cc} 1 & 0 \\ -mp^2 & 1 \end{array} \right]_i \left\{ \begin{array}{c} x \\ N \end{array} \right\}_i^{\mathrm{L}} \tag{15.9}$$

or

$$\{z\}_i^{\mathrm{R}} = [P]_i\,\{z\}_i^{\mathrm{L}} \tag{15.10}$$

The matrix $[P]_i$ is called a *point transfer matrix*. From Eqs. (15.6) and (15.10) we have

$$\{z\}_i^{\mathrm{R}} = [P]_i\,[F]_i\,\{z\}_{i-1}^{\mathrm{R}} = [T]_i\,\{z\}_{i-1}^{\mathrm{R}} \tag{15.11}$$

where

$$[T]_i = \left[\begin{array}{cc} 1 & 1/k \\ -mp^2 & 1 - mp^2/k \end{array} \right]_i \tag{15.12}$$

This matrix $[T]_i$ is called the *transfer matrix* of element i.

3) **Determination of the overall transfer matrix**: Multiplying the transfer matrices of each element $[T]_i$ successively, we can derive the overall transfer matrix $[T]$. From Eq. (15.11), we have

$$\{z\}_n^{\mathrm{R}} = [T]_n \cdots [T]_2\,[T]_1\,\{z\}_0^{\mathrm{R}} = [T]_n \cdots [T]_2\,[T]_1\,[P]_0\,\{z\}_0^{\mathrm{L}} \tag{15.13}$$

Therefore, the matrix

$$[T] = [T]_n \cdots [T]_2\,[T]_1\,[P]_0 \tag{15.14}$$

combines the state vector $\{z\}_n \left(= \{z\}_n^{\mathrm{R}}\right)$ on one end to the state vector $\{z\}_0 \left(= \{z\}_0^{\mathrm{L}}\right)$ on the other end in the form

$$\{z\}_n = [T]\,\{z\}_0 \tag{15.15}$$

4) **Application of the boundary conditions**: If the supporting condition at the end of the shaft is given, some of the state variables are determined correspondingly. For example, since the left and right ends of the bar in Figure 15.1a are fixed and free, respectively, the deflection is equal to zero in the former and the force is equal to zero in the latter. Substituting these conditions into Eq. (15.15), we have

$$\left\{ \begin{array}{c} x \\ 0 \end{array} \right\}_n = \left[\begin{array}{cc} T_{11} & T_{12} \\ T_{21} & T_{22} \end{array} \right] \left\{ \begin{array}{c} 0 \\ N \end{array} \right\}_0 \tag{15.16}$$

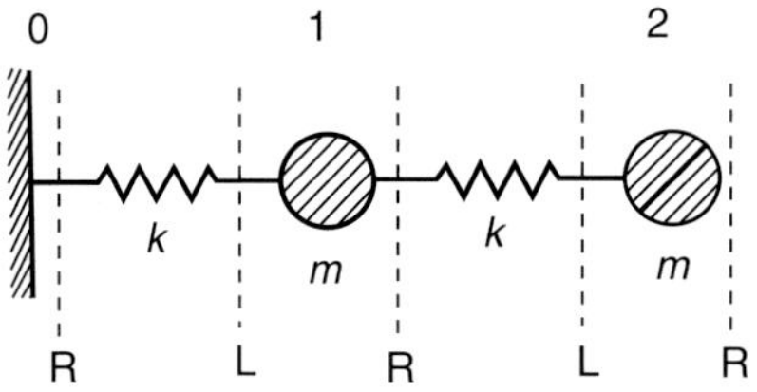

Figure 15.2 2 DOF of a spring–mass system.

5) **Determination of the natural frequencies**: From Eq. (15.16), we have

$$0 = T_{22} N_0 \tag{15.17}$$

Vibration solutions exist when

$$T_{22} = 0 \tag{15.18}$$

The left-hand side of this expression is a complex function of p. By solving it numerically, we can obtain natural frequencies.

6) **Determination of the mode shapes**: If we substitute one of the natural frequencies p_i into Eq. (15.12), we can determine the elements of matrix $[T]_i$. By giving appropriate values to the state variable N_0, we can determine the vector $\{z\}_0$. Then, using the relationship

$$\{\mathrm{z}\}_i^{\mathrm{R}} = [T]_i \cdots [T]_2 \, [T]_1 \, [P]_0 \, \{\mathrm{z}\}_0 \tag{15.19}$$

we can determine the state vectors $\{\mathrm{z}\}_i^{\mathrm{R}}, \{\mathrm{z}\}_2^{\mathrm{R}}, \cdots, \{\mathrm{z}\}_n^{\mathrm{R}}$ successively and, therefore, can know the mode shape for natural frequency p_i.

As an example, we consider the simple 2 DOF spring–mass system shown in Figure 15.2. Deriving the overall transfer matrix from Eq. (15.12) and applying the boundary conditions, we have

$$\begin{aligned}
\begin{Bmatrix} x \\ 0 \end{Bmatrix}_2^{\mathrm{R}} &= \begin{bmatrix} 1 & 1/k \\ -mp^2 & 1 - mp^2/k \end{bmatrix} \begin{bmatrix} 1 & 1/k \\ -mp^2 & 1 - mp^2/k \end{bmatrix} \begin{Bmatrix} 0 \\ N \end{Bmatrix}_0^{\mathrm{R}} \\
&= \begin{bmatrix} 1 - mp^2/k & 2/k - mp^2/k^2 \\ -2mp^2 + m^2p^4/k & -mp^2/k + (1 - mp^2/k)^2 \end{bmatrix} \begin{Bmatrix} 0 \\ N \end{Bmatrix}_0^{\mathrm{R}}
\end{aligned} \tag{15.20}$$

The frequency equation is obtained from $T_{22} = 0$, that is

$$-\frac{mp^2}{k} + \left(1 - \frac{mp^2}{k}\right)^2 = 0 \tag{15.21}$$

The roots of this equation give the natural frequencies as

$$p_1 = \sqrt{\left(\frac{3 - \sqrt{5}}{2}\right)\frac{k}{m}} \qquad p_2 = \sqrt{\left(\frac{3 + \sqrt{5}}{2}\right)\frac{k}{m}} \tag{15.22}$$

Next we derive the vibration mode for p_1. To determine the state vector of the left end $\{z\}_0^R$, we get $N_0^R = 1$. Then we get $\{z\}_1^R$ by

$$\left\{ \begin{array}{c} x \\ N \end{array} \right\}_1^R = \left[\begin{array}{cc} 1 & 1/k \\ -mp_1^2 & 1 - mp_1^2/k \end{array} \right] \left\{ \begin{array}{c} 0 \\ 1 \end{array} \right\} = \left\{ \begin{array}{c} 1/k \\ \left(-1 + \sqrt{5}\right)/2 \end{array} \right\} \tag{15.23}$$

Moreover, we get $\{z\}_2^R$ by

$$\begin{aligned} \left\{ \begin{array}{c} x \\ 0 \end{array} \right\}_2^R &= \left[\begin{array}{cc} 1 & 1/k \\ -mp_1^2 & 1 - mp_1^2/k \end{array} \right] \left\{ \begin{array}{c} 1/k \\ (-1 + \sqrt{5})/2 \end{array} \right\} \\ &= \left\{ \begin{array}{c} (1 + \sqrt{5})/2k \\ 0 \end{array} \right\} \end{aligned} \tag{15.24}$$

As a result, we get the mode shape of the first mode as follows:

$$\left\{ \begin{array}{c} x_1^R \\ x_2^R \end{array} \right\} = \left\{ \begin{array}{c} 1 \\ (1 + \sqrt{5})/2 \end{array} \right\} \tag{15.25}$$

The mode shape of the second mode can be obtained in a similar way.

15.2.2 Analysis of Forced Vibration

The steps for the analysis of forced vibrations are as follows:

1) **Division of the system**: Make a discrete model in the same procedure as step 1 for free vibration.
2) **Expression of the forces and displacements by complex numbers**: Let us consider a periodic external force $F_i = f_i \cos(\omega t + \alpha)$ acting on the ith mass. Here we represent this force and the associated displacement by complex numbers.
   As an example, let us consider the spring–mass system shown in Figure 15.3. The equations of motion are given by

   $$m\ddot{x} + c\dot{x} + kx = f\cos(\omega t + \alpha) \tag{15.26}$$

   We can represent the right-hand side by a complex number as follows:

   $$m\ddot{x} + c\dot{x} + kx = \mathrm{Re}\left(\bar{f}e^{j\omega t}\right) \tag{15.27}$$

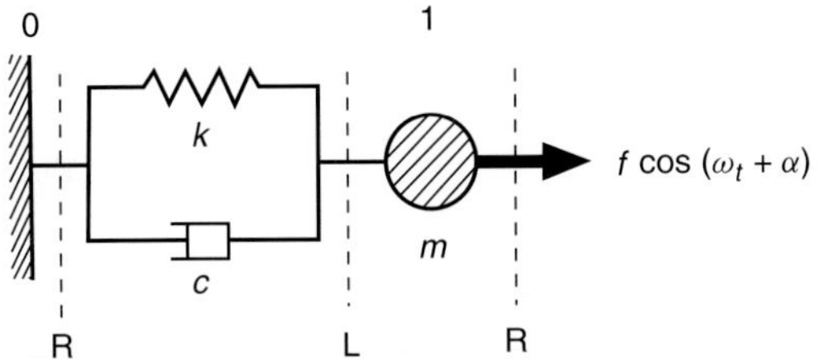

Figure 15.3 1 DOF of a spring–mass system with an external force.

where $\overline{f} = f e^{j\alpha}$ and "Re" indicates the real part. To avoid confusion with the element number, we use the notation "j" as an imaginary unit. This solution is represented by

$$x = \mathrm{Re}\left(\overline{x} e^{j\omega t}\right) \tag{15.28}$$

where $\overline{x} = a e^{j\beta}$ is a complex amplitude that includes phase delay. Since the 1:1 correspondence holds between the circular motion represented by a complex number and the rectilinear motion represented by its real part, we can eliminate the notation Re(·). Therefore, we have the following expression after substituting Eq. (15.28) into Eq. (15.27):

$$\left(-m\omega^2 + jc\omega + k\right) \overline{x} = \overline{f} \tag{15.29}$$

Then we have

$$\overline{x} = \frac{\overline{f}}{k - m\omega^2 + jc\omega} \tag{15.30}$$

This expression using complex numbers represents the amplitude a and the phase difference $(\beta - \alpha)$ as follows:

$$a = \frac{f}{\sqrt{\left(k - m\omega^2\right) + c^2\omega^2}}, \quad \tan(\beta - \alpha) = \frac{-c\omega}{k - m\omega^2} \tag{15.31}$$

3) **Determination of the extended transfer matrices of elements**: Next we derive the transfer matrix of the system shown in Figure 15.3. Considering the forces working in the spring and the damper, we have

$$N_0^{\mathrm{R}} = N_1^{\mathrm{L}} = c\left(\dot{x}_1^{\mathrm{L}} - \dot{x}_0^{\mathrm{R}}\right) + k\left(x_1^{\mathrm{L}} - x_0^{\mathrm{R}}\right) \tag{15.32}$$

This relationship is represented by complex numbers as follows:

$$\overline{N}_0^{\mathrm{R}} = \overline{N}_1^{\mathrm{L}} = (k + jc\omega)\left(\overline{x}_1^{\mathrm{L}} - \overline{x}_0^{\mathrm{R}}\right) \tag{15.33}$$

These can be written in matrix form as

$$\left\{ \begin{array}{c} \overline{x} \\ \overline{N} \end{array} \right\}_1^{\mathrm{L}} = \begin{bmatrix} 1 & 1/(k + jc\omega) \\ 0 & 1 \end{bmatrix} \left\{ \begin{array}{c} \overline{x} \\ \overline{N} \end{array} \right\}_0^{\mathrm{R}} \tag{15.34}$$

or

$$\{\overline{z}\}_1^{\mathrm{L}} = \left[\overline{F}\right] \{\overline{z}\}_0^{\mathrm{R}} \tag{15.35}$$

Concerning the mass, we have

$$\left.\begin{aligned} x_1^{\mathrm{L}} &= x_1^{\mathrm{R}} \\ m\ddot{x}_1^{\mathrm{L}} &= N_1{}^{\mathrm{R}} - N_1^{\mathrm{L}} + f\cos(\omega t + \alpha) \end{aligned}\right\} \tag{15.36}$$

These relationships are represented by complex numbers as follows:

$$\left.\begin{aligned} \overline{x}_1^{\mathrm{R}} &= \overline{x}_1^{\mathrm{L}} \\ \overline{N}_1^{\mathrm{R}} &= -m\overline{x}_1^{\mathrm{L}}\omega^2 + \overline{N}_1^{\mathrm{L}} - \overline{f} \end{aligned}\right\} \tag{15.37}$$

These can be written in matrix form as

$$\begin{Bmatrix} \overline{x} \\ \overline{N} \end{Bmatrix}_1^R = \begin{bmatrix} 1 & 0 \\ -m\omega^2 & 1 \end{bmatrix} \begin{Bmatrix} \overline{x} \\ \overline{N} \end{Bmatrix}_1^L + \begin{Bmatrix} 0 \\ -\overline{f} \end{Bmatrix} \tag{15.38}$$

The identical relationship $1 = 1$ can be written in the form

$$\{1\} = \lfloor\, 0 \quad 0 \,\rfloor \begin{Bmatrix} \overline{x} \\ \overline{N} \end{Bmatrix}_1^L + \{1\} \tag{15.39}$$

Using this relationship, we can rewrite Eqs. (15.34) and (15.38) as follows:

$$\begin{Bmatrix} \overline{x} \\ \overline{N} \\ 1 \end{Bmatrix}_1^L = \begin{bmatrix} 1 & 1/(k+jc\omega) & 0 \\ 0 & 1 & 0 \\ 0 & 0 & 1 \end{bmatrix} \begin{Bmatrix} \overline{x} \\ \overline{N} \\ 1 \end{Bmatrix}_0^R \tag{15.40}$$

$$\begin{Bmatrix} \overline{x} \\ \overline{N} \\ 1 \end{Bmatrix}_1^R = \begin{bmatrix} 1 & 0 & 0 \\ -m\omega^2 & 1 & -\overline{f} \\ 0 & 0 & 1 \end{bmatrix} \begin{Bmatrix} \overline{x} \\ \overline{N} \\ 1 \end{Bmatrix}_1^L \tag{15.41}$$

Since the system shown in Figure 15.3 is a 1 DOF system, the element transfer matrix and the overall transfer matrix are the same. However, in a general system, this relationship is obtained as a relationship concerning the ith element. Hence, replacing the parameters from k, c, m, and $\overline{f}$ to k_i, c_i, m_i, and $\overline{f}_i$, we can write Eqs. (15.40) and (15.41) as follows:

$$\left\{\tilde{\overline{z}}\right\}_i^L = [\tilde{\overline{F}}]_i \left\{\tilde{\overline{z}}\right\}_{i-1}^R \tag{15.42}$$

$$\left\{\tilde{\overline{z}}\right\}_i^R = \left[\tilde{\overline{P}}\right]_i \left\{\tilde{\overline{z}}\right\}_i^L \tag{15.43}$$

These quantities $\left\{\tilde{\overline{z}}\right\}_i$, $[\tilde{\overline{F}}]_i$, and $[\tilde{\overline{P}}]_i$ are called the *extended state vector*, *extended field transfer matrix*, and *extended point transfer matrix*, respectively. From Eqs. (15.42) and (15.43), we can define the extended transfer matrix $[\tilde{\overline{T}}]_i$ as follows:

$$\left\{\tilde{\overline{z}}\right\}_i^R = [\tilde{\overline{T}}]_i \left\{\tilde{\overline{z}}\right\}_{i-1}^R \tag{15.44}$$

where

$$\left[\tilde{\overline{T}}\right]_i = \begin{bmatrix} 1 & 1/(k_i+jc_i\omega) & 0 \\ -m_i\omega^2 & -m_i\omega^2/(k_i+jc_i\omega)+1 & -\overline{f}_i \\ 0 & 0 & 1 \end{bmatrix} \tag{15.45}$$

In this way, the analysis of forced vibrations is reduced to the analysis of free vibration by using extended state vectors and extended transfer matrices.

4) **Determination of the overall transfer matrix**: Similar to the case of free vibration, we can obtain overall extended transfer matrix $[\tilde{\overline{T}}]$ by multiplying extended transfer matrices of its elements successively as follows:

$$\left\{\tilde{\overline{z}}\right\}_n = [\tilde{\overline{T}}]_n \cdots [\tilde{\overline{T}}]_2 [\tilde{\overline{T}}]_1 \left\{\tilde{\overline{z}}\right\}_0 = [\tilde{\overline{T}}] \left\{\tilde{\overline{z}}\right\}_0 \tag{15.46}$$

where we put $\{\tilde{\overline{z}}\}_0^R = \{\tilde{\overline{z}}\}_0$.

5) **Application of the boundary conditions**: By applying boundary conditions, we can determine the vibration. For example, in the case of the fixed support at the left end $\left(\overline{x}_0^{\text{R}} = 0\right)$ and the free support at the right end $\left(\overline{N}_n^{\text{R}} = 0\right)$, we have

$$\begin{Bmatrix} \overline{x}_n^{\text{R}} \\ 0 \\ 1 \end{Bmatrix} = \begin{bmatrix} \overline{T}_{11} & \overline{T}_{12} & \overline{f}_1 \\ \overline{T}_{21} & \overline{T}_{22} & \overline{f}_2 \\ 0 & 0 & 1 \end{bmatrix} \begin{Bmatrix} 0 \\ \overline{N}_0^{\text{R}} \\ 1 \end{Bmatrix} \tag{15.47}$$

The second row gives

$$0 = \overline{T}_{22}\overline{N}_0^{\text{R}} + \overline{f}_2 \tag{15.48}$$

Knowing $\overline{N}_0^{\text{R}}$ from this, we can obtain the extended state vector $\left\{\tilde{\overline{z}}\right\}_0$. For example, in the case of Figure 15.3, since Eqs. (15.44) and (15.45) also represent overall transfer matrices, we have

$$\left(\frac{-m\omega^2}{k + jc\omega} + 1\right)\overline{N}_0^{\text{R}} - \overline{f} = 0 \tag{15.49}$$

and then

$$\overline{N}_0^{\text{R}} = \frac{(k + jc\omega)}{-m\omega^2 + (k + jc\omega)}\overline{f} \tag{15.50}$$

6) **Determination of a forced solution**: Multiplying successively the extended transfer matrix of each element with $\left\{\tilde{\overline{z}}\right\}_0$ obtained in step 5, we can determine the extended state vectors at each station and therefore the overall response of the system. For example, in the case of Figure 15.3, from $\left\{\tilde{\overline{z}}\right\}_1 = [\tilde{\overline{T}}]\left\{\tilde{\overline{z}}\right\}_0$, we have

$$\overline{x} = \overline{x}_1^{\text{R}} = \frac{\overline{f}}{k - m\omega^2 + jc\omega} \tag{15.51}$$

We can execute the same calculation using real numbers only. Now, let us assume that Eq. (15.44) is expressed in the form

$$\begin{Bmatrix} \tilde{\overline{z}} \\ 1 \end{Bmatrix}_i^{\text{R}} = \begin{bmatrix} \overline{T} & \overline{f} \\ 0 & 1 \end{bmatrix}_i \begin{Bmatrix} \tilde{\overline{z}} \\ 1 \end{Bmatrix}_{i-1}^{\text{R}} \tag{15.52}$$

We express each element showing explicitly the real part and the imaginary part as follows:

$$\overline{z}_i = z_i^1 + jz_i^2, \quad \overline{T}_i = T_i^1 + jT_i^2, \quad \overline{f}_i = f_i^1 + jf_i^2 \tag{15.53}$$

where the superscript 1 denotes the real part and 2 denotes the imaginary part. Substituting these expressions into $\overline{z}_i = \overline{T}_i\overline{z}_{i-1} + \overline{f}_i$, which is obtained from the first row of Eq. (15.52), and equating the real parts and the imaginary part on the right- and left-hand sides, respectively, we have

$$\left.\begin{aligned} z_i^1 &= T_i^1 z_{i-1}^1 - T_i^2 z_{i-1}^2 + f_i^1 \\ z_i^2 &= T_i^2 z_{i-1}^1 + T_i^1 z_{i-1}^2 + f_i^2 \end{aligned}\right\} \tag{15.54}$$

Using these relationships, Eq. (15.52) can be written in matrix form as follows:

$$\left\{\begin{array}{c} z^1 \\ z^2 \\ 1 \end{array}\right\}_i^{\mathrm{R}} = \left[\begin{array}{ccc} T^1 & -T^2 & f^1 \\ T^2 & T^1 & f^2 \\ 0 & 0 & 1 \end{array}\right]_i \left\{\begin{array}{c} z^1 \\ z^2 \\ 1 \end{array}\right\}_{i-1}^{\mathrm{R}} \tag{15.55}$$

Using this expression, we proceed to the analytical procedure using only real numbers.

15.3 Free Vibrations of a Rotor

15.3.1 State Vector and Transfer Matrix

Next, we discuss the rotor system shown in Figure 15.4. The rotor is rotating with rotational speed ω. We consider the gyroscopic moment acting on the rotor but neglect the moment acting on the shaft on the assumption that the latter is comparatively small compared to the former. The shaft diameter changes stepwise. First, we discretize this rotor. This shaft translates by u_i and v_i and inclines by ϕ_{xi} and ϕ_{yi} at station i. The shear forces V_{xi} and V_{yi} and the moments M_{xi} and M_{yi} work as the internal forces. The positive directions of these parameters are defined as shown in Figure 15.5a. For example, the sign of the internal forces are defined as follows. Concerning the faces of the edges of the element, the face is positive when its normal line points to the positive direction of the s-coordinate and is negative when it points to the negative direction. Similarly, when the shear force points to the positive direction of the coordinates on the positive face or when it points to the negative direction on the negative face, the force is defined as positive. The sign of a moment is defined in the same way after its vector is determined following the right-screw rule.

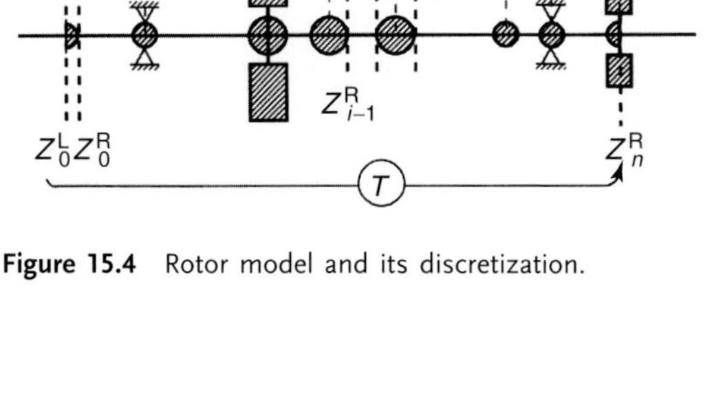

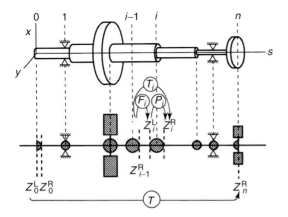

Figure 15.4 Rotor model and its discretization.

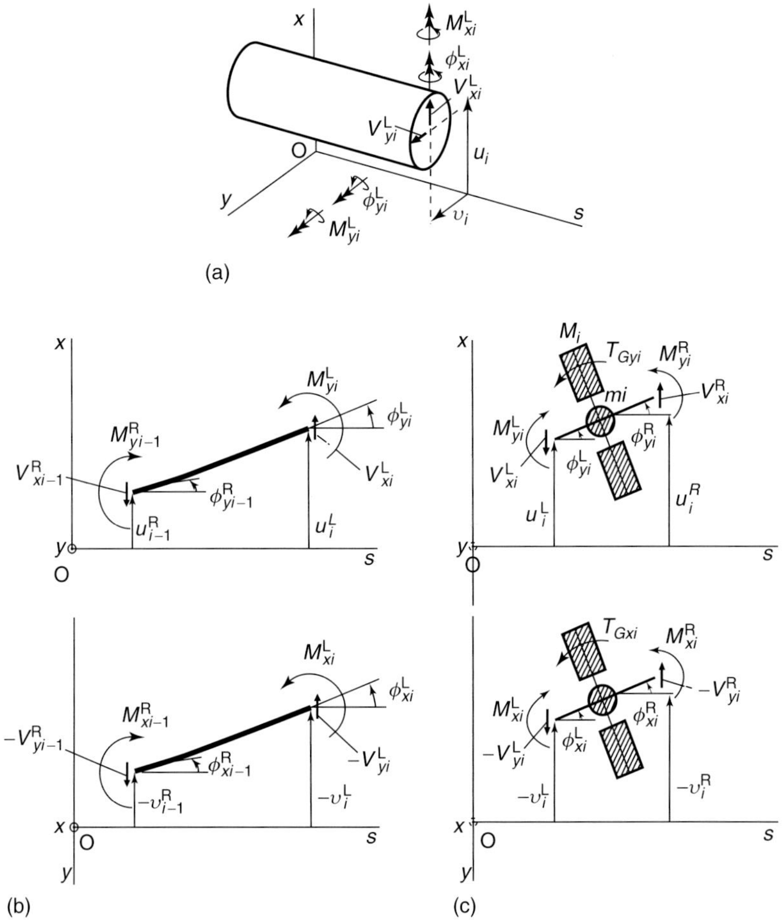

Figure 15.5 Forces and moments acting on an element. (a) Positive directions of parameters, (b) free-body diagrams of a beam, and (c) free-body diagrams of a mass.

Let us discuss the balancing of internal forces acting on an element. Figure 15.5b shows the components of displacements and internal forces in the xs- and ys-planes acting on the ith element with length l_i. From the figure on the xs-plane, the forces and the moments balance when

$$\left.\begin{aligned} V_{xi}^{\mathrm{L}} - V_{x\,i-1}^{\mathrm{R}} &= 0 \\ M_{yi}^{\mathrm{L}} - M_{y\,i-1}^{\mathrm{R}} + l_i V_{x\,i-1}^{\mathrm{R}} &= 0 \end{aligned}\right\} \tag{15.56}$$

From the beam theory of the strength of materials, we know that

$$\left.\begin{aligned} \phi_{yi}^{\mathrm{L}} &= \phi_{y\,i-1}^{\mathrm{R}} + \left(\frac{l_i}{EI_i}\right) M_{y\,i}^{\mathrm{L}} + \left(\frac{l_i^2}{2EI_i}\right) V_{x\,i}^{\mathrm{L}} \\ u_i^{\mathrm{L}} &= u_{i-1}^{\mathrm{R}} + l_i \phi_{y\,i-1}^{\mathrm{R}} + \left(\frac{l_i^2}{2EI_i}\right) M_{y\,i}^{\mathrm{L}} + \left(\frac{l_i^3}{6EI_i}\right) V_{x\,i}^{\mathrm{L}} \end{aligned}\right\} \tag{15.57}$$

From Eq. (15.56) and the expressions obtained by substituting Eq. (15.56) into Eq. (15.57), we have

$$\begin{Bmatrix} u \\ \phi_y \\ M_y \\ V_x \end{Bmatrix}_i^{\mathrm{L}} = \begin{bmatrix} 1 & l & l^2/2EI & -l^3/6EI \\ 0 & 1 & l/EI & -l^2/2EI \\ 0 & 0 & 1 & -l \\ 0 & 0 & 0 & 1 \end{bmatrix}_i \begin{Bmatrix} u \\ \phi_y \\ M_y \\ V_x \end{Bmatrix}_{i-1}^{\mathrm{R}} \tag{15.58}$$

This expression is represented by using notations as follows:

$$\{z\}_{xi}^{\mathrm{L}} = \left[F'\right]_i \{z\}_{x\,i-1}^{\mathrm{R}} \tag{15.59}$$

Next, we consider the components in the ys-plane. We derive the similar relationship utilizing the result above. In the figure on the xs-plane, vectors M_{yi}^{L} and ϕ_{yi}^{L} are perpendicular to the xs-plane and point upward. Therefore, to utilize the correspondence, we consider the shear forces in the reverse direction in the negative range of the y-axis. That is, by replacing $v_i \to -v_i$ and $V_{yi} \to -V_{yi}$, we have

$$\{z\}_{yi}^{\mathrm{L}} = \left[F'\right]_i \{z\}_{y\,i-1}^{\mathrm{R}} \tag{15.60}$$

where

$$\{z\}_{yi} = \begin{Bmatrix} -v \\ \phi_x \\ M_x \\ -V_y \end{Bmatrix}_i \tag{15.61}$$

Here we define a new state vector using $\{z\}_{xi}$ and $\{z\}_{yi}$ as follows:

$$\{z\}_i = \begin{Bmatrix} \{z\}_{xi} \\ \{z\}_{yi} \end{Bmatrix} \tag{15.62}$$

From Eqs. (15.59) and (15.60), we have

$$\{z\}_i^{\mathrm{L}} = [F]_i \{z\}_{i-1}^{\mathrm{R}} \tag{15.63}$$

where

$$[F]_i = \begin{bmatrix} [F']_i & [0] \\ [0] & [F']_i \end{bmatrix} \tag{15.64}$$

Next we consider the concentrated mass element and the rigid disk element. Figure 15.5c shows forces and moments acting on a mass m_i made by concentrating the distributed shaft mass and a rigid disk having mass M_i, the polar moment of inertia $I_{\mathrm{p}i}$, and the diametral moment of inertia I_i. In the xs-plane, the translation u_i and the inclination ϕ_{yi} of the rigid disk have the following relationships:

$$u_i^{\mathrm{L}} = u_i^{\mathrm{R}} = u_i, \quad \phi_{yi}^{\mathrm{L}} = \phi_{yi}^{\mathrm{R}} = \phi_{yi} \tag{15.65}$$

When the rotor is whirling with frequency p, the following equations of motion are obtained:

$$\left.\begin{aligned} (m_i + M_i)\,\ddot{u}_i &= (m_i + M_i)\left(-p^2 u_i\right) = V_{xi}^{\mathrm{R}} - V_{xi}^{\mathrm{L}} \\ I_i \ddot{\phi}_{yi} &= I_i\left(-p^2 \varphi_{yi}\right) = M_{yi}^{\mathrm{R}} - M_{yi}^{\mathrm{L}} + T_{Gyi} \end{aligned}\right\} \tag{15.66}$$

where T_{Gyi} is the gyroscopic moment. As explained in Section 2.4.2, the quantities $-I_{\mathrm{p}i}\omega\dot{\theta}_y$ and $I_{\mathrm{p}i}\omega\dot{\theta}_x$, which are obtained by changing the signs of the gyroscopic terms in Eq. (2.21), represent the gyroscopic moment. From the comparison between Figures 2.11 and 15.5a, we have $\theta_x = \phi_{yi}, \quad \theta_y = -\phi_{xi}$. Therefore, substituting these relationships into Eq. (2.21) gives

$$\left.\begin{aligned} I_i\ddot{\phi}_{yi} &= I_{\mathrm{p}i}\omega\dot{\phi}_{xi} + \delta\phi_{yi} \\ I_i\ddot{\phi}_{xi} &= -I_{\mathrm{p}i}\omega\dot{\phi}_{yi} + \delta\phi_{xi} \end{aligned}\right\} \tag{15.67}$$

From these expressions and Figure 15.3c, we have

$$T_{Gxi} = -I_{\mathrm{p}i}\omega\dot{\phi}_{yi}, \quad T_{Gyi} = I_{\mathrm{p}i}\omega\dot{\phi}_{xi} \tag{15.68}$$

If the rotor is whirling in the forms $\theta_x = \phi_{yi} = R\cos pt$ and $\theta_y = -\phi_{xi} = R\sin pt$, these expressions become

$$T_{Gxi} = -I_{\mathrm{p}i}\omega p\phi_{xi}, \quad T_{Gyi} = -I_{\mathrm{p}i}\omega p\phi_{yi} \tag{15.69}$$

Substituting these expressions into Eq. (15.66), we arrive at

$$\{z\}_{xi}^{\mathrm{R}} = \left[P'\right]_i \{z\}_{xi}^{\mathrm{L}} \tag{15.70}$$

where

$$\left[P'\right]_i = \begin{bmatrix} 1 & 0 & 0 & 0 \\ 0 & 1 & 0 & 0 \\ 0 & \left(I_{\mathrm{p}}\omega p - Ip^2\right) & 1 & 0 \\ -(m+M)p^2 & 0 & 0 & 1 \end{bmatrix}_i \tag{15.71}$$

In a similar way, we have

$$\{z\}_{yi}^{\mathrm{R}} = \left[P'\right]_i \{z\}_{yi}^{\mathrm{L}} \tag{15.72}$$

Moreover, we have

$$\{z\}_i^{\mathrm{R}} = [P]_i \{z\}_i^{\mathrm{L}} \tag{15.73}$$

where

$$[P]_i = \begin{bmatrix} [P']_i & 0 \\ 0 & [P']_i \end{bmatrix} \tag{15.74}$$

From Eqs. (15.63) and (15.73), we have

$$\{z\}_i^{\mathrm{R}} = [T]_i \{z\}_{i-1}^{\mathrm{R}} \tag{15.75}$$

where

$$[T]_i = [P]_i [F]_i \tag{15.76}$$

is the transfer matrix of the element.

In practical rotating machines, the supporting parts are often not as rigid. In such a case, the displacement of these parts should be considered. For simplicity, we consider that the rotor is supported elastically by springs (spring constant k)

with no directional difference in stiffness. The following forces act on the mass at the supporting position:

$$F_{xi} = -k_i u_i, \quad F_{yi} = -k_i v_i \tag{15.77}$$

For example, in the xs-plane in Figure 15.5c, since F_{xi} acts in the same direction as V_{xi}^{R}, Eq. (15.71) is modified as follows:

$$[P']_i = \begin{bmatrix} 1 & 0 & 0 & 0 \\ 0 & 1 & 0 & 0 \\ 0 & \left(I_{\mathrm{p}}\omega p - Ip^2\right) & 1 & 0 \\ k-(m+M)p^2 & 0 & 0 & 1 \end{bmatrix}_i \tag{15.78}$$

From Eqs. (15.64), (15.74), and (15.76), the transfer matrix of the ith element is given by

$$[T]_i = [P]_i [F]_i = \begin{bmatrix} [P']_i [F']_i & 0 \\ 0 & [P']_i [F']_i \end{bmatrix} \tag{15.79}$$

When the elasticity at the support is considered as shown in Figure 15.6, the transfer matrix is given by

$$[P']_i [F']_i = \begin{bmatrix} 1 & l & \frac{l^2}{2EI} & \frac{-l^3}{6EI} \\ 0 & 1 & \frac{l}{EI} & \frac{-l^2}{2EI} \\ 0 & I_{\mathrm{p}}\omega p - Ip^2 & \frac{1+(I_{\mathrm{p}}\omega p - Ip^2)l}{EI} & \frac{-l-\left(I_{\mathrm{p}}\omega p - Ip^2\right)l^2}{2EI} \\ k-(m+M)p^2 & \left\{k-(m+M)p^2\right\} l & \frac{\left\{k-(m+M)p^2\right\}l^2}{2EI} & \frac{1-\left\{k-(m+M)p^2\right\}l^3}{6EI} \end{bmatrix}_i \tag{15.80}$$

By applying the relationship in Eq. (15.75) repeatedly, we can obtain an expression that relates the state vector $\{z\}_i^{\mathrm{R}}$ of station i to the state vector $\{z\}_0^{\mathrm{L}} = \{z\}_0$ at the left end as follows:

$$\{z\}_i^{\mathrm{R}} = [T]_i [T]_{i-1} \cdots [T]_1 \{z\}_0^{\mathrm{R}} = [T]_i [T]_{i-1} \cdots [T]_1 [P]_0 \{z\}_0 \tag{15.81}$$

Further, by continuing this operation, we can relate the state vector $\{z\}_0$ at the left end to the state vector $\{z\}_n^{\mathrm{R}} = \{z\}_n$ at the right end as follows:

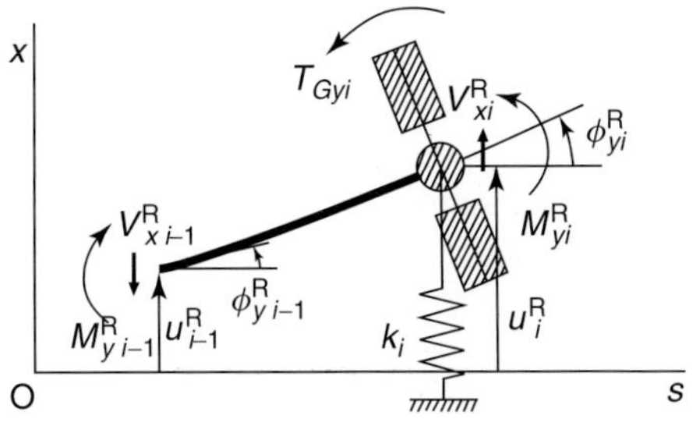

Figure 15.6 Element with elasticity at the support.

$$\{z\}_n = [T]\{z\}_0 \tag{15.82}$$

where

$$[T] = [T]_n \ldots [T]_2[T]_1[P]_0 \tag{15.83}$$

This matrix $[T]$ is called an *overall transfer matrix.*

15.3.2 Frequency Equation and the Vibration Mode

If the supporting conditions at both ends are given, values of some state variables at those ends are determined. For example, when the rotor is supported simply at both ends, we can put

$$u_0 = v_0 = M_{x0} = M_{y0} = 0, \quad u_n = v_n = M_{xn} = M_{yn} = 0 \tag{15.84}$$

and when it is supported freely, we have

$$V_{xo} = Y_{y0} = M_{x0} = M_{y0} = 0, \quad V_{xn} = V_{yn} = M_{xn} = M_{yn} = 0 \tag{15.85}$$

Suppose that four elements in each of the state vectors $\{z\}_n$ and $\{z\}_0$ become zero in a certain supporting condition. We represent these elements by $\{0\}$ and the rest of the elements by $\{z\}'_n$ and $\{z\}'_0$. Then Eq. (15.82) is written as

$$\begin{Bmatrix} \{z\}'_n \\ \{0\} \end{Bmatrix} = \begin{bmatrix} [T_{11}] & [T_{12}] \\ [T_{21}] & [T_{22}] \end{bmatrix} \begin{Bmatrix} \{z\}'_0 \\ \{0\} \end{Bmatrix} \tag{15.86}$$

The elements of this transfer matrix are functions of the natural frequency p. From this expression, we obtain

$$\{0\} = \left[T_{21}(p,\omega)\right]\{z\}'_0 \tag{15.87}$$

Since $\{z\}'_0 \neq \{0\}$ holds during vibration, the following relationship holds:

$$\left|T_{21}(p,\omega)\right| = 0 \tag{15.88}$$

This gives the frequency equation, and its roots $p = p_1, \ldots, p_n$ give natural frequencies.

Once the natural frequencies are obtained, we can determine the corresponding mode shapes. Now we substitute one of the natural frequencies p_i into Eq. (15.87). If we assume that one of the unknown components of $\{z\}_0$ is 1, we can determine the other components. Then if we determine $\{z\}_i^{\mathrm{R}}$ using the relationship in Eq. (15.81), we know the mode shape corresponding to the natural frequency p_i.

In the case of Eq. (15.79), we have

$$\begin{Bmatrix} \{z\}_{xi} \\ \{z\}_{yi} \end{Bmatrix} = \begin{bmatrix} [P']_i[F']_i & 0 \\ 0 & [P']_i[F']_i \end{bmatrix} \begin{Bmatrix} \{z\}_{x\,i-1} \\ \{z\}_{y\,i-1} \end{Bmatrix} \tag{15.89}$$

Here, quantities in the x-direction and those in the y-direction are not coupled. Therefore, we can use the following two 4×4 matrices instead of an 8×8 matrix to get natural frequencies and natural modes:

$$\{z\}_{xi} = \left[T'\right]_i \{z\}_{x\,i-1}, \quad \left[T'\right]_i = \left[P'\right]_i \left[F'\right]_i \tag{15.90}$$

In the following, we explain concretely how to obtain the natural frequencies and vibration modes in the case of a 4×4 matrix. When both ends are supported simply, we can put $u_0 = M_{y0} = u_n = M_{yn} = 0$. Corresponding to Eq. (15.86), we have

$$\begin{Bmatrix} 0 \\ \varphi_{yn} \\ 0 \\ V_{xn} \end{Bmatrix} = \begin{bmatrix} t_{11} & t_{12} & t_{13} & t_{14} \\ t_{21} & t_{22} & t_{23} & t_{24} \\ t_{31} & t_{32} & t_{33} & t_{34} \\ t_{41} & t_{42} & t_{43} & t_{44} \end{bmatrix} \begin{Bmatrix} 0 \\ \varphi_{y0} \\ 0 \\ V_{x0} \end{Bmatrix} \tag{15.91}$$

The first and third rows give

$$\left.\begin{aligned} 0 &= t_{12}\phi_{y0} + t_{14}V_{x0} \\ 0 &= t_{32}\phi_{y0} + t_{34}V_{x0} \end{aligned}\right\} \tag{15.92}$$

From its coefficient determinant, we get the following frequency equation:

$$t_{12}t_{34} - t_{14}t_{32} = 0 \tag{15.93}$$

Consider one natural frequency *pi*. If we assume that $\phi_i = 1$, we have $V_{x0} = -(t_{12}/t_{14})$ from the first equation of Eq. (15.92). Therefore, the mode shape is given by

$$\{z\}_0 = \begin{Bmatrix} 0 \\ 1 \\ 0 \\ -t_{12}/t_{14} \end{Bmatrix} \tag{15.94}$$

Substituting this into Eq. (15.90), we can determine $\{z\}_1, \{z\}_2, \ldots, \{z\}_n$ successively.

15.3.3
Examples

As examples, we calculate natural frequencies and mode shapes of a uniform continuous rotor (model I) and a disk–shaft system where the shaft diameter changes stepwise (model II). For comparison, the same models as those used in Chapter 14 for the finite element method are used here.

15.3.3.1 Model I: Uniform Continuous Rotor

The natural frequencies and the mode shapes of uniform continuous rotor (length 100 cm, diameter 2 cm) are shown in Figure 15.7a,b, respectively. These results are obtained by dividing the model into 20 equal elements. For simplicity, the results for higher modes are eliminated, although more frequencies and modes are obtained here. For comparison, the theoretical results are also depicted in these diagrams. These results and those of the transfer matrix analysis agree well.

The change of natural frequencies depending on the dividing number is shown in Table 15.1. Although the dividing number is small, a fairly accurate result is obtained, especially for lower modes. For example, concerning the second mode, the case for 15 elements gives a sufficiently accurate result, which does not change even for a larger number of elements.

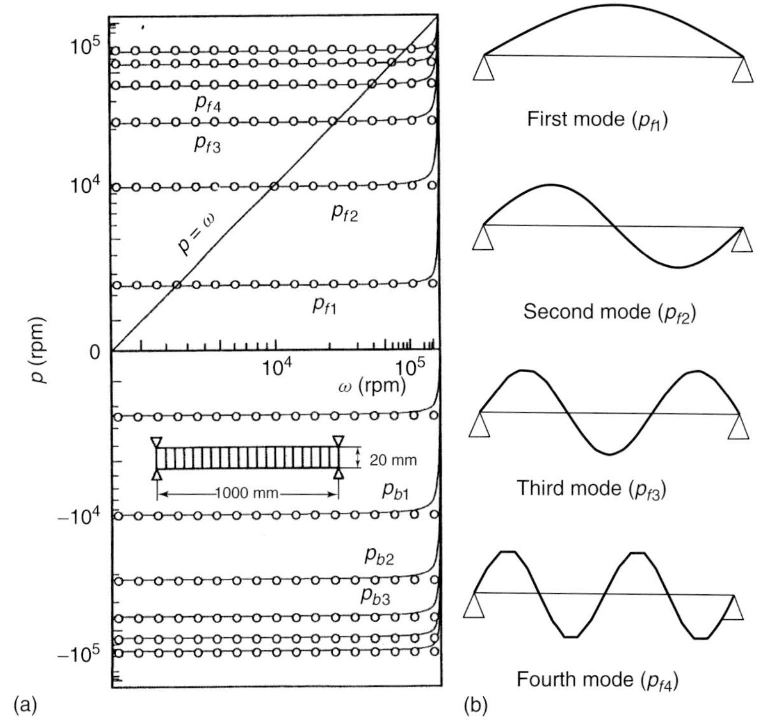

Figure 15.7 Model I. (a) Frequency diagram and (b) vibration modes at $\omega = 1000$ rpm.

Table 15.1 Natural frequencies for various divisions: model I ($\omega = 1000$ rpm).

Dividing number	**Natural frequency (rpm): with simple support**				
	First	**Second**	**Third**	**Fourth**	**Fifth**
5	2 412	9 625	21 309	35 086	–
10	2 412	9 648	21 696	38 508	59 881
15	2 412	9 650	21 711	38 579	60 249
20	2 412	9 650	21 711	38 591	60 299
Theoretical	2 412	9 650	21 712	38 600	60 312

15.3.3.2 **Model II: Disk–Shaft System**

The natural frequencies and the mode shapes of the disk-shaft system shown in Figure 14.3b are shown in Figure 15.8a,b, respectively. The dimensions of the disk and shaft are the same as that in Figure 14.6.

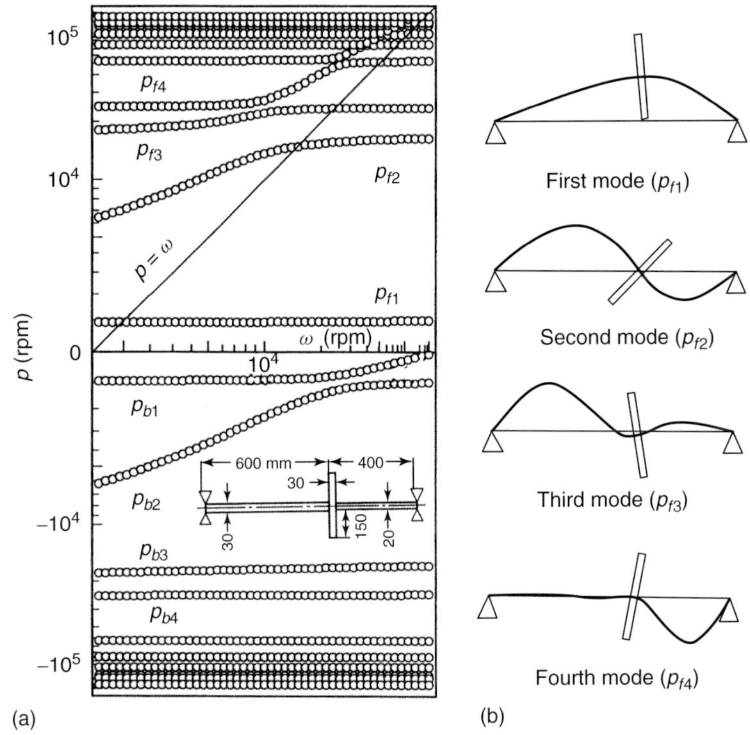

Figure 15.8 Model II. Frequency diagram and (b) vibration modes at $\omega = 1000$ rpm.

15.4 Forced Vibrations of a Rotor

15.4.1 External Force and Extended Transfer Matrix

We consider forced vibrations of a rotor. The external forces and damping forces are added to the rotor system used Section 14.3. First, we study external forces. Periodic forces due to unbalance act on the rotor. In the model shown in Figure 15.4, we consider the rotating coordinate system O-$x'y's$ whose x'-axis has an angle ωt relative to the x-axis of the static coordinate system. Let us assume that the disk has eccentricity e_i which is located at angle α_i from the x'-axis. The components of the unbalance in the x- and y-directions are given by

$$\left.\begin{aligned} F_{xi} &= M_i e_i \omega^2 \cos(\omega t + \alpha_i) = \mathrm{Re}\left(\bar{f}_i e^{j\omega t}\right) \\ F_{yi} &= M_i e_i \omega^2 \sin(\omega t + \alpha_i) = \mathrm{Re}\left(-j\bar{f}_i e^{j\omega t}\right) \end{aligned}\right\} \quad (15.95)$$

where $\bar{f}_i = M_i e_i \omega^2 e^{j\alpha_i}$.

Next we consider damping forces. Various types of damping work in rotating machines. Here we consider journal bearings. As explained in Section 10.2, the support by this bearing produces asymmetrical elastic forces and also has damping effect. Let us assume that this bearing is used at station i. The elastic force and the damping force are linearized in the equilibrium position of the rotor and expressed as follows:

$$\left.\begin{aligned} F_{xi} &= -\left(k_{11}u_i + k_{12}v_i + c_{11}\dot{u}_i + c_{12}\dot{v}_i\right) \\ F_{yi} &= -\left(k_{21}u_i + k_{22}v_i + c_{21}\dot{u}_i + c_{22}\dot{v}_i\right) \end{aligned}\right\} \tag{15.96}$$

We derive transfer matrices considering these unbalance forces as well as elastic and damping forces at the bearing. Considering these forces, we can obtain the following expressions corresponding to Eq. (15.36):

$$\left.\begin{aligned} & u_i^{\mathrm{R}} = u_i^{\mathrm{L}} \\ & \phi_{yi}^{\mathrm{R}} = \phi_{yi}^{\mathrm{L}} \\ & I_i\ddot{\phi}_{yi} = M_{yi}^{\mathrm{R}} - M_{yi}^{\mathrm{L}} + I_{pi}\omega\dot{\phi}_{xi} \\ & (m_i + M_i)\,\ddot{u}_i^{\mathrm{L}} = V_{xi}^{\mathrm{R}} - V_{xi}^{\mathrm{L}} - \left(k_{11}u_i + k_{12}v_i + c_{11}\dot{u}_i + c_{12}\dot{v}_i\right) \\ & \qquad\qquad + \mathrm{Re}\left(\overline{f}_i e^{j\omega t}\right) \\ & -v_i^{\mathrm{R}} = -v_i^{\mathrm{L}} \\ & \phi_{xi}^{\mathrm{R}} = \phi_{xi}^{\mathrm{L}} \\ & I_i\ddot{\phi}_{xi} = M_{xi}^{\mathrm{R}} - M_{xi}^{\mathrm{L}} - I_{pi}\omega\dot{\phi}_{yi} \\ & (m_i + M_i)\left(-\ddot{v}_i^{\mathrm{L}}\right) = \left(-V_{yi}^{\mathrm{R}}\right) - \left(-V_{yi}^{\mathrm{L}}\right) + \left\{k_{21}u_i - k_{22}\left(-v_i\right)\right. \\ & \qquad\qquad \left. + c_{21}\dot{u}_i - c_{22}\left(-\dot{v}_i\right)\right\} - \mathrm{Re}\left(-j\overline{f}_i e^{j\omega t}\right) \end{aligned}\right\} \tag{15.97}$$

First, we express variables by complex numbers as follows:

$$\left.\begin{aligned} & u = \mathrm{Re}\left(\overline{u}e^{j\omega t}\right), \quad v = \mathrm{Re}\left(\overline{v}e^{j\omega t}\right), \\ & \phi_x = \mathrm{Re}\left(\overline{\phi}_x e^{j\omega t}\right), \quad \phi_y = \mathrm{Re}\left(\overline{\phi}_y e^{j\omega t}\right), \\ & M_x = \mathrm{Re}\left(\overline{M}_x e^{j\omega t}\right), \quad M_y = \mathrm{Re}\left(\overline{M}_y e^{j\omega t}\right) \\ & V_x = \mathrm{Re}\left(\overline{V}_x e^{j\omega t}\right), \quad V_y = \mathrm{Re}\left(\overline{V}_y e^{j\omega t}\right) \end{aligned}\right\} \tag{15.98}$$

and substitute them into Eq. (15.97). Then, we define the extended state vectors and the extended point transfer matrices as follows:

$$\begin{Bmatrix} \overline{u} \\ \overline{\phi}_y \\ \overline{M}_y \\ \overline{V}_x \\ -\overline{v} \\ \overline{\phi}_x \\ \overline{M}_x \\ -\overline{V}_y \\ 1 \end{Bmatrix}_i^{\mathrm{R}} = \begin{bmatrix} 1 & 0 & 0 & 0 & 0 & 0 & 0 & 0 & 0 \\ 0 & 1 & 0 & 0 & 0 & 0 & 0 & 0 & 0 \\ 0 & a_{32} & 1 & 0 & 0 & a_{36} & 0 & 0 & 0 \\ a_{41} & 0 & 0 & 1 & a_{45} & 0 & 0 & 0 & -\overline{f} \\ 0 & 0 & 0 & 0 & 1 & 0 & 0 & 0 & 0 \\ 0 & 0 & 0 & 0 & 0 & 1 & 0 & 0 & 0 \\ 0 & a_{72} & 0 & 0 & 0 & a_{76} & 1 & 0 & 0 \\ a_{81} & 0 & 0 & 0 & a_{85} & 0 & 0 & 1 & -j\overline{f} \\ 0 & 0 & 0 & 0 & 0 & 0 & 0 & 0 & 1 \end{bmatrix}_i \begin{Bmatrix} \overline{u} \\ \overline{\phi}_y \\ \overline{M}_y \\ \overline{V}_x \\ -\overline{v} \\ \overline{\phi}_x \\ \overline{M}_x \\ -\overline{V}_y \\ 1 \end{Bmatrix}_i^{\mathrm{L}} \tag{15.99}$$

where

$$\left.\begin{aligned} & a_{32} = -I\omega^2, \quad a_{36} = -jI_p\omega^2, \quad a_{41} = -(m+M)\omega^2 + \left(k_{11} + jc_{11}\omega\right) \\ & a_{45} = -\left(k_{12} + jc_{12}\omega\right), \quad a_{72} = jI_p\omega^2, \quad a_{76} = -I\omega^2 \\ & a_{81} = -\left(k_{21} + jc_{21}\omega\right), \quad a_{85} = -(m+M)\omega^2 + \left(k_{22} + jc_{22}\omega\right) \end{aligned}\right\} \tag{15.100}$$

Equation (15.99) is represented by using notations as follows:

$$\left\{\tilde{\bar{z}}\right\}_i^{\mathrm{R}} = [\tilde{\bar{P}}]_i \left\{\tilde{\bar{z}}\right\}_i^{\mathrm{L}} \tag{15.101}$$

Similarly, we have

$$\begin{Bmatrix} \bar{u} \\ \bar{\phi}_y \\ \bar{M}_y \\ \bar{V}_x \\ -\bar{v} \\ \bar{\phi}_x \\ \bar{M}_x \\ -\bar{V}_y \\ 1 \end{Bmatrix}_i^{\mathrm{R}} = \begin{bmatrix} 1 & l & b_{13} & b_{14} & 0 & 0 & 0 & 0 & 0 \\ 0 & 1 & b_{23} & b_{24} & 0 & 0 & 0 & 0 & 0 \\ 0 & 0 & 1 & -l & 0 & 0 & 0 & 0 & 0 \\ 0 & 0 & 0 & 1 & 0 & 0 & 0 & 0 & 0 \\ 0 & 0 & 0 & 0 & 1 & l & b_{13} & b_{14} & 0 \\ 0 & 0 & 0 & 0 & 0 & 1 & b_{23} & b_{24} & 0 \\ 0 & 0 & 0 & 0 & 0 & 0 & 1 & -l & 0 \\ 0 & 0 & 0 & 0 & 0 & 0 & 0 & 1 & 0 \\ 0 & 0 & 0 & 0 & 0 & 0 & 0 & 0 & 1 \end{bmatrix} \begin{Bmatrix} \bar{u} \\ \bar{\phi}_y \\ \bar{M}_y \\ \bar{V}_x \\ -\bar{v} \\ \bar{\phi}_x \\ \bar{M}_x \\ -\bar{V}_y \\ 1 \end{Bmatrix}_{i-1}^{\mathrm{L}} \tag{15.102}$$

where

$$b_{13} = \frac{l^2}{2EI}, \quad b_{14} = -\frac{l^3}{6EI}, \quad b_{23} = \frac{l}{EI}, \quad b_{24} = -b_{13} \tag{15.103}$$

Using notations, Eq. (15.102) becomes

$$\left\{\tilde{\bar{z}}\right\}_i^{\mathrm{L}} = [\tilde{\bar{F}}]_i \left\{\tilde{\bar{z}}\right\}_{i-1}^{\mathrm{R}} \tag{15.104}$$

From Eqs. (15.101) and (15.104), we have

$$\{z\}_i^{\mathrm{R}} = [\tilde{\bar{P}}]_i[\tilde{\bar{F}}]_i \left\{\tilde{\bar{z}}\right\}_{i-1}^{\mathrm{R}} = [\tilde{\bar{T}}]_i \left\{\tilde{\bar{z}}\right\}_{i-1}^{\mathrm{R}} \tag{15.105}$$

where the extended transfer matrix $[\tilde{\bar{T}}]_i$ is given by

$$[\tilde{\bar{T}}]_i = \begin{bmatrix} 1 & l & b_{13} & b_{14} & 0 & 0 & 0 & 0 & 0 \\ 0 & 1 & b_{23} & b_{24} & 0 & 0 & 0 & 0 & 0 \\ 0 & a_{32} & t_{33} & t_{34} & 0 & a_{36} & t_{37} & t_{38} & 0 \\ a_{41} & t_{42} & t_{43} & t_{44} & a_{45} & t_{46} & t_{47} & t_{48} & -\bar{f} \\ 0 & 0 & 0 & 0 & 1 & l & b_{13} & b_{14} & 0 \\ 0 & 0 & 0 & 0 & 0 & 1 & b_{23} & b_{24} & 0 \\ 0 & a_{72} & t_{73} & t_{74} & 0 & a_{76} & t_{77} & t_{78} & 0 \\ a_{81} & t_{82} & t_{83} & t_{84} & a_{85} & t_{86} & t_{87} & t_{88} & -j\bar{f} \\ 0 & 0 & 0 & 0 & 0 & 0 & 0 & 0 & 1 \end{bmatrix} \tag{15.106}$$

where

$$\left.\begin{aligned} & t_{33} = a_{32}b_{23} + 1, \quad t_{34} = a_{32}b_{24} - l, \quad t_{37} = a_{36}b_{23} \\ & t_{38} = a_{36}b_{24}, \quad t_{42} = a_{41}l, \quad t_{43} = a_{41}b_{13} \\ & t_{44} = a_{41}b_{14} + 1, \quad t_{46} = a_{45}l, \quad t_{47} = a_{45}b_{13} \\ & t_{48} = a_{45}b_{14}, \quad t_{73} = a_{72}b_{23}, \quad t_{74} = a_{72}b_{24} \\ & t_{77} = a_{76}b_{23} + 1, \quad t_{78} = a_{76}b_{24} - l, \quad t_{82} = a_{81}l \\ & t_{83} = a_{81}b_{13}, \quad t_{84} = a_{81}b_{14}, \quad t_{88} = a_{85}l \\ & t_{87} = a_{85}b_{13}, \quad t_{88} = a_{85}b_{14} + 1 \end{aligned}\right\} \tag{15.107}$$

By applying this operation repeatedly, we can obtain the overall extended transfer matrix as follows:

$$\left\{\tilde{\bar{z}}\right\}_n = [\tilde{\bar{T}}]\left\{\tilde{\bar{z}}\right\}_0 \tag{15.108}$$

These calculations using complex numbers can be carried out using only real numbers in the manner shown in Eqs. (15.52–15.55). We can also calculate using only real numbers if we use the following transformation. When a phase shift occurs due to damping, we can observe the phenomena by decomposing it into the sine and cosine components. For example the displacement u in the xs-plane represented by

$$u = u_c \cos \omega t + u_s \sin \omega t \tag{15.109}$$

Then we can use u_c and u_s instead of u. By such transformation, we can derive a 9×9 matrix with the complex numbers in Eq. (11.106) into a 17×17 matrix with the real numbers.

15.4.2
Steady-State Solution

After we derive the transfer matrix of Eq. (15.108), we can obtain the steady-state solution as follows. First, substituting supporting conditions at the shaft ends, we have Eq. (15.108) in the form

$$\begin{Bmatrix} \left\{\tilde{\bar{z}}\right\}_n' \\ \{0\} \\ 1 \end{Bmatrix} = \begin{bmatrix} \left[\bar{T}_{11}\right] & \left[\bar{T}_{12}\right] & \left\{\bar{f}\right\}' \\ \left[\bar{T}_{21}\right] & \left[\bar{T}_{22}\right] & \left\{\bar{f}\right\}'' \\ [0] & [0] & 1 \end{bmatrix} \begin{Bmatrix} \{\tilde{z}\}_0' \\ \{0\} \\ 1 \end{Bmatrix} \tag{15.110}$$

Then we have

$$\left[\bar{T}_{21}\right]\left\{\tilde{\bar{z}}\right\}_0' = -\left\{\bar{f}\right\}'' \tag{15.111}$$

We can solve this and get

$$\{\tilde{z}\}_0' = -\left[\bar{T}_{21}\right]^{-1}\left\{\bar{f}\right\}'' \tag{15.112}$$

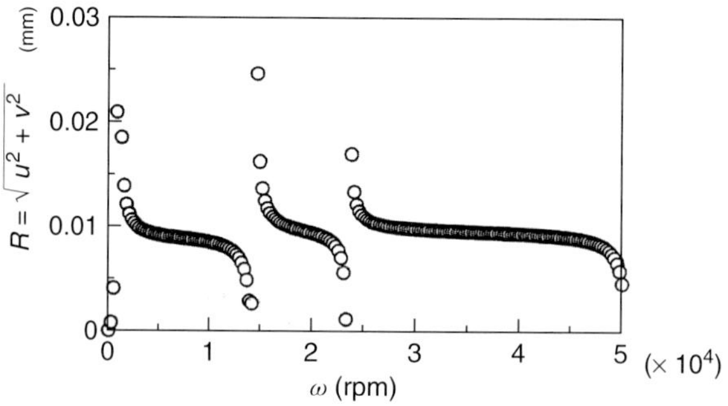

Figure 15.9 Resonance curves.

Adding elements with values 0 and 1, we have $\left\{\tilde{\bar{z}}\right\}_0$. Then we can determine the state vectors $\left\{\tilde{\bar{z}}\right\}_i^{\mathrm{R}} \left(= \left[\overline{T}\right]_i \cdots \left[\overline{T}\right]_1 \left[\overline{P}_0\right] \{\tilde{z}\}_0\right)$ one by one and are also able to determine the vibration shapes.

15.4.3 Example

Figure 15.9 shows a response to an unbalance in model II. Note the difference in scales in Figures 15.8 and 15.9. The resonance curves have peaks at the rotational speeds corresponding to the cross-points of frequency curves and the straight line $p = \omega$ in Figure 15.8.

References

Hahn, E.J. (1989) Rotordynamic analysis using transfer matrices. *Trans. IEAME*, **14**(3), 141–148.

Kikuchi, K. (1969) Vibration analysis of a rotor system by transfer matrix method. Proceedings of the 329th Seminar of the JSME, pp. 25–34. (in Japanese).

Myklestad, N.O. (1944) A new method for calculating natural modes of uncoupled bending vibrations of airplane wings and other types of beams. *J. Aeronaut. Sci.*, **II**(2), 153–162.

Pestal, E.C. and Leckie, F.A. (1963) *Matrix Methods in Elastomechanics*, McGraw-Hill, New York.

Prohl, M.A. (1945) A general method for calculating critical speeds of flexible rotors. *J. Appl. Mech.*, **12**(3), 142–148.

16
Measurement and Signal Processing

16.1
General Considerations

When we investigate the causes of vibrations, we first measure frequency spectra and investigate the relationship between the frequencies and the rotational speed. We can do such *spectrum analysis* easily using any fast Fourier transform (FFT)-analyzer in the market. Spectrum analyzers have various convenient functions, such as tracking analysis, Campbell diagram, and waterfall diagrams. In *tracking analysis*, dynamic characteristics of a rotating machine are investigated by changing the rotational speed. A *waterfall diagram* is a three-dimensional plot of spectra at various speeds. Examples of these diagrams are illustrated in Figure 16.1. However, to use these functions correctly, we must have some background knowledge of signal processing. Further, if we have to construct a specific data analysis system that fits our research, we must have sufficient understanding of the fundamental theories of signal processing.

As mentioned repeatedly, rotor vibration is a whirling motion and therefore not only the frequencies but also the directions of the whirling motion are important enough to pursue their causes. However, since the usual FFT *theory* gives information about magnitudes of frequencies and phases only, we cannot know the whirling direction using the conventional FFT analyzer. For this purpose, Ishida, Ikeda, Yamamoto, and Murakami (1989). Ishida, Yamamoto, and Murakami (1992) and Ishida, Yasuda, and Murakami (1997) proposed a signal processing method in which the whirling plane of a rotor is overlapped to the complex plane. This method, called the *complex-FFT method*, enables us to know the directions of whirling motions besides the magnitudes of the frequencies. They also used this method to extract a component from nonstationary time histories obtained in numerical simulations and experimental data and depicted the amplitude variations of the component.

In this chapter, fundamental ideas necessary for understanding signal processing by computers are explained. In addition, applications of the complex-FFT method of studying stationary and nonstationary vibrations are explained.

Linear and Nonlinear Rotordynamics: A Modern Treatment with Applications, Second Edition.
Yukio Ishida and Toshio Yamamoto.

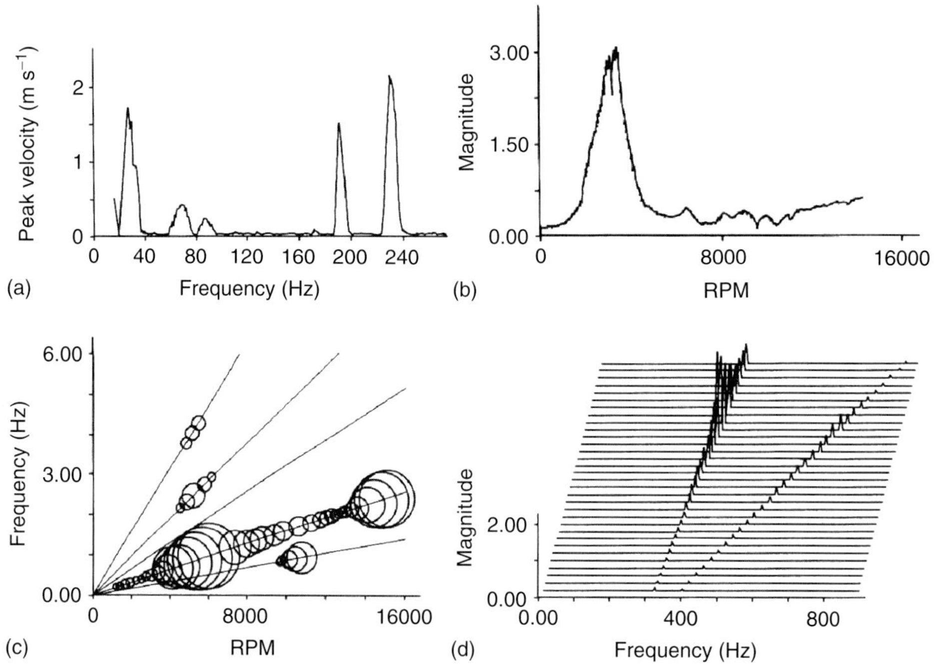

Figure 16.1 Functions of an FFT analyzer. (a) Spectrum diagram, (b) RPM tracking diagram, (c) Campbell diagram, and (d) waterfall diagram.

16.2 Measurement and Sampling Problem

16.2.1 Measurement System and Digital Signal

A measurement system is shown in Figure 16.2. Necessary data are detected from the vibrating structure by sensors. In a rotating machine, rotor displacements in two directions forming a right angle and rotational speed are detected as voltage variations. The output signal $x(t)$ from the sensor is an analog signal that is continuous with time. But the signal is digitized when it is acquired by computers through an interface. This digital signal x_n is a series of discrete data $\{x_n\}$ obtained by measuring (called *sampling*) an analog signal instantly at every time interval Δt, and is given as $x_n = x(n\Delta t)$, where n is an integer. This interval Δt is called a *sampling interval.* A digital signal is discrete in both time and magnitude. Discretization in magnitude is called *quantization,* and the magnitude is represented by binary numbers (unit: *bits*). Digital data in a personal computer are processed into various forms using software programs. In this operation, two representative processing are performed. One is *signal extraction,* where unnecessary signal components are abandoned in the acquired data, and the other is *data transformation,* where the data are converted to a convenient form.

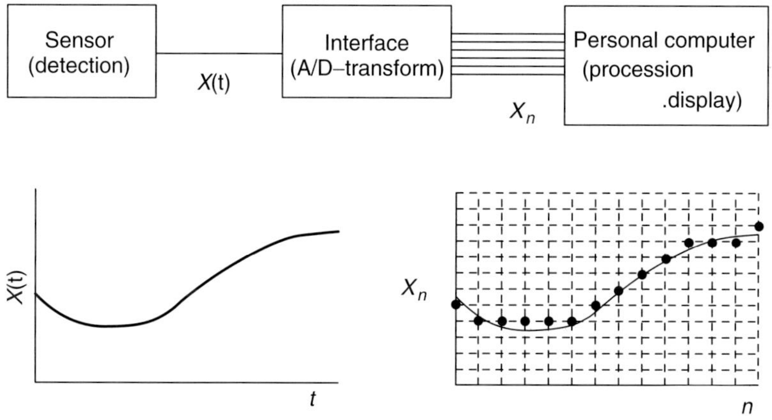

Figure 16.2 Measurement system and signals.

16.2.2
Problems in Signal Processing

When an analog signal $x(t)$ is changed into a sequence of digital data $\{x_n\}$ $(n = 0, 1, 2, \ldots, \mathrm{N})$, a virtual (or imaginary) wave is obtained if a frequent signal is sampled slowly. For example, when a signal illustrated by the full line is sampled as shown in Figure 16.3, a virtual signal wave illustrated by the dashed line appears, although it is not contained in the original signal. This phenomenon is called *aliasing*.

It is obvious that we must sample with a smaller sampling interval as the signal frequency increases. We can determine whether or not we have this aliasing by following the *sampling theorem*. It states that when a signal is composed of the components whose frequencies are all smaller than f_c, we must sample it with a frequency higher than $2f_c$ for the sake of not losing the original signal's information. The frequency $2f_c$ is called the *Nyquist frequency*. For example, if a sine wave with period T is sampled whenever $x(t) = 0$, that is, with sampling interval $T/2$, we have $x_n = 0$. Therefore, processes sampling twice in a period are clearly

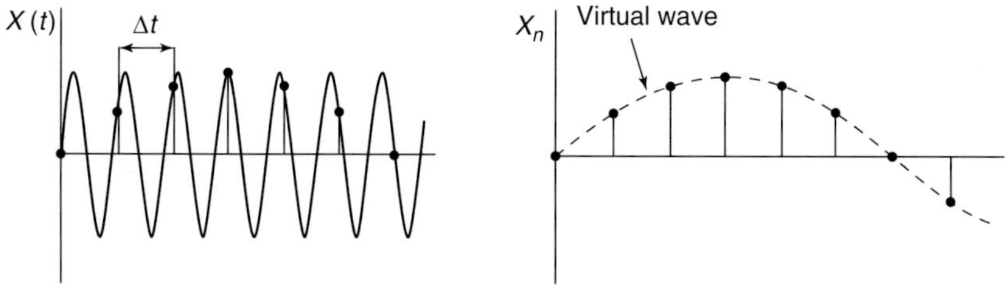

Figure 16.3 Aliasing.

insufficient. However, this theorem teaches us that digital data with more than two points during one period can express the original signal correctly. For example, if we sample the signal components of 1, 2, and 6 kHz with a sampling frequency of 10 kHz, we have an imaginary spectrum of 4 kHz, which does not exist practically. But, if we sample it with a frequency of more than 12 kHz (= 6 × 2 kHz), such an aliasing problem does not occur.

In practical measurements, we do not commonly determine the sampling frequency by trial measurement. Instead, we use a low-pass filter to eliminate the unnecessary high-frequency components in the signal and sample with the frequency higher than twice the cutoff frequency. By such a procedure, we can prevent aliasing.

16.3 Fourier Series

In data processing, we must first know the frequency components contained in a signal. The fundamental knowledge necessary for it is the Fourier series. Although it is assumed that the reader studying vibrations knows this theory well, here we summarize it briefly from the point of view of signal processing. One type of Fourier series is expressed by real numbers, while the other is expressed by complex numbers.

16.3.1 Real Fourier Series

A periodic function $x(t)$ with period T can be expanded by trigonometric functions that belong to the orthogonal function systems as follows:

$$x(t) = \frac{a_0}{2} + \sum_{n=1}^{\infty} (a_n \cos n\omega t + b_n \sin n\omega t) \tag{16.1}$$

where $\omega = 2\pi/T$. This series is called the *Fourier series* or *real Fourier series.* Its coefficients are given by

$$a_n = \frac{2}{T}\int_{-T/2}^{T/2} x(t) \cos n\omega t \, \mathrm{d}t, \quad b_n = \frac{2}{T}\int_{-T/2}^{T/2} x(t) \sin n\omega t \, \mathrm{d}t \tag{16.2}$$

16.3.2 Complex Fourier Series

Fourier series can be expressed by complex numbers using Euler's formula $\mathrm{e}^{i\theta} = \cos\theta + i \sin\theta$. Complex numbers make it easier to treat the expressions. As will be mentioned later, complex representation makes it possible to represent a whirling motion on the complex plane. Substituting Euler's formula into Eq. (16.1), we have

$$x(t) = \sum_{n=-\infty}^{\infty} C_n e^{in\omega t} \tag{16.3}$$

where the complex coefficients are given by

$$C_n = \frac{1}{T}\int_{-T/2}^{T/2} x(t)e^{-in\omega t}dt \quad (n = 0, \pm 1, \pm 2, \cdots) \tag{16.4}$$

Equation (16.3) is called the *complex Fourier series*. Between the real and complex Fourier coefficients, the relationship

$$C_n = \frac{a_n - ib_n}{2}, \quad C_0 = \frac{a_0}{2}, \quad C_{-n} = \frac{a_n + ib_n}{2} \tag{16.5}$$

hold, where $n > 0$. From this we know the following relationship:

$$C_{-n} = \overline{C}_n \tag{16.6}$$

Therefore, if these quantities are illustrated in the figure taking the order $n(n = 0, \pm 1, \cdots)$ as the abscissa, the real part is symmetric about the ordinate axis, and the imaginary part is skew-symmetric about the origin. These complex Fourier coefficients can also be represented by

$$C_n = |C_n|\, e^{-i\theta n} \tag{16.7}$$

where the absolute value $|C_n| = \sqrt{a_n^2 + b_n^2}/2$ is called an *amplitude spectrum*, the angle $\theta_n = \angle C_n = \tan^{-1}(b_n/a_n)$ a *phase spectrum*, and $|C_n|^2$ a *power spectrum*.

As an example, the complex Fourier coefficients of the square wave with period $T = 8$ are shown in Figure 16.4. This wave is defined as $x(t) = 1$ for $0 \le t \le 1, 7 \le t \le 8$, and $x(t) = 0$ for $1 \le t \le 7$. For this square wave, we have the following from Eq. (16.4):

$$C_0 = \frac{2}{T}, \quad C_n = \frac{2\sin n\omega}{Tn\omega} \text{ for } n \neq 0 \tag{16.8}$$

Since $x(t)$ is an even function, the imaginary part of C_n is zero.

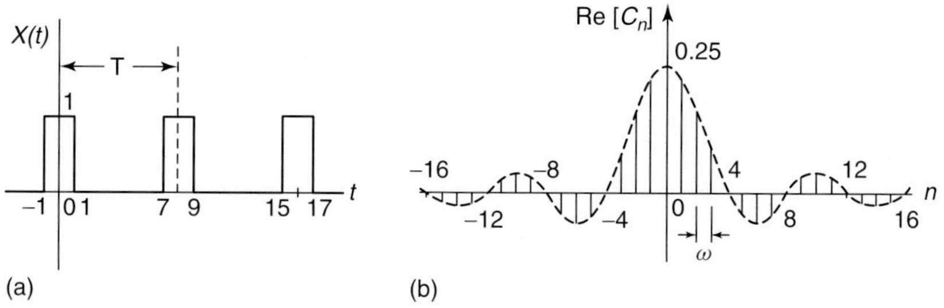

Figure 16.4 Spectrum of a square wave (complex form). (a) Time history and (b) spectrum.

16.4
Fourier Transform

When $x(t)$ is an isolated pulse, it cannot be converted to a discrete spectrum since it is not periodic. However, let us assume that these impulses appear repeatedly at the interval T and consider that this interval is extended to infinity. Then the spectra obtained will represent the spectra of the isolated pulse.

Substituting Eq. (16.3) into Eq. (16.4) gives

$$x(t) = \sum_{n=-\infty}^{\infty} \left[\frac{1}{2\pi} \int_{-T/2}^{T/2} x(t)\mathrm{e}^{-in\omega_0 t}\mathrm{d}t \right] \mathrm{e}^{in\omega_0 t} \frac{2\pi}{T} \tag{16.9}$$

where the frequency $\omega = 2\pi/T$ of the fundamental wave is denoted by ω_0. Here we represent the frequency of the nth order by $n\omega_0 = \omega_n$ and the difference in frequencies between the adjacent components by $\omega_{n+1} - \omega_n = \omega_0 = 2\pi/T = \Delta\omega$. If we make $T \to \infty$, we have

$$x(t) = \int_{-\infty}^{\infty} \left[\frac{1}{2\pi} \int_{-\infty}^{\infty} x(t)\mathrm{e}^{-i\omega t}\mathrm{d}t \right] e^{i\omega t}\mathrm{d}\omega \tag{16.10}$$

where ω_n, $\Delta\omega$, and $\sum$ are replaced by ω, $d\omega$, and $\int$, respectively. This can be expressed in separate forms as follows:

$$x(t) = \int_{-\infty}^{\infty} X(t)\mathrm{e}^{i\omega t}\mathrm{d}\omega \tag{16.11}$$

where

$$X(\omega) = \frac{1}{2\pi} \int_{-\infty}^{\infty} x(t)\mathrm{e}^{-i\omega t}\mathrm{d}t \tag{16.12}$$

Equation (16.12) is called the *Fourier transform* of $x(t)$, and Eq. (16.11) is called the *inverse Fourier transform* of $X(\omega)$.

As an example, Figure 16.5 shows a square pulse defined by $x(t) = 1$ for $-1 \le t \le +1$ and $x(t) = 0$ for a different t, and its continuous spectrum

$$X(\omega) = \frac{\sin \omega}{\pi \omega} \tag{16.13}$$

obtained after Fourier transformation by Eq. (16.12).

Now, let us compare the spectrum of a square wave of period Tshown in Figure 16.4 with that of a square pulse shown in Figure 16.5. As mentioned before, we see from Eqs. (16.8) and (16.13) that the Fourier coefficients in Figure 16.4 have the following relationship to $X(\omega)$:

$$\frac{TC_n}{2\pi} = X(n\omega_0) \tag{16.14}$$

where ω_0 is the fundamental frequency. Therefore, the envelope of the quantities obtained by multiplying $T/2\pi$ with the line spectra of the Fourier coefficients of

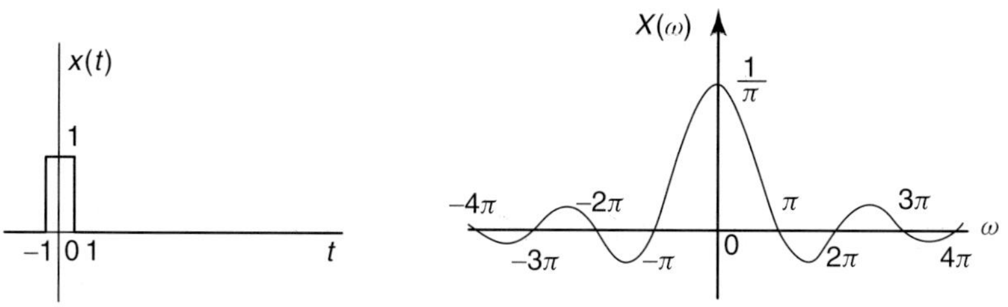

Figure 16.5 Spectrum for a square pulse (Fourier transform).

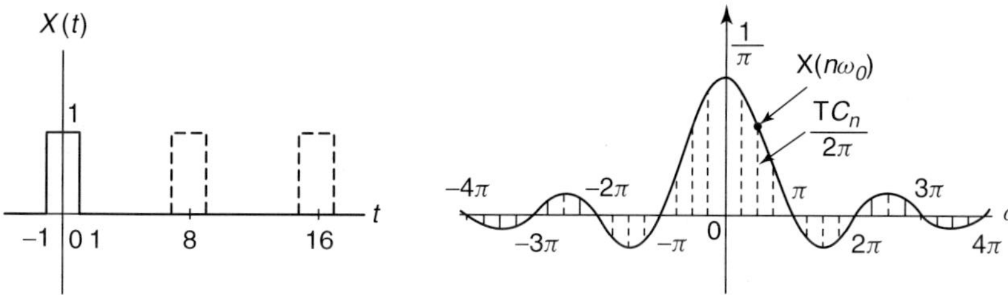

Figure 16.6 Comparison between spectra for a square wave and a square pulse.

the square wave gives the continuous spectra of the Fourier transform $X(\omega)$ of the square pulse. As shown in Figure 16.6, multiplying the spectrum for the square wave with period $T = 8$ in Figure 16.4 by $8/(2\pi)$ gives the spectrum for the square pulse in Figure 16.5.

16.5 Discrete Fourier Transform

So far, we have studied the Fourier series and Fourier transform with the assumption that we know a continuous signal wave in the infinite time duration. However, in practical experiments, the data acquired, converted from the data measured by an analog-to-digital converter, are the sequences of data $\{x_n\}$ $(n = 0, 1, \cdots, N-1)$ that are "discrete" and with "finite number." To perform spectrum analysis using these finite numbers of discrete data, we must use the *discrete Fourier transform* (DFT). DFT is defined as follows: Given N data sampled within the interval Δt, the *DFT* is defined as a series expansion on the assumption that the original signal is a periodic function with the period $N\Delta t$ (although the original signal is not necessarily periodic).

However, various problems occur in the course of this processing. The first is the aliasing problem. When the signal is sampled with interval Δt, information about the components with frequencies higher than $1/(2\Delta t)$ is lost. Therefore, we must pay attention to the valid range of the spectra obtained. The second is the

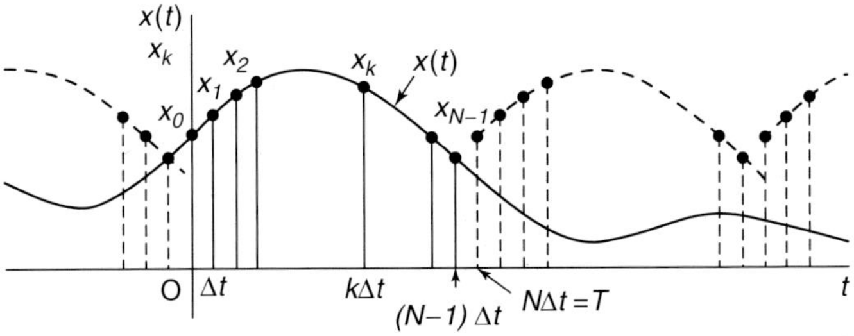

Figure 16.7 Sampled signal sequence.

problem of the coincidence of periods. It is impossible to know the correct period of the original signal before the measurement. Therefore, the period of the original signal and the period of DFT do not coincide, and this difference produces the leakage error. We explain this leakage error and its countermeasure later. The third is the problem of the length of measurement. In the case of an isolated signal $x(t)$, we cannot have data in an infinite time range. However, since the Fourier coefficients C_n and Fourier transform $X(\omega)$ coincide at discrete points as explained in Figure 16.6, we can obtain $X(\omega)$ by connecting the values of C_n smoothly.

In the following discussion, we explain how to compute DFT. Let us assume that we obtained data sequence $x_0, x_1, x_2, \ldots, -x_{N-1}$ by sampling. These data are extended periodically, as shown by the dashed curve in Figure 16.7. The fundamental period is $T = N\Delta t$, and the fundamental frequency is $\Delta\omega = 2\pi/T$. If this dashed curve is given as a continuous time function, its Fourier series expansion is given by the expressions obtained by replacing ω with $\Delta\omega$ in Eqs. (16.3) and (16.4). However, in the case of a discrete signal, the integral of Eq. (16.4) must be calculated by replacing t, T, $x(t)$, and $\int$ with $k\Delta t$, $N\Delta t$, x_k, and $\sum$, respectively. By such replacements, we have

$$C_n \approx \frac{1}{T}\sum_{k=0}^{N-1} x_k e^{-in\Delta\omega k\Delta t}\Delta t = \frac{1}{N}\sum_{k=0}^{N-1} x_k e^{-in(2\pi/T)k(T/N)} \tag{16.15}$$

We represent the right-hand side of this expression by X_n, that is,

$$X_n = \frac{1}{N}\sum_{k=0}^{N-1} x_k e^{-i2\pi nk/N} \quad (n = 0, 1, \ldots, N-1) \tag{16.16}$$

This expression is called the *discrete Fourier transform* of the discrete signal $x_0, x_1, \ldots, x_{N-1}$ Paired with this is the following expression, called the *inverse discrete Fourier transform* (IDFT):

$$x_k = \sum_{n=0}^{N-1} X_n e^{i2\pi nk/N} \quad (k = 0, 1, \ldots, N-1) \tag{16.17}$$

These transformations map the discrete signal of a finite number on the time axis to the discrete spectra of a finite number on the frequency axis, or vice versa. These expressions using complex numbers are called the *complex discrete Fourier transform* and the *complex inverse discrete Fourier transform*. We also have transformations using only real numbers. One is the *real discrete Fourier transform*, given by

$$A_n = \frac{1}{N}\sum_{k=0}^{N-1} x_k \cos\frac{2\pi nk}{N}$$

$$B_n = -\frac{1}{N}\sum_{k=0}^{N-1} x_k \sin\frac{2\pi nk}{N} \quad (n = 0, 1, \ldots, N-1) \tag{16.18}$$

where A_n and B_n are quantities defined by $X_n = A_n + iB_n$ Further, the *inverse real discrete Fourier transform* is given by

$$x_k = \sum_{n=0}^{N-1}\left(A_n \cos\frac{2\pi nk}{N} - B_n \sin\frac{2\pi nk}{N}\right) \quad (n = 0, 1, \ldots, N-1) \tag{16.19}$$

We explain the characteristics of the spectra obtained by DFT using an example in the following discussion. Figure 16.8a shows a square wave with period $T = 8$ and 16 sampled data

$$x_0 = \cdots = x_4 = 1, \quad x_5 = \cdots = x_{15} = 0 \tag{16.20}$$

obtained by sampling with interval $\Delta t = 0.5$. In this example, the signal is intentionally sampled in the range that coincides with the period of the original square wave to avoid the leakage error. Figure 16.8b–e shows spectra representing the real part of X_n, the imaginary part of X_n, the amplitude $|X_n|$, and the phase $\angle X_n$. These spectra have the following characteristics:

1) The spectra are periodic with a period N.
2) The same spectra as those of the negative order $n = -N/2, \ldots, -1$ also appear in the range $n = N/2, \ldots, (N-1)$
3) The spectra of the real part and those of the amplitude are both symmetric about $n = N/2$.
4) The spectra of the imaginary part and those of the phase are skew-symmetric about $n = N/2$.
5) The spectra in the left half of the zone $n = 0, \ldots, (N-1)$ are valid. The spectra in the right half are virtual and are too high compared to the sampling frequency. If the sampling interval is narrowed, the number of spectra increase, and therefore such a spectrum diagram written in the interval $\Delta\omega = 2\omega/T$ extends to the right. For comparison, the spectrum for $N = 32$ is depicted in Figure 16.8f. If the sampling period is shortened, the sampled data become substantially equal to a continuous wave, and therefore its spectra will approach those of the Fourier series shown in Figure 16.4.
6) The magnitude of X_0 is 0.313 in Figure 16.8a and X_0 is 0.281 in Figure 16.8f. This value approaches $C_0 = 0.25$ in Figure 16.4 as the number of data sampled increases.

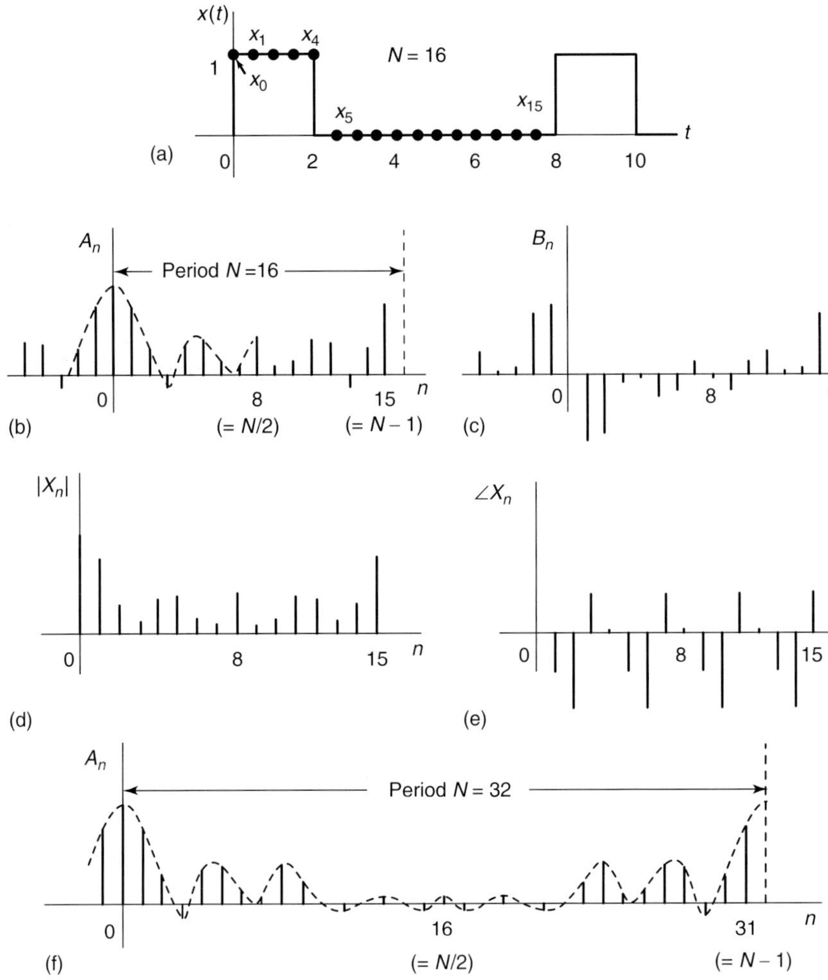

Figure 16.8 Examples of DFT. (a) Square wave and 18 sampled data, (b) real part of X_n, (c) imaginary part of X_n, (d) amplitude, (e) phase, and (f) case of $N = 32$.

Note

Different types of definitions of DFT and IDFT are used, depending on personal preference. Some use the following definitions, in which the magnitudes of X_n coincide with those of Fourier transform in Figure 16.5:

$$X_n = \Delta t \sum_{k=0}^{N-1} x_k e^{-i2\pi nk/N} \quad (n = 0, 1, \ldots, N-1) \tag{16.21}$$

$$x_n = \frac{1}{T} \sum_{n=0}^{N-1} X_n e^{i2\pi nk/N} \quad (k = 0, 1, \ldots, N-1) \tag{16.22}$$

Some use the following expressions, which have the coefficient $1/N$ in the counterpart expression.

$$X_n = \sum_{k=0}^{N-1} x_k e^{-i2\pi nk/N} \quad (n = 0, 1, \ldots, N-1) \tag{16.23}$$

$$x_n = \frac{1}{N} \sum_{n=0}^{N-1} X_n e^{i2\pi nk/N} \quad (k = 0, 1, \ldots, N-1) \tag{16.24}$$

Of course, every definition has the same function as mapping. However, we must be careful when we interpret the physical meaning of the magnitude of the spectra. For example, for $x(t) = \sin t$, it is Eq. (16.16) that gives a spectrum with magnitude 1.

16.6 Fast Fourier Transform

The vast computational task necessary for DFT prevented its practical utilization. In 1965, Cooley and Tukey (1965) proposed an algorithm that enables fast computation of DFT. This algorithm, called *fast Fourier transform*, has made real-time spectrum analysis a practical tool. In the calculation of DFT given by Eq. (16.16), we must perform many multiplications and additions. However, the same calculation appears repeatedly since the function $e^{-i2\pi nk/N} = \cos\{nk(2\pi/N)\} - i\sin\{nk(2\pi/N)\}$ has a periodic characteristic. The FFT algorithm eliminated such repetition and allowed the DFT to be computed with significantly fewer multiplications than direct evaluation of the DFT. For more information on this algorithm, refer to Kido (1985) and Newland (1975).

The FFT algorithm has the restriction that the number of data must be 2^n. When the number of data is $N = 2^m$, DFT needs N^2 multiplications and FFT needs $2Nm$ multiplications. For example, when $N = 2^{10} = 1024$, about 1 050 000 multiplications are necessary in DFT and about 20 480 in FFT. If N increases, this difference becomes extremely large. An FFT program is provided in Appendix G.

16.7 Leakage Error and Countermeasures

16.7.1 Leakage Error

In FFT or DFT, computations are based on the assumption that the data sampled over a finite period are repeated before and after data measurement. For example, Figure 16.9 shows the assumed signals and their spectra for two types

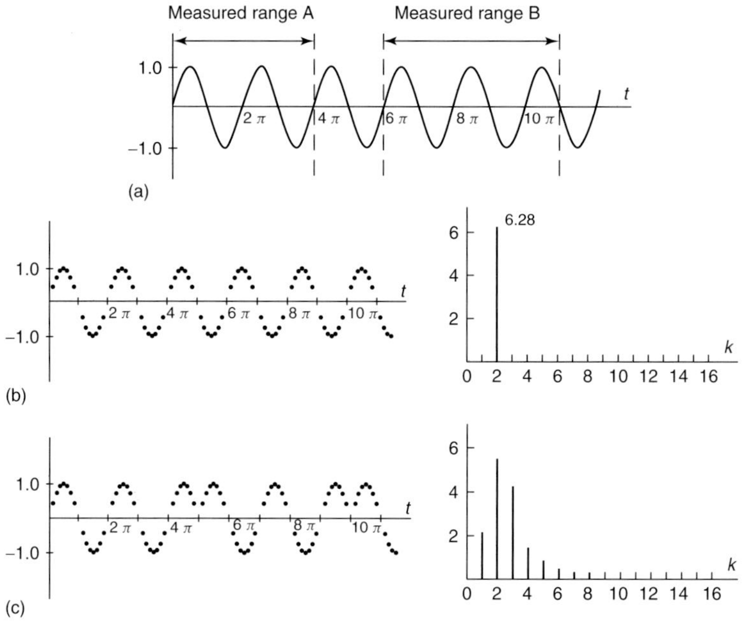

Figure 16.9 Leakage error. (a) Original signal and measured ranges, (b) assumed signal and its spectrum (case A), and (c) assumed signal and its spectrum (case B).

of measurement of a sinusoidal signal $x(t) = \sin t$. Both cases have 32 sampled data, but their sampling intervals are different. In case A, the sampling interval is $\Delta t = 4\pi/32 \approx 0.3926$ and the range measured is exactly twice the fundamental period. The computation of FFT or DFT is performed for the wave as shown by the dotted line. In this case, the assumed wave is the same as the original signal, and therefore we get a single spectrum. In case B, the sampling interval is $\Delta t = 5\pi/32 \approx 0.490$, and the range measured is about 2.5 times the period of the original signal. In this case, the assumed wave shown in Figure 16.9c is not smooth at the junction and differs from the original signal. As a result, the magnitude of the correct spectrum decreases and spectra that do not exist in the original signal appear. As seen in this example, if the time duration measured and the period of the original signal do not coincide, the magnitude of the spectrum decreases and spectra that do not exist in the original signal appear on both sides of the correct spectrum. This phenomenon is called *leakage error*.

16.7.2 Countermeasures for Leakage Error

16.7.2.1 Window Function

To decrease leakage error due to discrepancy between the time duration measured and the period of the original signal, we must connect the repeated wave smoothly.

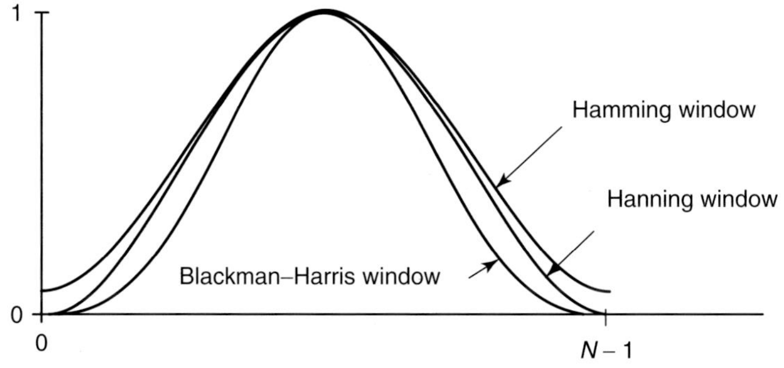

Figure 16.10 Window function.

For this purpose, we multiply a weighting function that decreases gradually at both sides. This weighting function is called a *time window*. Representative time windows, such as, the Hanning window, Hamming window, and Blackman–Harris window, are shown in Figure 16.10. These functions are defined in the range $0 \le n \le N-1$ as

$$\begin{aligned}
&\text{Hanning window: } w(n) = 0.5 - 0.5\cos(2\pi n/N)\\
&\text{Hamming window: } w(n) = 0.54 - 0.46\cos(2\pi n/N)\\
&\text{Blackman–Harris window: } w(n) = 0.423 - 0.498\cos(2\pi n/N)\\
&\quad +0.0792\cos(4\pi n/N)
\end{aligned}$$

and outside $0 \le n \le N-1$ as $w(n) = 0$. For a discussion of their characteristics and the effects of these window functions, refer to some reference books on signal processing.

16.7.2.2 **Prevention of Leakage by Coinciding Periods**

As mentioned above, we can obtain the correct result if the time duration measured coincides with the integer multiple of the period of the original signal. If we attain this by some means, it is better than the use of window functions, which distorts the original signal. For example, for numerical calculations that can be repeated in exactly the same way and whose sampling interval can be adjusted freely, we can determine the measurement duration after we know the period of the original signal by trial simulation and then execute the actual numerical simulation. On the contrary, for experiments, fine adjustment of sampling intervals is generally impossible using practical measuring instruments. However, if the phenomenon appears within a speed range, we can change the rotational speed little by little and adopt the best result where the period, often determined by the rotational speed, and the sampling interval fit.

16.8 Applications of FFT to Rotor Vibrations

In the investigation of rotor vibration, we must know the direction of a whirling motion as well as its angular velocity. In FFT (or DFT), elements of data sequence $\{x_k\}$ obtained by sampling are considered as real numbers and those of data sequence $\{X_n\}$ obtained by DFT are considered as complex numbers. In the following, we introduce a method that can distinguish between whirling directions utilizing the revised FFT. In this FFT, rotor whirling motion is represented by a complex number by overlapping the whirling plane on the complex plane and applying FFT to these complex sampled data.

Let us assume that a disk mounted on an elastic shaft is whirling in the xy-plane. We get sampled data $\{x_k\}$ and $\{y_k\}$ by measuring the deflections $x(t)$ and $y(t)$ in the x- and y- directions, respectively. Taking the x-axis as a real axis and the y-axis as an imaginary axis, we overlap the whirling plane on the complex plane. Using sampled data x_k and y_k, we define the complex numbers as follows:

$$z_k = x_k + iy_k \quad (k = 0, 1, 2, \ldots, N-1) \tag{16.25}$$

and apply FFT (DFT) to them. We call such a method the *complex-FFT method* (Ishida, Yasuda, and Murakami, 1997).

16.8.1 Spectra of Steady-State Vibration

We use the complex-FFT method to investigate steady-state vibrations of some of the nonlinear resonances explained in Chapter 6. Equations of motion are given by Eq. (6.9), that is,

$$\begin{aligned} \ddot{\theta}_x + i_p\omega\dot{\theta}_y + c\dot{\theta}_x + \theta_x + N_{\theta x} &= F\cos\omega t \\ \ddot{\theta}_y - i_p\omega\dot{\theta}_x + c\dot{\theta}_y + \theta_y + N_{\theta y} &= F\sin\omega t \end{aligned} \tag{16.26}$$

16.8.1.1 Subharmonic Resonance of Order 1/2 of a Forward Whirling Mode

In the vicinity of the cross-point of the natural frequency curve p_f and the straight line $p = \omega/2$ in Figure 6.6, a subharmonic component of order $(1/2)\omega$ occurs in addition to the harmonic component with frequency ω. Moreover, some components with higher frequencies coexist with these two major components. Here we investigate the magnitudes and whirling directions of these spectra using the complex-FFT method.

At the rotational speed $\omega = 3.2$, which is near the critical speed of the subharmonic resonance, Eq. (16.26) is integrated numerically using the Runge–Kutta method. The time history obtained is shown in Figure 16.11a. From such time histories in the θ_{xk}- and θ_{yk}-directions, $N = 2^{12} = 4096$ data are sampled with the interval $\Delta t = 0.01389$. Processing these data $\{\theta_{xk}\}$ and $\{\theta_{yk}\}$ using the FFT program in Appendix G, we obtained the amplitude spectra shown in Figure 16.11a. Only the low-frequency spectra are shown in this figure. Since X_n is the magnitude of the

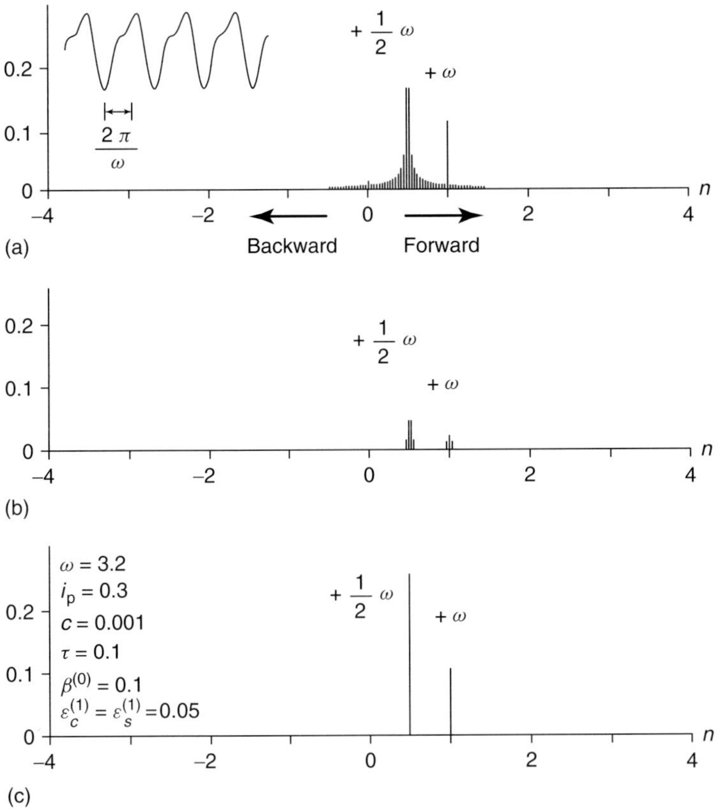

Figure 16.11 Spectra of the subharmonic resonance of order 1/2 of a forward whirling mode. (a) Spectrum obtained without using window function, (b) spectrum obtained by using window function, and (c) spectrum obtained by adjusting sampling interval.

spectrum of $e^{i2\pi nk/N}$ as shown in Eq. (16.17), the spectra for $n > 0$ represent forward whirling motions and those for $n < 0$ represent backward whirling motions. We see that spectra corresponding to a forward whirling motion of order 1/2 appear predominantly. In addition, spectra corresponding to a harmonic component and a constant component occur. In Figure 16.8, the amplitude spectra are symmetric about $N = 0$. This characteristic holds for real sampled data. In the case of Figure 16.11, where complex data are processed, the spectra are not symmetric. Much leakage error occurs on both sides of the spectrum $+(1/2)\omega$ in this case because data in the time duration of 29 periods of the harmonic component, that is, 14.5 periods of the subharmonic component, are sampled. To reduce this leakage error, a window function (a Blackman–Harris window) is multiplied and Figure 16.11b is obtained. Although the leakage error decreases, the necessary spectrum $+(1/2)\omega$ also decreases. Figure 16.11c shows the result obtained by the procedure explained in Section 16.7. In this procedure, the sampling interval is adjusted through trial sampling. By adjusting Δt, data in the time duration of just

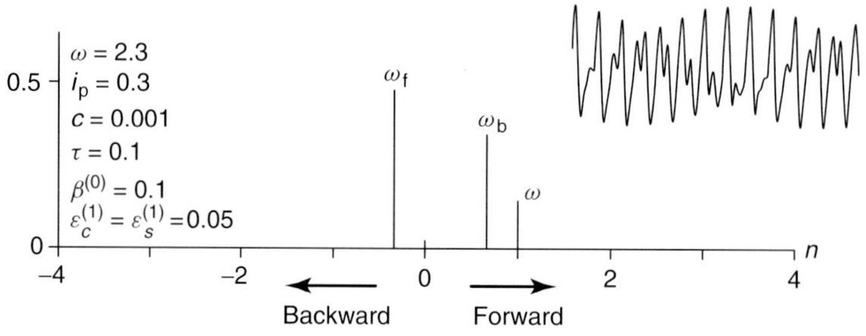

Figure 16.12 Spectra of the combination resonance of order $[p_f - p_b]$ obtained by the complex-FFT method.

30 periods of harmonic components are sampled. Spectra with no leakage error are obtained in this case.

16.8.1.2 **Combination Resonance**

Combination resonance occurs in the vicinity of the rotational speed, where the relationship $p_f - p_b = \omega$ holds in Figure 6.11. The result is shown in Figure 16.12, where a component with frequency $\omega_f(\approx p_b > 0$, forward) and a component with frequency $\omega_b(\approx p_b < 0$, backward) appear predominantly in addition to the harmonic component. The relationship $\omega_f = \omega_b = \omega$ holds in this resonance.

16.8.2 **Nonstationary Vibration**

A subhannonic resonance occurs at the rotational speed ω_0 in Figure 6.6. Let us investigate the amplitude variation when a rotor passes through this resonance with constant angular acceleration. The equations of motion for the 2 DOF inclination rotor system are given as follows (Ishida *et al.*, 1989):

$$\begin{aligned} \ddot{\theta}_x + i_p\ddot{\psi}\theta_y + i_p\dot{\psi}\dot{\theta}_y + c\dot{\theta}_x + \theta_x + N_{\theta x} &= F(\dot{\psi}^2 \cos\psi + \ddot{\psi}\sin\psi) \\ \ddot{\theta}_y - i_p\ddot{\psi}\theta_x - i_p\dot{\psi}\dot{\theta}_x + c\dot{\theta}_y + \theta_y + N_{\theta y} &= F(\dot{\psi}^2 \sin\psi + \ddot{\psi}\cos\psi) \end{aligned} \tag{16.27}$$

The angular position ψ of the unbalance is given by $\psi = 1/2\lambda t^2 + \omega_s t + \psi_0$, where λ is a constant angular acceleration, ω_s is the initial angular velocity, and ψ_0 is the initial angular position when the acceleration starts.

In Figure 16.13, resonance curves of the subharmonic resonance of order 1/2 of a forward whirling mode (amplitude R_0) and the radius of nonstationary vibration are shown. The resonance curve is of the hard spring type. The nonstationary vibration is obtained by accelerating with a constant acceleration from the angular velocity $\omega_s = 2.9$, which is a little lower than the resonance speed. The curve is obtained by integrating Eq. (16.27) numerically by the Runge–Kutta method and calculating $r(t) = \sqrt{\theta_x{}^2 + \theta_y{}^2}$ from the values of θ_x and θ_y obtained. Since the radius of the nonstationary vibration changes rapidly, it is difficult to know the variation of the subharmonic component. This is because, due to the coexistence of a subharmonic component and a harmonic component, the rotor traces a complicated orbit similar

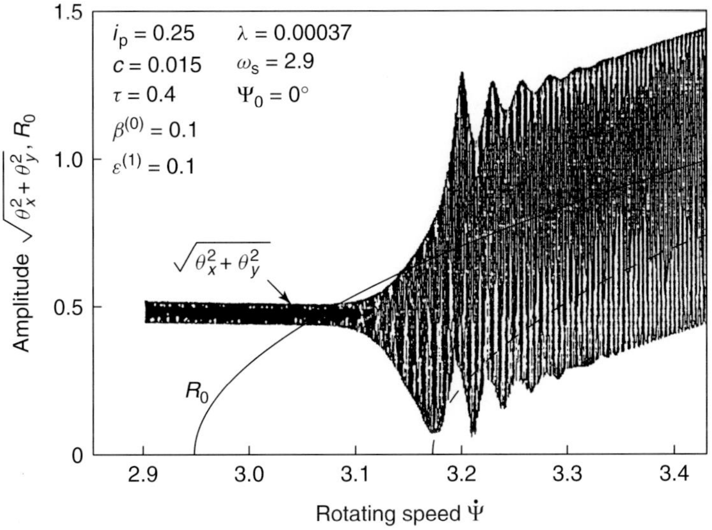

Figure 16.13 Variation of radius during passage through the subharmonic resonance of order 1/2.

to that in Figure 6.8. Therefore, we obtain the change of amplitude of a subharmonic component utilizing the complex-FFT method and signal extraction processing.

The processing is performed as follows:

1) Obtain the time histories of θ_x and θ_y by integrating Eq. (16.27) numerically. The result is shown in Figure 16.14a. The sampling interval is 1/24 of a period of a harmonic component at the initial angular velocity $\omega_s = 2.9$. The number of the data sampled is $N = 2^{14} = 16\,384$. The left half is a harmonic vibration of almost constant amplitude. At about the midposition of these time histories, a subharmonic vibration begins to occur. In this case, the amplitude of the subharmonic resonance increases rapidly along the resonance curve.
2) To reduce the leakage error, the Blackman–Harris window function is multiplied by these time histories. The result is shown in Figure 16.14b. Time histories that are narrow at both ends are obtained.
3) Make complex-numbered data $z_k = \theta_{xk} + i\theta_{yk}$ from the data sequences $\{\theta_{xk}\}$ and $\{\theta_{yk}\}$ $(k = 0, 1, \ldots, N-1)$ in the θ_x- and θ_y-directions. Apply FFT processing given by Eq. (16.16) to the data sequence $\{z_k\}$. The result is shown in Figure 16.14c. Only the comparatively low-frequency spectra are shown in this figure. We see that the backward subharmonic component of order 1/2 and the backward harmonic component appear simultaneously in addition to constant and forward components. Since these data are sampled during the time when angular velocity $\dot{\psi}$ varies from 2.9 to 3.5, the spectra for the forward harmonic components are distributed in the same frequency range. The spectra for the forward subharmonic component are distributed from 1.4 to 2.0. We eliminate all the spectra except the shaded ones corresponding to

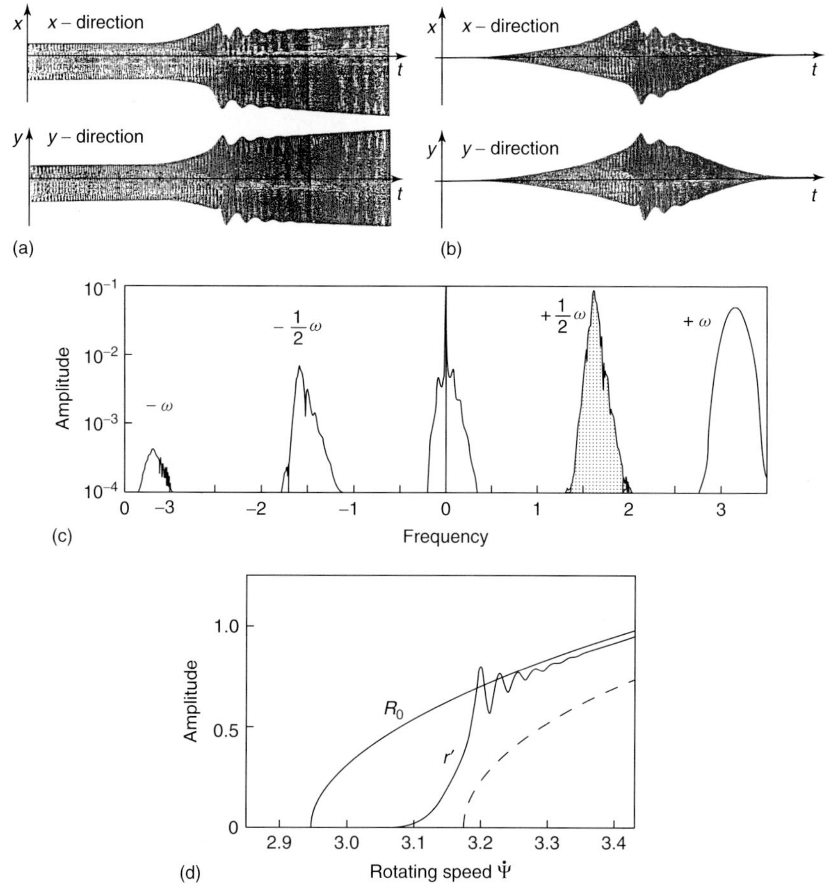

Figure 16.14 Nonstationary vibrations during passage through the subharmonic resonance of order 1/2. (a) Time histories, (b) time histories multiplied by a window function, (c) spectra obtained by the complex-FFT method, and (d) variation of amplitude of subharmonic component.

the forward subharmonic vibration of order 1/2 by the program called a *filtering process*.

4) Transform from the frequency range to the time range by applying inverse FFT.
5) Compensate for the magnitude changed in process 2 by dividing data by the window function.
6) The real and imaginary parts of the data sequence represent the components in the θ_x- and θ_y-directions, respectively. Therefore, the absolute value of these complex numbers gives the amplitude $r'(t)$ of the forward subharmonic component of order 1/2. The result is shown in Figure 16.14d. An amplitude variation curve is obtained clearly without noise.

References

Cooley, J.W. and Tukey, J.W. (1965) An algorithm for the machine calculation of complex Fourier series. *Math. Comput.*, **19**, 297–302.

Ishida, Y., Ikeda, T., Yamamoto, T., and Murakami, S. (1989) Nonstationary vibration acceleration through a critical speed (a critical speed of a l/2-order subharmonic oscillation). *JSME Int.J. Ser. 111*, **32** (4), 575–584.

Ishida, Y., Yamamoto, T., and Murakami, S. (1992) Nonstationary oscillation of a rotating shaft with nonlinear spring characteristics during acceleration through a critical speed (a critical speed of a 1/3-order subharmonic oscillation). *JSME Int.J. Ser. III*, **35** (3), 360–368.

Ishida, Y., Yasuda, K., and Murakami, S. (1997) Nonstationary vibration of a rotating shaft with nonlinear spring characteristics during acceleration through a major critical speed: a discussion by the asymptotic method and the FFT method. *Trans. ASME, J. Vib. Acoust. Stress Reliab. Des.*, **119** (1), 31–36.

Kido, K. (1985) *Fundamentals of Digital Signal Processing*, Maruzen Co. Ltd. (in Japanese).

Newland, D.E. (1975) *Random Vibrations and Spectral Analysis*, Longman Group, Harlow, Essex.

17
Active Magnetic Bearing

17.1
General Considerations

As mentioned before, journal bearings and rolling bearings are widely used. In both cases, the rotating parts have contact with the static parts via fluid and rolling bodies. As a result, the friction at the contact points causes energy dissipation. On the contrary, *active magnetic bearings* (AMBs) can support a rotor without contact. Figure 17.1 shows the features and the production output of various kinds of bearings (Kakuta, 2001). For a rotor that rotates at a low rotational speed, bearings in which the two faces of the rotor and the bearing have direct contact are used. Metal bush bearings, plastic bearings, oil-impregnated porous metal bearings, and so on, belong to this category. For a rotor that rotates at a very high speed, AMBs are used. An AMB supports a rotor without physical contact using magnetic force. AMBs have the following advantages: (i) they do not suffer from wear, (ii) they have low friction, (iii) they create minimal vibration and noise, (iv) they permit very high speed operation, and (v) they can be used in a vacuum atmosphere without contamination because they do not use lubricant. However, since the magnetic forces are controlled using various electronic devices, the price becomes unavoidably expensive. In addition, there is danger that an accident, such as the shutdown of the electric current, could occur. In this chapter, the fundamentals of magnetic levitation are explained along with an outline of AMBs.

17.2
Magnetic Levitation and Earnshaw's Theorem

When the same poles of two bar magnets face each other, a magnetic force works and they repel each other. Therefore, it seems that we can levitate a material by utilizing this characteristic. However, *Earnshaw's theorem*, outlined below, teaches us that it is impossible to levitate a magnet without contact in a static magnetic field.

This theorem was originally proved in the electrostatic field by Samuel Earnshaw in 1842. He states that "A collection of point charges cannot be maintained in a stable stationary equilibrium configuration solely by the electrostatic interaction of

Linear and Nonlinear Rotordynamics: A Modern Treatment with Applications, Second Edition.
Yukio Ishida and Toshio Yamamoto.

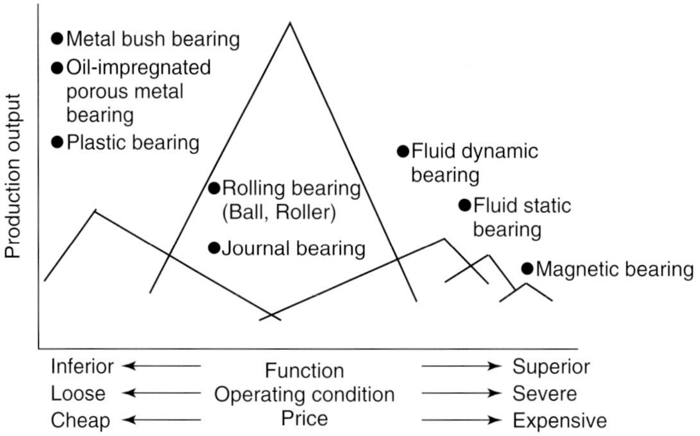

Figure 17.1 Characteristics of bearings and output. (Courtesy of Prof. Kakuta, 2001.)

the charges." If a point charge levitates stably in a static electric field, it means that the electric potential V is minimum at this position locally. In other words, when the particle deviates in any direction, a force works to push it back to the previous position. One theory of electromagnetics teaches us that Laplace's equation $\Delta V = \left(\partial^2/\partial x^2 + \partial^2/\partial y^2 + \partial^2/\partial z^2\right) V = 0$ holds in a field where no electric particle exists. In order for a location to be the local minimum point of the potential, the following two conditions must be satisfied. One is the condition $\partial V/\partial x = 0$, $\partial V/\partial y = 0$, $\partial V/\partial z = 0$, which means that the slope of the tangent is zero. The other is the condition $\partial^2 V/\partial x^2 > 0$, $\partial^2 V/\partial y^2 > 0$, $\partial^2 V/\partial z^2 > 0$, which indicates that the potential curved surface is convex downward. However, it is apparent that the second condition does not hold if Laplace's equation holds.

Earnshaw's theorem also holds in a static magnetic field produced by permanent magnets. Therefore, we must apply some active control methods to levitate a rigid body in a static magnetic field.

17.3
Active Magnetic Levitation

17.3.1
Levitation Model

In this section, a one-dimensional magnetic levitation system is explained (Matsumura, 1986; Schweitzer, Bleuler, and Traxier, 1994). Figure 17.2 illustrates a physical model in which a spherical iron ball is attracted by an electric magnet and is floating (levitating). The equilibrium point is the position where the gravitational force downward and the electric attractive force upward are balanced. However, this equilibrium position is unstable because the electric magnetic force becomes larger than the gravitational force when the ball moves upward and smaller when

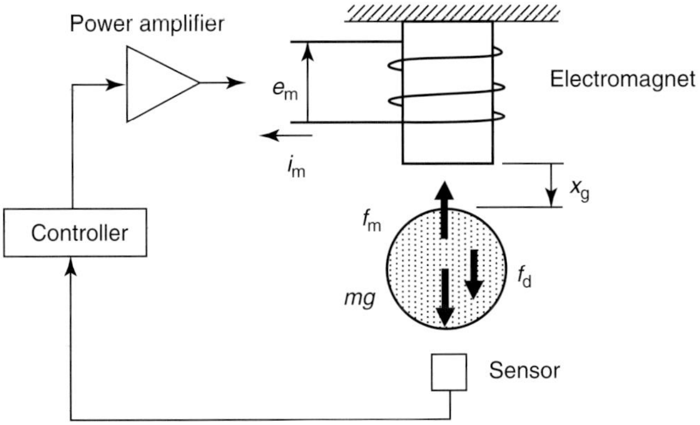

Figure 17.2 Magnetic levitation.

it moves downward. Therefore, in order to levitate the ball stably, we must stabilize the equilibrium position by controlling the magnetic force in the following steps: (i) the ball position is detected by a sensor, (ii) using this information control signal is determined by a controller, (iii) this control signal is amplified by a power amplifier, and (iv) the output voltage is applied to the electric magnet.

In Figure 17.2, let the downward deflection of the ball from the contact point be x_g. The voltage e_m is applied to the electric magnetic coil to produce the current, which in turn produces the magnetic attraction force f_m. This force f_m is inversely proportional to the square of the gap x_g and proportional to the square of the current i_m. The relationship between the magnetic force and the gap x_g, when the current i_m is constant, and that between the magnetic force and the current i_m, when the gap x_g is constant, are illustrated in Figure 17.3. In the former case, the force does not become infinite at $x_g = 0$ but approaches a constant value. Considering these characteristics, we represent the magnetic force approximately as follows:

$$f_m = -k\left(\frac{i_m}{x_g + \delta}\right)^2 \tag{17.1}$$

The positive direction is downward. Let the mass of the levitated body be m, the gravitational constant be g, and the force applied from the outside be f_d. Then the equation of motion of the ball is given as follows.

$$m\frac{d^2x_g}{dt^2} = mg - k\left(\frac{i_m}{x_g + \delta}\right)^2 + f_d \tag{17.2}$$

Suppose that the electromagnetic and gravitational forces balance each other when the current $i_m = i_0$ in the case of the force $f_d = 0$ and that the ball levitates at the position $x_g = x_0$ statically. From Eq. (17.2), we have

$$mg = k\left(\frac{i_0}{x_0 + \delta}\right)^2 \tag{17.3}$$

Next, we control the magnetic force to stabilize this position. First, the nonlinear relationships in Figure 17.3 are linearized in the vicinity of the equilibrium position

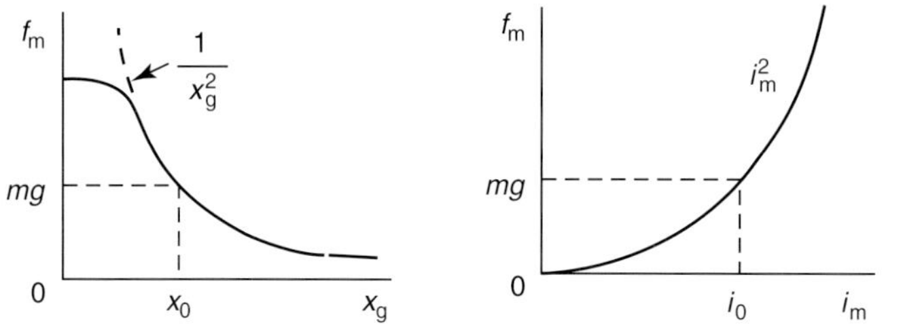

Figure 17.3 Electromagnetic force. (a) Change for the gap x_g at constant current and (b) change for the current i_m at constant gap.

x_0 and i_0. Considering small deviations x and i, we put

$$x_g = x_0 + x, \quad i_m = i_0 + i \tag{17.4}$$

Substituting these expressions into Eq. (7.2) and neglecting higher order terms of small quantities x/x_0 and i/i_0, we obtain

$$m\frac{d^2x}{dt^2} = \frac{2ki_0^2}{(x_0+\delta)^3}x - \frac{2ki_0}{(x_0+\delta)^2}i + f_d \tag{17.5}$$

where the relationship of Eq. (17.3) is used.

The electric current i necessary to levitate the ball stably is determined by the deflection x. Then the equations governing the motion of the ball are expressed as follows.

$$\left.\begin{aligned} \frac{dx}{dt} &= v \\ \frac{dv}{dt} &= ax - bi + df_d \\ i &= f\left(x, \dot{x}, \cdots, \int x dt, \cdots\right) \end{aligned}\right\} \tag{17.6}$$

where

$$a = \frac{2ki_0^2}{m(x_0+\delta)^3}, \quad b = \frac{2ki_0}{m(x_0+\delta)^2}, \quad d = \frac{1}{m} \tag{17.7}$$

The concrete expression of the third equation of Eq. (17.6) depends on the control method.

17.3.2
Current Control with PD-Control

In a feedback control system, a controller calculates a difference (namely, an error value) between a measured position x and a reference input (a desired set point) r, and attempts to minimize the error by adjusting the control input. In case of the PD (Proportional-Derivative) controller, the current is given by the following expression:

$$i = K_P\,(x - r) + K_D \frac{d\,(x - r)}{dt} \tag{17.8}$$

where K_P is the proportional gain and K_D is the derivative gain.

17.3.2.1 Physical Meanings of PD Control

The physical meanings of P control and D control are as follows. Substituting Eq. (17.8) into Eq. (17.5), we obtain

$$\begin{aligned} m\frac{d^2x}{dt^2} &+ \frac{2ki_0}{(x_0+\delta)^2}K_D\frac{dx}{dt} + 2k\left\{\frac{i_0 K_P}{(x_0+\delta)^2} - \frac{i_0^2}{(x_0+\delta)^3}\right\}x \\ &= \frac{2ki_0}{(x_0+\delta)^2}\left(K_P r + K_D\frac{dr}{dt}\right) + f_d \end{aligned} \tag{17.9}$$

This expression shows that K_P and K_D have effects such as stiffness and damping, respectively. From this expression, we see that this unstable system with a negative coefficient of the restoring force is stabilized if the coefficient of x is changed to a positive value.

17.3.2.2 Transfer Function and Stability Condition

Transforming Eqs. (17.6) and (17.8) by the Laplace transformation and adopting zero values as the initial values of x, v, and i, we obtain

$$\left.\begin{aligned} X(s) &= \frac{1}{s}V(s) \\ V(s) &= \frac{1}{s}\left(aX(s) - bI(s) + dF_d(s)\right) \\ I(s) &= \left(K_P + K_D s\right)\left(X(s) - R(s)\right) \end{aligned}\right\} \tag{17.10}$$

where $L[x] = X(s)$, $L[v] = V(s)$, $L[i] = I(s)$, $L[r] = R(s)$, and $L[f_d] = F_d(s)$. From this, we can derive the block diagram shown in Figure 17.4.

From Eq. (17.10), we can obtain the following relationship between the output $X(s)$, the input $R(s)$, and the disturbance $F_d(s)$.

$$X = \frac{bK_P + bK_D s}{s^2 + bK_D s + \left(bK_P - a\right)}R + \frac{d}{s^2 + bK_D s + \left(bK_P - a\right)}F_d \tag{17.11}$$

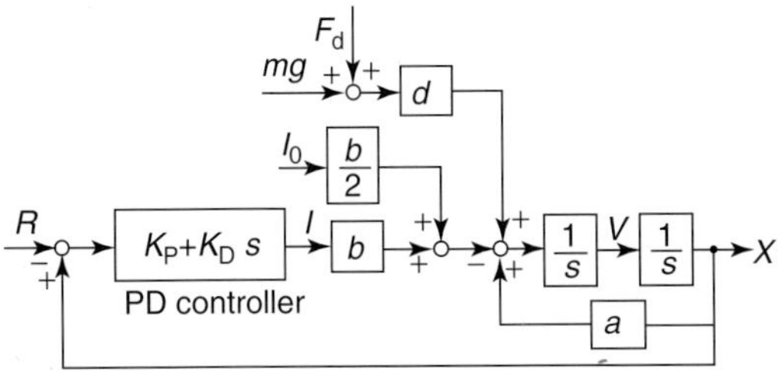

Figure 17.4 Block diagram of a PD controller.

The response in the time domain is given by $x(t) = \mathrm{L}^{-1}\left[X(s)\right]$. In order to be stable, all roots (eigenvalues) of the following characteristic equation must have negative real parts. (refer to "Note: Eigenvalues and the Stability".)

$$s^2 + bK_\mathrm{D}s + \left(bK_\mathrm{P} - a\right) = 0 \tag{17.12}$$

Applying the Routh–Hurwits stability criterion and considering that $b > 0$, we obtain the necessary and sufficient conditions to establish stability as follows:

$$K_\mathrm{D} > 0, \quad bK_\mathrm{P} - a > 0 \tag{17.13}$$

Note: Eigenvalues and the Stability

Equation (17.12) has real eigenvalues $s = \alpha_1, \alpha_2$ or conjugate complex eigenvalues $s = \beta \pm \omega i$. The denominator of Eq. (17.11) is represented by $(s - \alpha_1)(s - \alpha_2)$ in the former and $(s - \beta)^2 + \omega^2$ in the latter. The coefficient of R in the former case is decomposed into the partial fractions in the forms $1/(s - \alpha_1)$ and $1/(s - \alpha_2)$. As a result, the terms in the right-hand side of Eq. (17.11) have the form $1/(s - \alpha)$ or $1/\{(s - \beta)^2 + \omega^2\}$. When an impulse $R(s) = 1$ is given together with $F_d(s) = 0$, the response in the time domain is obtained by the inverse Laplace transformation as follows.

$$\mathrm{L}^{-1}\left[\frac{1}{s - \alpha}\right] = \mathrm{e}^{\alpha t}, \quad \mathrm{L}^{-1}\left[\frac{\omega}{(s - \beta)^2 + \omega^2}\right] = \mathrm{e}^{\beta t}\sin \omega t \tag{17.14}$$

This shows that the system is stable if the real parts of all eigenvalues are negative.

17.3.2.3 Determination of Gains

In Section 17.3.2.1 we derived the conditions that determine whether the system is stable or not. However, if we investigate the positions of the eigenvalues on the complex plane, we can find the degree of stability. In the design of the feedback control system, the positions of eigenvalues determine the characteristics of the transient property (namely, the speed of response and damping). The representative design methods for a controller are those called *pole assignment* and *Bode diagram*. Here we adopt the former. As an example, we discuss the case of the critical damping, which is the state that corresponds with the fastest damping. The arrangement of the roots (multiple roots) in this case is shown in Figure 17.5. Let the distance between the root and the origin be ω_n. Then the characteristic equation is represented by

$$(s + \omega_\mathrm{n})^2 = 0 \tag{17.15}$$

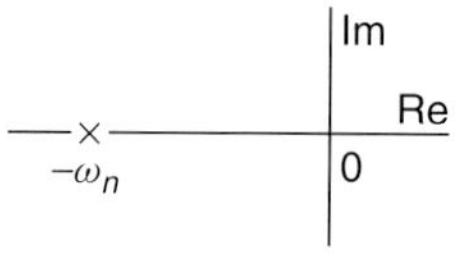

Figure 17.5 Pole assignment.

By comparing this with Eq. (17.12), we have

$$K_P = \frac{1}{b}\left(\omega_n^2 + a\right), \quad K_D = \frac{2}{b}\omega_n \tag{17.16}$$

When we use this theory, we first determine the value of ω_n appropriately and then determine the gains K_P and K_D using Eq. (17.16).

17.3.2.4 Case with a Static Load

If the reference input r and the disturbance force f_d are zero in Eq. (17.9), the ball comes to rest at $x = 0$, whereas if a static load (a step input) f_d is applied, it moves to a new rest point. For a magnetic bearing, it is desirable to keep the position at $x = 0$, namely at the bearing center, independent of the change in load. An integral feedback (*I control*) solves this problem. In I control, deviation of the position is integrated over time and the control force proportional to this value is added until the steady-state position becomes zero. In the next section, proportional-integral-derivative (PID) control is explained.

17.3.3 Current Control with PID-Control

In PID control, the electric current i is given as follows

$$i = K_P (x - r) + K_D \frac{d(x - r)}{dt} + K_I \int (x - r)\, dt \tag{17.17}$$

where K_I is the *integral gain*.

17.3.3.1 Transfer Function and Stability Condition

Similar to the previous case, taking the Laplace transform of Eqs. (17.6) and (17.17) yields the block diagram shown in Figure 17.6.

The following relationship holds among the output $X(s)$, the input $R(s)$, and the disturbance force $F_d(s)$.

$$X = \frac{bK_P s + bK_D s^2 + bK_I}{s^3 + bK_D s^2 + \left(bK_P - a\right) s + bK_I} R + \frac{ds}{s^3 + bK_D s^2 + \left(bK_P - a\right) s + bK_I} F_d \tag{17.18}$$

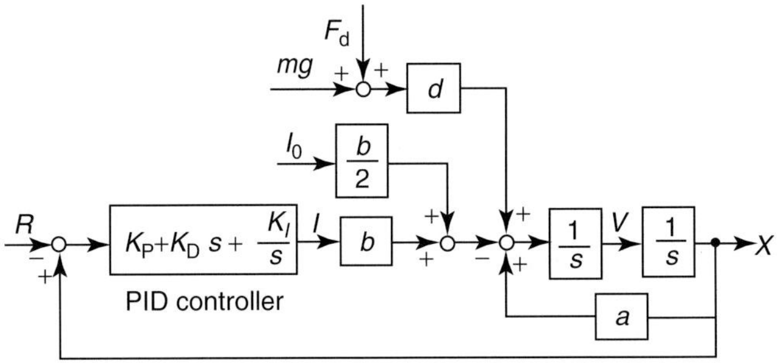

Figure 17.6 Block diagram (PID control).

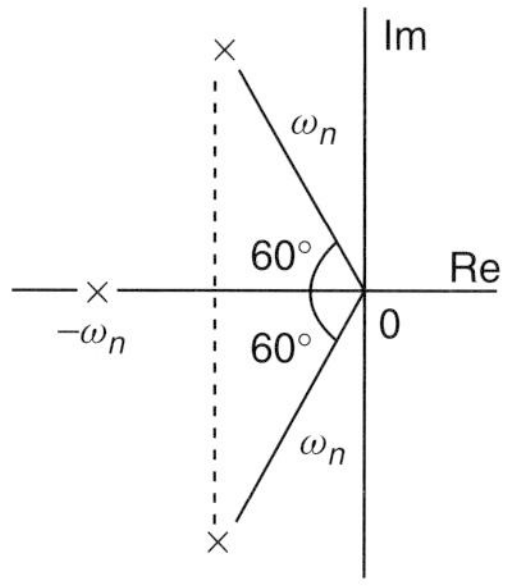

Figure 17.7 Pole assignment.

The characteristic equation is

$$s^3 + bK_Ds^2 + (bK_P - a)\,s + bK_I = 0 \tag{17.19}$$

From the Routh–Hurwits criteria, the necessary and sufficient conditions in order to be stable are given as follows:

$$K_D > 0,\;\; bK_P - a > 0,\;\; K_I > 0,\;\; K_D\,(bK_P - a) - K_I > 0 \tag{17.20}$$

where $b > 0$.

17.3.3.2 Determination of Gains

As an example, we adopt the pole assignment shown in Figure 17.7 (Matsumura, 1986). Let the distance from the origin to the poles be ω_n. Then these poles are given by

$$(s + \omega_n)\left(s + \frac{1}{2}\omega_n - \frac{\sqrt{3}}{2}\omega_n i\right)\left(s + \frac{1}{2}\omega_n + \frac{\sqrt{3}}{2}\omega_n i\right) = 0 \tag{17.21}$$

By comparing this with Eq. (17.19), we obtain

$$K_P = \frac{1}{b}\left(2\omega_n^2 + a\right),\quad K_I = \frac{1}{b}\omega_n^3,\quad K_D = \frac{2}{b}\omega_n \tag{17.22}$$

17.3.3.3 Case with a Static Load

Here we investigate the deflection when a static load is applied. In the case of PD control, we studied that a steady-state offset remains. This can be understood by using a transfer function. Suppose that a unit step function $f(t) = 1$ $(t \geq 0)$ is applied to the ball. Since its Laplace transformation is given by $F_d(s) = 1/s$, we can obtain the following equation by applying the *final value theorem* to Eq. (17.11):

$$\lim_{t\to\infty} x(t) = \lim_{s\to 0} s\frac{d}{s^2 + bK_Ds + (bK_P - a)}\left(\frac{1}{s}\right) = \frac{d}{(bK_P - a)} \tag{17.23}$$

This shows that an offset remains in this case.

In the case of PID control, we obtain the following equation by applying the final value theorem to Eq. (17.18):

$$\lim_{t\to\infty} x(t) = \lim_{s\to 0} s\frac{ds}{s^3 + bK_Ds^2 + (bK_P - a)\,s + bK_I}\left(\frac{1}{s}\right) = 0 \tag{17.24}$$

In this case, the static equilibrium position does not change.

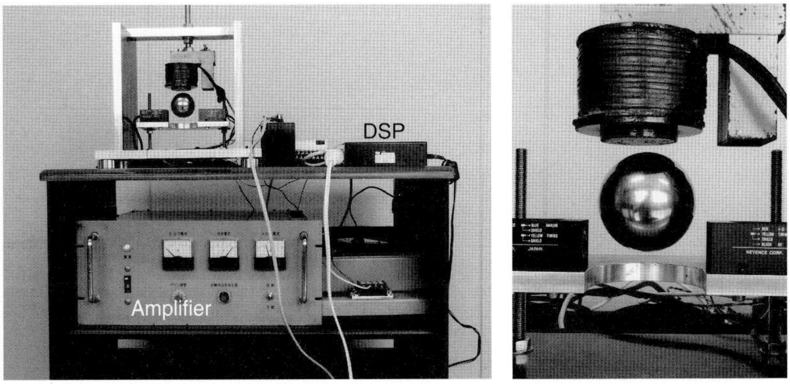

Figure 17.8 Experimental setup.

17.3.4
Practical Examples of Levitation

17.3.4.1 Identification of System Parameters

The experimental setup is illustrated in Figure 17.8. A steel ball of mass $m = 535$ g is pulled up by a magnet. In this setup, the controller sends a control signal to the power amplifier based on the signal from the sensor, and then the amplifier determines the current in the electric magnet following this control signal. Since it is assumed in this theory that there is no delay between the control signal and the current of the magnet, the amplifier must respond fast. In the sensor, a laser beam with a liner cross section is projected from the projector. The vertical width of this beam is 10 mm. As a part of this beam is interrupted by the ball, the quantity of the beam entering the receiver varies depending on the position of the ball.

We can determine the relationship between the magnetic attraction force f_m and the electric current i_m as follows. The ball is placed onto a set of scales, and the weight of the ball is measured at various distances $x_{g.}$ from the magnet (Figure 17.2). From this relationship, we obtain the value of the parameters k and δ in Eq. (17.1), which makes the square of the error between the measured value and the curve.

$$k \approx 5.06 \times 10^{-5}\mathrm{Nm^2\,A^{-2}}, \qquad \delta \approx 4.71 \times 10^{-3}\mathrm{m} \tag{17.25}$$

Figure 17.9 illustrates the measured values of the magnetic attraction force and the value given by Eq. (17.1) with these parameter values for various currents. The gap is fixed at 3 mm. The abscissa represents the electric current, and the ordinate represents the magnetic force. It shows that the magnetic force is a quadratic function of the current.

If the levitated equilibrium position is $x_0 = 2$ mm, the current given by Eq. (17.3) is $i_0 = 2.16$ A. Then, from Eq. (17.7), we obtain $a = 2.92 \times 10^3\ \mathrm{s^{-2}}$ and $b = 9.08$ m $\mathrm{As^{-2}}$.

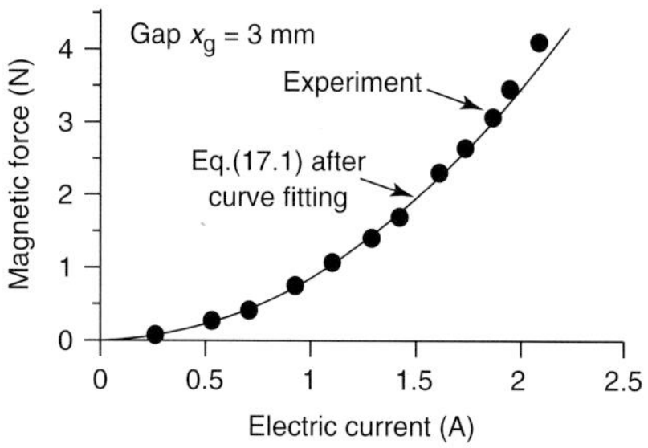

Figure 17.9 Measured values of the magnetic force and its approximation by Eq. (17.1).

17.3.4.2 Digital PD-Control with DSP

In the following, an iron ball is levitated by applying the PD control. DSP (digital signal processor) is used in this control system. Of course, it is possible to use a personal computer, micro computer, or operation amplifiers instead. The advantage of using DSP is that we can design and perform trial experiments easily using software. However, in a digital control system, the system becomes unstable if the sampling interval is not small enough. In this experiment, the sampling interval is 200 μs. Comparatively, in an analog control system in which operation amplifiers are used, the sampling problem does not occur but changing parameter values is troublesome. In addition, it is a little difficult to perform a highly sophisticated control.

In this PD control, $\omega_n = 8\pi$ rad s^{-1} (4 Hz) is adopted in the pole assignment. From Eq. (17.16) we obtain

$$K_P = 392\ \mathrm{Am}^{-1}, \quad K_D = 5.54\ \mathrm{sA\ m}^{-1} \tag{17.26}$$

In experiments, the deflection is measured and differentiation is performed approximately. This is because we cannot attain an exact differentiation using measured data. In addition, the differentiation of signal amplifies noise and cause the system unstable when the derivative gain is large. In the approximate differentiation, PD control is expressed by

$$K_P + \frac{K_D s}{\tau s + 1} \tag{17.27}$$

instead of $K_P + K_D s$. The time constant τ is a positive constant small enough to obtain sufficient approximation of $K_P + K_D s$. As τ approaches zero, the denominator of the second term in Eq. (17.27) approaches 1 and, as a result, the term represents correct differentiation. In the experiment, the inverse of time constant τ, that is, the corner frequency, is selected as 500 Hz. Therefore $\tau = 1/1000\pi$ s.

In the experiment, the DSP program is performed using MATLAB/Simulink (The MathWorks Inc.). In this software, we can make the program graphically by drawing a block diagram. Figure 17.10 illustrates a block diagram made using Simulink. Keep in mind that we can operate DSP by using a programing language such as C instead of Simulink.

Figure 17.8 shows a levitating iron ball. It is floating stably without any contact. The gap is about 5 mm. The difference from the design value of 2 mm is due to various errors that occurred in modeling, the amplifier, the sensor, and so on. If we use a PID controller, this difference will diminish.

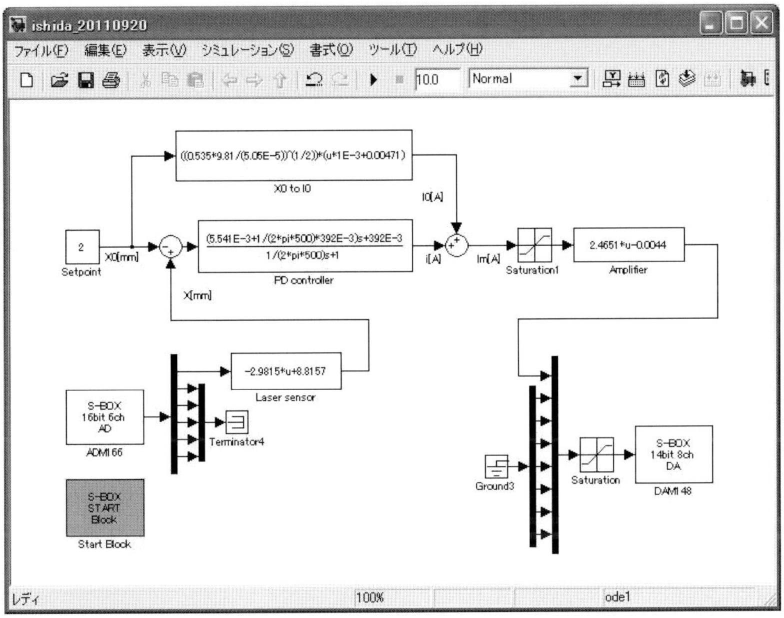

Figure 17.10 DSP program by Simulink.

17.3.5
Current Control with State Feedback Control

The above-mentioned frequency domain analysis is called the *classical control theory*. In contrast to this, the control theory utilizing time-domain state space representation is called the *modern control theory*. In practical rotor systems, the degrees of freedom (DOF) and the numbers of input and output increase from those of the levitation model mentioned above. In such a case, the following state variable expression is useful.

In the modern control theory, Eq. (17.6) is expressed by vectors and matrices as follows.

$$\dot{\boldsymbol{x}} = \mathbf{A}\boldsymbol{x} + \mathbf{B}\boldsymbol{u} + \mathbf{D}\boldsymbol{f} \tag{17.28}$$

where

$$\boldsymbol{x} = \begin{bmatrix} x \\ v \end{bmatrix}, \quad \mathbf{A} = \begin{bmatrix} 0 & 1 \\ a & 0 \end{bmatrix}, \quad \mathbf{B} = \begin{bmatrix} 0 \\ -b \end{bmatrix},$$

$$\boldsymbol{u} = [i], \quad \mathbf{D} = \begin{bmatrix} 0 \\ d \end{bmatrix}, \quad \boldsymbol{f} = [f_d] \tag{17.29}$$

The notation $\boldsymbol{x}$, which represents the state of the system, is called the *state variable*, and Eq. (17.28) is called the *state equation*.

All the components of the state variable cannot always be observed from outside. The vector $\boldsymbol{y}$ composed of observable variables is called the *output vector*. It is represented by

$$y = \mathbf{C}x \tag{17.30}$$

where

$$y = [y], \ \mathbf{C} = \begin{bmatrix} 1 & 0 \end{bmatrix} \tag{17.31}$$

Equation (17.30) is called the *output equation.*

Now, it is assumed that the reference value is $r = 0$. The system is designed so that the state variable $\boldsymbol{x}$ converges to 0 rapidly when the initial position is given. Since the input vector $\boldsymbol{u}$ in Eq. (17.8) is determined by the position and the velocity of a ball, we represent it as

$$\boldsymbol{u} = \begin{bmatrix} K_P & K_D \end{bmatrix} \begin{bmatrix} x \\ v \end{bmatrix} = \mathbf{F}\boldsymbol{x} \tag{17.32}$$

where F is called the gain matrix. In this setup, the velocity is not measured by a sensor. As mentioned previously, it is determined by differentiating the signal of the position. In control theory, the velocity is often estimated from the signal of the position. This method is called the *state observer*. This method's advantage is that we need not differentiate the signal of the position. From here onward, we continue the explanation with the assumption that the velocity v has been obtained ideally by some means.

When there is no disturbance, the following is obtained from Eqs. (17.32) and (17.28).

$$\dot{\boldsymbol{x}} = (\mathbf{A} + \mathbf{BF})\,\boldsymbol{x} \tag{17.33}$$

There are two means to determine the gain matrix $\mathbf{F}$, which stabilizes the system. One is the *pole assignment method,* which determines the positions of the eigenvalues of the system matrix $\mathbf{A}_0 = \mathbf{A} + \mathbf{BF}$ of the closed loop. This method is the same as that mentioned above in the classical control.

The other is the method based on the optimal control theory. The outline is summarized as follows. For further details regarding this control theory, refer to some reference book. In the *linear-quadratic regulator* (LQR) *theory,* which is one of the optimal control methods, the feedback control is performed so as to minimize the following cost function:

$$J = \int_0^\infty \left(\boldsymbol{x}^T\mathbf{Q}\boldsymbol{x} + \boldsymbol{u}^T\mathbf{R}\boldsymbol{u}\right) dt \tag{17.34}$$

This cost function, which is a quadratic function of the state variable and the control input variable, represents the integrals of the deviation and the input. In other words, minimizing this cost means that neither the state variable $\boldsymbol{x}$ nor the control variable $\boldsymbol{u}$ can be large. If the infinite integral of $\boldsymbol{x}$ is finite, this implies that $\boldsymbol{x}(t)$ converges to zero as t goes to infinity. Symmetric, nonnegative, semidefinite matrices $\mathbf{Q}$ and $\mathbf{R}$ are the weighting factors. The designers can select the ratio of these two matrices depending on if it is the deflection or the control force they hope to minimize. Choosing to make $\mathbf{Q}$ large means that the state variable $\mathbf{x}$ must be small (the state decays faster). On the other hand, making $\mathbf{R}$ large means that the control input force $\boldsymbol{u}$ must be small (less control force is necessary).

In order to get the feedback control force $\boldsymbol{u}$ that minimizes this cost function, we first solve the following *Riccati differential equation* for $\mathbf{P}$:

$$\mathbf{A}^T\mathbf{P} + \mathbf{P}\mathbf{A} - \mathbf{P}\mathbf{B}\mathbf{R}^{-1}\mathbf{B}^T\mathbf{P} + \mathbf{Q} = 0 \tag{17.35}$$

and then calculate

$$u = -\mathbf{R}^{-1}\mathbf{B}^T\mathbf{P}\boldsymbol{x} \tag{17.36}$$

(Meirovitch, 1990). Therefore, we can determine the gain matrix as follows:

$$\mathbf{F} = -\mathbf{R}^{-1}\mathbf{B}^T\mathbf{P} \tag{17.37}$$

17.4
Active Magnetic Bearing

AMB supports a rotor without physical contact utilizing the above-mentioned magnetic levitation technology. There are many kinds of magnetic bearings, such as an attractive type and a repelling type. The most widely used is the attractive type of AMB, which is outlined below.

17.4.1
Principle of an Active Magnetic Bearing

Figure 17.11 shows a principle of AMB. In the previous section, the attractive force that balances against gravity is given by one electromagnet. An AMB has two pairs of corresponding horseshoe-type electromagnets. One pair is in the horizontal direction and the other is in the vertical direction. Two electromagnets on opposite sides of the rotor compose one pair and attract the rotor. When the rotor deviates by a small amount from the center, the controller produces offset in the current in the two paired electromagnets by equal but opposite perturbation.

Figure 17.11 illustrates the control system for a pair of magnets in the vertical direction. The rotor position is detected by inductive gap sensors and this information determines the deviation of the rotor from the center. The controller, which is usually a microprocessor or DSP, sends a control signal to the power amplifier. The power amplifier supplies the necessary electric currents to the electromagnets.

17.4.2
Active Magnetic Bearings in a High-Speed Spindle System

Figure 17.12a illustrates an AMB unit for a super high speed spindle (Shimada, Horiuchi, and Shamoto, 1999). An end mill shown in Figure 17.12b is mounted

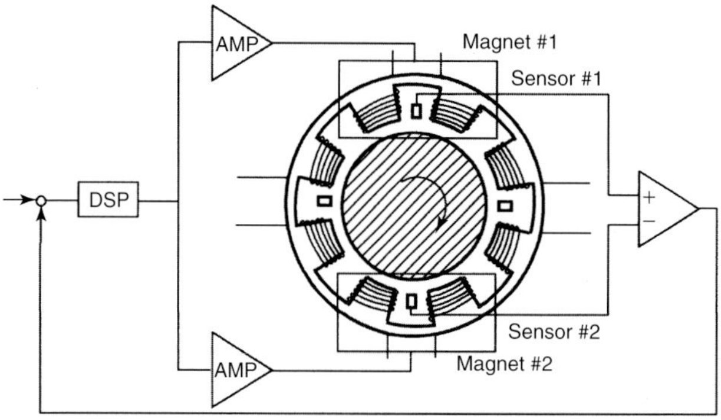

Figure 17.11 Construction of an electromagnet and a control system.

on the main shaft, and it cuts the material at a super high speed. The maximum rotational speed of the shaft is about 70 000 rpm. Figure 17.12c shows various parts of this unit. The cross section of this unit is shown in Figure 17.12d. The motor in the middle rotates the shaft. The upper and lower radial magnetic bearings (diameter of 49.1 mm) and the middle axial magnetic bearing support the rotor without contact. Since this rotor is vertical, the axial magnetic bearing not only controls the axial position of the shaft but also supports the rotor weight. Touchdown bearings are also used in order to backup the main bearings if they should fail due to a stoppage of electric current or a failure of the control system. The gap between the rotor and the backup bearing is about half of the air gap in the magnetic bearings.

The whole unit composed of magnetic bearings, motor, rotor, and so on, is often called a *magnetic bearing*. In the analysis of such a rigid rotor, we use the coordinate system shown in Figure 2.20. Since the rotation of the rotor around the Z_1-axis is determined by the rotational speed, the rotor position is expressed by the deflections of its geometrical center (x, y, z) and the inclination of the center line (θ_x, θ_y). Therefore, this system has 5 DOF, which is composed of 1 DOF in the thrust direction, 2 DOF representing a parallel up-and-down motion and a pitching motion in the vertical plane, and 2 DOF representing a parallel right-and-left motion and a yawing motion in the horizontal plane. These motions corresponding to these 5 DOF must be controlled simultaneously. Such a bearing is called a *five-axis magnetic bearing*.

In this system, the analog signals of deflections detected by eddy current sensors are changed to digital signals and then enter the PID controller, which contains the DSP. The control signals from the PID controller enter into the PWM power amplifier, and the output currents are supplied to the magnet coils.

17.4.3
Dynamics of a Rigid Rotor system

As mentioned above, P controller produces a spring force and D controller produces a damping force. Therefore, they can be represented by a spring with the spring

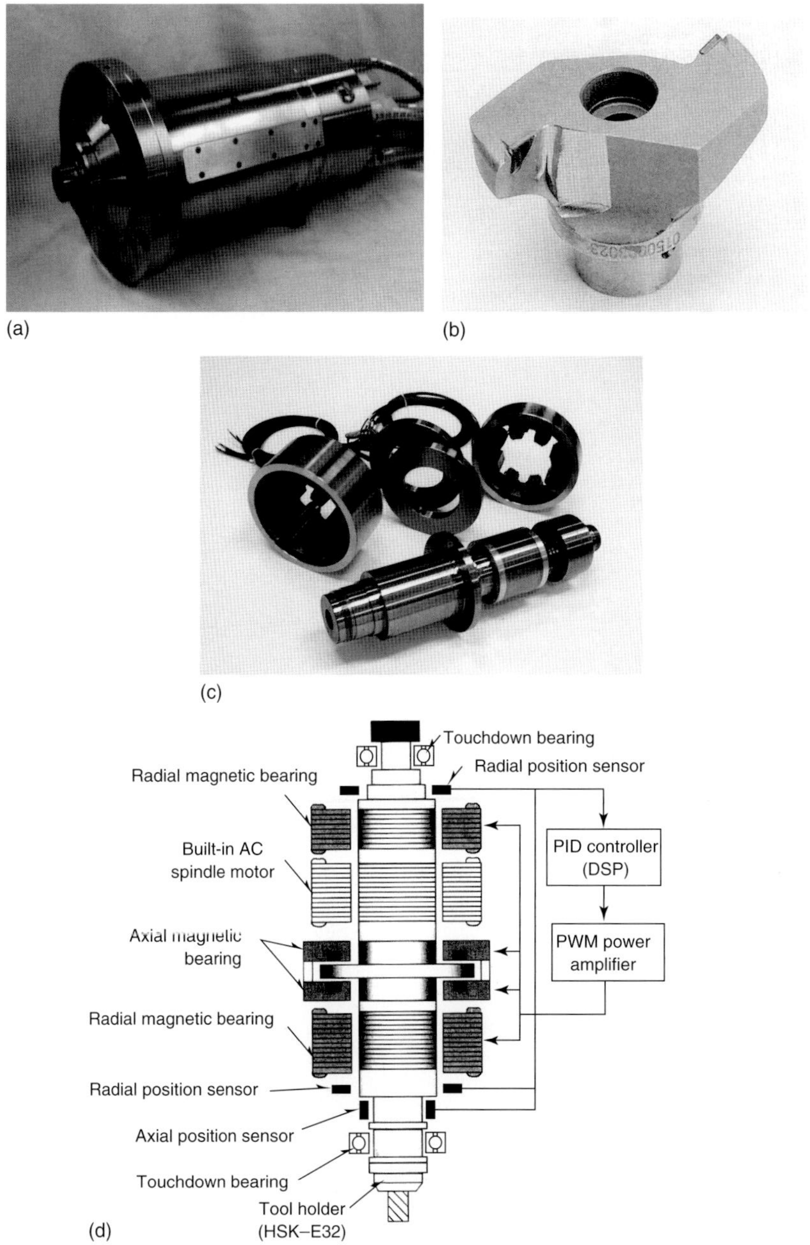

Figure 17.12 Example of a magnetic bearing. (a) Overview of AMB, (b) endmill, (c) various parts of AMB, and (d) cross section of a bearing. (Courtesy of Mitsubishi Electric Corp.)

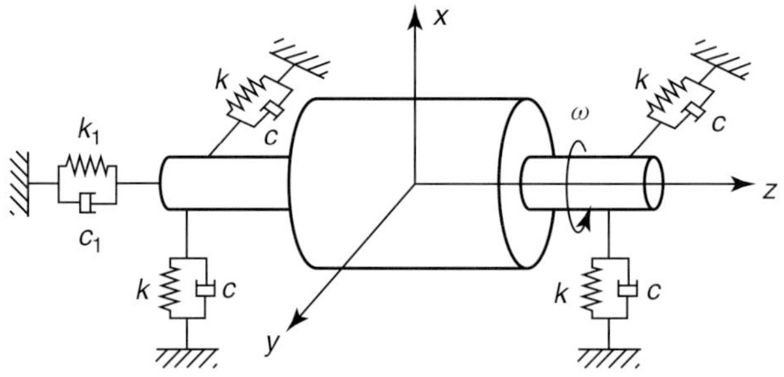

Figure 17.13 Model of a rigid rotor.

constant k and a damper with the damping coefficient c as shown in Figure 17.13. Let the rotational speed be ω. Then the equations of motion are represented by

$$\left.\begin{aligned} m\ddot{x} &= F_x \\ m\ddot{y} &= F_y \\ m\ddot{z} &= F_z \\ I\ddot{\theta}_x + I_p\omega\dot{\theta}_y &= M_{xz} \\ I\ddot{\theta}_y - I_p\omega\dot{\theta}_x &= M_{yz} \end{aligned}\right\} \tag{17.38}$$

The restoring forces F_x, F_y and the restoring moments M_{xz}, M_{yz} are produced by AMB. In the z-direction, the force is represented by

$$F_z = f_{zd} - k_z z - c_z\dot{z} \tag{17.39}$$

where f_{zd} is the disturbance force in the z-direction. These equations of motion can be analyzed in the same way as explained in Chapter 2.

References

Kakuta, K. (2001) Fundamentals of rolling bearings and the points for their selection. *Machine Des.*, **45** (9), 28–35. (in Japanese).

Matsumura, F. (1986) Fundamental theory and system construction of magnetic bearings. *Fujietsu Tech. Rep.*, **42** (1), 1–11. (in Japanese).

Meirovitch, L. (1990) Dynamics and Structures. *Wiley-Interscience*.

Schweitzer, G., Bleuler, H., and Traxier, A. (1994) *Active Magnetic Bearings*, Chapter 2, Hochschulverlag AG an der ETH Zürich.

Shimada, A., Horiuchi, Y., and Shamoto, K. (1999) A study on active magnetic bearings for machine tool's high speed spindle. 2nd International Symposium on Magnetic Bearing, Zurich, pp. 183–188.

Appendix A
Moment of Inertia and Equations of Motion

Let us consider a rigid body mounted on a rigid shaft as shown in Figure A.1. The rigid body is rotating with an angular velocity $\omega = d\varphi/dt$. A small particle of mass dm and at a distance r from the shaft has angular momentum $(dmr\omega)r = r^2\omega dm$. Summing this, we have the total angular momentum L of the rigid body as follows:

$$L = \int r^2 \omega dm = \omega \int r^2 dm = I\omega \tag{A.1}$$

where

$$I = \int r^2 dm \tag{A.2}$$

is the *moment of inertia* about the axis coinciding with the shaft. Equation (A.1) shows that the angular momentum of a rigid body is given by the product of the moment of inertia and the angular velocity. Corresponding to the equation of motion for linear motion $d(mv)/dt = F$, where v is the velocity of the center of gravity and F is the applied force, the equation of motion $dL/dt = M$ holds for rotational motion when a moment M applied. Then, substituting Eq. (A.1) into this expression, we have

$$I\frac{d^2\varphi}{dt^2} = M \tag{A.3}$$

This is the fundamental equation for the rotational motion of a rigid body. We consider the rectangular coordinates fixed to the rigid body to be O-$x'y'z'$, as shown in Figure A.2a. The moment of inertia about an arbitrary axis OA that passes through the point O is derived as follows. Let the direction cosines of OA with respect to the x'-, y'-, and z'- axes be λ, μ, and υ, respectively, the unit vector along OA be $\vec{e}\ (\lambda, \mu, \nu)$, the position vector of point P in the rigid body be $\vec{p}(x', y', z')$, and the angle between OP and OA be θ. Then, the inner product of $\vec{e}$ and $\vec{p}$ gives

$$p\cos\theta = \lambda x' + \mu y' + \nu z' \tag{A.4}$$

Linear and Nonlinear Rotordynamics: A Modern Treatment with Applications, Second Edition.
Yukio Ishida and Toshio Yamamoto.

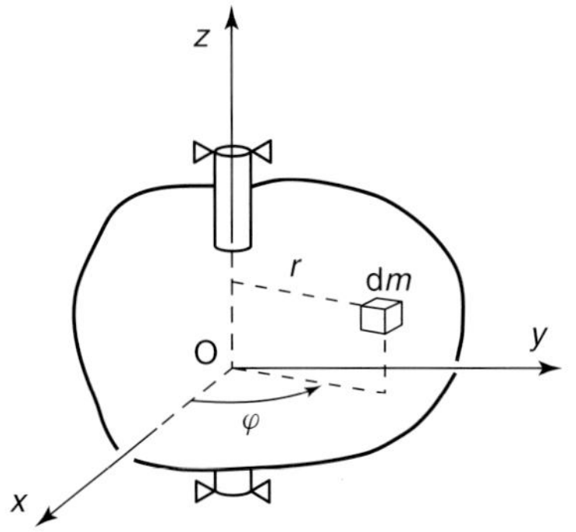

Figure A.1 Rigid rotor with a rigid shaft.

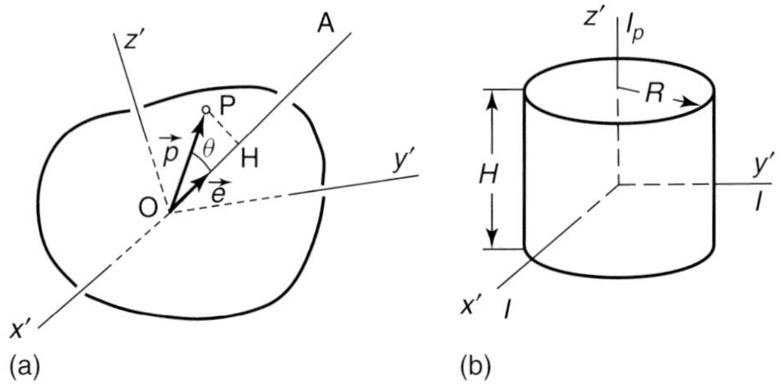

Figure A.2 Coordinate axis fixed to the rigid body. (a) General case and (b) principal axis.

where $p = \overline{\text{OP}}$. To calculate the moment of inertia about OA, we first calculate the length PH perpendicular to OA. It is given by

$$\begin{aligned}
\overline{\text{PH}^2} &= p^2 \sin^2\theta \\
&= (x'^2 + y'^2 + z'^2) - (\lambda x' + \mu y' + \nu z')^2 \\
&= (x'^2 + y'^2 + z'^2)(\lambda^2 + \mu^2 + \nu^2) - (\lambda x' + \mu y' + \nu z')^2 \\
&= (y'^2 + z'^2)\lambda^2 + (z'^2 + x'^2)\mu^2 + (x'^2 + y'^2)\nu^2 \\
&\quad -2\mu\nu y'z' - 2\nu\lambda z'x' - 2\lambda\mu x'y'
\end{aligned} \tag{A.5}$$

Here, in addition to the moment of inertia around the x'-, y'- and z'-axes

$$I_x = \int (y'^2 + z'^2)\, dm, \quad I_y = \int (z'^2 + x'^2)\, dm, \quad I_z = \int (x'^2 + y'^2)\, dm \tag{A.6}$$

we define the *product of inertia* as follows:

$$I_{xy} = \int x'y'\, dm, \quad I_{yz} = \int y'z'\, dm, \quad I_{zx} = \int z'x'\, dm \tag{A.7}$$

Then the moment of inertia about OA is given by

$$I_A = I_x\lambda^2 + I_y\mu^2 + I_z\nu^2 - 2I_{yz}\mu\nu - 2I_{zx}\nu\lambda - 2I_{xy}\lambda\mu \tag{A.8}$$

From the mathematical theory, we know that the right-hand side can be transformed into the form $A\lambda^2 + B\mu^2 + C\nu^2$ if we select the coordinate axes properly. In this case, the axes are called the *principal axes of moment of inertia* and A, B, and C are called the *principal moments of inertia*.

In a symmetric rigid body, the axes of symmetry coincide with the principal axes of the moment of inertia. For example, in a cylindrical rotor with radius R, height H, and mass m shown in Figure A.2b, the products of inertia about the axes x', y', and z' are all zero. The moment of inertia I_p about the z'-axis is called the *polar moment of inertia*, and the moments of inertia I about the x'- and y'-axes are called the *diametral moments of inertia*. They are given by

$$I_\mathrm{p} = \frac{mR^2}{2}, \quad I = m\left(\frac{R^2}{4} + \frac{H^2}{12}\right) \tag{A.9}$$

Appendix B
Stability above the Major Critical Speed

When the Jeffcott rotor executes a steady-state whirling motion as shown in Figure 2.4, the centrifugal force $C = m(r \pm e)\omega^2$ and the restoring force $F = kr$ in Fig. 2.8 balance and the following relationships hold:

$$\left.\begin{aligned} m(r+e)\omega^2 = kr \qquad & \mathit{for}\ \omega < \omega_c \\ m(r-e)\omega^2 = kr \qquad & \mathit{for}\ \omega > \omega_c \end{aligned}\right\} \tag{B.1}$$

Figure B.1 illustrates these relationships. From Figure B.1b, it seems that such balancing cannot be attained in the postcritical range. Namely, if the radius r becomes a little smaller than this equilibrium position r_0 because of some disturbance, the radius will decrease further because F is larger than C. On the contrary, if the radius r becomes a little larger, the radius will increase further because C is larger than F. Therefore, this position seems to be unstable above the major critical speed. However, this interpretation is insufficient because it does not take into account the Coriolis force. We can prove that this position is stable as discussed below.

Let us assume that points O, M, and G are aligned in the equilibrium state as shown in Figure B.2. Let the radius of the center of gravity be r_G and the angular position be φ. The coordinates of the center of gravity are given by

$$x_G = r_G \cos\varphi, \quad y_G = r_G \sin\varphi \tag{B.2}$$

Let us assume that because of a disturbance, these positions shift and become

$$r_G = r_{G0} + \xi, \quad \varphi = \omega t - \psi \tag{B.3}$$

If the shaft is not twisted, the direction MG is not influenced by the disturbance and is given by ωt. Substituting Eq. (B.2) into Eqs. (2.2) and (2.3), we have

$$\left.\begin{aligned} & \ddot{r}_G - r_G\dot{\varphi}^2 + \omega_c^2 r_G = p_0^2 e\cos\psi \\ & 2\dot{r}_G\dot{\varphi} + r_G\ddot{\varphi} = -p_0^2 e\sin\psi \end{aligned}\right\} \tag{B.4}$$

Substituting Eq. (B.3) into these expressions and neglecting small terms of second order or more of small quantities ξ, ψ, and e, we have

$$\left.\begin{aligned} & \ddot{\xi} + (\omega_c^2 - \omega^2)\xi + 2r_{G0}\omega\dot{\psi} = 0 \\ & -2\omega\dot{\xi} + r_{G0}\ddot{\psi} = 0 \end{aligned}\right\} \tag{B.5}$$

Linear and Nonlinear Rotordynamics: A Modern Treatment with Applications, Second Edition.
Yukio Ishida and Toshio Yamamoto.

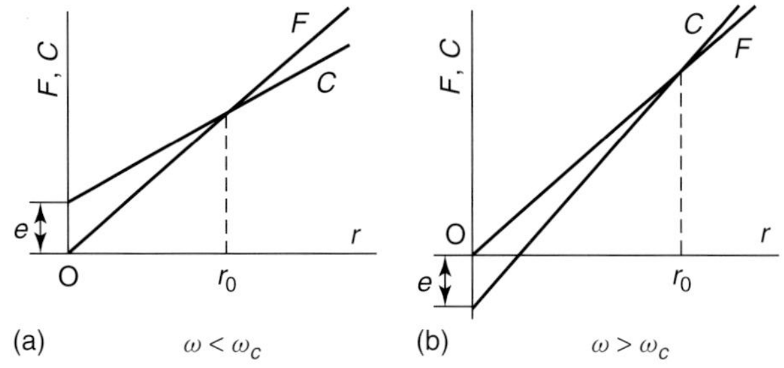

Figure B.1 Centrifugal force C and restoring force F.

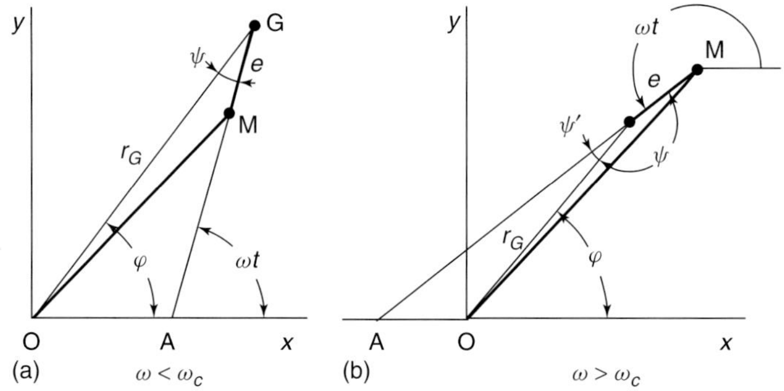

Figure B.2 Locations after disturbance.

Assuming the solution as

$$\xi(t) = Ae^{st}, \quad \psi(t) = Be^{st} \tag{B.6}$$

and substituting them into Eq. (B.5), we have

$$\left.\begin{array}{l} \left\{s^2 + (\omega_c^2 - \omega^2)\right\} A + 2r_{G0}\omega s B = 0 \\ -2\omega s A + r_{G0}s^2 B = 0 \end{array}\right\} \tag{B.7}$$

Then the characteristic equation is given by

$$r_{G0}s^2\left\{s^2 + \left(\omega_c^2 + 3\omega^2\right)\right\} = 0 \tag{B.8}$$

The roots of this equation are

$$s = 0, \quad \pm i\sqrt{\omega_c^2 + 3\omega^2} \tag{B.9}$$

If the real part of one of these roots is positive, the quantities ξ and ψ increase and the system becomes unstable. However, since these roots are zero and purely imaginary in this case, they do not increase. (The damping is not considered above, for simplicity. If it exists, the real parts of s become negative and the small vibrations due to disturbance dampen.)

Appendix C
Derivation of Equations of Motion of a 4 DOF Rotor System by Using Euler Angles

In Section 2.5.1, the equations of motion of a 4 DOF system are derived. We can also obtain them by the Euler angles, which are introduced in the "Note" in Section 2.5.1. The coordinate systems and the Euler angles are shown in Figure C.1.

The equations of motion for the deflection can be derived easily in a manner similar to that explained in Section 2.3. Let the position of the center of gravity G be (x_G, y_G) and that of the geometrical center be $(x.y)$. We get the following equations:

$$\begin{aligned} m\ddot{x}_G &= -(\alpha x + \gamma\theta_x) \\ m\ddot{y}_G &= -(\alpha y + \gamma\theta_y) \end{aligned} \tag{C.1}$$

Let the angular position of the center of gravity, measured from the X-axis in the counterclockwise direction, be ωt (refer to Figures 2.20 and C.1). Then the relationships

$$x_G = x + e\cos\omega t, \quad y_G = y + e\sin\omega t \tag{C.2}$$

hold approximately when the inclination angle θ_1 is small. Substituting this relationship into Eq. (C.1) yields the equations of motion for the lateral oscillations:

$$\begin{aligned} m\ddot{x} + \alpha x + \gamma\theta_x &= me\omega^2\cos\omega t \\ m\ddot{y} + \alpha y + \gamma\theta_y &= me\omega^2\sin\omega t \end{aligned} \tag{C.3}$$

Next, the equations of motion for the inclination can be derived by the following procedure:

1) Derive expressions for the angular velocity of the disk in terms of the Euler angles.
2) Obtain the components of the angular velocity in each direction of the principal moment of inertia.
3) Derive the components of the angular momentum in the directions of the principal moment of inertia by taking products of the angular velocity and the principal moment.
4) Derive the changes in the angular momentum per unit time in the direction of each principal axis.
5) Divide these variations into the x-, y-, and z-directions of the static coordinate system.
6) Derive the equations of motion in the x-, y-, and z-directions.

Linear and Nonlinear Rotordynamics: A Modern Treatment with Applications, Second Edition.
Yukio Ishida and Toshio Yamamoto.

Following this procedure, we analyze as follows:

1) Let the Euler angles of the coordinate system M-$X_1Y_1Z_1$ fixed to the rotor be θ_1, φ_1, and ψ_1 (Figure C.1). The angular speeds for these angles are given by $\dot{\theta}_1$, $\dot{\psi}_1$, and $\dot{\psi}_1$. Let the line of node of the disk and the horizontal plane (xy-plane) be MK. The vectors $\dot{\boldsymbol{\theta}}_1$, $\dot{\boldsymbol{\varphi}}_1$, and $\dot{\boldsymbol{\psi}}_1$ defined in Fig. 2.2.2 are in the same directions as MK, MZ, and MZ_1, respectively.
2) The directions of MK, MZ, and MZ_1 relative to M-$X_1Y_1Z_1$ whose axis coincide with the principal axes are given by the direction cosines shown in Table 2.1. For example, the direction cosines of MZ are obtained by deriving the components in the X_1-, Y_1-, and Z_1-directions of the unit vector in the MZ direction. By referring to this table, the components of the angular velocity in

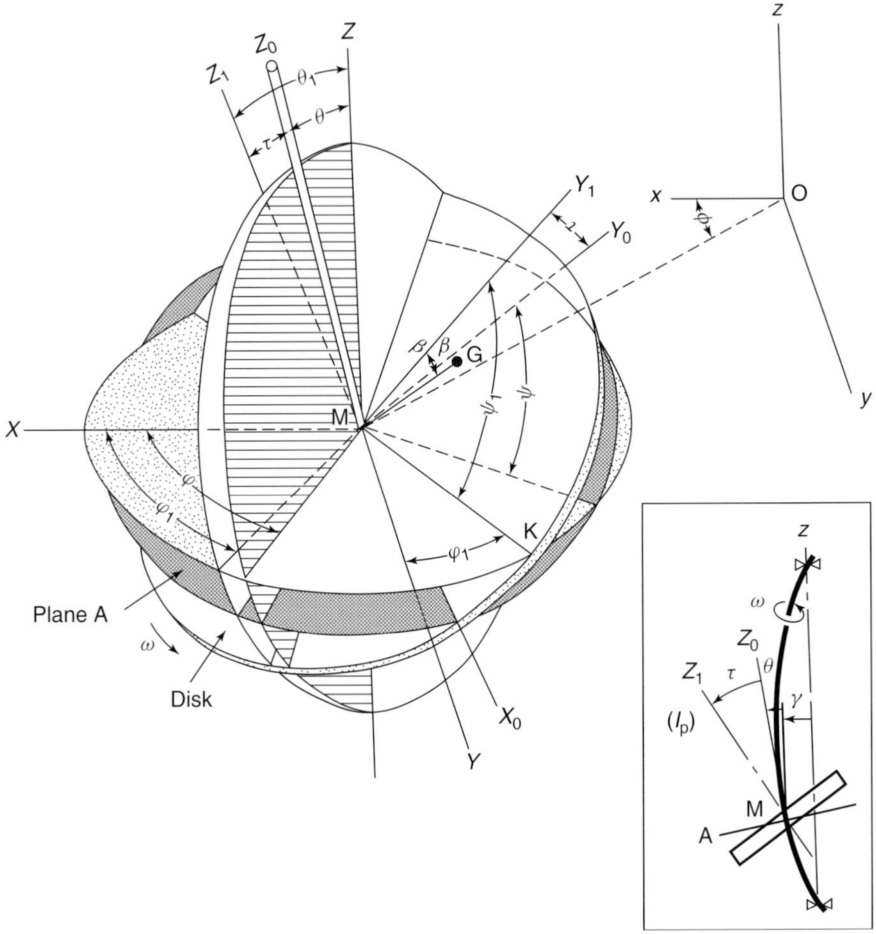

Figure C.1 Rotor and the coordinate systems.

the directions of the principal axes MX_1, MY_1, and MZ_1 become

$$\left.\begin{aligned} \omega_{X1} &= \dot{\theta}_1 \sin\psi_1 - \dot{\varphi}_1 \sin\theta_1 \cos\psi_1 \\ \omega_{Y1} &= \dot{\theta}_1 \cos\psi_1 + \dot{\varphi}_1 \sin\theta_1 \sin\psi_1 \\ \omega_{Z1} &= \dot{\psi}_1 \cos\theta_1 + \dot{\psi}_1 \end{aligned}\right\} \tag{C.4}$$

3) The angular momentums L_{X1}, L_{Y1}, and L_{Z1} about each principal axis of moment of inertia are given by

$$\left.\begin{aligned} L_{X1} &= I\omega_{X1} \\ I_{Y1} &= I\omega_{Y1} \\ I_{Z1} &= I_p\omega_{Z1} \end{aligned}\right\} \tag{C.5}$$

4) From Eq. (C.5) and Figure C.2, we get the expressions for H_{X1}, H_{Y1}, and H_{Z1} which are the variations of the angular momentums in the directions of the X_1-, Y_1-, and Z_1-axes per unit time, respectively:

$$\left.\begin{aligned} H_{X1} &= I\dot{\omega}_{X1} - (I - I_p)\omega_{Y1}\omega_{Z1} \\ H_{Y1} &= I\dot{\omega}_{Y1} - (I_p - I)\omega_{Z1}\omega_{X1} \\ H_{Z1} &= I_p\dot{\omega}_{Z1} \end{aligned}\right\} \tag{C.6}$$

Substituting Eq. (C.4) into Eq. (C.6) gives

$$\left.\begin{aligned} H_{X1} &= I\frac{\mathrm{d}}{\mathrm{d}t}(\dot{\theta}_1 \sin\psi_1 - \dot{\varphi}_1 \sin\theta_1 \cos 2\psi_1) \\ &\quad -(I - I_p)(\dot{\theta}_1 \cos\psi_1 + \dot{\varphi}_1 \sin\theta_1 \sin\psi_1)(\dot{\varphi}_1 \cos\theta_1 + \dot{\psi}_1) \\ H_{Y1} &= I\frac{\mathrm{d}}{\mathrm{d}t}(\dot{\theta}_1 \cos\psi_1 + \dot{\varphi}_1 \sin\theta_1 \sin\psi_1) \\ &\quad -(I_p - I)(\dot{\theta}_1 \sin\psi_1 - \dot{\varphi}_1 \sin\theta_1 \cos\psi_1)(\dot{\varphi}_1 \cos\theta_1 + \dot{\psi}_1) \\ H_{Z1} &= I_p\frac{\mathrm{d}}{\mathrm{d}t}(\dot{\varphi}_1 \cos\theta_1 + \dot{\psi}_1) \end{aligned}\right\} \tag{C.7}$$

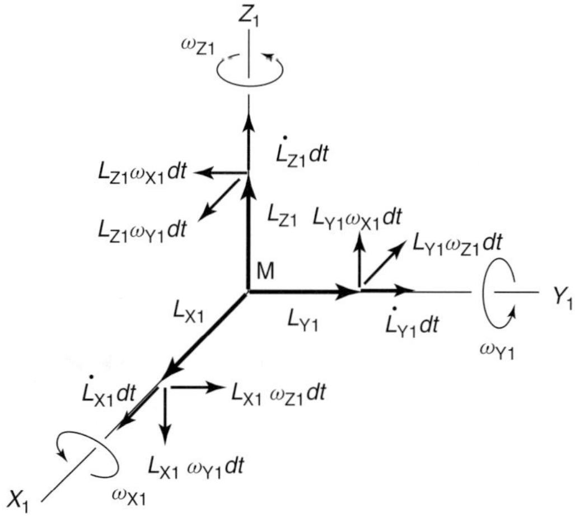

Figure C.2 Variations of angular momentum during time *dt*.

5) Next, we obtain H_x, H_y, and H_z which are the changes in angular momentum along the static coordinate axes x, y, and z, respectively. These changes are the same as H_X, H_Y, and H_Z along the X-, Y-, and Z-axes, respectively (Figure 2.22). To transform H_{X1}, H_{Y1}, and H_{Z1} into H_X, H_Y, and H_Z, we need direction cosines $l_1', m_1', n_1', l_2', m_2', n_2', l_3', m_3,'$ and n_3', which give the positions of the X_1, Y_1, and Z_1-axes with respect to the X-, Y-, and Z-axes. They are given in Table C.1.
These direction cosines can be derived as follows. Suppose that there is a unit vector a in the X_1-direction in Figure C.3. This vector is decomposed into the components $a_1 = \cos\psi_1$ and $a_2 = \sin\psi_1$ in the plane P_1 defined by the X_1- and Y_1-axes. The component a_1 is then decomposed into $a_3 = a_1 \cos\theta_1 = \cos\psi_1 \cos\theta_1$ and $a_{1z} = a_1 \times \sin\theta_1 = \cos\psi_1 \sin\theta_1$ in the ZZ_1-plane defined by the Z- and Z_1-axes, while the component a_2 is decomposed into $a_{2Y} = a_2 \times \cos\varphi_1 = \sin\psi_1 \cos\varphi_1$ and $a_{2X} = a_2 \times \sin\varphi_1 = \sin\psi_1 \sin\varphi_1$ in the horizontal plane P (XY-plane). The component a_3 is further decomposed into $a_{3X} = a_3 \times \cos\varphi_1 = \cos\psi_1 \cos\theta_1 \cos\varphi_1$ and $a_{3Y} = a_3 \times \sin\varphi_1 = \cos\psi_1 \cos\theta_1 \sin\varphi_1$ in plane P. Summing up the components in the X-, Y-, and Z-directions in Figure C.3, we get the direction cosines l_1', m_1', and n_1' in Table C.1, respectively.

Table C.1 Direction cosines for X-, Y-, and Z-axes.

	X_1	Y_1	Z_1
X	$l_1' = \cos\theta_1 \cos\varphi_1 \cos\psi_1 - \sin\varphi_1 \sin\psi_1$	$l_2' = -\cos\theta_1 \cos\varphi_1 \sin\psi_1 - \sin\varphi_1 \cos\psi_1$	$l_3' = \sin\theta_1 \cos\varphi_1$
Y	$m_{1'} = \cos\theta_1 \sin\varphi_1 \cos\psi_1 + \cos\varphi_1 \sin\psi_1$	$m_2' = -\cos\theta_1 \sin\varphi_1 \sin\psi_1 + \cos\varphi_1 \cos\psi_1$	$m_3' = \sin\theta_1 \sin\varphi_1$
Z	$n_1' = -\sin\varphi_1 \cos\psi_1$	$n_2' = -\sin\theta_1 \sin\psi_1$	$n_3' = \cos\theta_1$

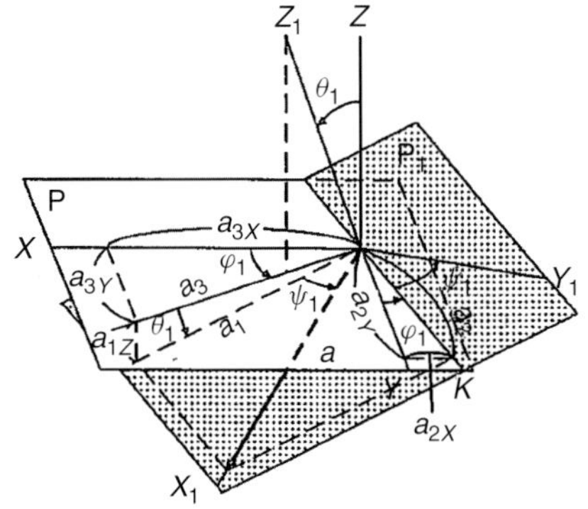

Figure C.3 Derivations of direction cosines.

After deriving such direction cosines, we get the following relationship between the angular momentum changes H_X, H_Y, H_Z and H_{X1}, H_{Y1}, H_{Z1}:

$$\left.\begin{aligned} H_X &= l_1{}'H_{X1} + l_2{}'H_{Y1} + l_3{}'H_{Z1} \\ H_Y &= m_1{}'H_{X1} + m_2{}'H_{Y1} + m_3{}'H_{Z1} \\ H_Z &= n_1{}'H_{X1} + n_2{}'H_{Y1} + n_3{}'H_{Z1} \end{aligned}\right\} \tag{C.8}$$

Substituting Eq. (C7) into this expression, we get the following expressions after some rearrangement:

$$\begin{aligned} H_X = {} & I\left(-\ddot{\theta}_1 S_{\varphi 1} - \ddot{\varphi}_1 C_{\theta 1} C_{\varphi 1} S_{\theta 1} - 2\dot{\theta}_1 \dot{\psi}_1 C_{\theta 1}^2 C_{\varphi 1} + \dot{\psi}_1^2 S_{\varphi 1} S_{\theta 1} C_{\theta 1}\right) \\ & + I_p\Big\{\dot{\theta}_1 \dot{\psi}_1 \left(C_{\theta 1}^2 C_{\varphi 1} - S_{\theta 1}^2 C_{\varphi 1}\right) - \dot{\psi}_1^2 S_{\varphi 1} C_{\theta 1} S_{\theta 1} + \dot{\theta}_1 \dot{\psi}_1 C_{\theta 1} C_{\varphi 1} \\ & - \dot{\psi}_1 \dot{\psi}_1 S_{\varphi 1} S_{\theta 1} + \ddot{\varphi}_1 S_{\theta 1} C_{\varphi 1} C_{\theta 1} + \ddot{\psi}_1 S_{\theta 1} C_{\varphi 1}\Big\} \end{aligned}$$

$$\begin{aligned} H_Y = {} & I(\ddot{\theta}_1 C_{\varphi 1} - \ddot{\varphi}_1 C_{\theta 1} S_{\varphi 1} S_{\theta 1} - 2\dot{\theta}_1 \dot{\psi}_1 C_{\theta 1}^2 S_{\varphi 1} - \dot{\psi}_1^2 C_{\varphi 1} S_{\theta 1} C_{\theta 1}) \\ & + I_p \left\{\dot{\theta}_1 \dot{\psi}_1 \left(C_{\theta 1}^2 S_{\varphi 1} - S_{\theta 1}^2 S_{\varphi 1}\right) + \dot{\psi}_1^2 C_{\varphi 1} S_{\theta 1} C_{\theta 1}\right. \\ & \left. + \dot{\theta}_1 \dot{\psi}_1 C_{\theta 1} S_{\varphi 1} + \dot{\psi}_1 \dot{\psi}_1 C_{\varphi 1} S_{\theta 1} + \ddot{\varphi}_1 S_{\theta 1} S_{\varphi 1} C_{\theta 1} + \ddot{\psi}_1 S_{\theta 1} S_{\varphi 1}\right\} \end{aligned} \tag{C.9}$$

$$\begin{aligned} H_Z = {} & I\left(\ddot{\varphi}_1 S_{\theta 1}^2 + 2\dot{\theta}_1 \dot{\varphi}_1 S_{\theta 1} C_{\theta 1}\right) \\ & + I_p\left(-2\dot{\theta}_1 \dot{\varphi}_1 S_{\theta 1} C_{\theta 1} - \dot{\theta}_1 \dot{\psi}_1 S_{\theta 1} + \ddot{\varphi}_1 C_{\theta 1}^2 + \ddot{\psi}_1 C_{\theta 1}\right) \end{aligned}$$

Here, for simplicity, we adopt the notations $C_{\theta 1} = \cos\theta_1$, $S_{\theta 1} = \sin\theta_1$, $C_{\varphi 1} = \cos\varphi_1$, and $S_{\varphi 1} = \sin\varphi_1$.

As the angle θ_1 is very small in practical rotating machineries, we simplify these expressions by using some approximations. From here onward, we use a small parameter ε and the symbol $O(\varepsilon)$, which represents the same order as ε. The angles θ_{1x} and θ_{1y}, which are the projections of the inclination angle θ_1 onto the XZ- and YZ-planes, are given by

$$\theta_{1x} \approx \theta_1 \cos\varphi_1, \quad \theta_{1y} \approx \theta_1 \sin\varphi_1 \tag{C.10}$$

approximately in the accuracy of $O(\varepsilon)$. Substituting these relationships into Eq. (C.9) and neglecting small quantities of $O(\varepsilon^2)$ and below, we obtain simple expressions for H_X, H_Y, and H_Z, and therefore those for H_X, H_Y, and H_Z as well, which gives us

$$\left.\begin{aligned} H_x &= -I\ddot{\theta}_{1y} + I_p \frac{d}{dt}\left(\dot{\Theta}_1 \theta_{1x}\right) \\ H_y &= I\ddot{\theta}_{1x} + I_p \frac{d}{dt}\left(\dot{\Theta}_1 \theta_{1y}\right) \\ H_z &= I_p \ddot{\Theta}_1 \end{aligned}\right\} \tag{C.11}$$

where $\Theta_1 = \varphi_1 + \psi_1$.

When the shaft is rotating with the constant angular velocity ω, the relationships

$$\dot{\Theta}_1 = \omega, \quad \ddot{\Theta}_1 = 0 \tag{C.12}$$

hold. Then Eq. (C.11) become

$$\left.\begin{array}{l} H_x = -I\ddot{\theta}_{1y} + I_p\omega\dot{\theta}_{1x} \\ H_y = I\ddot{\theta}_{1x} + I_p\omega\dot{\theta}_{1y} \\ H_z = 0 \end{array}\right\} \tag{C.13}$$

6) The moments acting on the disk are obtained from Eq. (2.51). When the moments $M_{xz}{}'$ and $M_{yz}{}'$ given by Eq. (2.51) are applied to the shaft, the restoring moments $M_{xz}\left(= -M_{xz}{}'\right)$ and $M_{yz}\left(= -M_{yz}{}'\right)$ appear as a reaction. These restoring moments act on the disk during whirling motion. The signs of M_{xz} and M_{yz} are defined as positive when they act so as to increase θ_x and θ_y, respectively. For example, when $x > 0, y > 0$ and $\theta_x > 0, \theta_y > 0$ hold, the restoring moments are $M_{xz} = -\gamma x - \delta\theta_x < 0$, $M_{yz} = -\gamma y - \delta\theta_y$ and therefore they work in the direction to decrease θ_x and θ_y, respectively. These moments are represented by vectors similar to the way shown in Figure 2.22. The vectors representing positive M_{xz} and positive M_{yz} point to the positive direction of the y-axis and the negative direction of the x-axis, respectively.
If we represent a restoring moment vector by $\boldsymbol{M} = M_x\boldsymbol{i} + M_y\boldsymbol{j}$, where $\boldsymbol{i}$ and $\boldsymbol{j}$ are unit vectors in the x- and y-directions, respectively, these components are given by

$$M_x = \gamma y + \delta\theta_y \left(= -M_{yz}\right), M_y = -\gamma x - \delta\theta_x \left(= M_{xz}\right) \tag{C.14}$$

The equations of motion are derived by the following relationships:

$$H_x = M_x, \quad H_y = M_y, \quad H_z = M_z \tag{C.15}$$

From the first and second equations, we get

$$\left.\begin{array}{l} -I\ddot{\theta}_{1y} + I_p\omega\dot{\theta}_{1x} = \gamma y + \delta\theta_y \\ I\ddot{\theta}_{1x} + I_p\omega\dot{\theta}_{1y} = -\gamma x - \delta\theta_x \end{array}\right\} \tag{C.16}$$

These expressions are the same as Eq. (2.53). Therefore, in the same manner, we can obtain the equations of motion for 4 DOF rotor system.

$$\left.\begin{array}{l} m\ddot{x} + c_{11}\dot{x} + c_{12}\dot{\theta}_x + \alpha x + \gamma\theta_x = me\omega^2\cos\omega t \\ m\ddot{y} + c_{11}\dot{y} + c_{12}\dot{\theta}_y + \alpha y + \gamma\theta_y = me\omega^2\sin\omega t \\ I\ddot{\theta}_x + I_p\omega\dot{\theta}_y + c_{21}\dot{x} + c_{22}\dot{\theta}_x + \gamma x + \delta\theta_x = \left(I_p - I\right)\tau\omega^2\cos\left(\omega t + \beta_\tau\right) \\ I\ddot{\theta}_y - I_p\omega\dot{\theta}_x + c_{21}\dot{x} + c_{22}\dot{\theta}_y + \gamma y + \delta\theta_y = \left(I_p - I\right)\tau\omega^2\sin\left(\omega t + \beta_\tau\right) \end{array}\right\} \tag{C.17}$$

Appendix D
Asymmetrical Shaft and Asymmetrical Rotor with Four Degrees of Freedom

Equations of motion and frequency diagrams of a 4 DOF asymmetrical shaft and a 4 DOF asymmetrical rotor are summarized in this appendix. Refer to Yamamoto and Ota (1963) for the details.

D.1
4 DOF Asymmetrical Shaft System

Let us consider the coordinate system M-$X_2Y_2Z_2$, in which the X_2- and Y_2-axes coincide with the axes of symmetry of a cross section as shown in Figure D.1. Let the angular positions of the static unbalance e and the dynamic unbalance τ measured from the direction of minimum stiffness (the X_2-direction) in the clockwise direction be ξ and η, respectively. Similar to Figure 5.7, the position of this coordinate system is represented by the Eulerian angles θ_2, φ_2, and ψ_2. To make correspondence to Figure 5.2 for 2 DOF system, we represent the rotational angle as $\Theta_2(=\varphi_2+\psi_2)=\omega t$. If the inclination angle θ_2 is small, Θ_2 approximately represents the angular position of the X_2-axis relative to the x-axis. The equations of motion are given by

$$\left.\begin{array}{l}
m\ddot{x}+c_{11}\dot{x}+c_{12}\dot{\theta}_x+\alpha x+\gamma\theta_x-\Delta\alpha(x\cos 2\omega t+y\sin 2\omega t)\\
\quad -\Delta\gamma\left(\theta_x\cos 2\omega t+\theta_y\sin 2\omega t\right)=me\omega^2\cos(\omega t+\xi)\\
m\ddot{y}+c_{11}\dot{y}+c_{12}\dot{\theta}_y+\alpha y+\gamma\theta_y-\Delta\alpha(x\sin 2\omega t-y\cos 2\omega t)\\
\quad -\Delta\gamma\left(\theta_x\sin 2\omega t-\theta_y\cos 2\omega t\right)=me\omega^2\sin(\omega t+\xi)\\
I\ddot{\theta}_x+I_{\mathrm{p}}\omega\dot{\theta}_y+c_{12}\dot{x}+c_{22}\dot{\theta}_x+\gamma x+\delta\theta_x-\Delta\gamma(x\cos 2\omega t+y\sin 2\omega t)\\
\quad -\Delta\delta\left(\theta_x\cos 2\omega t+\theta_y\sin 2\omega t\right)=\left(I_{\mathrm{p}}-I\right)\tau\omega^2\cos(\omega t+\eta)\\
I\ddot{\theta}_y-I_{\mathrm{p}}\omega\dot{\theta}_x+c_{12}\dot{y}+c_{22}\dot{\theta}_y+\gamma y+\delta\theta_y-\Delta\gamma(x\sin 2\omega t-y\cos 2\omega t)\\
\quad -\Delta\delta\left(\theta_x\sin 2\omega t-\theta_y\cos 2\omega t\right)=\left(I_{\mathrm{p}}-I\right)\tau\omega^2\sin(\omega t+\eta)
\end{array}\right\}\quad\text{(D.1)}$$

In the shaft with uniform cross section, the following relationship holds:

$$\frac{\Delta\alpha}{\alpha}=\frac{\Delta\gamma}{\gamma}=\frac{\Delta\delta}{\delta}=\frac{I_{x2}-I_{y2}}{I_{x2}+I_{y2}}=\Delta s\quad\text{(D.2)}$$

Linear and Nonlinear Rotordynamics: A Modern Treatment with Applications, Second Edition.
Yukio Ishida and Toshio Yamamoto.

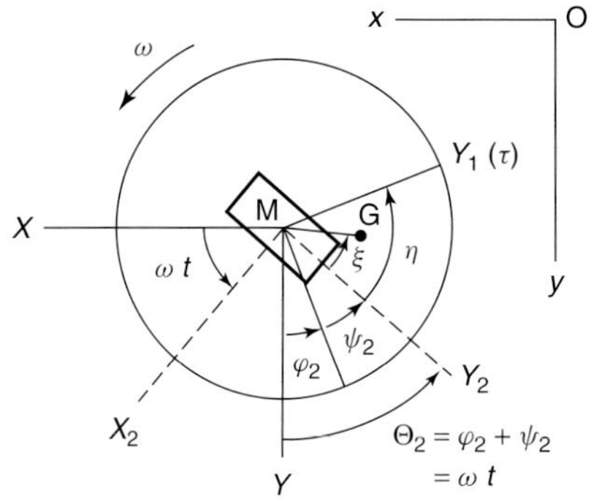

Figure D.1 Asymmetrical shaft system ($I_p/I \approx 2$).

where I_{x2} and $I_{y2}(I_{x2} > I_{y2})$ are the second moments of area about the neutral axes X_2 and Y_2, respectively. Here we define the following dimensionless quantities using the static deflection $e_{\rm st} = mg/\alpha$:

$$\left.\begin{array}{l} x' = x/e_{\rm st}, \quad y' = y/e_{\rm st}, \quad \theta'_x = \theta_x\sqrt{I/m}/e_{\rm st}, \quad \theta'_y = \theta_y\sqrt{I/m}/e_{\rm st}, \\ i_{\rm p} = I_{\rm p}/I, \quad t' = t\sqrt{\alpha/m}, \quad \omega' = \omega/\sqrt{\alpha/m}, \quad \gamma' = \gamma\sqrt{m/I}/\alpha, \\ \delta' = m\delta/(\alpha I), \quad c'_{11} = c_{11}/\sqrt{m\alpha}, \quad c'_{12} = c_{12}\sqrt{\alpha I}, \\ c_{22}{}' = c_{22}/(I\sqrt{\alpha/m}), \quad e' = e/e_{\rm st}, \quad \tau' = \tau\sqrt{I/m}/e_{\rm st} \end{array}\right\} \tag{D.3}$$

The equations of motion are represented by these dimensionless quantities as follows:

$$\left.\begin{array}{l} \ddot{x} + c_{11}\dot{x} + c_{12}\dot{\theta}_x + x + \gamma\theta_x - \Delta_{\rm s}(x\cos 2\omega t + y\sin 2\omega t) \\ \quad - \gamma\Delta_{\rm s}\left(\theta_x\cos 2\omega t + \theta_y\sin 2\omega t\right) = e\omega^2\cos(\omega t + \xi) \\ \ddot{y} + c_{11}\dot{y} + c_{12}\dot{\theta}_y + y + \gamma\theta_y - \Delta_{\rm s}(x\sin 2\omega t - y\cos 2\omega t) \\ \quad - \gamma\Delta_{\rm s}\left(\theta_x\sin 2\omega t - \theta_y\cos \omega t\right) = e\omega^2\sin(\omega t + \xi) \\ \ddot{\theta}_x + i_{\rm p}\omega\dot{\theta}_y + c_{12}\dot{x} + c_{22}\dot{\theta}_x + \gamma x + \delta\theta_x - \gamma\Delta_{\rm s}(x\cos 2\omega t + y\sin 2\omega t) \\ \quad - \delta\Delta_{\rm s}\left(\theta_x\cos 2\omega t + \theta_y\sin 2\omega t\right) = \left(i_{\rm p} - 1\right)\tau\omega^2\cos(\omega t + \eta) \\ \ddot{\theta}_y - i_{\rm p}\omega\dot{\theta}_x + c_{12}\dot{y} + c_{22}\dot{\theta}_y + \gamma y + \delta\theta_y - \gamma\Delta_{\rm s}(x\sin 2\omega t - y\cos 2\omega t) \\ \quad - \delta\Delta_{\rm s}\left(\theta_x\sin 2\omega t - \theta_y\cos 2\omega t\right) = \left(i_{\rm p} - 1\right)\tau\omega^2\sin(\omega t + \eta) \end{array}\right\} \tag{D.4}$$

where the primes are eliminated for simplicity.

A natural frequency diagram is shown in Figure D.2. In the 4 DOF symmetrical rotor system discussed in Section 2.5, four natural frequencies $p_i(i = 1 \sim 4)$ exist as shown in Figure 2.23. However, as shown in Figure D.2, this asymmetrical shaft has four other natural frequencies $\tilde{p}_i(i = 1 \sim 4)$, which have the relationship

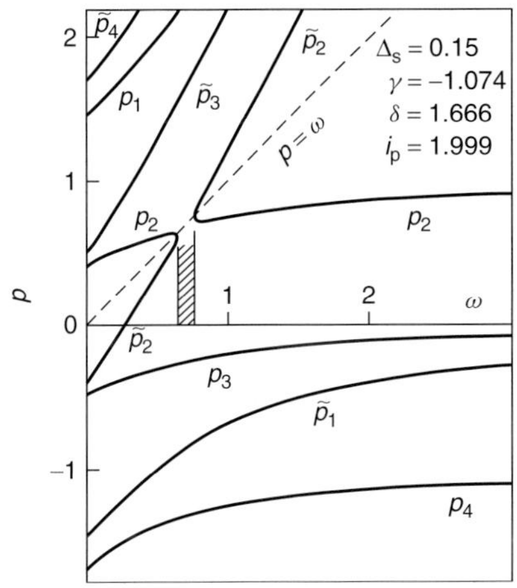

Figure D.2 Natural frequency diagram of an asymmetrical shaft system ($I_p/I \approx 2$).

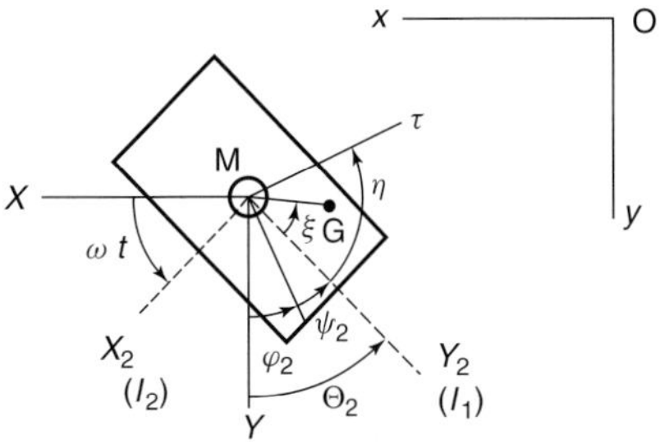

Figure D.3 Asymmetrical rotor system.

$\tilde{p}_i = 2\omega - p_i$ in addition to the four natural frequencies corresponding to these in a symmetrical shaft system. In the vicinities of the rotational speeds where $p_i = \tilde{p}_i$ hold, the natural frequencies become complex numbers and unstable ranges appear. In Figure D.2, since the rotor has the shape of a short cylinder ($I_p > I$), there exists only one unstable range.

D.2
4 DOF Asymmetrical Rotor System

As shown in Figure D.3, let us consider the coordinate system $M{-}X_2Y_2Z_2$ in which the axes coincide with the principal axes of the moments of inertia of the asymmetrical rotor. Let the principal moments of inertia about the X_2-, Y_2-, Z_2-axes be I_2, $I_1 (I_2 \neq I_1)$, and I_p, respectively. Let the angular positions of the static unbalance e and the dynamic unbalance τ be ξ and η. We represent the rotational angle by $\Theta_2 (= \varphi_2 + \psi_2) = \omega t$. The equations of motion are given as follows:

$$
\left.
\begin{aligned}
& m\ddot{x} + c_{11}\dot{x} + c_{12}\dot{\theta}_x + \alpha x + \gamma\theta_x = me\omega^2\cos(\omega t + \xi) \\
& m\ddot{y} + c_{11}\dot{y} + c_{12}\dot{\theta}_y + \alpha y + \gamma\theta_y = me\omega^2\sin(\omega t + \xi) \\
& I\ddot{\theta}_x + I_p\omega\dot{\theta}_y + c_{12}\dot{x} + c_{22}\dot{\theta}_x + \gamma x + \delta\theta_x \\
& \quad - \Delta I\left\{\left(\ddot{\theta}_x\cos 2\omega t + \ddot{\theta}_y\sin 2\omega t\right) - 2\omega\left(\dot{\theta}_x\sin 2\omega t - \dot{\theta}_y\cos 2\omega t\right)\right\} \\
& = \tau\omega^2\left\{\left(I_p - I\right)\cos(\omega t + \eta) - \Delta I\cos(\omega t - \eta)\right\} \\
& I\ddot{\theta}_y - I_p\omega\dot{\theta}_x + c_{12}\dot{y} + c_{22}\dot{\theta}_y + \gamma y + \delta\theta_y \\
& \quad - \Delta I\left\{\left(\ddot{\theta}_x\sin 2\omega t - \ddot{\theta}_y\cos 2\omega t\right) + 2\omega\left(\dot{\theta}_x\cos 2\omega t + \dot{\theta}_y\sin 2\omega t\right)\right\} \\
& = \tau\omega^2\left\{(I_p - I)\sin(\omega t + \eta) - \Delta I\sin(\omega t - \eta)\right\}
\end{aligned}
\right\} \quad \text{(D.5)}
$$

In addition to the dimensionless quantities of Eq. (D.3), we use

$$
\Delta = \Delta I / I \quad \text{(D.6)}
$$

The dimensionless equations of motion are given by

$$
\left.
\begin{aligned}
& \ddot{x} + c_{11}\dot{x} + c_{12}\dot{\theta}_x + x + \gamma\theta_x = e\omega^2\cos(\omega t + \xi) \\
& \ddot{y} + c_{11}\dot{y} + c_{12}\dot{\theta}_y + y + \gamma\theta_y = e\omega^2\sin(\omega t + \xi) \\
& \ddot{\theta}_x + i_p\omega\dot{\theta}_y + c_{12}\dot{x} + c_{22}\dot{\theta}_x + \gamma x + \delta\theta_x \\
& \quad - \Delta\left\{\left(\ddot{\theta}_x\cos 2\omega t + \ddot{\theta}_y\sin 2\omega t\right) - 2\omega\left(\dot{\theta}_x\sin 2\omega t - \dot{\theta}_y\cos 2\omega t\right)\right\} \\
& = \tau\omega^2\left\{\left(i_p - 1\right)\cos(\omega t + \eta) - \Delta\cos(\omega t - \eta)\right\} \\
& \ddot{\theta}_y - i_p\omega\dot{\theta}_x + c_{12}\dot{y} + c_{22}\dot{\theta}_y + \gamma y + \delta\theta_y \\
& \quad - \Delta\left\{\left(\ddot{\theta}_x\cos 2\omega t - \ddot{\theta}_y\cos 2\omega t\right) + 2\omega\left(\dot{\theta}_x\cos 2\omega t + \dot{\theta}_y\sin 2\omega t\right)\right\} \\
& = \tau\omega^2\left\{(i_p - 1)\sin(\omega t + \eta) - \Delta\sin(\omega t - \eta)\right\}
\end{aligned}
\right\} \quad \text{(D.7)}
$$

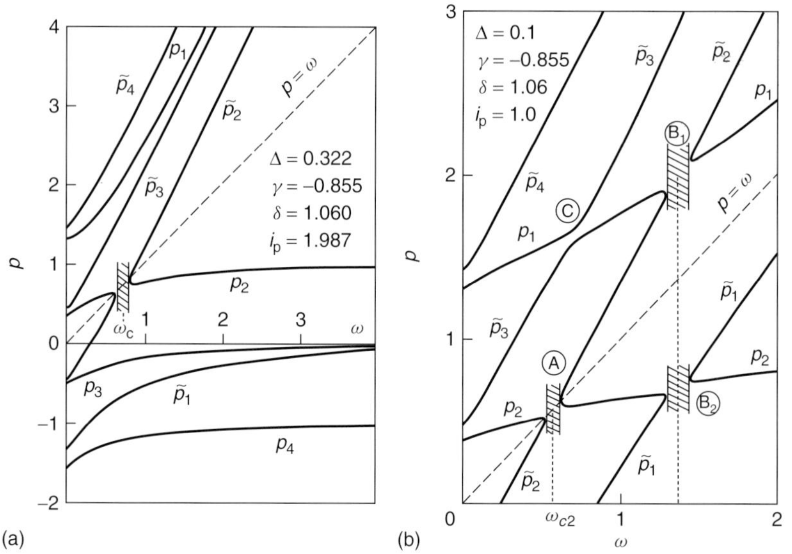

Figure D.4 Natural frequency diagrams of asymmetrical rotors. (a) $i_p = I_p/I \approx 2$ and (b) $i_p = I_p/I \approx 1$.

Natural frequency diagrams are shown in Figure D.4. Figure D.4a is the case of a thin asymmetrical rotor. The ratio of the moments of inertia $i_p(= I_p/I)$ of this rotor is about 2. In this case, similar to Figure D.2 for an asymmetrical shaft, one unstable range appears at the rotational speed where $p_2 = \omega$ holds. Figure D.4b shows a thick asymmetrical rotor with a ratio $i_p \approx 1$. In addition to the unstable range at the major critical speed ω_{c2}, an unstable range appears at the rotational speed ω_d where $p_1 = \tilde{p}_2$ and $p_2 = \tilde{p}_1$ (i.e., $p_1 + p_2 = 2\omega$) hold. In this speed range, two vibration components with frequencies $\omega_1(\approx p_1)$ and $\omega_2(\approx p_2)$ occur simultaneously and the relationship $\omega_1 + \omega_2 = 2\omega$ holds. This type of vibration, called a *dynamically unstable vibration*, occurs at $p_1 + p_2 = 2\omega$. The values p_1 and p_2 approach $i_p\omega$ and 1 asymptotically, respectively, as the rotational speed increases (refer to Section 2.5.2). Therefore, when the value of i_p approaches 2, this resonance speed ω_d moves toward the higher-speed side and the diagram changes from the type of Figure D.4b to that of Figure D.4a.

An unstable range does not necessarily appear at the cross point of two natural frequency curves. For example, an unstable range does not appear in the vicinity of point C, where $p_1 = \tilde{p}_3$ and $p_3 = \tilde{p}_1$ hold.

Reference

Yamamoto, T. and Ota, H. (1963) On the vibrations of a shaft carrying an unsymmetrical rotating body. *Bull. JSME*, **6** (21), 29–36.

Appendix E

Transformation of Equations of Motion to Normal Coordinates: 4 DOF Rotor System

Expressions of nonlinear spring characteristics and transformation to normal coordinates in a 4 DOF rotor system where the deflection and the inclination couple with each other are summarized here. For details, refer to Yamamoto (1957) and Yamamoto and Ishida (1975).

E.1
Transformation of Equations of Motion to Normal Coordinates

Equations of motion of a 4 DOF symmetrical rotor system are given by Eq. (2.59). The following dimensionless quantities are introduced:

$$\left.\begin{array}{llll} x^* = \dfrac{x}{\sqrt{I/m}}, & y^* = \dfrac{y}{\sqrt{I/m}}, & e^* = \dfrac{e}{\sqrt{I/m}}, & c_{11}^* = \dfrac{c_{11}}{\sqrt{\alpha m}} \\ c_{12}^* = c_{21}^* = \dfrac{c_{12}}{\sqrt{\alpha I}} = \dfrac{c_{21}}{\sqrt{\alpha I}}, & & c_{22}^* = \dfrac{c_{22}}{I\sqrt{\alpha/m}}, & \omega^* = \dfrac{\omega}{\sqrt{\alpha/m}} \\ t^* = t\sqrt{\alpha/m}, & \gamma^* = \dfrac{\gamma}{\alpha\sqrt{I/m}}, & \delta^* = \dfrac{\delta m}{\alpha I}, & i_p = \dfrac{I_p}{I} \end{array}\right\} \tag{E.1}$$

With these notations, the equations of motion can be rewritten as

$$\left.\begin{array}{l} \ddot{x} + c_{11}\dot{x} + c_{12}\dot{\theta}_x + x + \gamma\theta_x + N_x = e\omega^2\cos\omega t \\ \ddot{y} + c_{11}\dot{y} + c_{12}\dot{\theta}_y + y + \gamma\theta_y + N_y = e\omega^2\sin\omega t \\ \ddot{\theta}_x + i_p\omega\dot{\theta}_y + c_{21}\dot{x} + c_{22}\dot{\theta}_x + \gamma x + \delta\theta_x + N_{\theta x} = \left(i_p - 1\right)\tau\omega^2\cos(\omega t + \beta) \\ \ddot{\theta}_y - i_p\omega\dot{\theta}_x + c_{21}\dot{y} + c_{22}\dot{\theta}_y + \gamma y + \delta\theta_y + N_{\theta y} = \left(i_p - 1\right)\tau\omega^2\sin(\omega t + \beta) \end{array}\right\} \tag{E.2}$$

where N_x, N_y, $N_{\theta x}$, and $N_{\theta y}$ are nonlinear terms. In the following, the asterisks are eliminated for convenience.

The frequency equation corresponding to Eq. (2.75) is given by

$$f(p) \equiv \left(1 - p^2\right)\left(\delta + i_p\omega p - p^2\right) - \gamma^2 = 0 \tag{E.3}$$

Linear and Nonlinear Rotordynamics: A Modern Treatment with Applications, Second Edition.
Yukio Ishida and Toshio Yamamoto.

This has four roots: p_1, p_2, p_3, and p_4. We can transform Eq. (E.2) to normal coordinates $X_1, \cdots, X_4$ by the following transformation.

$$\left.\begin{aligned} x &= X_1 + X_2 + X_3 + X_4 \\ y &= -\frac{\dot{X}_1}{p_1} - \frac{\dot{X}_2}{p_2} - \frac{\dot{X}_3}{p_3} - \frac{\dot{X}_4}{p_4} \\ \theta_x &= \kappa_1 X_1 + \kappa_2 X_2 + \kappa_3 X_3 + \kappa_4 X_4 \\ \theta_y &= -\frac{\kappa_1 \dot{X}_1}{p_1} - \frac{\kappa_2 \dot{X}_2}{p_2} - \frac{\kappa_3 \dot{X}_3}{p_3} - \frac{\kappa_4 \dot{X}_4}{p_4} \end{aligned}\right\} \tag{E.4}$$

where

$$\kappa_s = \frac{p_s^2 - 1}{\gamma} \tag{E.5}$$

Substituting Eq. (E.4) into Eq. (E.2) and reforming by using the procedure described in Section 6.3, we arrive at the following equations of motion expressed in normal coordinates:

$$\begin{aligned} \ddot{X}_s + \omega_s^2 X_s = (\omega_s^2 - p_s^2) X_s + n_s \Bigg[& c_{11} \left(\frac{-\dot{x} + p_s y}{\kappa_s} \right) \\ & + c_{12} \left(-\dot{x} + p_s y - \frac{\dot{\theta}_x}{\kappa_s} + \frac{p_s \theta_y}{\kappa_s} \right) \\ & + \left(-\frac{N_x}{\kappa_s} - N_{\theta x} + \frac{p_s}{\kappa_s} \int N_y \mathrm{d}t + p_s \int N_{\theta y} \mathrm{d}t \right) \\ & + \omega (\omega + p_s) \left\{ \frac{e}{\kappa_s} \cos \omega t + (i_p - 1) \tau \cos(\omega t + \beta) \right\} \Bigg] \\ & (s = 1, \cdots, 4) \end{aligned} \tag{E.6}$$

where

$$\left.\begin{aligned} n_s &= \frac{\gamma p_s}{(p_s - p_i)(p_s - p_j)(p_s - p_k)} \\ (s &\neq i \neq j \neq k, i, j, k, s = 1, 2, 3, 4) \end{aligned}\right\} \tag{E.7}$$

E.2
Nonlinear Terms

Corresponding to Eq. (6.2), the potential energy of the 4 DOF rotor system is given as follows:

$$\begin{aligned} V = & \frac{1}{2}(x^2 + y^2) + \gamma (x\theta_x + y\theta_y) + \frac{1}{2}\delta \left(\theta_x^2 + \theta_y^2 \right) \\ & + \sum_{\substack{a,b,c,d=0 \\ (a+b+c+d=3)}}^{3} \varepsilon_{abcd} x^a y^b \theta_x^c \theta_y^d + \sum_{\substack{a,b,c,d=0 \\ (a+b+c+d=4)}}^{4} \beta_{abcd} x^a y^b \theta_x^c \theta_y^d \end{aligned} \tag{E.8}$$

The restoring force and the restoring moment are given by partially differentiating this potential energy by each coordinate. Nonlinear terms N_x, N_y, $N_{\theta x}$, and $N_{\theta y}$ are obtained by this calculation.

$$\left.\begin{array}{ll} F_x = -\dfrac{\partial V}{\partial x} = -x - \gamma\theta_x - N_x, & F_y = -\dfrac{\partial V}{\partial y} = -y - \gamma\theta_y - N_y \\ M_{\theta x} = -\dfrac{\partial V}{\partial \theta_x} = -\gamma x - \delta\theta_x - N_{\theta x}, & M_{\theta y} = -\dfrac{\partial V}{\partial \theta_y} = -\gamma y - \delta\theta_y - N_{\theta y} \end{array}\right\} \tag{E.9}$$

By the transformation

$$x = r\cos\varphi_r, \quad y = r\sin\varphi_r, \quad \theta_x = \theta\cos\varphi_\theta, \quad \theta_y = \theta\sin\varphi_\theta \tag{E.10}$$

we transform the nonlinear terms into the polar coordinate expressions. Substituting this into Eq. (E.8), we can express nonlinear terms in the polar coordinates. The outline of this analysis is as follows (refer to Yamamoto and Ishida, 1975). From Eqs. (E.8) and (E.10), we have

$$\begin{aligned} V = {} & \frac{1}{2}r^2 + \gamma r\theta\cos(\varphi_r - \varphi_\theta) + \frac{1}{2}\delta\theta^2 \\ & + (\varepsilon_{30c}^{(1)}\cos\varphi_r + \varepsilon_{30s}^{(1)}\sin\varphi_r)r^3 \\ & + \Big\{\varepsilon_{21c}^{(1)}\cos\varphi_\theta + \varepsilon_{21s}^{(1)}\sin\varphi_\theta + \varepsilon_{21c}^{(1)\prime}\cos(2\varphi_r - \varphi_\theta) \\ & + \varepsilon_{21s}^{\prime(1)}\sin(2\varphi_r - \varphi_\theta)\, r^2\theta\Big\} + (\ldots)r\theta^2 + (\ldots)\theta^3 \\ & + (\ldots)r^4 + (\ldots)r^3\theta + (\ldots)r^2\theta^2 + (\ldots)r\theta^3 + (\ldots)\theta^4 \end{aligned} \tag{E.11}$$

Since the angles φ_r and φ_θ increase with the same speed during a whirling motion, we know that the quantities $(2\varphi_r - \varphi_\theta)$, φ_r, and φ_θ change their values once, while the shaft whirls once with the same radius (i.e., these terms are $N(1)$ component). Similarly, the terms $(\varphi_r - \varphi_\theta)$, $(3\varphi_r - \varphi_\theta)$, $(2\varphi_r + \varphi_\theta)$, and $(3\varphi_r + \varphi_\theta)$, which appear in the omitted part in Eq. (E.11), belong to the $N(0)$, $N(2)$, $N(3)$, and $N(4)$ components. Relationships hold between the coefficient ε_{abcd} in Eq. (E.8) in the rectangular coordinate expression and the coefficients $\varepsilon_{30c}^{(1)}, \ldots$ in Eq. (E.11) in the polar coordinate expression. For example,

$$\varepsilon_{30c}^{(1)} = \frac{1}{4}(3\varepsilon_{3000} + \varepsilon_{1200}), \quad \varepsilon_{30s}^{(1)} = \frac{1}{4}(3\varepsilon_{0300} + \varepsilon_{2100}), \cdots \tag{E.12}$$

References

Yamamoto, T. (1957) On the vibrations of a rotating shaft, Chapter V: on subharmonic oscillations and on "summed and differential harmonic oscillations" in non-Iinear systems having multiple degrees of freedom. *Mem. Fac. Eng., Nagoya Univ.*, **9** (1), 53–71.

Yamamoto, T. and Ishida, Y. (1975) Theoretical discussions of a rotating shaft with nonlinear spring characteristics. *Trans. JSME*, **4J** (345), 1374–1384. (in Japanese)

Appendix F
Routh–Hurwitz Criteria for Complex Expressions

In Chapter 7, we represented the motion of a rotor with complex numbers and derived equations of motion with complex coefficients. In such a case, the characteristic equation is an algebraic equation with complex coefficients. The Routh–Hurwitz coefficients for a complex expression are given as follows (Tondl, 1965; Morris, 1962).

Substituting a solution for free oscillation

$$z = Ae^{i\mu t} \tag{F.1}$$

into the equations of motion, we get

$$(a_0 + ib_0)\mu^n + (a_1 + ib_1)\mu^{n-1} + \cdots + (a_n + ib_n) = 0 \tag{F.2}$$

The solution (Eq. (F.1)) becomes stable when the real part of $i\mu$ is negative. This corresponds to the case where all roots μ of Eq. (F.2) have positive coefficients of imaginary parts. It is known that the necessary and sufficient condition for this is given as follows:

$$\left.\begin{aligned}
&(-1)\Delta_2 = -\begin{vmatrix} a_0 & a_1 \\ b_0 & b_1 \end{vmatrix} > 0, \\
&(-1)^2\Delta_4 = \begin{vmatrix} a_0 & a_1 & a_2 & 0 \\ b_0 & b_1 & b_2 & 0 \\ 0 & a_0 & a_1 & a_2 \\ 0 & b_0 & b_1 & b_2 \end{vmatrix} > 0, \\
&\cdots\cdots\cdots\cdots\cdots\cdots\cdots\cdots\cdots\cdots \\
&(-1)^{(n)}\Delta_{2n} = (-1)^n \begin{vmatrix} a_0 & a_1 & \cdots & a_n & 0 & \cdots & \cdots & 0 \\ b_0 & b_1 & \cdots & b_n & 0 & \cdots & \cdots & 0 \\ 0 & a_0 & \cdots & a_{n-1} & a_n & 0 & \cdots & 0 \\ 0 & b_0 & \cdots & b_{n-1} & b_n & 0 & \cdots & 0 \\ \cdots & \cdots & \cdots & \cdots & \cdots & \cdots & \cdots & \cdots \\ 0 & \cdots & \cdots & 0 & a_0 & \cdots & a_{n-1} & a_n \\ 0 & \cdots & \cdots & 0 & b_0 & \cdots & b_{n-1} & b_n \end{vmatrix} > 0
\end{aligned}\right\} \tag{F.3}$$

Linear and Nonlinear Rotordynamics: A Modern Treatment with Applications, Second Edition.
Yukio Ishida and Toshio Yamamoto.

References

Morris, J. (1962) The Routh and Routh-Hurwitz stability criteria. *Aircraft Eng.*, **34** (395), 25–27.

Tondl, A. (1965) *Some Problems of Rotor Dynamics*, Czechoslovak Academy of Sciences, Prague.

Appendix G
FFT Program

A Fortran program of Fast Fourier Transform (FFT) is provided below as a subroutine program. In this program, characteristic processing techniques for the FFT algorithm, such as *bit reversal* and the *FFT butterfly*, appear. For details of these techniques, refer to Newland (1975) and Kido (1985). The program given below is composed of a part of the main program for the analysis of rotor whirling motion and a subroutine program for FFT. This program is used in both FFT and inverse FFT.

To use this program as FFT, we set DIR = 1.0. X(I) and Y(I) are sampled values in the x- and y-directions, and ZN(I) represents complex data composed of X(I) and Y(I), considering the xy-plane as a complex plane. (When we use this program for ordinary FFT, we may put data into X(I) and zero into Y(I).) The value of N is taken as a power of 2 (NF order of 2). A certain value larger than N is assigned to the asterisk of the array A(*). When this subroutine is called, the result is assigned to ZN(I). The N-times real part of Fourier coefficient is given to the real part of ZN(1) and the N-times imaginary part is given to the imaginary part of ZN(1) (refer to Eq. (16.16)).

To use this program as inverse FFT, we set DIR = −1.0. If the real and imaginary parts of Fourier coefficients are assigned to X(I) and Y(I), respectively, the deflections in the x- and y-directions are assigned, respectively, to the real and imaginary parts as output (refer to Eq. (16.16)).

(A part of the main program for FFT)

```
....................................
    DO 100 I=1,N
    ZN=CMPLX(X(I),Y(I))
100 CONTINUE
    NF=...
    DIR = 1.0D0
    CALL FFT(ZN,NF,DIR)
    DO 200 1,N
    FFTS(I-1)=CDABS(ZN(I))/DBLE(N)
200 CONTINUE
....................................
```

Linear and Nonlinear Rotordynamics: A Modern Treatment with Applications, Second Edition.
Yukio Ishida and Toshio Yamamoto.

```
999 STOP
    END
```

(A part of the main program for inverse FFT)

```
...................................
    DO 100 I=1,N
    ZN=CMPLX(X(I),Y(I))
100 CONTINUE
    NF=...
    DIR = -1.0D0
    CALL FFT(ZN,NF,DIR)
...................................
999 STOP
    END
```

(Subroutine program)

```
C ************************************************************
C     Fast Fourier Transformation
C ************************************************************
      SUBROUTINE FFT(A,NF,DIR)
      IMPLICIT REAL8*(A-H,O-Y)
      COMPLEX*16 A(*),TMP,U,W
      DPAI=4.0D0*DATAN(1.0D0)
      N=2**NF
C ***** Bit reversal *****
      J=1
      DO 10 L=1,N-1
        IF(L.LT.J) THEN
           TMP = A(J)
           A(J) = A(L)
           A(L) = TMP
      ENDIF
      N2=N/2
20 IF(N2.LT.J) THEN
           J=J-N2
           N2=N2/2
           GOTO 20
      ENDIF
      J=J+N2
10 CONTINUE
```

```
C ***** Butterfly *****
    DO 30 M=1,NF
      K=2**(M-1)
      U=CMPLX(1.0D0,0.0D0)
      W=CMPLX(DCOS(DPAI/DBLE(K)),DIR*(-1.0D0)*DSIN(DPAI/DBLE(K))
      DO 40 J=1,K
       L= J
50       IF(L.LE.N) THEN
          N0=L+K
          TMP=A(N0)*U
          A(N0)=A(L)-TMP
          A(L)=A(L)+TMP
          L=L+2**M
          GOTO 50
        ENDIF
     U=U*W
40    CONTINUE
30 CONTINUE
C
    RETURN
    END
```

References

Kido, K. (1985) *Fundamentals of Digital Signal Processing*, Maruzen Co. Ltd. (in Japanese).

Newland, D.E. (1975) *Random Vibrations and Spectral Analysis*, Longman Group, Harlow, Essex.

Index

Linear and Nonlinear Rotordynamics: A Modern Treatment with Applications, Second Edition.
Yukio Ishida and Toshio Yamamoto.

S